Communications in Computer and Information Science 1452

More information about this series at http://www.springer.com/series/7899

Jianchao Zeng · Pinle Qin ·
Weipeng Jing · Xianhua Song ·
Zeguang Lu (Eds.)

Data Science

7th International Conference of Pioneering Computer Scientists, Engineers and Educators, ICPCSEE 2021 Taiyuan, China, September 17–20, 2021 Proceedings, Part II

Springer

Editors
Jianchao Zeng
North University of China
Taiyuan, China

Pinle Qin
North University of China
Taiyuan, China

Weipeng Jing
Northeast Forestry University
Harbin, China

Xianhua Song
Harbin University of Science
and Technology
Harbin, China

Zeguang Lu
National Academy of Guo Ding
Institute of Data Science
Beijing, China

ISSN 1865-0929 ISSN 1865-0937 (electronic)
Communications in Computer and Information Science
ISBN 978-981-16-5942-3 ISBN 978-981-16-5943-0 (eBook)
https://doi.org/10.1007/978-981-16-5943-0

This Springer imprint is published by the registered company Springer Nature Singapore Pte Ltd.
The registered company address is: 152 Beach Road, #21-01/04 Gateway East, Singapore 189721, Singapore

Preface

As the program chairs of the 7th International Conference of Pioneer Computer Scientists, Engineers and Educators 2021 (ICPCSEE 2021, originally ICYCSEE), it is our great pleasure to welcome you to the conference proceedings. ICPCSEE 2021 was held in Taiyuan, China, during September 17–20, 2021, and hosted by the North University of China and the National Academy of Guo Ding Institute of Data Science, China. The goal of this conference series is to provide a forum for computer scientists, engineers, and educators.

This year's conference attracted 256 paper submissions. After the hard work of the Program Committee, 81 papers were accepted to appear in the conference proceedings, with an acceptance rate of 31%. The major topic of this conference series is data science. The accepted papers cover a wide range of areas related to basic theory and techniques for data science including mathematical issues in data science, computational theory for data science, big data management and applications, data quality and data preparation, evaluation and measurement in data science, data visualization, big data mining and knowledge management, infrastructure for data science, machine learning for data science, data security and privacy, applications of data science, case studies, multimedia data management and analysis, data-driven scientific research, data-driven bioinformatics, data-driven healthcare, data-driven management, data-driven eGovernment, data-driven smart city/planet, data marketing and economics, social media and recommendation systems, data-driven security, data-driven business model innovation, and social and/or organizational impacts of data science.

We would like to thank all the Program Committee members, a total of 215 people from 102 different institutes or companies, for their hard work in completing the review tasks. Their collective efforts made it possible to attain quality reviews for all the submissions within a few weeks. Their diverse expertise in each research area helped us to create an exciting program for the conference. Their comments and advice helped the authors to improve the quality of their papers and gain deeper insights.

We thank Lanlan Chang and Jane Li from Springer, whose professional assistance was invaluable in the production of the proceedings. A big thanks also to the authors and participants for their tremendous support in making the conference a success.

Besides the technical program, this year ICPCSEE offered different experiences to the participants. We hope you enjoyed the conference.

July 2021

Pinle Qin
Weipeng Jing

Organization

The 7th International Conference of Pioneering Computer Scientists, Engineers and Educators (http://2021.icpcsee.org) was held in Taiyuan, China, during September 17–20, 2021, and hosted by the North University of China and the National Academy of Guo Ding Institute of Data Science, China.

General Chair

Jianchao Zeng	North University of China, China

Program Chairs

Pinle Qin	North University of China, China
Weipeng Jing	Northeast Forestry University, China

Program Co-chairs

Yan Qiang	Taiyuan University of Technology, China
Yuhua Qian	Shanxi University, China
Peng Zhao	Taiyuan Normal University, China
Lihu Pan	Taiyuan University of Science and Technology, China
Alex kou	University of Victoria, Canada
Hongzhi Wang	Harbin Institute of Technology, China

Organization Chairs

Juanjuan Zhao	Taiyuan University of Technology, China
Fuyuan Cao	Shanxi University, China
Donglai Fu	North University of China, China
Xiaofang Mu	Taiyuan Normal University, China
Chang Song	Institute of Coal Chemistry, CAS, China

Organization Co-chairs

Rui Chai	North University of China, China
Yanbo Wang	North University of China, China
Haibo Yu	North University of China, China
Yi Yu	North University of China, China
Lifang Wang	North University of China, China
Hu Zhang	Shanxi University, China
Wei Wei	Shanxi University, China
Rui Zhang	Taiyuan University of Science and Technology, China

Publication Chair

Guanglu Sun	Harbin University of Science and Technology, China

Publication Co-chairs

Xianhua Song	Harbin University of Science and Technology, China
Xie Wei	Harbin University of Science and Technology, China

Forum Chairs

Haiwei Pan	Harbin Engineering University, China
Qiguang Miao	Xidian University, China
Fudong Liu	Information Engineering University, China
Feng Wang	RoarPanda Network Technology Co., Ltd., China

Oral Session and Post Chair

Xia Liu	Sanya Aviation and Tourism College, China

Competition Committee Chairs

Peng Zhao	Taiyuan Normal University, China
Xiangfei Cai	Huiying Medical Technology (Beijing) Co., Ltd., China

Registration and Financial Chairs

Chunyan Hu	National Academy of Guo Ding Institute of Data Science, China
Yuanping Wang	Shanxi Jinyahui Culture Spreads Co., Ltd., China

Steering Committee Chair

Hongzhi Wang	Harbin Institute of Technology, China

Steering Committee Vice Chair

Qilong Han	Harbin Engineering University, China

Steering Committee Secretary General

Zeguang Lu	National Academy of Guo Ding Institute of Data Science, China

Steering Committee Vice Secretary General

Xiaoou Ding	Harbin Institute of Technology, China

Steering Committee Secretaries

Dan Lu	Harbin Engineering University, China
Zhongchan Sun	National Academy of Guo Ding Institute of Data Science, China

Steering Committee

Xiaoju Dong	Shanghai Jiao Tong University, China
Lan Huang	Jilin University, China
Ying Jiang	Kunming University of Science and Technology, China
Weipeng Jing	Northeast Forestry University, China
Min Li	Central South University, China
Junyu Lin	Institute of Information Engineering, CAS, China
Xia Liu	Hainan Province Computer Federation, China
Rui Mao	Shenzhen University, China
Qiguang Miao	Xidian University, China
Haiwei Pan	Harbin Engineering University, China
Pinle Qin	North University of China, China
Xianhua Song	Harbin University of Science and Technology, China
Guanglu Sun	Harbin University of Science and Technology, China
Jin Tang	Anhui University, China
Ning Wang	Xiamen Huaxia University, China
Xin Wang	Tianjin University, China
Yan Wang	Zhengzhou University of Technology, China
Yang Wang	Southwest Petroleum University, China
Shengke Wang	Ocean University of China, China
Yun Wu	Guizhou University, China
Liang Xiao	Nanjing University of Science and Technology, China
Junchang Xin	Northeastern University, China
Zichen Xu	Nanchang University, China
Xiaohui Yang	Hebei University, China
Chen Ye	Hangzhou Dianzi University, China
Canlong Zhang	Guangxi Normal University, China
Zhichang Zhang	Northwest Normal University, China
Yuanyuan Zhu	Wuhan University, China

Program Committee

Witold Abramowicz	Poznan University of Economics and Business, Poland
Chunyu Ai	University of South Carolina Upstate, USA
Jiyao An	Hunan University, China

Ran Bi	Dalian University of Technology, China
Zhipeng Cai	Georgia State University, USA
Yi Cai	South China University of Technology, China
Zhao Cao	Beijing Institute of Technology, China
Richard Chbeir	LIUPPA Laboratory, France
Wanxiang Che	Harbin Institute of Technology, China
Wei Chen	Beijing Jiaotong University, China
Hao Chen	Hunan University, China
Xuebin Chen	North China University of Science and Technology, China
Chunyi Chen	Changchun University of Science and Technology, China
Yueguo Chen	Renmin University of China, China
Zhuang Chen	Guilin University of Electronic Technology, China
Siyao Cheng	Harbin Institute of Technology, China
Byron Choi	Hong Kong Baptist University, China
Vincenzo Deufemia	University of Salerno, Italy
Xiaofeng Ding	Huazhong University of Science and Technology, China
Jianrui Ding	Harbin Institute of Technology, China
Hongbin Dong	Harbin Engineering University, China
Minggang Dong	Guilin University of Technology, China
Longxu Dou	Harbin Institute of Technology, China
Pufeng Du	Tianjin University, China
Lei Duan	Sichuan University, China
Xiping Duan	Harbin Normal University, China
Zherui Fan	Xidian University, China
Xiaolin Fang	Southeast University, China
Ming Fang	Changchun University of Science and Technology, China
Jianlin Feng	Sun Yat-sen University, China
Yongkang Fu	Xidian University, China
Jing Gao	Dalian University of Technology, China
Shuolin Gao	Harbin Institute of Technology, China
Daohui Ge	Xidian University, China
Yu Gu	Northeastern University, China
Yingkai Guo	National University of Singapore, Singapore
Dianxuan Gong	North China University of Science and Technology, China
Qi Han	Harbin Institute of Technology, China
Meng Han	Georgia State University, USA
Qinglai He	Arizona State University, USA
Tieke He	Nanjing University, China
Zhixue He	North China Institute of Aerospace Engineering, China
Tao He	Harbin Institute of Technology, China
Leong Hou	University of Macau, China

Yutai Hou	Harbin Institute of Technology, China
Wei Hu	Nanjing University, China
Xu Hu	Xidian University, China
Lan Huang	Jilin University, China
Hao Huang	Wuhan University, China
Kuan Huang	Utah State University, USA
Hekai Huang	Harbin Institute of Technology, China
Cun Ji	Shandong Normal University, China
Feng Jiang	Harbin Institute of Technology, China
Bin Jiang	Hunan University, China
Xiaoyan Jiang	Shanghai University of Engineering Science, China
Wanchun Jiang	Central South University, China
Cheqing Jin	East China Normal University, China
Xin Jin	Beijing Electronic Science and Technology Institute, China
Chao Jing	Guilin University of Technology, China
Hanjiang Lai	Sun Yat-sen University, China
Shiyong Lan	Sichuan University, China
Wei Lan	Guangxi University, China
Hui Li	Xidian University, China
Zhixu Li	Soochow University, China
Mingzhao Li	RMIT University, Australia
Peng Li	Shaanxi Normal University, China
Jianjun Li	Huazhong University of Science and Technology, China
Xiaofeng Li	Sichuan University, China
Zheng Li	Sichuan University, China
Mohan Li	Jinan University, China
Min Li	South University, China
Zhixun Li	Nanchang University, China
Hua Li	Changchun University of Science and Technology, China
Rong-Hua Li	Shenzhen University, China
Cuiping Li	Renmin University of China, China
Qiong Li	Harbin Institute of Technology, China
Qingliang Li	Changchun University of Science and Technology, China
Wei Li	Georgia State University, USA
Yunan Li	Xidian University, China
Hongdong Li	Central South University, China
Xiangtao Li	Northeast Normal University, China
Xuwei Li	Sichuan University, China
Yanli Liu	Sichuan University, China
Hailong Liu	Northwestern Polytechnical University, China
Guanfeng Liu	Macquarie University, Australia
Yan Liu	Harbin Institute of Technology, China

Dong Wang	Hunan University, China
Yongheng Wang	Hunan University, China
Chunnan Wang	Harbin Institute of Technology, China
Jinbao Wang	Harbin Institute of Technology, China
Xin Wang	Tianjin University, China
Peng Wang	Fudan University, China
Chaokun Wang	Tsinghua University, China
Xiaoling Wang	East China Normal University, China
Jiapeng Wang	Harbin Huade University, China
Qingshan Wang	Hefei University of Technology, China
Wenfeng Wang	CAS, China
Shaolei Wang	Harbin Institute of Technology, China
Yaqing Wang	Xidian University, China
Yuxuan Wang	Harbin Institute of Technology, China
Wei Wei	Xi'an Jiaotong University, China
Haoyang Wen	Harbin Institute of Technology, China
Huayu Wu	Institute for Infocomm Research, Singapore
Yan Wu	Changchun University of Science and Technology, China
Huaming Wu	Tianjin University, China
Bin Wu	Institute of Information Engineering, CAS, China
Yue Wu	Xidian University, China
Min Xian	Utah State University, USA
Sheng Xiao	Hunan University, China
Wentian Xin	Xidian University, China
Ying Xu	Hunan University, China
Jing Xu	Changchun University of Science and Technology, China
Jianqiu Xu	Nanjing University of Aeronautics and Astronautics, China
Qingzheng Xu	National University of Defense Technology, China
Yang Xu	Harbin Institute of Technology, China
Yaohong Xue	Changchun University of Science and Technology, China
Mingyuan Yan	University of North Georgia, USA
Yu Yan	Harbin Institute of Technology, China
Cheng Yan	Central South University, China
Yajun Yang	Tianjin University, China
Gaobo Yang	Hunan University, China
Lei Yang	Heilongjiang University, China
Ning Yang	Sichuan University, China
Xiaochun Yang	Northeastern University, China
Shiqin Yang	Xidian University, China
Bin Yao	Shanghai Jiao Tong University, China
Yuxin Ye	Jilin University, China
Xiufen Ye	Harbin Engineering University, China

Minghao Yin	Northeast Normal University, China
Dan Yin	Harbin Engineering University, China
Zhou Yong	China University of Mining and Technology, China
Jinguo You	Kunming University of Science and Technology, China
Xiaoyi Yu	Peking University, China
Ye Yuan	Northeastern University, China
Kun Yue	Yunnan University, China
Yue Yue	SUTD, Singapore
Xiaowang Zhang	Tianjin University, China
Lichen Zhang	Shaanxi Normal University, China
Yingtao Zhang	Harbin Institute of Technology, China
Yu Zhang	Harbin Institute of Technology, China
Wenjie Zhang	University of New South Wales, Australia
Dongxiang Zhang	University of Electronic Science and Technology of China, China
Xiao Zhang	Renmin University of China, China
Kejia Zhang	Harbin Engineering University, China
Yonggang Zhang	Jilin University, China
Huijie Zhang	Northeast Normal University, China
Boyu Zhang	Utah State University, USA
Jin Zhang	Beijing Normal University, China
Dejun Zhang	China University of Geosciences, China
Zhifei Zhang	Tongji University, China
Shigeng Zhang	Central South University, China
Mengyi Zhang	Harbin Institute of Technology, China
Yongqing Zhang	Chengdu University of Information Technology, China
Xiangxi Zhang	Harbin Institute of Technology, China
Meiyang Zhang	Southwest University, China
Zhen Zhang	Xidian University, China
Jian Zhao	Changchun University, China
Qijun Zhao	Sichuan University, China
Bihai Zhao	Changsha University, China
Xiaohui Zhao	University of Canberra, Australia
Peipei Zhao	Xidian University, China
Bo Zheng	Harbin Institute of Technology, China
Jiancheng Zhong	Hunan Normal University, China
Jiancheng Zhong	Central South University, China
Fucai Zhou	Northeastern University, China
Changjian Zhou	Northeast Agricultural University, China
Min Zhu	Sichuan University, China
Yuanyuan Zhu	Wuhan University, China
Yungang Zhu	Jilin University, China
Bing Zhu	Central South University, China
Wangmeng Zuo	Harbin Institute of Technology, China

Contents – Part II

Social Media and Recommendation Systems

Data Security and Privacy

Applications of Data Science

Contents – Part I

Basic Theory and Techniques for Data Science

Machine Learning for Data Science

Social Media and Recommendation Systems

Natural Language Inference Using Evidence from Knowledge Graphs

Boxuan Jia[1], Hui Xu[1(✉)], and Maosheng Guo[2]

[1] School of Computer Science and Technology, Heilongjiang University, Harbin, China
xuhui@hlju.edu.cn

[2] Harbin Institute of Technology, Harbin, China
msguo@ir.hit.edu.cn

Abstract. Knowledge plays an essential role in inference, but is less explored by previous works in the Natural Language Inference (NLI) task. Although traditional neural models obtained impressive performance on standard benchmarks, they often encounter performance degradation when being applied to knowledge-intensive domains like medicine and science. To address this problem and further fill the knowledge gap, we present a simple Evidence-Based Inference Model (EBIM) to integrate clues collected from knowledge graphs as evidence for inference. To effectively incorporate the knowledge, we propose an efficient approach to retrieve paths in knowledge graphs as clues and then prune them to avoid involving too much irrelevant noise. In addition, we design a specialized CNN-based encoder according to the structure of clues to better model them. Experiments show that the proposed encoder outperforms strong baselines, and our EBIM model outperforms other knowledge-based approaches on the SciTail benchmark and establishes a new state-of-the-art performance on the MedNLI dataset.

Keywords: Knowledge graphs · Natural language processing · Natural Language Inference · Neural networks

1 Introduction

Natural Language Inference (NLI) is a fundamental task in the Natural Language Understanding (NLU) area, with the goal to determine the reasoning relationship between two texts, namely premise and hypothesis, from {*Entailment*, *Neutral*, *Contradiction*}[5]. For human beings, knowledge plays an essential role when performing inference. For example, we can infer the hypothesis 'Joseph Biden was born in the United States.' from the premise 'Joseph Biden is the 46th president of the U.S.' only if we have the knowledge 'A presidential candidate must be a natural born citizen of the United States.'

This work is supported by Basic Research Funds for Higher Education Institution in Heilongjiang Province (Fundamental Research Project, Grant No.2020-KYYWF-1011).

J. Zeng et al. (Eds.): ICPCSEE 2021, CCIS 1452, pp. 3–15, 2021.
https://doi.org/10.1007/978-981-16-5943-0_1

However, previous NLI approaches often only involve textual information, by using techniques such as sentence encoding or matching, while ignoring the knowledge related to them when conducting inference on natural language texts. Although obtained impressive performance on standard NLI benchmarks, e.g., SNLI [5] and MultiNLI [22], those methods encounter performance degradation when being applied to some specialized, knowledge-intensive areas such as the clinical domain, e.g., MedNLI [18], where the data are derived from the past medical history of patients and annotated by doctors using their clinical knowledge.

Table 1. An example of sentence pairs from the MedNLI dataset, with relevant clues extracted from knowledge graphs.

Example from the MedNLI dataset	
P:	History of present illness : This 66 year old white male has a five day history of chest tightness with exertion.
H:	The patient has angina .
Related evidence from UMLS	
1)	chest tightness $\xrightarrow{\text{finding site of}}$ chest pain $\xrightarrow{\text{belongs to}}$ angina
Related evidence from ConceptNet	
2)	chest $\xrightarrow{\text{RelatedTo}}$ angina
3)	illness $\xrightarrow{\text{AtLocation}}$ sick person $\xrightarrow{\text{IsA}}$ patient

The knowledge of medicine is often necessary for inference in this domain. For example, Table 1 shows an example of sentence pairs describing the situation of a patient, where the entailment relation depends on whether the symptom, i.e., chest tightness, indicates the illness, i.e., angina, which is hard to learn only from textual information, especially when involving clinical terminologies.

To fill this knowledge gap, we propose a novel method, namely, Evidence-Based Inference Model (EBIM), to explicitly integrate relevant evidence from Knowledge Graphs (KGs) to help natural language inference.

In this example, we retrieve professional medical clues from the UMLS (Unified Medical Language System, [3]), a specialized knowledge graph built for clinical domain, e.g., chest tightness $\xrightarrow{\text{finding site of}}$ chest pain $\xrightarrow{\text{belongs to}}$ angina, where colored-boxes represent concepts from the two sentences separately, and the plain text, i.e., 'chest pain', is an internal node from KG, while the labeled-arrows indicating the relationship among them are edges from the KG. A clue is a path consisting of edges and nodes from a KG, which connects the concepts from the two sentences, playing an essential role in inference on clinical statements. In addition, we also retrieve clues from general KGs as supplements, e.g., chest $\xrightarrow{\text{RelatedTo}}$ angina from a commonsense knowledge graph, i.e., ConceptNet [19], which helps inference on this example too.

We also propose a specialized convolution-based encoder to embed those relevant clues retrieved from KGs. Experiments not only indicate that the clues from KGs contribute a lot to the accuracy of inference, but also demonstrate the proposed encoders outperforms other existing encoders, showing the superiority of our EBIM model.

The contribution of this paper focuses on the integration of evidence from KGs. To the best of our knowledge, it is the first work that explicitly exploits the semantics of paths in KGs connecting sentence pairs as evidence in NLI tasks. Experiments show that the proposed EBIM approach establishes a new state-of-the-art performance on the MedNLI benchmark, where the evidence from KGs brings about 2% improvement of accuracy. Additional experiment on the SciTail NLI benchmark [14], which comes from another knowledge-intensive domain, i.e., science question answering, shows that the performance of EBIM is better than other knowledge-based approaches, also indicating it could be generalized to more areas besides medicine.

2 Related Work

Numerous efforts [8,11,20] have been dedicated to the NLI tasks. However, most of them focused on the improvement of sentence encodings or the interaction between the premise and hypothesis. Although obtaining impressive accuracy of inference in normal domains, they encountered performance degradation when applied to knowledge-intensive area.

To encourage the research of NLI on the knowledge-intensive domains, Khot [14] and Romanov et al. [18] proposed textual entailment benchmarks from science question answering and medicine areas, i.e., SciTail and MedNLI, respectively.

To the best of our knowledge, only a few works on NLI involve knowledge from WordNet [16], ConceptNet, PPDB [10], by means of using embeddings derived from KGs [21] or heuristic features [6,18] or hand-crafted rules [12,13].

Chen et al. [6] tried to improve the attention mechanism by heuristic features to exploit lexical knowledge from WordNet. Their experiments showed that those lexical information is helpful to improve the accuracy of inference. However, WordNet could not provide enough knowledge for NLI tasks in professional areas.

Kang et al.[13] proposed an approach to utilize external knowledge to rewrite the premise and hypothesis to augment training data. They implicitly used PPDB as knowledge source for inference. PPDB [10] is an automatically-built dictionary for paraphrase, which could be treated as bidirectional entailment and is therefore useful in NLI. However, it is not as robust as WordNet, which is compiled by experts.

Wang et al. [21] explored the feasibility of incorporating knowledge graphs to fill the knowledge gap in inference tasks. They used KGs to learn word embeddings, and then fed them to NLI models. They also compared the contribution of different KGs, i.e., WordNet, DBPeida [13] and ConceptNet, and found models involving ConceptNet performs the best. Although re-training word embeddings using KGs could improve the performance, it still could not exploit knowledge explicitly.

Inspired by those previous approaches, EBIM is the first work focusing on extracting relevant clues from KGs as evidence for inference, which is our main contribution.

3 Model Architecture

The input of the EBIM model is a pair of sentences, namely, premise p and hypothesis h, and the output of the EBIM model is the prediction of reasoning relation label l between them, from {*Entailment, Neutral, Contradiction*}.

The overall architecture is depicted in Fig. 1. As the figure shows, it first extracts related clues from KGs as evidence, and then employs CNN-based clue encoders and HBMP encoders to embed the evidence and the input sentences respectively. Then an attention mechanism is introduced to integrate the evidence into the encodings of premise and hypothesis. Finally, it utilizes an MLP (Multi-Layer Perceptron) classifier to make the final predictions.

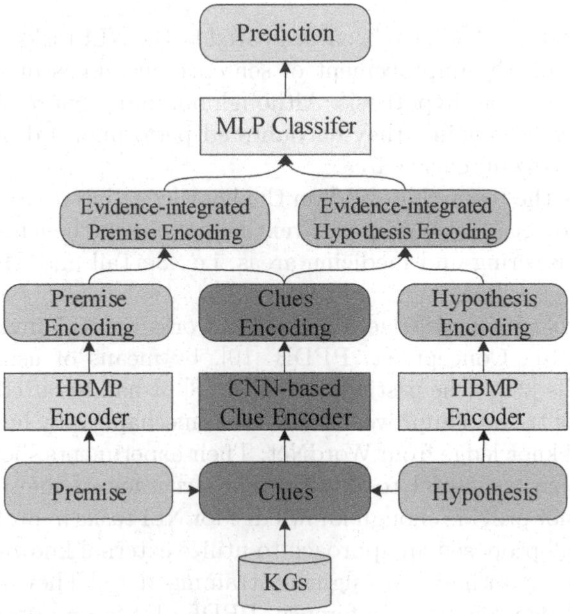

Fig. 1. The overall architecture of EBIM.

We describe the EBIM approach in the following parts: clue extraction (Sect. 3.1), evidence and sentence encoding (Sect. 3.2), evidence integration (Sect. 3.3), and the final classifier (Sect. 3.4).

3.1 Clue Extraction

To extract paths from KGs as clues for inference, we first need to map the mentions in sentence pairs to concepts/entities in KGs. Firstly, we employ Metamap [2] to detect medical concepts from texts and map them to standard terminologies in the UMLS. For example, 'BP' is mapped to 'Blood Pressure'. For commonsense entities, we use the n-grams matching between tokens in sentences and the vocabulary of concepts in ConceptNet. For instance, 'living room' is linked to '/c/en/living_room'.

After concept mapping and entity linking, we get two sets of concepts $C_p = \{c_1^p, c_2^p, ...\}$, i.e., concepts mentioned in the premise, and $C_h = \{c_1^h, c_2^h, ...\}$, i.e., concepts mentioned in the hypothesis, for the pair of sentences. Then we extract the paths connecting concepts from $C_p \times (C_h - C_p)$ and $(C_p - C_h) \times C_h$, where '$\times$' represents Cartesian product, and '$-$' indicates set difference, making the clues only focus on the connections between sentences instead of within them. Take the sentences in Table 1 as an example, we have $C_p = \{$ illness , chest tightness $\}$ and $C_h = \{$ patient , angina $\}$, and we only retrieve the clues connecting concepts from two sentences, e.g., chest tightness $\xrightarrow{\text{finding site of}}$ chest pain $\xrightarrow{\text{belongs to}}$ angina , but do not care about paths only connecting concepts within one sentence, e.g., chest tightness $\xrightarrow{\text{RelatedTo}}$ illness , because usually only the former ones are related to the reasoning relationship between the two sentences.

To reduce noise and save computation, it is necessary to prune the number of clues. We limit the length of paths within m hops, i.e., number of edges in the path. For example, illness $\xrightarrow{\text{AtLocation}}$ sick person $\xrightarrow{\text{IsA}}$ patient is a 2-hop clue. To avoid introducing noise, we only keep the most n confident clues for each concept pair, where the confidence is calculated by the geometric mean of weights of edges in a clue, which are pre-defined in KGs.

We finally get the set of clues $E = \{e_0, e_1, ..., e_N\}$ as the evidence for inference, where N represents the number of clues retrieved from KGs. Please note that $e_1, ...e_N$ are paths extracted from KGs, but e_0 is a special dummy clue Null $\xrightarrow{\text{NoRelation}}$ Null to make the clue set always not empty, which is also useful in the evidence integration step (Sect. 3.3).

3.2 Evidence and Sentence Encoding

After retrieving the related clues from KGs, we need to encode them to feed the neural networks. Please note the structure of a clue, i.e., $e = \{c_1, r_1, c_2, r_2, c_3, ...\}$, which could be viewed as a chain of meaningful triplets $\{t_1, t_2, ...\}$, where $t_1 = (c_1, r_1, c_2)$, $t_2 = (c_2, r_2, c_3)$, ... For example, illness $\xrightarrow{\text{AtLocation}}$ sick person $\xrightarrow{\text{IsA}}$ patient could be viewed as a chain of (illness, AtLocation, sick person), (sick person, isA, patient).

Based on this observation, we propose a specialized convolution-based encoder. It models these triplets first and then aggregate them together to represent the clue. In detail, it is implemented as a 1D convolution with kernel size

3 and stride 2, followed by a max-over-time pooling:

$$\{t_1, t_2, ...\} = Conv\_1D^{size=3}_{stride=2}(\{c_1, r_1, c_2, r_2, c_3, ...\}), \quad (1)$$
$$v = Max\_pooling(\{t_1, t_2, ...\}), \quad (2)$$

where $t_1, t_2, ...$ are the encodings of triplets, and v is the representation of clue e. $c_1, c_2, ...$ and $r_1, r_2, ...$ represent concepts (nodes in the clue) and relations (edges in the clue), whose embeddings are computed by taking average over the embedding of their tokens, e.g., the vector representation of 'sick person' is the average of the embeddings of word 'sick' and 'person'. Experiments in Sect. 4 show the proposed CNN-based clue encoder outperforms other baselines.

Besides the extracted evidence, we also need to encode the premise and hypothesis into vectors p and h. Since it is not our main contribution in this work, we just employ the best-performing sentence encoders on both the SNLI [5] and SciTail [14] benchmarks here, i.e., HBMP [20] encoders. It consists of a max-pooling layer on a stack of 3 BiLSTMs. Now we have the encoding of premise p and hypothesis h by passing their token sequences $\{w^p_1, ..., w^p_{l_p}\}$ and $\{w^h_1, ..., w^h_{l_h}\}$ to the HBMP encoders:

$$p = HBMP(\{w^p_1, ..., w^p_{l_p}\}), \quad (3)$$
$$h = HBMP(\{w^h_1, ..., w^h_{l_h}\}), \quad (4)$$

where word vectors w are initialized from pre-trained fastText [4] embeddings, l represents lengths of sequences.

3.3 Evidence Integration

In fact, not all extracted clues are close related to the sentence pairs. Therefore, we need to select those most relevant clues according their compatibilities with the sentence pairs.

In the EBIM model, we soft-select the relevant clues and integrate them into the encodings of sentences using weighted attention mechanism:

$$u_p = \sum_{i=1}^{N} \frac{pWv_i}{\sum_{j=1}^{N} pWv_j} v_i, \quad (5)$$
$$u_h = \sum_{i=1}^{N} \frac{hWv_i}{\sum_{j=1}^{N} hWv_j} v_i, \quad (6)$$

where v, p and h are the encodings of clues and sentence pairs from Sect. 3.2, W is the parameter matrix of attention, which is initialized randomly and trained by backpropagation, u_p and u_h are the evidence-involved premise and hypothesis representation. Please note that, if there are no relevant clues or most clues are not compatible, the attention mechanism will select the dummy clue introduced in the clue extraction step (Sect. 3.1), which is useful especially for the neural and contradiction reasoning relationships.

3.4 Final Classifier

As we discussed before, the reasoning relationship depends on both textual information and evidence from KGs. So we combine them together as the new sentence encoding,

$$\hat{p} = [p; u_p], \; \hat{h} = [h; u_h], \tag{7}$$

where $[\cdot, \cdot]$ represents concatenation. Then we employ the heuristic features, i.e., $|\hat{p} - \hat{h}|, \hat{p} * \hat{h}$, introduced by Mou et al. [17], and a 3-layered MLP (Multi-Layer Perceptron) with softmax classifier to make final decisions as below,

$$f = [\hat{p}, \hat{h}, |\hat{p} - \hat{h}|, \hat{p} * \hat{h}], \tag{8}$$

$$l = Softmax(MLP(f)), \tag{9}$$

where f is the features used for classification, and $l \in$ {*Entailment, Neutral, Contradiction*} indicates the inference label of sentences pairs.

4 Experiments

4.1 Datasets and Knowledge Resources

We evaluate the proposed EBIM model on two knowledge-intensive NLI benchmarks, i.e., MedNLI and SciTail, which are derived from the clinical and science question answering domains, consisting of about 14k and 27k sentence pairs, respectively. We obtain medical concepts from the UMLS knowledge base, consisting of 3.8M concepts, which is developed by the US National Library of Medicine. For commonsense knowledge, the EBIM approach lookups up the ConceptNet, a publicly available knowledge graph created by crowd-sourcing, contains 8.8M concepts, which has been proved as the most reliable KG in the previous work on SciTail [21].

4.2 Effectiveness of Different KGs and Clue Encoders

We first test whether the introduction of the evidence from KGs helps the inference on natural texts in knowledge-intensive domains, and conduct an ablation study on the MedNLI dataset to examine the importance of the evidence extracted from KGs, and list the results in Table 2, where Model (1) indicates the proposed EBIM model with evidence from both UMLS and ConceptNet, and Model (2)–(4) are variations of EBIM models which removes clues from ConceptNet, UMLS, and both of them, respectively.

Table 2. Ablation study on the effectiveness of evidence from KGs, including accuracy (%) on the development and test set of different configurations of EBIM.

Model	Dev / Test Acc.
(1) EBIM w/ UMLS & ConceptNet	**80.22 / 78.69**
Effect of different KGs	
(2) EBIM w/ UMLS	80.14 / 77.92
(3) EBIM w/ ConceptNet	80.00 / 77.07
(4) EBIM w/o evidence from KGs	78.57 / 76.65

It is obvious that Model (1) performs the best, showing that removing any part of evidence hurts the performance, which demonstrates effectiveness of knowledge in NLI, which brings a 2% improvement of accuracy compared with the textual information only model, i.e., Model (4). It also shows that Model (2), i.e., the EBIM model with evidence only from UMLS, performs better than Model (3), i.e., the one with only ConceptNet, indicating clues from UMLS play a more critical role, since UMLS as a knowledge graph built for clinical domain, could provide more professional medical knowledge.

We examine the contribution of different KGs by two ways: besides comparing the improvement of accuracy brought by them, we also test their coverage, i.e., how many sentence pairs could find at least one clue from a KG, and list them in Table 3. As shown in the table, although UMLS is a more professional KG and contributes more to the inference on clinical texts, its coverage is much lower than that of ConceptNet, which makes ConceptNet a useful complement.

Table 3. Coverage (%) of clues from different KGs.

UMLS	ConceptNet	UMLS + ConceptNet
47.68	70.46	80.15

Next, we keep all the clues and inspect the effect of the convolutional clue encoder introduced in Sect. 3.3 by comparing its performance with strong baselines. The accuracy of different encoders are listed in Table 4.

Model (5) is an EBIM variation which replaces the proposed clue encoder with an LSTM-based clue encoder, but gets a lower performance. It is probably because the LSTM encoder could not model the structure of clues, i.e., a chain of meaningful triplets, as good as the proposed convolutional encoder. Also, we try reusing the HBMP encoder, which is used for encoding the text part, i.e., the premise and hypothesis, to encode clues from KGs, and get an even worse accuracy, indicated by Model (6). We guess the different structures of sentences (natural language) and clues (chains of triplets) causes the shared encoder could not model them well simultaneously. In summary, the proposed CNN-based clue

Table 4. Control experiment on different clue encoders, including accuracy (%) on the development and test set of different configurations of EBIM.

Model	Dev / Test Acc.
(1) EBIM w/ proposed clue encoder	**80.22 / 78.69**
Performances of other clue encoders	
(5) EBIM w/ LSTM clue encoder	80.07 / 77.99
(6) EBIM w/ HBMP clue encoder	79.86 / 77.85

encoder outperforms other baseline encoders due to its compatibility with the structure of extracted clues.

4.3 Experiments on MedNLI

To compare with previous state-of-the-art models, we report the performance of EBIM on the test set of MedNLI, and list the results in Table 5.

Table 5. Performance on the test set of MedNLI.

Model	Test Acc. (%)
(1) EBIM (ours)	**78.69**
(2) BERT	<u>77.71</u>
(3) InferSent w/ K Att.	76.8
(4) HBMP	76.65
(5) InferSent	76.6
(6) ESIM w/ K Att.	75.2
(7) ESIM	74.9

Among them, Model (5) and (7) are traditional neural approaches without introducing knowledge [7,8], while Model (3) and (6) are their corresponding variations by adding a knowledge-directed attention to utilize knowledge in an implicit way [18]. Comparing with those previous works, the proposed EBIM model is able to explicitly integrate evidence from KGs, and therefore establishes a new state-of-the-art performance.

It worth noting that the proposed EBIM approach also outperforms BERT [9], i.e., Model (2), which is a very strong baseline pre-trained by tons of corpus and months of time and then fine-tuned on the MedNLI data. We guess the reason for EBIM outperforming BERT is that there are lots of medical terms and abbreviations in MedNLI, making the data largely different than the corpus used for training BERT. Also, inference on lots of cases in clinical domain needs professional medical knowledge, which BERT has never seen during pre-training. On the contrast, the proposed EBIM method could resolve this by Metamap and further exploit professional knowledge in UMLS to inference effectively.

4.4 Experiments on SciTail

To test whether the proposed EBIM model could generalize to knowledge-intensive domains other than medicine, we also report the accuracy of EBIM on SciTail benchmark, comparing with previous knowledge-based approaches in Table 6.

The sentence pairs in SciTail dataset come from school-level science exams, making it another knowledge-intensive inference task. Since most of statements in this data are irrelevant with medicine, we didn't involve UMLS in this experiment. Model (1) indicates our EBIM method, but with only use the evidence from ConceptNet. Model (2) and Model (3) are two baselines which also involves knowledge in implicit ways during inference proposed by Wang et al. [21] and Kang et al. [13].

As shown in the table, our proposed approach EBIM performs the best among those knowledge-based NLI approaches, by explicitly exploiting clues from KGs as evidence for inference. Moreover, it shows EBIM also works on other knowledge-intense area, and not limited to the clinical domain.

Table 6. Accuracy (%) of knowledge-based NLI methods on the test set of SciTail.

Model	Test Acc.
(1) EBIM (ours)	**86.74**
(2) ConSeqNet [21]	<u>85.20</u>
(3) AdvEntuRe [13]	79.00

4.5 Implementation Details

The EBIM model is implemented using Tensorflow [1] and trained using AdamWR [15], with $lr \in [4\text{e}{-5}, 5\text{e}{-4}]$, $W_{norm} = 1/900$, $\mathrm{T} = 10$, batch size = 64. Word vectors are initialized using pre-trained 300d fastText embeddings [4]. The BERT model is fine-tuned using the base version on the training set. The hidden size of HBMP encoder and MLP are 600. The number of filters in the clue encoder is 300. m is chosen to 2, n to 1, by preliminary experiments.

5 Conclusion and Future Works

Traditional neural NLI models often encounter performance degradation when being applied to knowledge intensive domains. To address this issue and further fill the knowledge gap in natural language inference tasks, we presented a novel Evidence-Based Inference Model, i.e., EBIM, which could explicitly exploit relevant clues from KGs as evidence to support inference in professional area.

To effectively incorporate the knowledge, we proposed an efficient way to retrieve clues and then prune them to avoid involving too much irrelevant noise.

In addition, we designed a specialized CNN-based encoder according to the structure of clues to better model them. Experiments show that the proposed encoder outperforms strong baselines, and our EBIM model establishes a new state-of-the-art accuracy on the MedNLI benchmark and a strong performance on the SciTail dataset.

In the future, we will explore the feasibility to further incorporate knowledge graphs to other knowledge intensive tasks.

References

1. Abadi, M., et al.: Tensorflow: a system for large-scale machine learning. In: Keeton, K., Roscoe, T. (eds.) 12th USENIX Symposium on Operating Systems Design and Implementation, OSDI 2016, Savannah, GA, USA, 2–4 November 2016, pp. 265–283. USENIX Association (2016). https://www.usenix.org/conference/osdi16/technical-sessions/presentation/abadi
2. Aronson, A.R., Lang, F.: An overview of metamap: historical perspective and recent advances. JAMIA **17**(3), 229–236 (2010). https://doi.org/10.1136/jamia.2009.002733
3. Bodenreider, O.: The unified medical language system (UMLS): integrating biomedical terminology. Nucleic Acids Res. **32**(Database-Issue), 267–270 (2004). https://doi.org/10.1093/nar/gkh061
4. Bojanowski, P., Grave, E., Joulin, A., Mikolov, T.: Enriching word vectors with subword information. TACL **5**, 135–146 (2017). https://transacl.org/ojs/index.php/tacl/article/view/999
5. Bowman, S.R., Angeli, G., Potts, C., Manning, C.D.: A large annotated corpus for learning natural language inference. In: Màrquez, L., Callison-Burch, C., Su, J., Pighin, D., Marton, Y. (eds.) Proceedings of the 2015 Conference on Empirical Methods in Natural Language Processing, EMNLP 2015, Lisbon, Portugal, 17–21 September 2015, pp. 632–642. The Association for Computational Linguistics (2015). http://aclweb.org/anthology/D/D15/D15-1075.pdf
6. Chen, Q., Zhu, X., Ling, Z., Inkpen, D., Wei, S.: Neural natural language inference models enhanced with external knowledge. In: Gurevych, I., Miyao, Y. (eds.) Proceedings of the 56th Annual Meeting of the Association for Computational Linguistics, Volume 1: Long Papers, ACL 2018, Melbourne, Australia, 15–20 July 2018, pp. 2406–2417. Association for Computational Linguistics (2018). https://aclanthology.info/papers/P18-1224/p18-1224
7. Chen, Q., Zhu, X., Ling, Z., Wei, S., Jiang, H., Inkpen, D.: Enhanced LSTM for natural language inference. In: Barzilay, R., Kan, M. (eds.) Proceedings of the 55th Annual Meeting of the Association for Computational Linguistics, Volume 1: Long Papers, ACL 2017, Vancouver, Canada, 30 July–4 August, pp. 1657–1668. Association for Computational Linguistics (2017). https://doi.org/10.18653/v1/P17-1152
8. Conneau, A., Kiela, D., Schwenk, H., Barrault, L., Bordes, A.: Supervised learning of universal sentence representations from natural language inference data. In: Palmer, M., Hwa, R., Riedel, S. (eds.) Proceedings of the 2017 Conference on Empirical Methods in Natural Language Processing, EMNLP 2017, Copenhagen, Denmark, 9–11 September 2017, pp. 670–680. Association for Computational Linguistics (2017). https://aclanthology.info/papers/D17-1070/d17-1070

9. Devlin, J., Chang, M., Lee, K., Toutanova, K.: BERT: pre-training of deep bidirectional transformers for language understanding. CoRR abs/1810.04805 (2018). http://arxiv.org/abs/1810.04805
10. Ganitkevitch, J., Durme, B.V., Callison-Burch, C.: PPDB: the paraphrase database. In: Vanderwende, L., III, H.D., Kirchhoff, K. (eds.) Human Language Technologies: Conference of the North American Chapter of the Association of Computational Linguistics, Proceedings, Westin Peachtree Plaza Hotel, Atlanta, Georgia, USA, 9–14 June 2013, pp. 758–764. The Association for Computational Linguistics (2013). http://aclweb.org/anthology/N/N13/N13-1092.pdf
11. Guo, M., Zhang, Y., Liu, T.: Gaussian transformer: a lightweight approach for natural language inference. Proc. AAAI Conf. Artif. Intell. **33**(01), 6489–6496 (July 2019). https://doi.org/10.1609/aaai.v33i01.33016489. https://ojs.aaai.org/index.php/AAAI/article/view/4614
12. Kang, D., Khot, T., Sabharwal, A., Clark, P.: Bridging knowledge gaps in neural entailment via symbolic models. In: Riloff, E., Chiang, D., Hockenmaier, J., Tsujii, J. (eds.) Proceedings of the 2018 Conference on Empirical Methods in Natural Language Processing, Brussels, Belgium, 31 October–4 November 2018, pp. 4940–4945. Association for Computational Linguistics (2018). https://aclanthology.info/papers/D18-1535/d18-1535
13. Kang, D., Khot, T., Sabharwal, A., Hovy, E.H.: Adventure: adversarial training for textual entailment with knowledge-guided examples. In: Gurevych, I., Miyao, Y. (eds.) Proceedings of the 56th Annual Meeting of the Association for Computational Linguistics, Volume 1: Long Papers, ACL 2018, Melbourne, Australia, 15–20 July 2018, pp. 2418–2428. Association for Computational Linguistics (2018). https://aclanthology.info/papers/P18-1225/p18-1225
14. Khot, T., Sabharwal, A., Clark, P.: Scitail: a textual entailment dataset from science question answering. In: McIlraith, S.A., Weinberger, K.Q. (eds.) Proceedings of the 32nd AAAI Conference on Artificial Intelligence (AAAI-18), The 30th innovative Applications of Artificial Intelligence (IAAI-18), and The 8th AAAI Symposium on Educational Advances in Artificial Intelligence (EAAI-18), New Orleans, Louisiana, USA, 2–7 February 2018, pp. 5189–5197. AAAI Press (2018). https://www.aaai.org/ocs/index.php/AAAI/AAAI18/paper/view/17368
15. Loshchilov, I., Hutter, F.: Fixing weight decay regularization in Adam. CoRR abs/1711.05101 (2017). http://arxiv.org/abs/1711.05101
16. Miller, G.A.: WordNet: a lexical database for English. Commun. ACM **38**(11), 39–41 (1995). https://doi.org/10.1145/219717.219748. http://doi.acm.org/10.1145/219717.219748
17. Mou, L., et al.: Natural language inference by tree-based convolution and heuristic matching. In: Proceedings of the 54th Annual Meeting of the Association for Computational Linguistics, Volume 2: Short Papers, ACL 2016, Berlin, Germany, 7–12 August 2016. The Association for Computer Linguistics (2016). http://aclweb.org/anthology/P/P16/P16-2022.pdf
18. Romanov, A., Shivade, C.: Lessons from natural language inference in the clinical domain. In: Riloff, E., Chiang, D., Hockenmaier, J., Tsujii, J. (eds.) Proceedings of the 2018 Conference on Empirical Methods in Natural Language Processing, Brussels, Belgium, 31 October–4 November 2018, pp. 1586–1596. Association for Computational Linguistics (2018). https://aclanthology.info/papers/D18-1187/d18-1187

19. Speer, R., Havasi, C.: Representing general relational knowledge in conceptnet 5. In: Calzolari, N., et al. (eds.) Proceedings of the 8th International Conference on Language Resources and Evaluation, LREC 2012, Istanbul, Turkey, 23–25 May 2012, pp. 3679–3686. European Language Resources Association (ELRA) (2012). http://www.lrec-conf.org/proceedings/lrec2012/summaries/1072.html
20. Talman, A., Yli-Jyrä, A., Tiedemann, J.: Sentence embeddings in NLI with iterative refinement encoders. Nat. Lang. Eng. **25**(4), 467–482 (2019)
21. Wang, X., et al.: Improving natural language inference using external knowledge in the science questions domain. CoRR abs/1809.05724 (2018). http://arxiv.org/abs/1809.05724
22. Williams, A., Nangia, N., Bowman, S.R.: A broad-coverage challenge corpus for sentence understanding through inference. In: Walker, M.A., Ji, H., Stent, A. (eds.) Proceedings of the 2018 Conference of the North American Chapter of the Association for Computational Linguistics: Human Language Technologies, Volume 1 (Long Papers), NAACL-HLT 2018, New Orleans, Louisiana, USA, 1–6 June 2018, pp. 1112–1122. Association for Computational Linguistics (2018). https://aclanthology.info/papers/N18-1101/n18-1101

Construction of Multimodal Chinese Tourism Knowledge Graph

Jiawang Xie[1], Zhenhao Dong[2], Qinghua Wen[1], Hongyin Zhu[1](✉), Hailong Jin[1], Lei Hou[1], and Juanzi Li[1]

[1] Tsinghua University, Beijing 100084, China
{xiejw18,zhuhongyin2020,jinhl15}@mails.tsinghua.edu.cn, houlei@tsinghua.edu.cn

[2] Beijing University of Chemical Technology University, Beijing 100084, China

Abstract. With the development of tourism knowledge graphs (KGs), recommendation, question answering (QA) and other functions under its support enable various applications to better understand users and provide services. Existing Chinese tourism KGs do not contain enough entity information and relations. Besides, the knowledge storage usually contains only the text modality but lacks other modalities such as images. In this paper, a multi-modal Chinese tourism knowledge graph (MCTKG) is proposed based on Beijing tourist attractions to support QA and help tourists plan tourism routes. An MCTKG ontology was constructed to maintain the semantic consistency of heterogeneous data sources. To increase the number of entities and relations related to the tourist attractions in MCTKG, entities were automatically expanded belonging to the concepts of building, organization, relic, and person based on Baidu Encyclopedia. In addition, based on the types of tourist attractions and the styles of tourism route, a tourism route generation algorithm was proposed, which can automatically schedule the tourism routes by incorporating tourist attractions and the route style. Experimental results show that the generated tourist routes have similar satisfaction compared with the tourism routes crawled from specific travel websites.

Keywords: Knowledge graph · Tourism ontology · Entity expansion · Route generation algorithm

1 Introduction

With the prosperity of large-scale knowledge graphs (KGs) such as DBpedia [1], YAGO [18] and Wikidata [19], various knowledge-based applications, such as semantic search, recommendation system [12], question answering [11], etc., can benefit from the rich and precious knowledge in the KGs, and achieve significant performance improvement. Recently, in the tourism field, KGs are used to recommend travel attractions or travel plans [6, 13]. The rapid development of the Internet has given birth to many travel websites, e.g. Ctrip, Meituan, Mafengwo, etc. Some companies have also built their own KGs to support their downstream services. However, the heterogeneity of information is still

J. Zeng et al. (Eds.): ICPCSEE 2021, CCIS 1452, pp. 16–29, 2021.
https://doi.org/10.1007/978-981-16-5943-0_2

the bottleneck of downstream applications. Besides, existing tourism KGs usually lack multimodal knowledge and effective arrangement of knowledge from various tourism data sources. Therefore, it is an urgent need to construct a comprehensive multimodal tourism KG.

In recent years, academic researches on Chinese tourism KGs [14, 15] attract much attention. For example, Zeng et al. [7] construct a Chinese tourist attractions KG and a recommendation system CASIA-TAR, whose knowledge is extracted from Baidu Encyclopedia[1] and Hudong Encyclopedia[2]. Geng et al. [16] construct a Mongolian-Chinese bilingual knowledge graph for tourism by extracting entities and attributes through neural networks from text. Unlike existing tourism KG, our MCTKG focuses on travel itinerary and people-oriented information in Beijing, covering restaurant knowledge and rich image-modal knowledge, which can directly meet the needs of tourists for querying information and planning itineraries. We mainly solve the following key problems for constructing MCTKG:

- Tourism ontology construction. We construct a tourism ontology with consideration of the practical application and the conceptual hierarchy of data.
- Entity expansion. We extract 0.2M entities from the Baidu Encyclopedia webpages corresponding to specific tourism attraction through Xlink [17], then filter out entities belonging to person, building, organization, and relic.
- Attribute normalization. Entities obtained from heterogeneous data sources have many attributes. We classify the attribute names with the same suffix into the same set and then judge whether the attributes in the attribute set are object type.
- Tourism routes generation. We propose an automatic generation algorithm of tourist routes, which comprehensively considers the attraction level and the geographical location between attractions to generate a reasonable and effective tourist route. The generated routes are also defined in MCTKG through the RDF container element rdf:Seq.

To query and display the multimodal knowledge in MCTKG, we construct the application platform of MCTKG and provide the data access and query interface. In summary, the main contributions are as follows:

- Knowledge graph. We propose a tourism ontology and build the MCTKG which integrates multiple heterogeneous data sources. We introduce rich image-modal knowledge into our KG to understand specific tourism attractions more intuitively. The MCTKG contains 251,532 entities, 8,883 properties, 1,368,544 triples.
- Pipeline. We propose a pipeline for constructing MCTKG. The information can be accurately extracted from multiple travel websites and organized in an appropriate form.
- Platform. An application platform is built based on the MCTKG, which can effectively display the text and image modal knowledge in the MCTKG.

[1] https://baike.baidu.com/.
[2] https://www.hudong.com/.

2 Related Work

KGs have been used extensively to enhance the results provided by popular search engines at first. These KGs are typically powered by structured data repositories such as Freebase, DBpedia, Yago, and Wikidata. However, since KG construction is a particularly complex and time-consuming systematic project, it has become one of the bottlenecks in developing graph construction technology.

In the last few years, KGs have been used in many domains. In the financial domain, Sarah Elhammadi et al. [8] focus on constructing a KG automatically by developing a high precision knowledge extraction pipeline tailored for the financial domain. Miao et al. [9] construct a novel dynamic financial KG, which utilizes time information to capture data changes and trends over time. In the tourism domain, some research starts to build KGs to solve tourists' problems during the trip. DBtravel [6] is a tourism-oriented KG generated from the collaborative travel site Wikitravel. Lu et al. [2] constructed a rich world-scale travel KG from existing large KGs, namely Geonames, DBpedia, Wikidata and the underlying ontology contain more than 1,200 classes to describe attractions.

The research on construction technology in China is still in its infancy. In terms of domain KG construction in china, although there have been many important research results, such as Sogou's Knowledge Cube and Baidu's Zhixin. GDM Laboratory of Fudan University designed a Chinese KG [3] for book reading. Wang et al. [4] constructed the bilingual films KG, including the construction of films ontology base, entity link, entity matching and built the application platform and open data access interface. E et al. [5] proposed an end-to-end automatic construction scheme of Chinese KG based on Chinese encyclopedia data and developed a user-oriented Chinese KG system. But in general, there are few domains involved, which can not meet the needs of practical application in terms of scale and quality. Especially in the tourism domain with wide application prospects, there is no high-quality KG in Chinese.

3 MCTKG Construction

The construction process of MCTKG is shown in Fig. 1, which generally consists of the following parts.

- Tourism ontology is constructed semi-automatically by optimizing and simplifying the concept hierarchy relationship abstract from data resources and combining with tourists' actual needs and application scenarios. In addition, we also normalize the attributes to improve the ontology.
- We extract semi-structured knowledge from multiple data sources such as entities and their attributes and align it to the constructed tourism ontology.
- Based on the attraction entities and the concepts of tourism styles in KG, we propose an automatic algorithm for generating tourism routes. The generated route can be recommended to tourists as a recommended route in the needs of tourists.
- We extract entities from the encyclopedia page of attraction entities and filter entities that belong to concepts such as people, buildings, organizations, relics, etc., as expanded entities.

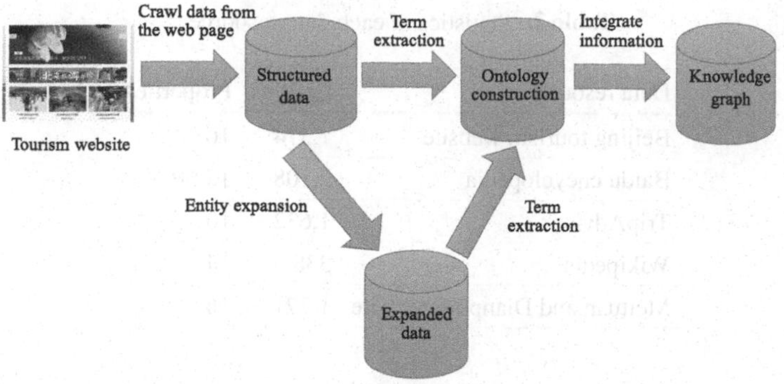

Fig. 1. MCTKG generation pipeline.

3.1 Semi-structured Knowledge Extraction

Multiple Heterogeneous Data Sources. The data used to construct MCTKG are semi-structured data obtained from major tourism websites on the Internet, mainly including the Beijing tourism website, Baidu Encyclopedia, TripAdvisor, Meituan website. A detailed description of these data resources is shown in Table 1.

Table 1. Description of each data resource

Data resource	Description
Beijing tourism website	Largest tourism portal in Beijing, covers all the tourist attractions, hotels, hot springs and resorts in Beijing and its surrounding areas
Baidu Encyclopedia	Baidu Encyclopedia, largest Chinese encyclopedia, has been significantly improved its data in terms of scale and quality in tourism recently
TripAdvisor	TripAdvisor is a tourism review website, which provides tourism-related information about the world's attractions and their nearby hotels and restaurants
Wikipedia	Wikipedia is a multi-language encyclopedia collaboration project based on Wiki technology
Meituan website	China's leading local life information and trading platforms

Text and Image Modal Knowledge Extraction. Text modal knowledge extraction involves extracting knowledge from heterogeneous data sources, analyzing data of multiple formats, unifying semantics and structure, and roughly includes the following two modules:

Table 2. Statistics of each data resource.

Data resource	Entities	Properties
Beijing tourism website	1,574	16
Baidu encyclopedia	2,108	1,060
TripAdvisor	1,632	10
Wikipedia	338	14
Meituan and Dianping website	1,771	16

- Web page analysis. This module mainly obtains web pages through a crawler tool and then extracts entity and entity attribute information from web pages by the regular expression.
- Attribute value completion. This module's main task is to align entities extracted from heterogeneous data sources and then complete the attribute values of the same attributes of the two entities after alignment to achieve the purpose of knowledge fusion.

After the above two steps, the extracted text modal knowledge is transformed into structured JSON format data. Statistics about semi-structured knowledge extracted from the above data resources are shown in Table 2.

To acquire image modal knowledge from Baidu images, we apply python selenium webdriver to simulate user scroll, click actions, etc., by using the entity name extracted from the Beijing tourism website. We save the image modal knowledge of each entity in the form of a link. After extracting each attraction's image modal information from Baidu images and filtering out low-quality images by manual filtering, we fuse them to MCTKG.

3.2 Entity Expansion

In this subsection, to increase the number of entities and enhance the abundance of entities in MCTKG, we give a detailed description of how to expand entities based on existing entities.

Given all the attraction entities as seed entities, use the python toolkit request to obtain each attraction entity's encyclopedia page text. Since part of knowledge has nothing to do with corresponding attractions in obtained text, we use regular expressions to extract abstracts and articles from the obtained text as valid text related to corresponding attractions.

To obtain the entities related to the attraction entity from the valid text, we adopt Xlink to find out all entities in the valid text. But many entities unrelated to tourism in the expanded entities. Therefore, it's necessary to filter these entities out.

We retrieved expanded entities to get their labels from XLORE [10], a large-scale English-Chinese bilingual KG. Since each entity has several labels, we set that once a label contains a keyword, the entity's class can set to the corresponding concept. For example, after retrieving, the entity "Guan Yu," who is a very famous general in the Warring States period, has two corresponding labels, namely "person" and "general." Since both labels contain keywords, the entity "Guan Yu" can be directly classified as person.

After filtering, entities belonging to class such as person, building, organization, relic are used as seed entities, then repeat the above steps for the second round iterations to get more entities.

Finally, except for the entities belonging to the concepts of people, buildings, organizations, cultural relics, etc., in the expanded entity, some entities in the remaining entities are almost irrelevant to tourism. Therefore, we need to filter out these entities. We count the frequency of all entities' occurrence and then perform statistics according to the set low-frequency interval, medium-frequency interval, and high-frequency interval. Statistics of the number of entities in each interval are shown in Table 3. We can see that most entities fall in the low-frequency interval, and low-frequency words are basically meaningful entities, so we mainly focus on processing high-frequency words, and the first two thousand high-frequency words are selected for manual filtering. And after observing the expanded entities, entities with a long length are generally misidentified or meaningless entities, so we count all entities' length and delete the obviously longer entities to support entity filtering. We lastly filtered out 1,279 entities from 7937 high-frequency entities.

Table 3. Number of entities in each interval.

The entity of each interval	Quantity
Low-frequency entity (<=3)	182,866
Medium-frequency entity (4–20)	64,796
high-frequency entity (>20)	7,937

3.3 Ontology Construction

Ontology is an abstract model for describing the objective world, which clarifies concepts and their relations in a formal way. It consists of some concepts and attributes. The common approaches are divided into three types: semi-automatic construction, automatic construction, and manual construction. MCTKG needs to directly participate in solving the actual problems in the tourism background. Therefore, it is challenging to adapt to this demand through automatic construction technology. This paper uses the approach of semi-automatic ontology construction, through comprehensive consideration of multiple factors to complete the ontology construction. And it mainly includes the following steps.

- Initial ontology and terms are generalized from obtained data. Firstly, according to the knowledge extracted from the Beijing tourism website, we can get an accurate concept hierarchy relationship of attractions by refining the relationship between attractions and their corresponding concept. Combined with the practical problems and

the specific application scenarios, a new concept hierarchy is added. We summarized the actual problems and defined the hierarchical relationship of tourism style, time-honored brands, time-honored stores, etc.
- Ontology iteratively update. In the initial stage, the construction of ontology requires lots of manual intervention, which is time-consuming. After the construction is completed, we will evaluate the ontology's rationality from multiple aspects and iteratively update the ontology version based on the feedback in the data.

The illustration of the part concept classification system is shown in Fig. 2.

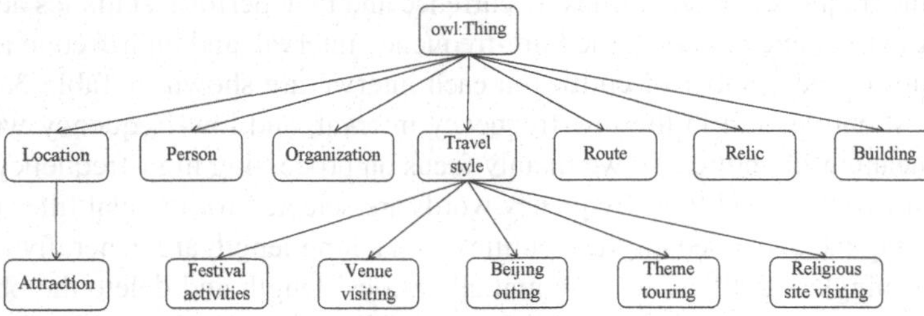

Fig. 2. Illustration of part concept classification system.

3.4 Attribute Normalization

Most of the attributes of the entities in MCTKG are extracted from information boxes of Baidu Encyclopedia web pages. These attributes include two types of numeric attributes and object attributes. To normalize the attributes, it is necessary to determine which attributes are object attributes.

Since the quantity of attributes in MCTKG is as high as 8,883, it can be time-consuming to determine whether each attribute is an object-type attribute. We are based on the fact that attributes with the same suffix usually have the same features. Therefore, we merge the attributes with the same suffix into the same set, and the suffix is the central word of all the attributes in the set. We only need to judge whether the attributes in the set are object-type attributes according to the 4,490 central words instead of judging one by one. This method greatly improves efficiency.

After a manual judge of 4,490 central words, we finally get 3,772 object-type attributes out of 8,883 attributes. Before normalizing the attribute, we treat each attribute value as a string instead of entities. But after normalization, the attribute values of object-type attributes should be treated as expanded entities.

To obtain the attribute information of these entities from Baidu Encyclopedia, 31,145 entities out of 92,442 expanded entities have their own links when they are obtained, so we can use the link to obtain the attribute information of the entity's Baidu Baike page through a crawler program. But the remaining 61,297 expanded entities cannot directly obtain the attribute information on their Baidu Encyclopedia page through the link. Therefore, we obtained 18,821,853 Baidu Baike entities and their webpage links in advance, and then we matched the remaining expanded entities with Baidu Baike entities. We found that 17,286 expanded entities can be completely matched with Baidu Baike entities through string matching. For the remaining 44,011 entities, we use the Levenshtein distance algorithm to calculate the distance between a specific expanded entity and each Baidu Baike entity.

Levenshtein distance refers to the minimum number of editing operations required to convert two strings from one to the other. Allowed editing operations include: replacing a character with another character, inserting a character, or deleting a character. Generally, the smaller the Levenshtein distance, the greater the similarity between the two strings. For the case where the distance is greater than 3, we treat it as a fail match. Since many expanded entities correspond to several Baidu Encyclopedia entities with Levenshtein distance less than or equal to 3, we manually judge each candidate entity to select the entity that best matches the expanded entity from several candidates. Finally, 1569 entities can match with Baidu Encyclopedia entities. Then we can get the attributes of these matched entities on the encyclopedia page by obtaining the links of the corresponding Baidu Encyclopedia entities.

3.5 Automatic Generation Algorithm of Tourist Routes

For enabling MCTKG to support subsequent practical applications such as planning tourist tourism routes for tourists, we need to generate specific tourism routes according to the attraction entities obtained from multiple data sources and add them into MCTKG. To automatically generate tourism routes, we propose an algorithm for the automatic generation of tourism routes. For each travel style, several routes are generated. The pseudo-code of the algorithm is shown in Algorithm 1.

Several travel styles have been added to MCTKG, and each travel style corresponds to several recommended attractions. Therefore, it is necessary to select several tourism routes from these attractions. Considering that popular attractions should be added to the route first, and the generated route should be as short as possible, so the order of the attractions should be arranged reasonably, so as shown in the pseudo-code, the function Permutation_and_CalculateShortestDistance is used to calculate the shortest route distance among all permutations and combinations of several attractions. According to the attraction level, set a weight for each attraction. The higher the attraction level, the larger the weight. Those with a weight larger than 7 are treated as hot attractions and will be given priority in generating routes. Besides, we use the function calculate_total_time to calculate the duration of the entire route and determine whether the route is a one-day tourism route or a half-day tourism route based on the duration.

```
Algorithm 1 Automatic generation algorithm of tourist routes
Input: – Travel style set: A={A1, A2, ..., Am}
       – Candidate attractions set corresponding to each travel style, Bi={C1, C2, ...,
         Cn} (i=1, 2, ..., m)
       – Weight dictionary d: Each attraction level corresponds to a weight, the higher
         attraction level is, the greater the weight is, and the highest weight is 10
       – quantity of cycles threshold Q
Output: – List of one day tour route: One_day_tour_routes – List of half day tour route:
Half_day_tour_routes
function Generating_Tourist_Routes(A, B, Q, d)
    for i = 0 → m do
        candidate ← []
        count ← 0
        for j = 0 → n do
            if d[B[j]] > 7 then
                candidate.append(B[j])
            end if
        end for
        while count < Q do
            count ← count + 1
            Attractions ← np.random.choice(candidate, np.random.choice([2, 3, 4, 5, 6]))
            T ← calculate_total_time(Attractions)
            if T > 6 and T <= 10 then
```

3.6 MCTKG Sharing Platform

KG is a kind of network of knowledge relationships constructed by information visualization technology. The purpose of establishing this KG sharing platform[3] is to display MCTKG from concept, instance, and attribute perspectives. Figure 3 shows the sharing platform of MCTKG. The platform is developed by an open-source framework React and uses Virtuoso as the database server, which mainly provides two functions: (1)

3 http://166.111.68.66:9083/.

basic information of tourism ontology, which can provide much statistical information of concept hierarchy and KG, (2) data query interface, includes SPARQL terminal query interface, classified index query interface, and the composite query interface.

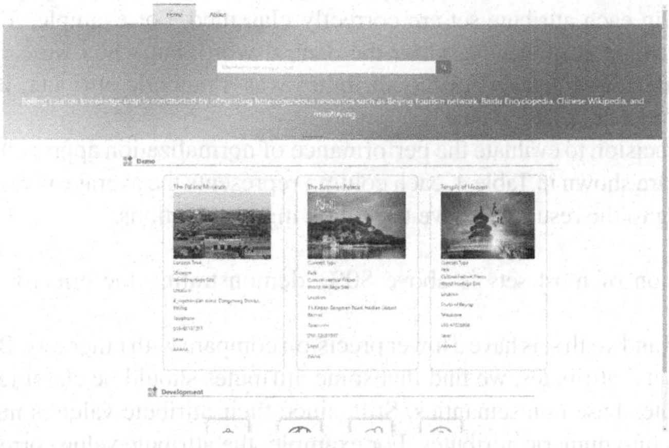

Fig. 3. The website of the MCTKG sharing platform.

In addition to the shared platform's information, Table 4 shows the quantity of instances, image links, properties, and triples of each concept.

Table 4. Statistics of each concept.

Concept	Instances	ImageLinks	Properties	Triples
Attraction	2,106	44,359	1,069	96,493
Time-honored restaurant	1,771	10,798	25	52,127
Brand	60	1,799	3	4,145
Route	232	0	6	1,392
Travel style	22	0	2	22
Person	37,290	194,564	4,102	584,477
Building	3,501	99,850	2,813	161,557
Relic	705	20,810	860	30,436
Organization	10,643	315,698	5,742	442,062

4 Experiments

To verify the attribute standardization method's effectiveness and evaluate the quality of the generated tourist routes, we conducted the following experiments.

4.1 Attribute Normalization Evaluation

We obtain 4,490 central words of 8,883 attributes. Each central word has a corresponding attributes set. We randomly selected ten sets of data from 4,490 central words. Each set includes ten central words and the corresponding attribute sets. We need to judge whether the attributes in each attribute set are correctly classified. For example, a attribute set was all object-type attributes based on the central word, but when we carried out the experiment and found that one of the attributes was a numeric attribute, it means this attribute was misclassified.

We use precision to evaluate the performance of normalization approach. The experiment results are shown in Table 4, each column represents the average precision of each set. According to the result, We have the following observations:

- The precision of most sets is above 80%, demonstrating the effectiveness of our approach.
- The second and sixth sets have a lower precision compare with other sets. By observing these two sets' attributes, we find that some attributes should be classified as object-type attributes based on semantics. Still, since their attribute value is numeric, they are divided into numeric attributes. For example, the attribute value corresponding to the attribute "existing member" should be a specific person, but the attribute value is the member's quantity (Fig. 4).

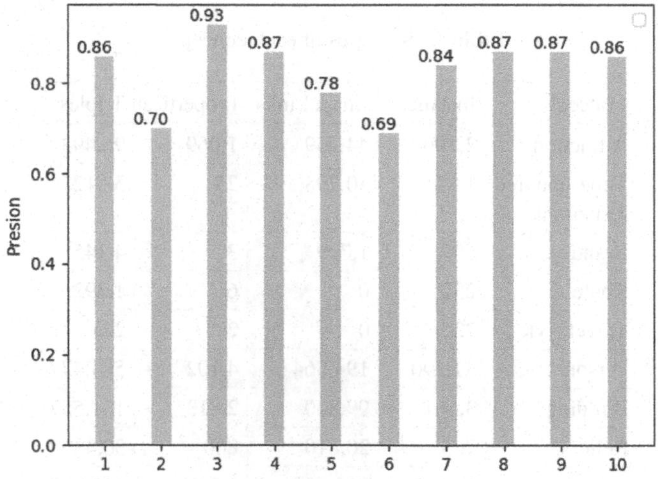

Fig. 4. Average precision of each set.

4.2 Route Evaluation

We obtained 189 tourist routes generated by the automatic route generation algorithm and carried out experiments on these data. To verify the automatic generation algorithm's effectiveness, we compare the generated routes with online routes crawled from travel websites. Since online routes are classic and popular, which are recommended to tourists who visit the website, they usually have a high evaluation and reliability. Therefore, it's reasonable to use online routes as references to evaluate the generated routes' quality.

We use the approach of scoring the generated routes according to some evaluation indicators to judge whether the generated routes are as satisfactory as those obtained from travel websites, including the Cellular website[4], Qunar[5] and so on.

To make the scoring result more reasonable and accurate, according to the proportion of the two types of routes' overall quantity, we generated 10 sets of route data. Each set of data randomly selects 6 and 30 routes from online routes and automatically generated routes. That is, each set of data contains 36 tourist routes.

We recruited some experienced tourists to score these routes. The score ranges from 1.0 to 5.0, and the evaluation indicators include the rationality of travel arrangement, the rationality of time planning, and the comfort of tourism experience. The scoring results are shown in Fig. 5.

From the scoring results, we can see that the score of the generated routes is slightly lower than online routes, but the score gap is not too large, which indicates that the generated routes can meet the needs of tourists to a certain extent. This illustrates the effectiveness of the automatic route generation algorithm.

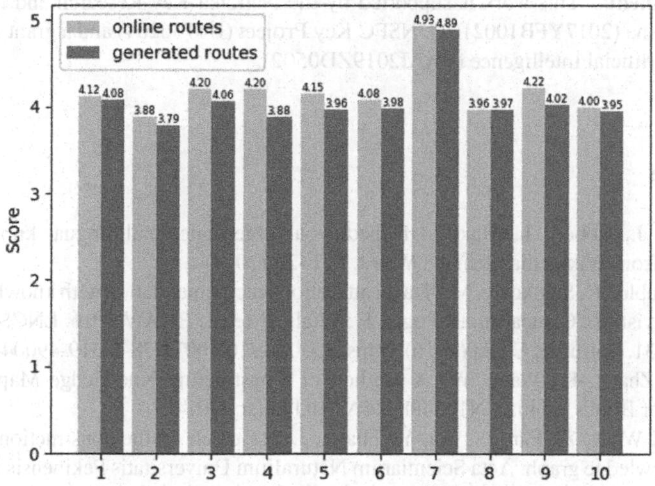

Fig. 5. Route scoring results.

5 Conclusion

In this paper, the construction process of MCTKG based on multiple heterogeneous data sources is proposed. We first extract semi-structured data from various heterogeneous

[4] https://www.mafengwo.cn/.

[5] https://www.qunar.com/.

data sources. Then we extract terms from these data and abstract the initial ontology, and then iteratively update it. The attractions entities are used as seed entities, and expanded entities are extracted from these entities' encyclopedia pages to increase the number of entities. We use the suffix method to normalize the attributes. We have designed an automatic tourist route generation algorithm to generate tourist routes with similar quality to online routes.

Our MCTKG is a high-quality RDF tourism ontology KG that integrates 6 heterogeneous high-quality data sources and fills the Chinese tourism ontology KG gap in China. KG provides an important basis for mining and utilizing tourism-related information and has an important significance to expand the international influence of Chinese tourism information.

The construction of KG is a long-term, systematic complex work, which needs continuous improvement. And there are still many aspects that require to be improved in the construction process of MCTKG. For example, (1) seeking higher quality and more complete data sources to complete some attraction entities' missing attribute values. (2) Establish the relational link between the concept of people and attractions. (3) The automatic updating mechanism of KG is established. The method proposed in this paper has a certain reference significance for the construction of domain KG, which needs to integrate multiple heterogeneous data sources and entity alignment in a specific domain.

Acknowledgements. This work is supported by the National Key Research and Development Program of China (2017YFB1002101), NSFC Key Project (U1736204) and a grant from Beijing Academy of Artificial Intelligence (BAAI2019ZD0502).

References

1. Lehmann, J., Robert, I., Max, J.: Dbpedia—a large-scale, multilingual knowledge base extracted from Wikipedia. Semant. Web J. **5**, 1–29 (2014)
2. Lu, C., Laublet, P., Stankovic, M.: Travel attractions recommendation with knowledge graphs. In: Blomqvist, E., Ciancarini, P., Poggi, F., Vitali, F. (eds.) EKAW 2016. LNCS, vol. 10024, pp. 416–431. Springer, Cham (2016). https://doi.org/10.1007/978-3-319-49004-5_27
3. Xiao, Y., Zhang, K., Wang, W.: A Method of Constructing Knowledge Map of Reading Domain for Books, China CN103488724A, 01 January 2014
4. Wang, W., Wang, Z., Pan, L., Liu, Y., Zhang, J.: Research on the construction of bilingual movie knowledge graph. Acta Scientiarum Naturalium Universitatis Pekinensis **52**(1), 25–34 (2016)
5. Shijia, E., Lin, P., Xing, Y.: Automatical construction of Chinese knowledge graph system. Comput. Appl. **36**(4), 992–996 (2016)
6. Calleja, P., Priyatna, F., Mihindukulasooriya, N., Rico, M.: DBtravel: a tourism-oriented semantic graph. In: Pautasso, C., Sánchez-Figueroa, F., Systä, K., Murillo Rodríguez, J. (eds.) ICWE 2018. LNCS, vol. 11153, pp. 206–212. Springer, Cham (2018). https://doi.org/10.1007/978-3-030-03056-8_19
7. Zeng, Y., Zhang, T., Hao, H.: Active recommendation of tourist attractions based on visitors interests and semantic relatedness. In: Ślęzak, D., Schaefer, G., Vuong, S.T., Kim, Y.S. (eds.) AMT 2014. LNCS, vol. 8610, pp. 263–273. Springer, Cham (2014). https://doi.org/10.1007/978-3-319-09912-5_22

8. Elhammadi, S., Lakshmanan, L.V., Ng, R., Simpson, M., Huai, B., Wang, Z., Wang, L.: A high precision pipeline for financial knowledge graph construction. In: Proceedings of the 28th International Conference on Computational Linguistics, pp. 967–977, December 2020
9. Miao, R., Zhang, X., Yan, H., Chen, C.: A dynamic financial knowledge graph based on reinforcement learning and transfer learning. In 2019 IEEE International Conference on Big Data (Big Data), pp. 5370–5378. IEEE, December 2019
10. Wang, Z., et al.: XLore: a large-scale English-Chinese bilingual knowledge graph. In: International Semantic Web Conference (Posters Demos), vol. 1035, pp. 121–124, October 2013
11. Qu, Y., Liu, J., Kang, L., Shi, Q., Ye, D.: Question answering over freebase via attentive RNN with similarity matrix based CNN (2018)
12. Wang, H.: Knowledge graph convolutional networks for recommender systems (2019)
13. Jorro-Aragoneses, J.L., Bautista-Blasco, S.: Adaptation process in context-aware recommender system of accessible tourism plan. In: Pautasso, C., Sánchez-Figueroa, F., Systä, K., Murillo Rodríguez, J.M. (eds.) ICWE 2018. LNCS, vol. 11153, pp. 292–295. Springer, Cham (2018). https://doi.org/10.1007/978-3-030-03056-8_29
14. Li, Y.U., Huang, Z., Gao, J., et al.: An interpretable attraction recommendation method based on knowledge graph. Scientia Sinica Informationis **50**(7), 1055 (2020)
15. Feng, X., Zhao, X.: A Chinese-Tibetan bilingual knowledge graph system in tourism domain. J. Chin. Inf. Process.**33**(11) (2019)
16. Geng: Research and Construction of Mongolain-Chinese Bilingual Knowledge Graph for Tourism (2019)
17. Zhang, J., Cao, Y., Hou, L., Li, J., Zheng, H.T.: XLink: an unsupervised bilingual entity linking system. In: Sun, M., Wang, X., Chang, B., Xiong, D. (eds.) NLP-NABD 2017, CCL 2017. LNCS, vol. 10565, pp. 172–183. Springer, Cham (2017). https://doi.org/10.1007/978-3-319-69005-6_15
18. Suchanek, F.M., Kasneci, G., Weikum, G.: YAGO: a core of semantic knowledge unifying WordNet and Wikipedia. In: International Conference on World Wide Web (2007)
19. Vrandecic, D., Krtoetzsch, M.: Wikidata: a free collaborative knowledgebase. Commun. ACM **57**(10), 78–85 (2014)

Study on Binary Code Evolution with Concrete Semantic Analysis

Bing Xia[1,2(✉)], Jianmin Pang[1(✉)], Jun Wang[1], Fudong Liu[1], and Feng Yue[1]

[1] Key Laboratory of Mathematical Engineering and Advanced Computing, Zhengzhou 450007, China
zztixiabing@sina.com

[2] ZhongYuan University of Technology, Zhengzhou 450007, China

Abstract. The study on binary code evolution is very crucial for understanding vulnerability repair and malicious code variants. Researchers on code evolution focus on the source code level, whereas very few works have been done to tackle this problem at the binary code level. In this paper, a binary code evolution analysis framework is proposed to automatically locate evolution area and identify evolution semantic with concrete semantic difference. Difference of binary function domain was applied based on function similarity. Trace alignment was used to find evolution blocks, instruction classification semantic was utilized to identify evolution operation, and evolution semantic was extracted combined with function domain elements. The experimental results show that binary code evolution analysis framework can correctly locate binary code evolution area and identify all concrete semantic evolution.

Keywords: Binary code · Code evolution · Evolution area · Semantic analysis

1 Introduction

Binary code evolution is fundamental in scenarios where the program source code is not available such as understand vulnerability repair and malicious code variants. However, researchers at present mainly focus on the source code evolution analysis [1] and source code clone detection [2], very few works have been done to tackle this problem at the binary-level, which often requires a significant amount of human efforts and expertise to understand the semantics of binary code instructions.

Binary code evolution is based on binary code similarity. Binary code similarity approaches compare two or more pieces of binary code e.g., basic blocks, function, or whole programs, to identify their similarities. According to the paper [3], based on open-source Java projects, 92% of the patches change only one file, almost 30% of the patches contain only addition of lines, the top-3 most applied repair actions are addition of method calls, conditionals, and assignments, occurring in 77% of the patches. So, binary code similarity approaches only detect the similarity, but it can't identify where the changes are and what the semantics have been changed. In this paper, evolution is

J. Zeng et al. (Eds.): ICPCSEE 2021, CCIS 1452, pp. 30–43, 2021.
https://doi.org/10.1007/978-981-16-5943-0_3

different from long-term evolution, but refers to the semantic evolution between original version and patched version of a binary program.

To address this problem, we propose a **Bin**ary **C**ode **E**volution **A**nalysis framework named **BinCEA**, to automatically locate and identify the concrete semantic difference. In particular, given the original version and patched version of a binary program, BinCEA can detect the changes of patched version to capture the function-level changes. There are many applications, e.g., finding repair actions, addition of method calls and conditionals, to identify semantic difference of patch analysis. We evaluated BinCEA on cross version binary programs. The experiment results showed that BinCEA can locate evolution area and identify all concrete semantic evolution.

This paper makes the following contributions.

1. We proposed a binary-level semantic analysis framework BinCEA, which can locate evolution area.
2. We have implemented a prototype analysis engine for binary code, which can identify function semantic difference with a combination of instruction classification semantic vector, machine learning and control flow sub-graphs.
3. Experiments show that BinCEA can extract binary code evolution actions such as add variables, add conditionals and assignments.

2 Approach Overview

2.1 Motivation Example

Figure 1(a) is a original version source code, Fig. 1(b) and Fig. 1(c) are patched version of Fig. 1(a). Note that Fig. 1(a) and Fig. 1(b) have the same format string vulnerability (*printf* (*optionalMsg*)). Despite Fig. 1(a) and Fig. 1(b) have high similarity at the source-code level, there are some difference indeed. For example the line 3 adds a variable in Fig. 1(b) which does not have a corresponding variable in Fig. 1(a). The line 11–13 adds a conditional process in Fig. 1(b) which does not have a corresponding process in Fig. 1(a). Compared with Fig. 1(b), Fig. 1(c) and Fig. 1(b) have higher similarity. The only difference is that the vulnerability is fixed by introducing the security function *puts()*, as shown in the green part of Fig. 1(c).

Challenges. Try to extract semantic difference between two cross version functions at the binary level, we face two challenges: evolution area and evolution semantic. In particular:

The control flow graph structure is different, which lead to the uncertainty of evolution area. For example, after compilation, the original version of the binary code has 5 basic blocks in Fig. 2(a) and the patched version of the binary code has 8 basic blocks in Fig. 2(b).

The semantic of instruction is different, which make evolution semantic extraction difficult. For example, *var_10* is introduced into basic block No. 4 in Fig. 2(b), but there is no corresponding variable in Fig. 2(a), and *xor ebx, ebx* as a new basic block No. 3 in Fig. 2(b) while the corresponding block in Fig. 2(a) does not use *xor*.

```
int doCommand(int cmd,char *optionalMsg, char *logPath){
    int counter=1;
    FILE *f=fopen(logPath,"w");
    if(cmd==1){
        printf("%d HELLO",counter);
    }else if(cmd==2){
        printf(optionalMsg);
    }
    fprintf(f,"CMD %d DONE",counter,bytes);
    return counter;
}
```

(a)

```
int doCommand(int cmd,char *optionalMsg, char *logPath){
    int counter=1;
    int bytes=0;
    FILE *f=fopen(logPath,"w");
    if(cmd==1){
        printf("%d HELLO",counter);
        bytes+=4;
    }else if(cmd==2){
        printf(optionalMsg);
        bytes+=strlen(optionalMsg);
    }else if(cmd==3){
        printf("(%d) BYE",counter);
        bytes+=3;
    }
    fprintf(f,"CMD %d DONE",counter, bytes);
    return counter;
}
```

(b)

```
int doCommand(int cmd,char *optionalMsg, char *logPath){
    int counter=1;
    int bytes=0;
    FILE *f=fopen(logPath,"w");
    if(cmd==1){
        printf("%d HELLO",counter);
        bytes+=4;
    }else if(cmd==2){
        puts(optionalMsg);
        bytes+=strlen(optionalMsg);
    }else if(cmd==3){
        printf("(%d) BYE",counter);
        bytes+=3;
    }
    fprintf(f,"CMD %d DONE",counter,bytes);
    return counter;
}
```

(c)

Fig. 1. Original function and Patched function

Furthermore, the code shown in the Fig. 2 assumes that the functions were compiled by the same compiler and optimization level for a special platform. Had the functions been compiled under different optimization level, the generated assembly would differ even more e.g., address or offset which have relation with location.

2.2 Problem Statement

Problem Definition. Given the original and patched versions of a binary program, our goals are to locate evolution area between two cross version binary program and to identify tiny evolution semantic. For the convenience of description, we introduce the following basic definitions.

Definition 1: *Binary code* is machine code that is produced by the compilation process and can by run directly by a CPU. The standard compilation process takes the source code files of a program as inputs. It compiles them with a chosen compiler and optimization level and for a special platform (defined by the architecture, word size, and OS) producing object files. Those object files are then linked into a binary program, either a stand-alone executable or a library. Each binary program contains a number of functions. In this work, we assume that binary programs are in the X86 format.

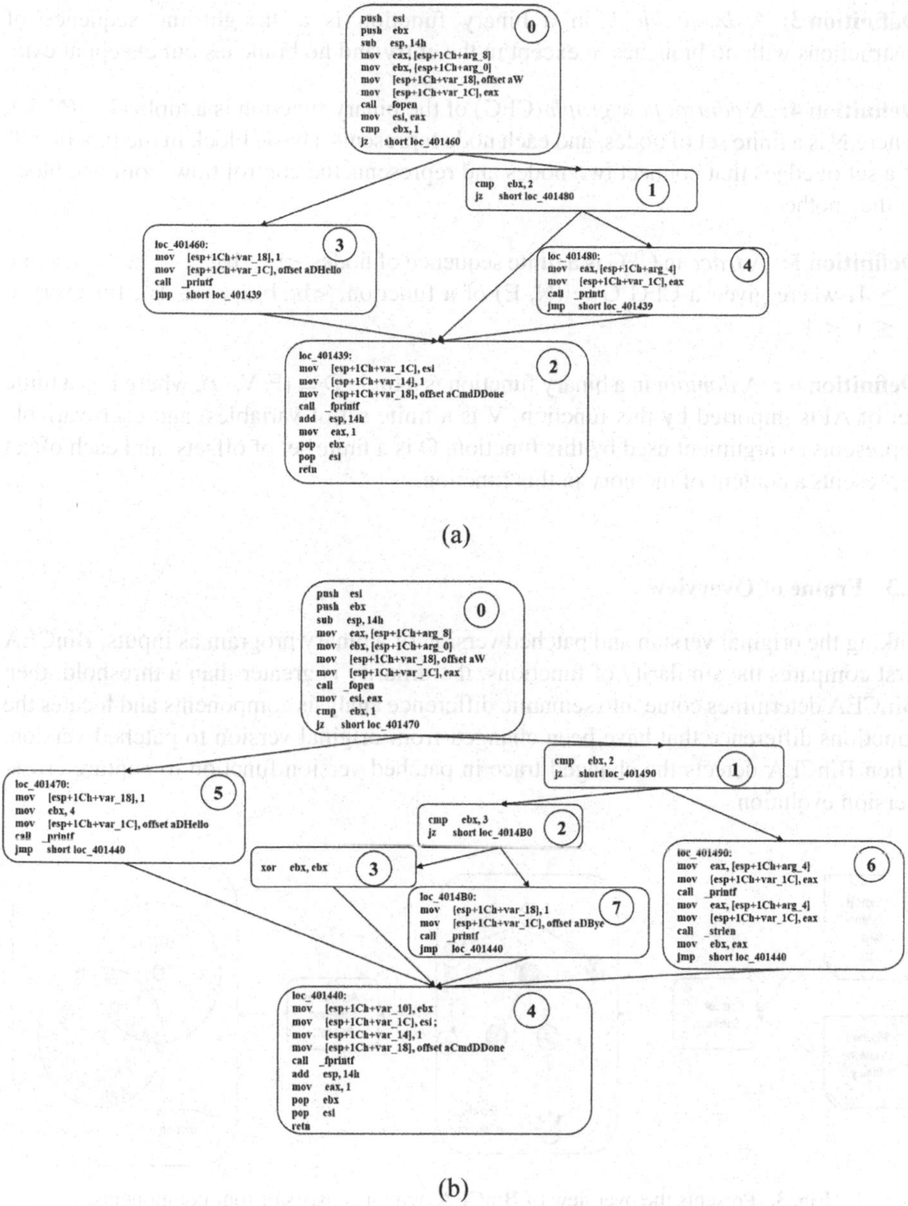

Fig. 2. Original function CFG and Patched function CFG

Definition 2: An assembly *instruction* consists of a mnemonic and up to 3 operands. Each operand can be one argument (register or offset) or a group of arguments used to address a certain memory offset.

Definition 3: A *basic block* in a binary function is a straight-line sequence of instructions with no branches in except to the entry and no branches out except at exit.

Definition 4: A *control flow graph* (CFG) of the binary function is a tuple G = (N, E), where N is a finite set of nodes, and each node represents a basic block in the function, E is a set of edges that connect two nodes and represents the control flow from one block to the another.

Definition 5: A *trace* in CFG is a finite sequence of nodes $<b_1, b_2, \ldots, b_k>$ for some $k \geq 1$, where given a CFG G = (N, E) of a function, $<b_i, b_{i+1}> \in E$, for every i: $1 \leq i < k$.

Definition 6 : A *domain* in a binary function is a tuple D = (F, V, O), where F is a finite set of APIs imported by this function, V is a finite set of variables, and each variable represents an argument used by this function, O is a finite set of offsets, and each offset represents a content of memory in this function.

2.3 Frame of Overview

Taking the original version and patched version of a binary program as inputs, BinCEA first compares the similarity of functions, if similarity is greater than a threshold, then BinCEA determines come into semantic difference analysis components and locates the functions difference that have been changed from original version to patched version. Then BinCEA detects the changed trace in patched version function to capture cross-version evolution.

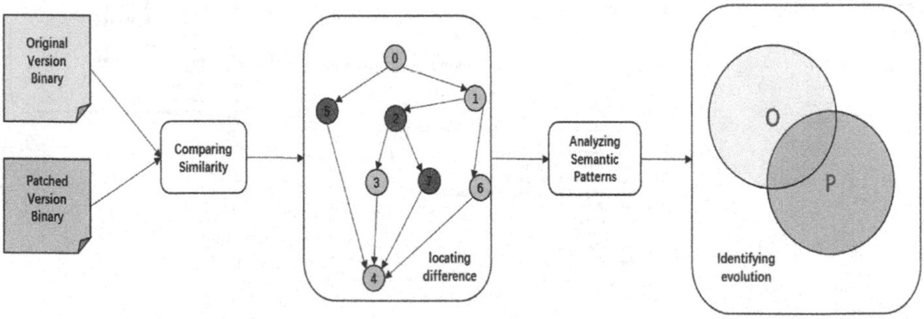

Fig. 3. Presents the overview of BinCEA, which consists of four components.

Comparing Similarity. Given the original version and patched version of a binary function which has been disassembled by IDA Pro [9], the component leverages the technique in paper [7] to compare similarity of function pairs in the original version and patched version. If similarity result is greater than the threshold, our process into the next component.

Locating Difference. This component leverages the pairwise basic block matching with Word2Vec [10] to identify the patched basic blocks. Then, it identifies the relationships among these basic block in terms of trace that capture the locality of a patch. If pairwise result does not equality, we can get the locations of difference between the cross-version functions. As shown in the middle of Fig. 3, the red numbers in CFG of patched version binary are the evolution area.

Analyzing Semantic Patterns. This component first categorizes assemble instructions with similar functionality into different semantic classes, which is robust against syntactical obfuscations such as junk insertion, instruction reordering, and instruction replacement. Then, this component uses Word2Vec to capture an instruction semantic vector, and plus all instruction vector of different semantic classes, so we can capture semantic difference of functions. Finally, component decides whether the changes are caused by patched version.

Identifying Evolution. For a tiny change caused by patched version, this component leverages the pairwise domain elements match to identify semantic difference.

Assumptions. BinCEA has a few underlying assumptions, which may limit its application. First, we focus on cross version function in which only corresponding function is modified. Second, we assume that the functions are compiled by the same compiler and the same optimization level. Third, we focus on X86 32bit binary format.

3 Methodology

3.1 Comparing Similarity

In this component, we use disassembler tool IDA Pro to extract the assembly instructions and construction the CFG for each function. Then, we use Trace to measure the function similarity between two functions by coverage rate of the matching trace.

Extraction Function. Input original version and patched version of a binary program, we leverage disassembler tool IDA Pro and python script to extract follow function information of CFG e.g., assembly instructions sequence of a block, the in-nodes and out-nodes of a block. So we can get all function lists of a cross version binary program.

Function Similarity. First, decomposing a function into a set of 3-traces. Second, proceeding pairwise comparison of traces with an LCS variation to score. Third, if their similarity score is above threshold α, calculate the cover rate used by all match result. Finally, two functions are considered similar if their similarity score is above the threshold β.

3.2 Locating Difference

Patches usually change a small part of the whole binary function, so we use embedding to identify tiny difference of two blocks.

Normalization. We normalize each basic block. The normalization process takes a basic block as an input and returns the normalized basic block. First literately normalizes each assembly instruction in the basic block. Second, according definition 2, an assembly instruction consists an opcode and up to 3 operands, where opcode represents the special action that has semantic information, we replace instruction operands with symbols. Finally, we get a normalized basic block.

Find Difference. Fetch all instruction of a block, compute instruction vector by Word2Vec technology and summation. Once the vector has been extracted, a similarity metric between semantic vectors is used to compute the similarity. We use the strongest constraint, that is if measure is not equal to 1, means that pairwise comparison of blocks are difference. After finishing this component, BinCEA get location difference between two blocks.

3.3 Analyzing Semantic Patterns

In this component, instruction is classified into 13 classes e.g., logical, data transfer, stack, arithmetic. BinCEA leverages Word2Vec to get every instruction semantic vector and summation all instruction vector of a class, so we get 13 vectors of different semantic, and compare the corresponding vector in cross version functions used by Chebyshev distance. If the distance between same instruction classification is not equal to 0, there is a semantic difference of cross version functions. Algorithm 1 gives this procedure.

Algorithm 1: Identify Semantic Difference

```
Input  :
         original function FO, patched function FP
Output : set of semantic difference cross-version
Oclass = Φ   //set of original instruction classification
Pclass = Φ   //set of original instruction classification
Result =Φ   //set of semantic difference cross-version
foreach instr io in FO do
    InstrType = GetInstrType(io)
    Oclass.append(InstrType, io)
end
foreach instr ip in FP do
    InstrType = GetInstrType(ip)
    Pclass.append(InstrType, ip)
end
foreach o,p in zip(Oclass, Pclass) do
    Ovecter = GetVecByWord2Vec(o, io)
    Pvecter = GetVecByWord2Vec(p, ip)
    similarity = ChebyshevDistance(Ovecter, Pvecter)
    Result.append(similarity)
end
return Result
```

In detail, we leverage the technique in paper [18] to generate the semantic summary from cross version functions, describing the semantic classes result in Table 1, which classify 13 classes within two functions according to the semantic type of opcode (Lines 4–11).

Next, fetch all instruction of every class, compute instruction vector by Word2Vec technology and summation (Line 12–14). Once the vector has been extracted, a similarity metric between semantic vectors is used to compute the similarity. Common similarity metrics are Euclidean or Cosine distance. In this component, we use Chebyshev Distance because we find Chebyshev Distance can identify tiny difference (Line 15). By such procedure, cross version concrete semantic difference are constructed.

3.4 Identifying Evolution

Once the differences are located, semantic differences are identified, BinCEA proceeds to summarize evolution patterns from the original and patched function.

BinCEA can identify the newly-added variables, arguments, functions and conditional sensitive instructions in patched version. In general, newly-added arguments are more likely to identify the addition of method call, which can reduce triggered relevant semantic changes in patched version, e.g., arithmetic and data transfer. Newly-added test instructions, especially the ones that depend on test type of instruction such as *cmp*, *test*, and logic type of instruction such as *xor*, *and*, are of the interest for conditional judgement of security patches analysis. Algorithm 2 describes the identification of the evolution process.

```
Algorithm 2: Identifying Evolution
   Input    :F  // binary function
   Output   :Domain  // {F, V, O}
1    Domain = Φ
2    foreach instruction in F do
3       opcode = GetOpcode(instruction)
4       if opcode == call  then
5           Domain.add(f, GetOperands(instruction) ) //type {api}
6    else
7           foreach operand in instruction  do
8               type = GetPrefix(operand)  //type {variable, offset}
9               Domain.add (type, operand)
10          end
11   end
12   return Domain
```

As shown in algorithm 2, it takes function as an input and returns domain (see Definition 6). First, this component gets the opcode (Line 3), if opcode is call, then fetches operand as an API (Line 4–5). An operand can be a register, an immediate value, or a variable, an offset, we uses *GetPrefix* operation to analysis all operands type of an instruction, then add to domain (Line 7–9).

After doing the above procedure, we can get two domains of functions. Accord to Definition 6, each tuple in domain is a set, we can use intersection of the two corresponding sets to extract different elements. Then, with different type element and different semantic classification result, we can identify evolution patterns, e.g., method calls, conditionals, and assignments.

Table 1. Semantic classification of instruction.

Type	X86 Instruction
DataTransfer	mov, cmov, cwd, cdq
String	movs, cmps, sca, lod, stos, rep
Stack	push, pop
Arithmetic	sub, add, shr, shl, sbb, sar, adc, xchg, imul, idiv, mul, div, inc, dec, xadd, ror, rol, p*
Call	call
LEA	lea
IO	in, out
Test	cmp, test, comiss, ucomi
Jump	jmp, j*
SKIP	nop, hlt, mfence, sldt, ret
Logic	and, xor, not, or, bt, bsr, neg
Flags	set, sahf, cld
Float	fld, cvt, vcvt, maxss, pun, shufps, sqrt, unpck, f

4 Evaluation

In this section, we conduct an experiment study on Fig. 1 to answer the following research questions.

RQ1: Where is the semantic difference in function?
RQ2: What code semantics have been evolved?

The experiments were conducted on DELL with 16 GB RAM and Intel Core i7-8650 1.90 GHz. We used Fig. 1 as our evaluation testing.

V1.0 is an open source program with a function, which has 3 arguments, one variable and has format string vulnerability. V1.1 add a variable and a conditional process in function, but it has format string vulnerability too. V1.2 updates a security API to patch format string vulnerability. After disassembled, the CFG of the three versions are shown in Fig. 4.

4.1 Locating Difference

Based on algorithm 1, We use the same colors to represent the same blocks as shown in Fig. 4, e.g., block No. 1 in V1.0 is the same as blocks No. 1 and No. 2 in V1.1. From Table 2 and Fig. 4, it can be seen that BinCEA successfully locates the difference basic blocks in cross-version, e.g., block No. 2 in V1.0 and block No. 4 in V1.1 are the last block, they are difference, which indicate evolution area based on blocks.

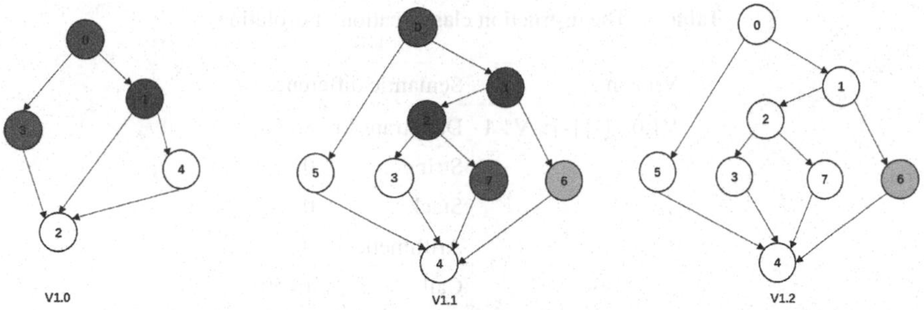

Fig. 4. CFG of cross-version

Table 2. Located same blocks in cross-versions

Cross-version	Same Blocks
V1.0-V1.1	0--->0 1--->1 1--->2 3--->7
V1.1-V1.2	0--->0 1--->1 1--->2 2--->1 2--->2 3--->3 4--->4 5--->5 6--->6 7--->7

4.2 What Semantics Are Evolved

After we capture the locations of semantic difference, we need to describe which semantics had been evolved. Table 3 gives what instruction classification had happened in cross version functions.

In Table 3, column 1 is the information of binary code version, column 2 refers to the differences of semantic behavior. Based on column 2, we know the concrete semantic difference between the cross-version, and know what changed and unchanged. Bigger the number, bigger the gap, 0 means no difference. For example, between V1.0 version and V1.1 version, semantic difference of Data Transfer value is 88.56, Call value is 13.59.

Normalization could lead to high false positives rate. For example, as shown in Table 4, between V1.1 version and V1.2 version, all semantics are not changed, but V1.2 updates a security API to patch format string vulnerability. So, we leverage algorithm 2 to compare elements of domain to identify evolution, experiment shows that BinCEA identifies update a API.

Based on the results form Table 2, 3 and 4, we used BinCEA to compare the cross-version from binary-level analysis, which successfully locate the difference of basic block in patch version and identify concrete semantic evolution in cross-version. The experiment shows that our model has the capability to capture binary code evolution in binary function.

Table 3. The instruction classification of evolution

Version	Semantic difference	
V1.0<{-}{-}>V1.1	Data transfer	88.56
	String	0
	Stack	0
	Arithmetic	0
	Call	13.59
	LEA	0
	IO	0
	Test	8.012
	Jump	0
	SKIP	0
	Logic	7.554
	Flags	0
	Float	0
V1.1<{-}{-}>V1.2	*	0

Table 4. The evolution in function domain

Function Domain			Element Difference	
V1.0	API	"_fopen","_fprintf","_printf"	V1.0 & V1.1	'_strlen' 'var_10' 'off-setaDBye'
	Variable	"var_10","var_1C","var_18","var_14"		
	Memory	"offsetaDHello","offsetaW","offsetaCmdDDone"		
V1.1	API	"_fopen","_fprintf","_strlen","_printf"		
	Variable	"var_10","var_1C","var_18","var_14"		
	Memory	"off-setaDBye","offsetaDHello","offsetaW","offsetaCmdDDone"	V1.1 & V1.2	'_puts'
V1.2	API	"_fopen","_fprintf","_strlen","_printf",'_puts'		
	Variable	"var_10","var_1C","var_18","var_14"		
	Memory	"off-setaDBye","offsetaDHello","offsetaW","offsetaCmdDDone"		

5 Related Work

Our work attempts to locate the differences and capture the concrete semantic changes, then try to explain why evolution happened. In this section, we briefly survey additional related work.

Audit Library Function Methods. The author of [5] propose a method for auditing the semantic differences to tell whether a function is vulnerability-related. However, this method simply judge the function by searching the calls towards dangerous library functions, such as *strcpy* and *strncpy*, which only able to audit buffer overflows. Our method can find more evolutions, including not only dangerous functions, but also variables and memory.

Dynamic Methods. BISSAM [8] uses binary patch diffing information to extract signature used by multiple sandboxes running different versions, e.g., open the document using the appropriate application, log the execution path, and automatically identify embedded malicious code using dynamic binary instrumentation. That is different from our approach, BinCEA uses static analysis to identify cross version evolution which can traverse all paths and recognize all changes.

Control Flow Methods. BinHunt [4] bases on the control flow of the programs using graph isomorphism technique, symbolic execution, and theorem proving. The theory proving and symbolic execution are too complex to find all semantic changes, we use classified instructions and distributed representation learning, which provide exactly semantic difference describing.

Classified Methods. BMM2006 [13], BEAGLE [14], FOSSIL [15] and SIGMA [16] which classified instructions into classes, but they are using semantic similarity comparison, we use classified instruction to find the differences and identify semantic evolution.

Similarity Methods. Safe [11] uses word2vec method to generate function embedding and send it into self attention neural network. The biggest advantage of this scheme is that it can automatically identify encryption function, but encryption and decryption have higher similarity. Safe only detect the similarity, but it can't identify where the changes are and what the semantics have been changed. word2vec method was used in the literature [17].

6 Conclusion

In this paper, we proposed a binary code evolution analysis framework BinCEA to automatically locate different blocks between cross version functions and identify concrete semantic differences, which understand vulnerability repair actions. This paper is based on strong constraint of semantic vector and identifies evolution by comparing elements in function domain, so BinCEA not only identify add new binary code, but also delete or change of the original binary code.

BinCEA is only based on X86 instructions, different optimization levels, different instruction sets, different compilers and stripped binaries have not yet been taken account. In the future, based on binary code evolution, we plan to extend this framework to analysis malicious code variants.

Acknowledgment. We would like to thank the anonymous reviewers for their comments that helped us improve the paper.

The research leading to these results has received founding from the Advanced Industrial Internet Security Platform Program of Zhijiang Laboratory (No. 2018FD0ZX01), the National Natural Science Foundation of China (Nos. 61802435), and the Key Research Projects of Henan college (No. 21A520054).

References

1. Ghazarian, A.: A case study of source code evolution. In: 13th European Conference on Software Maintenance and Reengineering (2009)
2. Saha, R.K., Roy, C.K., Schneider, K.A.: gCad: a near-miss clone genealogy extractor to support clone evolution analysis. In: 2013 29th IEEE International Conference on IEEE Software Maintenance (ICSM) (2013)
3. Sobreira, V., Durieuxy, T., Madeiral, F.: Dissection of a bug dataset: anatomy of 395 patches from Defects4J. In: SANER 2018 (2018)
4. Gao, D., Reiter, M.K., Song, D.: BinHunt: automatically finding semantic differences in binary programs. In: Chen, L., Ryan, M.D., Wang, G. (eds.) ICICS 2008. LNCS, vol. 5308, pp. 238–255. Springer, Heidelberg (2008). https://doi.org/10.1007/978-3-540-88625-9_16
5. Song, Y., Zhang, Y., Sun, Y.: Automatic vulnerability locating in binary patches. In: International Conference on Computational Intelligence and Security. IEEE (2009)
6. Xu, Z., Chen, B., Chandramohan, M., Liu, Y., Song, F.: SPAIN: security patch analysis for binaries towards understanding the pain and pills. In: International Conference on Software Engineering. IEEE Press (2017)
7. David, Y., Yahav, E.: Tracelet-based code search in executables. In: Conference on Programming Language Design and Implementation. ACM (2014)
8. Schreck, T., Berger, S., Göbel, J.: BISSAM: automatic vulnerability identification of office documents. In: Flegel, U., Markatos, E., Robertson, W. (eds.) DIMVA 2012. LNCS, vol. 7591, pp. 204–213. Springer, Heidelberg (2013). https://doi.org/10.1007/978-3-642-37300-8_12
9. IDA (2021). https://www.hex-rays.com/products/ida/
10. Mikolov, T., Chen, K., Corrado, G., et al.: Efficient estimation of word representations in vector space. Comp. Sci. (2013)
11. Massarelli, L., Di Luna, G.A., Petroni, F., Baldoni, R., Querzoni, L.: SAFE: self-attentive function embeddings for binary similarity. In: Perdisci, R., Maurice, C., Giacinto, G., Almgren, M. (eds.) DIMVA 2019. LNCS, vol. 11543, pp. 309–329. Springer, Cham (2019). https://doi.org/10.1007/978-3-030-22038-9_15
12. Xu, X., Liu, C., Feng, Q., Yin, H., Song, L., Song, D.: Neural network-based graph embedding for cross-platform binary code similarity detection. In: ACM Conference on Computer and Communication Security (2017)
13. Bruschi, D., Martignoni, L., Monga, M.: Detecting self-mutating malware using control-flow graph matching. In: Detection of Intrusions and Malware and Vulnerability Assessment (2006)
14. Lindorfer, M., Di Federico, A., Maggi, F., Comparetti, P.M., Zanero, S.: Lines of malicious code: insights into the malicious software industry. In: Annual Computer Security Applications Conference. ACM (2012)
15. Alrabaee, S., Shirani, P., Wang, L., Debbabi, M.: FOSSIL: a resilient and efficient system for identifying FOSS functions in malware binaries. In: IEEE Transactions on Privacy and Security (2018)

16. Alrabaee, S., Shirani, P., Wang, L., Debbabi, M.: Sigma: a semantic integrated graph matching approach for identifying reused functions in binary code. In: The International Journal of Digital Forensics & Incident Response (2015)
17. Massarelli, L., Luna, G., Petroni, F., et al.: Investigating graph embedding neural networks with unsupervised features extraction for binary analysis. In: Workshop on Binary Analysis Research (2019)
18. Kruegel, C., Kirda, E., Mutz, D., Robertson, W., Vigna, G.: Polymorphic worm detection using structural information of executables. In: Valdes, A., Zamboni, D. (eds.) RAID 2005. LNCS, vol. 3858, pp. 207–226. Springer, Heidelberg (2006). https://doi.org/10.1007/11663812_11

Research on COVID-19 Internet Derived Public Opinions Prediction Based on the Event Evolution Graph

Xu Chen[1,2], Feng Pan[1,2], Yiliang Han[1,2](✉), and Riming Wu[1,2]

[1] College of Cryptographic Engineering, Engineering University of PAP, Xi'an 710086, China
[2] Key Laboratory of PAP for Cryptology and Information Security, Xi'an 710086, China

Abstract. As one of the ways to reflect the views of the masses in modern society, online reviews have great value in public opinion research. The analysis of potential public opinion information from online reviews has a certain value for the government to clarify the next work direction. In this paper, the event evolution graph is designed to make COVID-19 network public opinion prediction. The causal relationship was extracted in the network reviews after the COVID-19 incident to build an event evolution graph of COVID-19 and predict the possibility of the occurrence of the derivative public opinion. The research results show the hot events and evolution direction of COVID-19 network public opinion in a clear way, and it can provide reference for the network regulatory department to implement intervention.

Keywords: Online reviews · Causal relationship · COVID-19

1 Introduction

The event evolution graph is based on the knowledge graph. Knowledge graph is a visualization technology to show the relationship between knowledge development process and structure. It provides an efficient and convenient way for data expression, organization and management under the background of big data. At present, it is widely used in intelligent search, deep question answering, social network and other fields. However, knowledge graph is only based on entity knowledge base, which is not enough to describe the evolution law between events. The emergence of event evolution graph solves this problem to a certain extent. The event evolution graph is essentially a logic knowledge base, which describes the evolution rules and patterns between events. In terms of structure, event evolution graph is a graph network with events as nodes and relationships between events as edges. Unlike most entities and relationships in knowledge map are stable, most relationships in reason map are uncertain and transfer with a certain probability. The emergence of event evolution graph provides a new idea for

Foundation Items: The National Natural Science Foundation of China (No. 61572521), Engineering University of PAP Innovation Team Science Foundation (No. KYTD201805), The National Social Science Fund of China (No. 20XTQ007, 2020-SKJJ-B-019).

J. Zeng et al. (Eds.): ICPCSEE 2021, CCIS 1452, pp. 44–54, 2021.
https://doi.org/10.1007/978-981-16-5943-0_4

event prediction, which can better show the coupling relationship between events and realize the prediction of event causality. Generally speaking, event evolution graph can automatically predict other events that may occur after the occurrence of an event, and give the probability of the occurrence of the events.

The concept of the event evolution graph was put forward by Professor Liu Ting's team of Harbin Institute of technology at China National Computer Congress in 2017 [1]. Since it was proposed, its construction technology and application methods have attracted extensive attention in various research fields. Zhao [2] and Ding [3] realized the construction of Chinese event evolution graph, but there are some difficulties in event generalization. In September 2018, the scir team of Harbin Institute of technology, combined with the financial news text released by the media, constructed and released financial event evolution graph Demo, which proved the practicability of the event evolution graph in the financial field. Zhou Jingyan et al. [4] revealed the law and logic of event evolution by analyzing the successive and causal relationship between events, and preliminarily discussed the value of event evolution graph in the field of information from the theoretical level. Bailu [5] put forward a set of automatic event evolution graph construction framework for political texts. It shows the relationship network and logical trend among the major events, which can promote the future research in the field of political science. Zhu Han et al. [6] extracted causal event pairs from aviation safety accident text data, and constructed aviation safety field event evolution graph, which provided data and method support for aviation safety accident prediction, early warning and decision management. Shan Xiaohong and others [7–9] tried to explore and improve the event generalization method in the construction of event evolution graph, and applied event evolution graph to the event prediction in the field of network public opinion, which provided a new method and idea for the prediction of network public opinion, and made up for the shortcomings of the existing event prediction methods.

At present, new local cases appear scattered all over the country, and the government has implemented corresponding control measures, which has caused extensive comments on the Internet. Under the background of high popularity of the Internet, rapid development of media and the reduction of crowd contact by COVID-19's prevention and control, network commentary has become the main way for the masses to express their aspirations and appeals, which has great public opinion value. It is necessary and reasonable to predict the derivative public opinion by building COVID-19 event evolution graph. First, from the present point of view, it is very likely that COVID-19 will become normal in a certain period of time. The focus of epidemic prevention and control and the concerns of the masses are highly similar after the occurrence of local cases. The public opinion after carrying out the relevant work has a certain reference value for the follow-up work of the local government. Second, it can predict the possibility of subsequent events according to the events that have happened, which is highly consistent with the prediction of derivative public opinion based on hot events in the field of online public opinion. Third, it has the function of automatization and visualization to show causal relationship events. Under the background of COVID-19, it can assist government departments to quickly and comprehensively determine the focus of epidemic prevention measures. In this context, this paper constructs an event evolution graph of COVID-19 network public opinion. The construction steps are shown in Fig. 1.

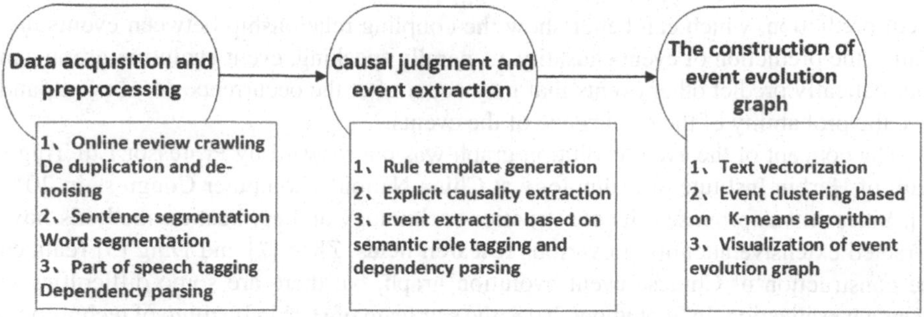

Fig. 1. Flow chart of COVID-19 event evolution graph

2 The Method and Steps of Building the Event Evolution Graph

The core and key to the construction of a domain event evolution graph is to extract causal events, and then construct the reasoning map with generalized events as nodes and causality as edges between nodes. The common steps of building the event evolution graph include the following aspects:

2.1 Data Acquisition and Preprocessing

In order to improve data quality and lay a good foundation for subsequent data processing, the crawled data should be preprocessed, including de duplication, de-noising, sentence segmentation, word segmentation, part of speech tagging and dependency parsing. The main purpose of de duplication is to remove repetitive comments in crawling data, especially when web crawler software Octopus is used to crawl secondary comments, the data is adulterated with repeated primary comments. Denoising mainly includes removing inherent impurities in online data, removing emoticons, rewriting traditional Chinese characters and so on. Sentence segmentation, word segmentation, part of speech tagging, semantic role tagging and dependency parsing are common natural language processing methods. This paper mainly uses Jieba word segmentation module, natural language processing tool HarvestText and natural language processing tool LTP developed by Harbin Institute of technology for processing, and the processing results are ready for subsequent text processing.

2.2 Causal Judgment and Event Extraction

Now that the Internet is highly popular, every small event reported by we media and commented by netizens may evolve into a major public opinion event. If two events occur successively or the former event leads to the latter event, it indicates that there is a causal relationship between them. Hayne's law points out: behind every serious accident, there must be 29 minor accidents, 300 near miss omens and 1000 potential accidents. Similarly, in the field of Internet public opinion, every major public opinion event is inevitably accompanied by some symptoms, and the causal sentence in netizen comments is one of the manifestations of this symptom. This paper extracts causal event pairs from netizens' comments to construct an event evolution graph, so as to predict the possibility of derivative public opinion.

According to whether the sentence contains causal cues, it can be divided into explicit causality and implicit causality [10]. The common methods of causality extraction include pattern matching based method and machine learning based method [11]. This paper uses pattern matching method and python programming to extract causal relationship.

The reason clause and result clause obtained by pattern matching need to extract events in a structured way. The common way to express events is [subject, predicate, object] triple, Ref. [14–16], or gerund double, Ref. [3]. Simply relying on triples or binary to extract events can easily lead to incomplete event extraction, incomplete semantics and even the opposite. Therefore, on the basis of extracting events from triples, this paper uses binary as a supplement and retains negative words.

2.3 Text Vectorization and Event Clustering

The node of the event evolution graph is a kind of causality events, so after extracting a single event, the events need to be vectorized and clustered. TF-IDF is a commonly used method of text vectorization, which is convenient and widely used. However, it only considers the frequency of each word, and ignores the meaning of the word. In this paper, Word2vec model is used to represent the text vector. The model can realize the vectorization of words by deeply learning the segmentation results of a certain field statement, and fully considering the semantics and context in the learning process. The vectorization of events is based on the average of word vectors.

$$model(A) = \frac{1}{n}(model(word_1) + model(word_2) + \ldots + model(word_n)) \tag{1}$$

Where event, $A = \{word_1, word_2, \ldots, word_n\}$ model (A) represents the vector of event A represented by Word2vec model, and model (word_n) represents the vector of word word_n represented by Word2vec model.

After using woc2vec model to vectorize events, this paper uses K-means clustering algorithm to cluster events. K-means clustering algorithm is an unsupervised clustering algorithm, which can realize vector self-clustering based on the number of clusters determined in advance. The number of clusters is determined by the evaluation value SSE of K-means clustering.

Using Visio as a tool, clustering events as nodes, and the relationship between nodes as edges, the event evolution graph is constructed. The nodes of the event evolution graph can be expressed as a set $Nodes = \{e_1, e_2, e_3, \ldots, e_n\}$. $count(e_i, e_j)$ is the number of times node e_i transfers to e_j. Then the weight $count(e_i, e_j)$ transferred from node e_i to e_j can be calculated by the following formula:

$$w(e_i, e_j) = \frac{count(e_i, e_j)}{\sum_k count(e_i, e_k)} \tag{2}$$

3 An Empirical Study of COVID-19 Network Public Opinion Event Evolution Graph

The COVID-19 incident is a global hot issue since 2019. Since the outbreak of the COVID-19 incident in Wuhan, new city crowns have been reported in other cities throughout China, which has aroused wide discussion and concern. This paper mainly selects micro-blog's Chinese commentary in the short term after the emergence of new crowns in various parts of China as corpus, and constructs a Chinese event evolution graph of public opinion on COVID-19.

3.1 COVID-19 Event Data Acquisition and Preprocessing

As a tool to predict the development of events, event evolution graph needs to select data for a specific field or event. The quality of data directly affects the effect of building event evolution graph. As one of the main representatives of new media social software, microblog has great popularity and attention. It can push messages and get netizens' comments at the first time when hot events occur, and has high public opinion research value. Therefore, this paper selects microblog as the corpus source and uses Octopus crawler software to crawl comments to construct the event evolution graph.

The city of Wuhan, Dalian and Qingdao, which once had local cases in China from January 2020 to January 2021, are selected as the foothold. The Octopus software is used to crawl the micro-blog reviews of the local residents in the short term after the local COVID-19 appeared in these cities. For the crawled data, the Octopus software is used for de duplication, and then the NLP tool HarvestText is used to remove the impurities "@", punctuation, emoticons, and simplified Chinese characters in the microblog text, and sentence segmentation is carried out. Then, Jieba word segmentation module is used for word segmentation, and LTP is used for part of speech tagging, semantic role tagging, dependency parsing and other processing to prepare for the subsequent operation (Table 1).

Table 1. COVID-19 event comment data (in Chinese)

Serial number	Place of event	Data volume (unit: sentence)	Sample data
1	Wuhan	6046	Wuhan, come on! I hope it won't be too late to overcome the epidemic.
2	Dalian	3636	Imported food should be strengthened epidemic prevention and control, and imported aquatic products should be appropriately prohibited.

(*continued*)

Table 1. (*continued*)

Serial number	Place of event	Data volume (unit: sentence)	Sample data
3	Qingdao	946	Don't take it lightly before the epidemic is over. You should wear a mask.
4	Shanghai	447	Tears are on the eye. It's too difficult for medical staff to be OK.
5	Chengdu	4664	I hope that the girl will recover soon. In the future, she will have to wear a mask when she goes to dense places, otherwise it will bring unnecessary trouble to the staff.
6	Fuyang	2056	They all wear masks. Have they ever touched the same kind of goods together?
7	Shenyang	3803	She is malicious spread, so this old lady is really not worthy of sympathy!
8	Shijiazhuang	5439	The more dangerous you are, the more you can't have a holiday. If a student is diagnosed, isn't it all over the country.
9	Jilin	2258	It's very nice to say that the health club is actually a pyramid marketing organization. We must intensify our efforts to crack down on it!

3.2 COVID-19 Event Data Acquisition and Preprocessing

Referring to Liu [12] for the construction of causal relationship knowledge base, according to the position and number of relationship cues, the matching syntactic patterns are divided into "at the beginning", "in the middle" and "matching type". According to the relative position of cause event and result event, the syntactic patterns are divided into "from effect to cause" and "from cause to effect". At the same time, referring to the construction of causality knowledge base in literature [5, 6] and the generation of Rule Template in literature [13], the causality syntax extraction template is constructed, as shown in Table 2.

Using Python language, 1338 causal sentences (669 cause sentences and 669 result sentences) are matched according to the template of causal sentences. On this basis, the stop words are removed with the help of stop words list, and then the event is extracted by the combination of triple and binary. The process of event extraction is based on the combination of semantic role tagging and dependency parsing in LTP tools, while retaining the negative words to ensure the semantic integrity. An example of the event is shown in Table 3.

Table 2. Causal syntactic pattern and corresponding clue words (in Chinese)

<table>
<tr><th>Syntactic pattern</th><th>Types of causal cues</th><th>Causal cues</th></tr>
<tr><td>P1: From cause to effect (matching type)</td><td><Cue1, Cue2></td><td><because| due to| since, so| with the result that| so that| thus| therefore| consequently| for this reason> |⟨only| unless | exactly, just⟩|⟨if, so| then⟩</td></tr>
<tr><td>P2: From result to cause (matching type)</td><td><Cue3, Cue4></td><td><cause| result in| lead to| spur| give rise to| the reason why| bring about, the possible reason is| as a result of| on account of| due to></td></tr>
<tr><td>P3: From cause to effect (at the beginning)</td><td>Cue5</td><td>due to| because| since| if| as long as| in order to</td></tr>
<tr><td>P4: From cause to effect (in the middle)</td><td>Cue6</td><td>lead to| result in| spur| facilitate| cause| therefore| consequently| thus| bring about| so that| give rise to| so then| be helpful to| not all</td></tr>
<tr><td>P5: From result to cause (in the middle)</td><td>Cue7</td><td>result of| due to| as a result of| attributed to| interfering factor</td></tr>
</table>

Table 3. Causal syntactic pattern and corresponding clue words (in Chinese)

Causal event	Result event
It's useful to do personal protection	Insist on being ourselves
COVID-19 outbreak	The examination is postponed
Divulging information is a crime	She should be sentenced
There is an epidemic	Hoarding masks
Hospitals should be responsible	Revocation of medical institution license

3.3 The Construction of Event Evolution Graph

For text quantization, we use Woc2vec model training to get COVID-19 domain text vector representation model. Woc2vec word vector training method is divided into two steps: training data preprocessing and word vector training. Here, micro-blog COVID-19 event media articles and user reviews are selected as training sets, and the training data amount is about one hundred thousand sentences. The data preprocessing uses the tool HarvestText to remove the impurities in the microblog text, and uses the Jieba word segmentation module for word segmentation. Input the corpus after word segmentation, set vector dimension size = 200, minimum word frequency min_count = 1 for vector training.

According to the trained vector representation model, the word vector is generated, and then the word vector of each event is calculated by formula (1), and the vector of 1338 events is obtained, which is expressed as a 1338 × 100 matrix to prepare for event clustering. The traditional K-means clustering algorithm is used for event clustering. K-means clustering algorithm is an unsupervised machine learning method, the algorithm needs to set the number of clusters K in advance, the selection of K value directly affects the clustering effect. In this paper, the K value is determined by the evaluation value SSE of K-means clustering. SSE is to calculate the average distance from all points to the corresponding cluster center. Generally, the larger the K value is, the smaller the SSE is. Therefore, through the change rule of SSE with K value, we can find the K value with the smallest reduction of SSE. At this time, K is a relatively reasonable value. As shown in Fig. 2, when k changes from 12 to 13, SSE decreases the least. Therefore, the value of K is set to 12.

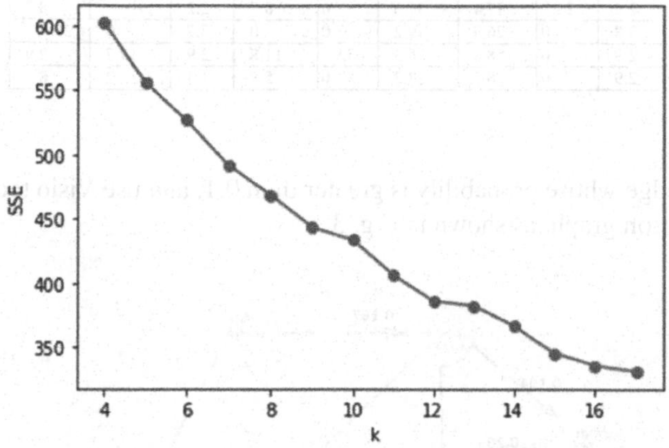

Fig. 2. The change of SSE with K value

After clustering the events with k-means algorithm, the word frequency of 12 categories of text is counted to determine the labels of 12 nodes. For example, for category 5 text, the statistical results of word frequency are shown in Table 4. According to the statistical results of word frequency, "epidemic outbreak" is selected as the label of node 5. The transition probability between 12 nodes is calculated by formula (2), and the results are shown in Table 5.

Table 4. Word frequency of category 5 event (in Chinese)

Key word	Word frequency
Epidemic	55
Burst	31
Serious	28
Hebei	12
Wuhan	10

Table 5. Statistical table of node transfer probability (%)

Target node / Source node	1	2	3	4	5	6	7	8	9	10	11	12
1		16.7	8.3	8.3	33.3	0	0	8.3	8.3	8.3	0	0
2	1.4		0	19.4	25.0	0	9.7	1.4	4.2	4.2	0	23.6
3	0	5.1		7.7	38.5	0	5.1	2.6	12.8	20.5	0	7.7
4	4.1	6.1	1.0		34.7	1.0	2.0	2.0	3.1	5.1	1.0	6.1
5	5.7	2.4	1.4	15.8		1.0	4.3	3.3	6.2	11.5	2.4	8.6
6	0	0	0	21.7	47.8		0	8.7	4.3	8.7	0	4.3
7	0	4.5	0	36.4	31.8	4.5		0	9.1	4.5	0	0
8	5.3	0	0	26.3	36.8	0	0		0	5.3	5.3	15.8
9	4.3	2.2	2.2	17.4	37.0	0	8.7	2.2		8.7	4.3	4.3
10	3.8	3.8	0	26.9	19.2	0	0	7.7	0		0	7.7
11	0	2.9	0	8.8	35.3	5.9	11.8	2.9	14.7	2.9		8.8
12	1.4	2.9	0	8.7	23.2	0	5.8	1.4	23.2	8.7	0	

Select the edge whose probability is greater than 0.1, and use Visio tool to visualize the event evolution graph, as shown in Fig. 3

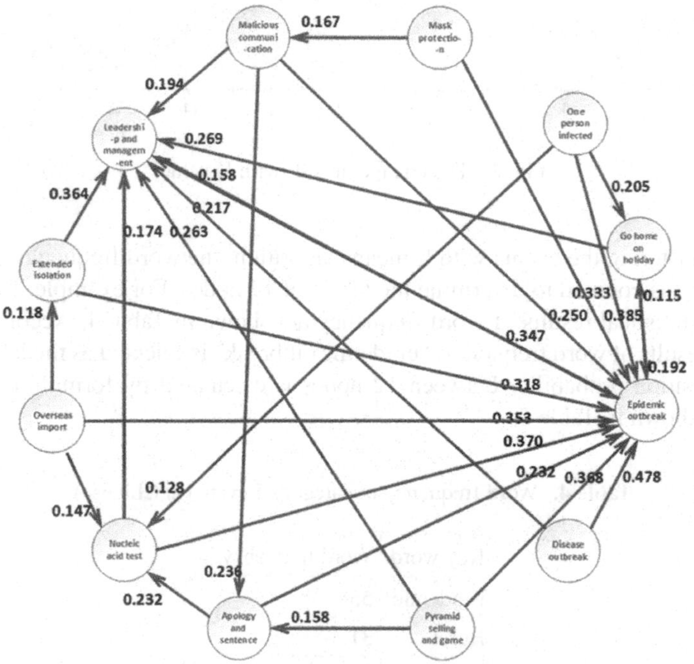

Fig. 3. The event evolution graph of COVID-19's public opinion (in Chinese)

4 Analysis and Discussion

Based on the text information of public opinion in a certain field, the construction of event evolution graph can reflect the possible direction of public opinion evolution in this field to a certain extent. The large-scale corpus in a certain field can fully reflect the hot spots of public opinion, and it can be presented in a clear visual way by the event evolution graph. According to the nodes of the event evolution graph, we can determine the hot spots of public opinion in a certain field. At the same time, the node with large degrees indicates that the event, which is represented by the node, has great attention, and many other events in this field are related to it. On the other hand, the event evolution graph can reflect the probability of transferring from one event to another, which provides a feasible means for event prediction in the field of network public opinion.

In this paper, the COVID-19 domain text is used as an object to construct an event evolution graph. It can clearly show 12 COVID-19 events with high attention and predict the possible direction and probability size of COVID-19 event evolution. For example, according to Fig. 3, the common concerns of COVID-19 during the period of COVID-19 include twelve aspects, such as epidemic outbreak, nucleic acid detection, mask protection, overseas import, extended isolation and so on. In particular, we pay the most attention to the "Epidemic outbreak" and "Leadership and management" events, the degree of which is large. At the same time, the evolution of public opinion events is multi-directional. According to the event evolution graph, we can know the probability of any two events transferring to each other. In the future, when any event occurs in the event evolution graph, we can predict its evolution direction and possibility. For example, when there is malicious communication in a certain place, the probability of 0.194 will evolve into the attention to the management of government and community leaders; the probability of 0.236 will evolve into the masses demanding the malicious communicators to apologize and sentence them; the probability of 0.250 will evolve into the masses' worry about epidemic outbreak. The evolution direction and possibility of public opinion events are of great significance for network regulatory departments to take intervention measures to guide, and government departments to determine the focus of epidemic prevention and control measures.

5 Conclusion

This paper applies the event evolution graph to COVID-19 network public opinion prediction. First, the Internet reviews of micro-blog COVID-19 events were crawled through Octopus software, then the Jieba segmentation tool, natural language processing tool LTP and HarvestText were used to preprocess the text. Then, the causal relation template is constructed to extract causal relation sentences, and the event is comprehensively extracted based on the combination of LTP semantic role annotation and dependency parsing. On this basis, word vectors are trained by Woc2vec model, and then events are represented by average word vectors. After determining K by SSE value, we use K-means algorithm to realize event clustering, and then use generalized events as nodes and Visio as the tool to construct event evolution graph of COVID-19 network public opinion. The research results can show the hot events and evolution direction of COVID-19 network public opinion in a clear way, and it can provide reference for the network regulatory department to implement intervention and the government to do a good job of continuous epidemic prevention and control in the next step.

References

1. Liu, T.: From Knowledge Graph to Event Logic Graph (2017). (in Chinese). https://blog.csdn.net/tgqdt3ggamdkhaslzv/article/details/78557548. Accessed 15 Nov 2017
2. Zhao, S., Wang, Q., Massung, S., et al.: Constructing and embedding abstract event causality networks from text snippets. In: Proceedings of the Tenth ACM International Conference on Web Search and Data Mining, pp. 335–344 (2017)
3. Ding, X., Li, Z., Liu, T., et al.: ELG: an event logic graph. arXiv preprint arXiv.1907.08015 (2019)
4. Zhou, J.Y., Liu, R., Li, J.Y., et al.: Study on the concept and value of intelligence event evolutionary graph. J. Intell. **37**(05), 31–36 (2018). (in Chinese)
5. Bai, L.: Event evolution graph construction in political field. Doctor, University of International Relations (2020). (in Chinese)
6. Zhu, H.: Research on causality of aviation safety accident based on event evolutionary graph. Doctor, Civil Aviation University of China (2019). (in Chinese)
7. Shan, X.H., Pang, S.H., Liu, X.Y., et al.: Analysis on the evolution path of internet public opinions based on the event evolution graph: taking medical public opinions as an example. Inf. Stud. Theory Appl. **42**(09), 99–103 (2019). (in Chinese)
8. Shan, X.H., Pang, S.H., Liu, X.Y., et al.: Analysis and empirical study of policy impact based on event evolution graph. Complex Syst. Complex. Sci. **16**(1), 74–82 (2019). (in Chinese)
9. Shan, X.H., Pang, S.H., Liu, X.Y., et al.: Research on internet public opinion event prediction method based on event evolution graph. Inf. Stud. Theory Appl. **43**(10), 165–170 (2020). (in Chinese)
10. Qiu, J.N.: Research on emergency causality extraction from Chinese corpus. Doctor, Dalian University of Technology (2011). (in Chinese)
11. Li, P.F., Huang, Y.L., Zhu, Q.M.: Global optimization to recognize causal relations between events. J. Tsinghua Univ. **50**(10), 1042–1047 (2017). (in Chinese)
12. Hirschman, L.: The evolution of evaluation: lessons from the message understanding conferences. Comput. Speech Lang. **12**(4), 281–305 (1998)
13. Qiu, J., Du, Y., Wang, Y.: Extraction and representation of feature events based on a knowledge model. In: 2008 IEEE/WIC/ACM International Conference on Web Intelligence and Intelligent Agent Technology, pp. 219–222 (2008)
14. Girju, R.: Toward social causality: an analysis of interpersonal relationships in online blogs and forums. In: Proceedings of the International AAAI Conference on Web and Social Media, vol. 4, no. 1 (2010)
15. Pichotta, K., Mooney, R.J.: Using sentence-level LSTM language models for script inference. arXiv preprint arXiv.1907.08015 (2019)
16. Modi, A.: Event embeddings for semantic script modeling. In: Proceedings of The 20th SIGNLL Conference on Computational Natural Language Learning, pp. 75–83 (2016)

A Text Correlation Algorithm for Stock Market News Event Extraction

Jiachen Wu[1] and Yue Wang[2(✉)]

[1] School of Information, Renmin University of China, Beijing, China
[2] School of Information, Central University of Finance and Economics, Beijing, China

Abstract. To extract effective information in massive financial news, this paper proposes a method to calculate the correlation between text and text set by extracting structured events in the stock market news text, and provides more detailed and interpretable information. First, the structured event triplet was extracted from the text set, and the trained word vector was used to represent the event triplet as an event vector. Event vectors were clustered, the cosine distances were calculated for the cluster centers, and the correlation between the text sets was determined by matching. Finally, the event triplets with the highest correlation between the text sets were selected to provide explanation information for the calculation results. Experimental results show that this method effectively measures the correlation between text and text set.

Keywords: Text relevance · Structured event · Word embedding

1 Introduction

Nowadays, the rapid development of the Internet has brought us a huge amount of information sources. In the financial and other related fields, most investors will use to check the news of the corresponding stock field in the network to learn about market changes. The network news has greatly affected how people view the market orientation and how to make investment decisions. However, it is not realistic to obtain the information in these reports in front of massive news artificially, which also greatly increases the time cost for investors and managers to obtain effective information. Therefore, automatic computing systems such as intelligent investment advisor and news hotspot extraction emerge as the times require. Pioneering work can find news hotspots and extract hot keywords by clustering, but these functions cannot capture structured relationships.

Using event triples for structured event representation can capture the initiator and executed object of an event better. For example, an event triples can be (initiator = CSRC, action = release, executed object = regulatory information public directory). More semantic information can be obtained by using structured event representation.

J. Wu and Y. Wang—Contributed equally to this work. This work is supported by: Engineering Research Center of State Financial Security, Ministry of Education, Central University of Finance and Economics, Beijing, 102206, China; Program for Innovation Research in Central University of Finance and Economics.

J. Zeng et al. (Eds.): ICPCSEE 2021, CCIS 1452, pp. 55–68, 2021.
https://doi.org/10.1007/978-981-16-5943-0_5

However, one disadvantage of using structured representations is that they increase the sparsity and limit the analysis ability. We can use word vector to train event triples into dense event vectors to solve this problem.

Simultaneously, the occurrence of natural events is not isolated. In a certain period of time, multiple events in different fields often have a certain correlation. With the development of China's stock market, there are more and more economic exchanges between different fields and industries. The interaction between similar fields and events may have a linkage effect on market changes. Investors can make investment decisions and guard against investment risks better according to the relevance between their areas of interest and news texts in recent years.

In this paper, we use natural language processing and machine learning methods. We can select domain text sets from a large number of macro stock market news based on keywords provided by investors, such as "epidemic", "medical", etc., calculate the correlation between these texts and domain text sets, and give more detailed explanatory information, including related events, related articles, etc., to facilitate investors' access. The overall research idea of the article is shown in Fig. 1 below. It remedies the information asymmetry problem of the whole market faced by investors, saves the time cost of acquiring effective information, and provides guidance for investors to obtain interested financial information, analyze market trends and make rational investments. It also provides references and suggestions for listed companies to maintain their public image, explore their potential and predict their development direction.

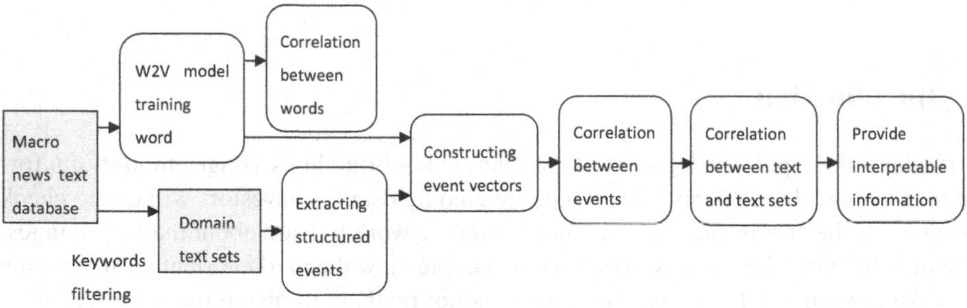

Fig. 1. Overall research idea.

2 Related Work

2.1 Extraction of Stock Market News Information

The efficient market hypothesis (EMH) proposes that the price of securities reflects all available information in the whole market. For a long time, researchers have tried many methods to extract effective information from the stock market to assist prediction and decision-making. A key source of market volatility is financial news. Pioneering work mainly uses simple functions in news documents, such as bag of words model, noun phrase extraction and named entity recognition [1].

Sentiment analysis of news is an entry point for semantic analysis of news, and many reliable sentiment analysis methods have been developed. Pei-Yi Hao et al. extracted hidden topic models and sentiment information from news articles and developed a novel fuzzy dual support vector machine to merge large amounts of information from online news and used it to predict stock price trends [2].

With the development of computing power and NLP technology, the extraction of text information is more accurate and effective. Structured event representation can capture the participants and actions of events in text [3]. This kind of structured representation will increase the sparsity and limit the prediction ability. In recent years, research has focused on deep learning. Benjamin Warner et al. used a hybrid genetic algorithm/support vector regression and BERT to predict the daily closing price of the Dow Jones Industrial Average [4]. The word vectors trained from the neural network model can transform words into dense vectors, and make the vector distances of words with similar semantic environment similar [5]. Zhang Bin et al. used word vector to improve the performance and accuracy of text similarity calculation [6]. However, the representation of events in this method is still a little fuzzy, and it is impossible to study the correlation between the extracted events and other texts or other events. Therefore, this study uses the trained word vector to represent structured events as dense vectors [7, 8]. Even if the events with similar semantics don't have the same vocabulary, similar vectors can be calculated to better obtain and express the semantic information in the news text.

2.2 Text Correlation Algorithm

Many researchers have tried a lot of work in the field of text correlation, and more and more scholars combine a variety of methods to calculate text similarity. Based on the word frequency inverse word frequency system calculation method, Wang Xiaolin et al. added information gain and entropy, and then added semantic weighting factor, and obtained more realistic text similarity calculation results [9]. Atoum et al. first used the algorithm combining information content and distance to measure the similarity between words, and then weighted the word similarity and combined with the text sentence length information to obtain the text similarity [10]. Jinfeng Rao et al. used a text similarity modeling method combining relevance matching and semantic matching [11]. Jiaqi Yang et al. proposed a short text similarity calculation method based on semantic and syntactic information, which used knowledge and corpus to express word meaning to solve the problem of polysemy, and used the selection analysis tree to obtain the syntactic structure of short text [12]. Farouk used discourse representation structure and word order similarity to improve accuracy [13]. Some hybrid methods can optimize the calculation effect of text similarity, and the influence factors and interpretability of each method are different, so we can develop ideas and combine calculation methods according to the needs of project application.

To sum up, this paper proposes a method based on matching and clustering to measure the relevance between text and text set by combining the word vector model of neural network model and extracting event triples by syntactic analysis. Although this method is relatively simple compared with many of the above algorithms, the innovation of this method is to obtain more text semantic information, effectively calculate the relevance, and provide some interpretable information for the calculated relevance.

3 Our Proposed Method

3.1 Word Vector Model

Word vector is to express words as dense vectors, so that the corresponding word vectors of two similar words often appear in the same semantic environment are also similar. A neural network language model with single hidden layer is obtained by unsupervised learning from a large number of unlabeled ordinary text data. The weight of hidden layer learned in this model is called "word vector". To get the word vector, the neural network needs to be able to input a word in a sentence, check the words around the word and select another word randomly, and output the possibility that the word selected in the established vocabulary list is the "nearby word" of the previous words. The probability of output is related to the probability of finding every other word near the input word in our training text. The first step is to create a vocabulary based on the corpus, which has n independent words. Then, the corresponding n dimensional vector is established for each word, which is marked as "1" in the corresponding place of the word and "0" in other places. The input of the network is a single heat vector representing the input words, and the output is also an n dimensional vector. This vector represents the probability (a pile of floating-point values) that each word in our vocabulary is randomly selected around it.

In 2013, Google announced the related tools of word vector. In practical application, we can directly call the jieba tool based on Python to segment the news text, and then use the tools in gensim package to establish the word vector model and obtain the word vector.

3.2 Structured Event and Event Vector

Event triples are usually defined as $E = (O_1, P, O_2)$, where P is the action of the event, O_1 is the initiating object of the event, O_2 is the executed object of the event action. For example, as mentioned in a news article "no one has thought novel coronavirus pneumonia will aggravate the liquidity crisis of HNA". The event e_1 was extracted as (O_1= "sudden novel coronavirus pneumonia", P = "aggravate", O_2 = "the liquidity crisis of HNA").

This paper uses LTP tools based on Python to extract structured events from free text. Language technology platform (LTP) is an open Chinese natural language processing system [14] developed by the laboratory of Harbin Institute of technology, which includes six efficient Chinese language processing functions, such as morphology, syntax and semantics.

In order to extract structured events, we need to do some basic processing such as Chinese word segmentation, part of speech tagging, named entity recognition, dependency parsing and so on. After that, we use semantic role tagging to locate the key predicate in the sentence, and then conduct structural analysis to mark other phrases in the sentence as the semantic roles of the chosen predicate, including the agent, the patient, the place and time of the event [15], to obtain the subject and object of the sentence. We assume that the three elements O_1, P and O_2 in the event should contain subject, predicate and object respectively, so that the event triples that meet the requirements can be screened out from the sentence.

In order to solve the problem of event sparsity, our goal is to express event $E = (O_1, P, O_2)$ as an event vector, so that even if events do not share the same action, agent or patient, we can establish a more basic relationship between events. Because the three elements of the event triplet have specific meanings and the sequence is not interchangeable, an easy representation is to splice the vectors representing the three elements into event vectors three times the length of the word vector. Most event elements are composed of multiple words, so we represent each element as the average of its word vector after word segmentation, so that we can share the statistical strength among the words describing each component.

3.3 Text Correlation Algorithm Based on Matching and Clustering

After obtaining the event triples representing the text, an idea to calculate the text correlation is to calculate the correlation coefficient between all triples of two texts and take the average value. The problem with this idea is that the denominator of the average is the product of the number of triples of two texts, which will dilute the calculation results to a great extent. Therefore, the matching based algorithm is used to optimize the computing performance, that is, for each triplet in the text, the most matching triplet in another text (the triplet with the highest correlation coefficient) is found, and the final text correlation is the sum average of the correlation coefficients of each pair of matching. The denominator of the average is the sum value of the triplet number of two texts.

In the actual calculation, because the extraction algorithm of event triples is to find the subject and object according to the predicate in the sentence, it will lead to the extraction of multiple triples according to the length of each news text, resulting in a large number of event triples corresponding to the text set and low computational efficiency. We use the K-means algorithm to improve the computational efficiency, and generate the cluster center of the event triplet of the text set, which is used to represent the distribution of the event features of the text set. At the same time, the contour coefficient method is used to measure the clustering effectiveness, so as to determine the optimal K value. The contour coefficient is used to evaluate whether the data is suitable for the selected number of clusters. The value is between -1 and 1. The closer the value is to 1, it proves that the average distance within clusters is far less than the average distance between clusters, and the clustering effect is better. In a certain range of K value clustering experiments, the K value with the largest contour coefficient is the optimal K value for this data sample. Here, we assume that under the premise of using the optimal K value in K-means algorithm training, the distribution of clustering centers can roughly represent the distribution characteristics of the data itself. Finally, the operation process of the text correlation algorithm based on matching and clustering is as follows.

Algorithm 1 Text correlation algorithm based on matching and clustering

1 Using the contour coefficient method to find the optimal *K* value of each event triple clustering.
2 Clustering two text sets *A* and *B* with their respective optimal *K* values.
3 Calculate the pairwise correlation degree of *A* and *B* cluster centers, which is recorded as *s(a,b)* and stored in the correlation matrix *S*.
4 Set the total similarity *total_s = 0*, total weight *total_w = 0*.
5 Traverse the matrix *S*, for each cluster *a* in *A*, find the cluster *b* with the highest matching (similarity) in *B*, and record it as *b_a*. *total_s += s(a, b_a)**|*a*|*|*b_a*|, *total_w* += |*a*|*|*b_a*|, |*a*| is the size of cluster *a* (number of triples).
6 Traverse matrix *S* and repeat step 5 for each cluster *b* corresponding to cluster *A* in *B*.
7 Similarity between two text sets *S(A,B)* is *S(A,B) = total_s / total_w*.

End

4 Experiments

4.1 Data Collection and Preprocessing

This paper uses web crawler program to obtain a large number of macro financial news from top financial websites from January 1, 2019 to March 31, 2020, with a total of 22797 news texts. The text processing program is used to remove the header information, meaningless characters and other useless information to prevent them from interfering with the main information.

In order to calculate the correlation between domain text sets, this paper proposes three groups of keywords that investors may be interested in. Each group is composed of one main keyword and three relevant keywords that need to measure the correlation. The main keywords are "epidemic", "5G" and "petroleum". The domain keywords are filtered by month according to the title to form a domain text set, and there is no cross between different domain text sets.

4.2 Relevance Between Words

The bottom layer of text correlation is the correlation between words. This paper uses the word vector model to obtain the vector representation of words, and uses the cosine distance of the word vector as the correlation degree between words.

In the experiment, we use jieba word segmentation device to segment 22797 texts in the crawled text set. It should be noted that the stock market news involves many fields proper nouns. Adding the relevant user-defined dictionary can greatly improve the recognition accuracy of segmentation and make the system have good domain adaptability. The user-defined dictionary used in this paper is composed of several Sogou thesaurus, including: A-share abbreviation thesaurus, such as Angang Steel, Antai group, etc.; financial thesaurus, such as mortgage loan, market value, unfair profit, etc.; economic thesaurus, such as business cycle, production structure, joint venture, etc.

Then the word vector model is built by using gensim package, and the parameter is set to 200 dimensions. The following Table 1 shows the six most relevant words of each keyword measured by word vector and their correlation degree, taking three simulated key words "epidemic", "5G" and "petroleum" as examples. According to the results in the table, we can see that the word vector model can get the semantic information well.

Table 1. Calculation results of relevance of word vector measurement.

Words most related to epidemic situation and their relevance		The most relevant words of 5G and their relevance		The most relevant words of petroleum and their relevance	
Novel coronavirus	0.794	4G	0.815	Saudi Arabia	0.806
Fight against the epidemic	0.771	Commercial	0.813	Aluminum	0.792
Virus	0.770	Base station	0.802	Crude oil	0.786
Pneumonia	0.765	PCB	0.795	Sinopec Group	0.780
Epidemic prevention	0.759	Huawei	0.791	Coke	0.779
Coronavirus	0.753	Localization	0.790	Oil	0.773

4.3 Correlation Between Event Triples

The middle layer of text correlation is the correlation between triples. In this paper, LTP tool is used to extract event triples in each text set, and each element of the first three elements extracted is expressed as the average value of the word vector after word segmentation, which is spliced into 600-dimensional event vector to represent events. In order to debug sampling, this experiment randomly selected 100 from the extracted four tuples to calculate the correlation between the three tuples. The three results with the highest correlation are shown in Table 2 below.

Table 2. Three calculation results with the highest correlation of event triples.

No.	Correlation degree	Subject	Predicate	Object
1	0.892	The 5G users in China	Exceed	200 million
		China Tower 5g base station	Exceed	110 thousand
2	0.860	First three quarters	Realize	The operating revenue was 785 million yuan
		168 listed companies in Shandong	Realize	The operating revenue was 69.4 billion yuan
3	0.839	National Bureau of statistics data	Show	China produced 1.907 billion tons of cement, a year-on-year increase of 5.8% in the first 10 months of this year
		The latest data released by the Chinese Academy of information and communication	Show	The shipment volume of 5G mobile phones in September was 497000, accounting for more than 60% of the first three quarters

According to Table 2, we can see that this algorithm for calculating the correlation of triples can measure the correlation degree of triples well. The most relevant triples are those with similar keywords, semantic and grammatical structures. However, we also find that this algorithm can give higher scores to simple triples with similar syntax structure. For example, in the third example, the correlation between cement and 5G is not high, but these two events are describing some data showing the change of product production in a certain period of time, so the correlation is very high. In fact, we hope that we can give a higher similarity to semantically similar triples. At the same time, the way to get every element vector of event vector is to take the average value of every word vector of every element word segmentation. This average calculation method actually obliterates part of the information, making the calculation result not accurate enough. Here, improvements can be made in the future.

4.4 Relevance Between Texts

This paper uses the algorithm based on matching and clustering to calculate the correlation between texts. The experiment selected a news text from the epidemic text set, which was titled "How much is the impact of the epidemic? Pharmaceutical enterprises return to full capacity production in an emergency", 128 event triples are extracted from the text. Then 100 different news texts are randomly selected from different domain text sets to calculate the correlation with this text.

In order to prove that using K-means algorithm under the premise of optimal *K* value, the distribution of clustering centers can roughly represent the distribution characteristics of the data itself, and the results of matching methods using clustering and not using clustering are almost the same, this paper uses two methods to test the same data. In the clustering experiment, the method of selecting the optimal *K* value is as follows: first, set *K* in a possible interval to experiment respectively, and draw the contour coefficient variation graph, in the graph, select the *K* value of the highest contour coefficient as the optimal *K* value. Figure 2 below takes the initial text to be calculated as an example.

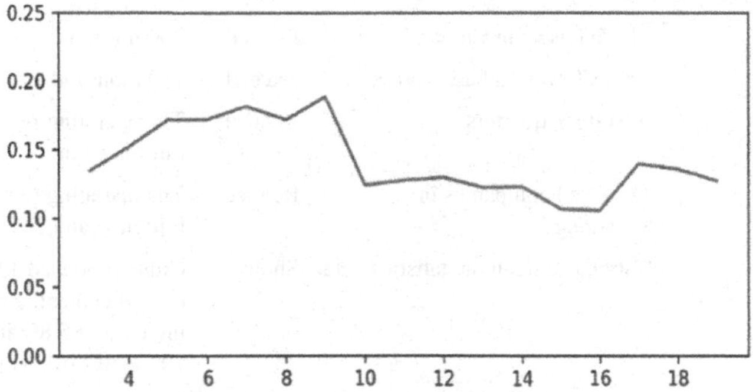

Fig. 2. Initial text contour coefficient diagram.

It is obvious from the figure that for this text, the clustering effect is the best when the *K* value is 9. If several contour coefficients are large and the difference is not big, the smaller value is selected considering the calculation efficiency.

Simultaneously, in order to compare the effect of similarity calculation, the gensim package is used for each text, the Doc2Vec model is used to build the text vector, and the cosine value calculated by the text vector is used as the contrast value to mark the text similarity. Due to the limitation of the length of the article, the following Table 3 randomly selects 5 out of 100 articles and lists the specific results of the text similarity experiment. The statistical results of all 100 experiments are shown in Table 3.

Table 3. Specific similarity experimental results of text relevance between five randomly selected texts and news text the initial text.

Manual sorting	Text name	Matching algorithm	Matching and clustering algorithm	Doc2Vec model
1	All living things in catering industry under epidemic situation	0.450	0.467	0.525
2	March Chinese mainland IELTS test cancelled	0.451	0.448	0.496
3	Set up national efforts to ensure the demand of medical materials	0.439	0.470	0.478
4	Sichuan Baijia food a round financing of 110 million yuan	0.418	0.447	0.444
5	The concept of 5G hot money game mask	0.394	0.381	0.291

The first column in Table 3 is the similarity order sorted manually according to the keywords and semantics of the text content. The smaller the serial number is, the more relevant it is to the original text. We can see that the ranking of the three methods of calculating similarity is similar to that of manual ranking, only the method with clustering gives a higher similarity in the third text, which is inconsistent with the other two methods. In terms of numerical value, the calculation method based on matching gives the most concentrated values, the range is small, the similarity difference is small, and it is more average. After adding the clustering algorithm, the range of similarity becomes larger, which proves that the vector distribution feature represented by the clustering center is more obvious. Similarly, the data in the table can prove that we can use the band clustering method to improve the computational efficiency without affecting the computational results.

Doc2Vec model itself can obtain the semantic information in the text and calculate the text similarity. The problem is that the vector information of the text cannot provide a reasonable explanation for the similarity calculation results. The method comparison proves that this method can measure the text similarity well, and the advantage of our algorithm is that we can also show the triads that are most helpful to calculate the similarity, so as to provide better explanatory information.

Next, the statistical results of all 100 experiments are shown in Table 4 below.

Table 4. Statistical results of text correlation experiment between 100 randomly selected texts and original news texts.

Text source domain	No. of texts	Maximum	Minimum	Average	Total similarity
Epidemic	25	0.668	0.088	0.380	0.019
Medical	25	0.518	0.067	0.349	0.012
Food	25	0.518	0.070	0.287	0.019
5G	25	0.468	0.13	0.296	0.010

According to the statistics of 100 results in Table 4, we can see that the calculation of the correlation with the text from the epidemic text set shows that the epidemic situation and medical treatment have a relatively high average correlation, and there is no significant difference between them, while the food and 5G have a relatively low average correlation, which is in line with the reality.

4.5 Correlation Between Text Sets

Finally, we use the algorithm based on matching and clustering to calculate the correlation between texts. As mentioned in the previous article, this paper proposes three groups of keywords that are interested in by simulated investors, each of which is composed of one main keyword and three related keywords that need to measure the relevance. The following is the result of correlation calculation of these three groups of text sets.

The key words interested of "5G" are "chip", "chemical industry" and "infrastructure", and the data of November 2019 are selected. Among them, 123 texts are obtained from 5G text set, and 1736 triples are extracted. The optimal K value obtained from contour coefficient graph is 29. The results of correlation calculation are shown in Table 5.

The key words interested of "petroleum" are "steel", "electricity" and "new energy", and the data of December 2019 are selected. Among them, 60 texts are obtained from the petroleum text set, 770 triples are extracted, and the optimal K value obtained from the contour coefficient graph is 28. The results of correlation calculation are shown in Table 6.

The keywords of "epidemic" are "medical", "biological" and "food", and the data in February 2020 are selected. 1033 texts and 20857 triples were extracted from the epidemic text set, and the optimal K value obtained from the contour coefficient graph was 29. The results of correlation calculation are shown in Table 7.

Table 5. Calculation results of relevance of domain text sets related to 5G.

Domain text set	No. of texts	No. of triples extracted	No. of clusters	Correlation with 5G text set
Chip	35	709	37	0.785
Chemical industry	46	995	27	0.671
Infrastructure	11	385	18	0.606

Table 6. Calculation results of relevance of text sets in petroleum related fields.

Domain text set	No. of texts	No. of triples extracted	No. of clusters	Correlation with petroleum text set
Steel	36	321	27	0.781
Electricity	194	2283	21	0.751
New energy	37	873	24	0.725

Table 7. Results of correlation calculation of epidemic related domain text sets.

Domain text set	No. of texts	No. of triples extracted	No. of clusters	Correlation with epidemic text set
Medical	264	3634	16	0.836
Biological	163	2447	14	0.746
Food	24	323	36	0.720

According to the calculation results of these three groups of text sets, we can see that the calculation method in this paper is ideal, which is in line with the actual situation. It proves that our algorithm can calculate the correlation of text sets well, and can be recommended to investors as an effective method.

4.6 Interpretable Information of Correlation Calculation

Compared with other methods, such as word matching method, vector space model, document embedding model and so on, the text similarity calculation method based on structured event extraction can provide interpretable information while containing more semantic information. For stock market news texts, when investors know that there is a high degree of correlation between two fields or two texts, they can have a deeper understanding of the structured events from which the high degree of correlation comes. Through in-depth reading of the extracted information, investors can better comb the stock market development.

When calculating the most relevant event triples, considering that there is only one or two words of the same event in Chinese, two different event triples are formed.

Therefore, the two event triples with high similarity should be filtered out in a certain range. Through experiments, the better parameter for the data set in this paper is to filter out the cosine distance between 0.99 and 1 A pair of event triples between. At the same time, in order to obtain more effective results, it is recommended to filter out triples with the same subject and predicate. Finally, taking 5G and the five most relevant triples of the chip text set as examples, we show the interpretable information of the text correlation calculation, as shown in Table 8 below.

Table 8. The most relevant event pairs of 5G and chip text set.

No	Correlation degree	Subject	Predicate	Object
1	0.972	Shunwang Technology	Said	The company has launched a client-side cloud game testing platform
		Runxin Technology	Said	The company provides TWS wireless Bluetooth headset chip design
2	0.968	Bohai Securities	Point out	Three operators announced the official launch of 5G business
		Chuqing	Point out	Mobile phone manufacturers are over prepared in 5G period
3	0.962	Huawei	Is	In the leading position in the industry
		General purpose the chip	Is	At the advanced level in the industry
4	0.949	Glory	Release	Its first 5G mobile phone series
		Huixinchen	Release	The first self-developed chip
5	0.931	Fiberhome Communications	Is	China's only scientific research center of three strategic technologies in the field of optical communication
		Coprocessor	Is	An auxiliary computing chip

According to the results in Table 8, because the object is relatively simple, the most relevant event pairs often have the same object. The most relevant groups of triples are similar in keyword, semantic and grammatical structure. The measurement method in this paper can provide some explanatory information, but the triad shown here has only been screened, and the effect needs to be improved.

5 Conclusion

In view of the background of extracting effective event information from massive financial news, this paper proposes a method to calculate the correlation between texts or text sets by extracting structured events from stock market news texts. Through empirical analysis, it is proved that the method can effectively extract structured events and express them as event vectors, so that the text representation contains more semantic information and improves the efficiency of information collection and collation; the algorithm of calculating text relevance can take into account both computational efficiency and text information. At the same time, it gives explanatory information to a certain extent, which is convenient for investors to consult. However, the calculation method of the correlation degree of triples in this paper also tends to give the triples with similar grammatical results higher scores and the average calculation method obliterates part of the information, which can be improved in the future.

References

1. Schumaker, R.P., Chen, H.: Textual analysis of stock market prediction using breaking financial news. ACM Trans. Inf. Syst. (TOIS) **27**(2), 12 (2009)
2. Hao, P.Y., Kung, C.F., Chang, C.Y., et al.: Predicting stock price trends based on financial news articles and using a novel twin support vector machine with fuzzy hyperplane. Appl. Soft Comput. **98**, 106806 (2021)
3. Ding, X., Zhang, Y., Liu, T., et al.: Using structured events to predict stock price movement: an empirical investigation. In: Proceedings of the 2014 Conference on Empirical Methods in Natural Language Processing (EMNLP), pp. 1415–1425 (2014)
4. Warner, B.: Predicting the DJIA with news headlines and historic data using hybrid genetic algorithm/support vector regression and BERT. In: Proceedings of the International Conference on Big Data – BigData 2020, pp. 23–37 (2020)
5. Kilimci, Z.H., Duvar, R.: An efficient word embedding and deep learning based model to forecast the direction of stock exchange market using Twitter and financial news sites: a case of Istanbul stock exchange (BIST 100). IEEE Access **8**, 188186–188198 (2020)
6. Zhang, B., Hu, L.M., Hou, L., et al.: Chinese event detection and representation based on word vector. Pattern Recognit. Artif. Intell. **3**, 275–282 (2018)
7. Ding, X., Zhang, Y., Liu, T., et al.: Deep learning for event-driven stock prediction. In: Proceedings of the International Joint Conferences on Artificial Intelligence Organization (IJCAI), pp. 2327–2333 (2015)
8. Lei, Y.X., Zhou, K.Y., Liu, Y.C.: Multi-category events driven stock price trends prediction. In: Proceedings of the 5th IEEE International Conference on Cloud Computing and Intelligence Systems (CCIS), pp. 497–501 (2018)
9. Wang, X.L., Xiao, H., Tai, W.P.: Research on text similarity detection system based on hadoop platform. Comput. Technol. Dev. **25**(8), 90–93 (2015)
10. Atoum, I., Otoom, A.: Efficient hybrid semantic text similarity using Wordnet and a corpus. Int. J. Adv. Comput. Sci. Appl. **7**(9), 124–130 (2016)

11. Rao, J., Liu, L., Tay, Y., et al.: Bridging the gap between relevance matching and semantic matching for short text similarity modeling. In: Proceedings of the 2019 Conference on Empirical Methods in Natural Language Processing and the 9th International Joint Conference on Natural Language Processing (EMNLP-IJCNLP), pp. 5369–5380 (2019)
12. Yang, J., Li, Y., Gao, C., et al.: Measuring the short text similarity based on semantic and syntactic information. Future Gener. Comput. Syst. **114**, 169–180 (2021)
13. Farouk, M.: Measuring text similarity based on structure and word embedding. Cogn. Syst. Res. **63**, 1–10 (2020)
14. Li, Z.: Research on Automatic Extraction of Key Words and Text Summaries in Text Mining. Qingdao University of Technology (2018)
15. Yang, X.: Research on Answer Optimization Method for Question Answering System. Harbin Institute of Technology (2017)

Citation Recommendation Based on Community Merging and Time Effect

Liang Xing[(✉)], Lina Jin, Yinshan Jia, and Chunxu Wu

Liaoning Petrochemical University, Fushun, Liaoning, China

Abstract. The accuracy of information network partition is not high and the characteristics of metapath cannot represent the attributes of network nodes in the existing academic citation recommendation algorithms. In order to solve the problems, a similarity measurement algorithm, community merging and time effect PathSim (CMTE-PathSim), based on community merging and time effect is proposed. On the premise of dividing heterogeneous information network (HIN) effectively, the algorithm considers the influence of node information on the characteristics of metapath. The results of Top-k query verify the effectiveness of CMTE-PathSim on real datasets and improve the quality of citation recommendation.

Keywords: Literature information network · Meta path · Community merging · Similarity measure · Time effect

1 Introduction

With the rapid growth of information containing secret connections, HIN emerges as the times require [1]. The information network is complex, and the appropriate division method is conducive to the processing of data. References [2] and [3] use hierarchical mutual information and hierarchical mapping equation to divide information network respectively. Reference [4] uses potential topic model to find related papers. A large number of papers can share a topic, which makes topic similarity not work well. Some researchers have proposed the graph model [5] and the meta path [6], which a variety of citation features can be obtained in the literature network. Personalized PageRank (abbreviated as P-PageRank) [7] and [8] are well researched similarity measurement algorithms on the meta path. In reference [9], a method is proposed with symmetric element path, which captures the similarity between peers. At present, the similarity measurement algorithms [10], [11] and [12] waste time searching the whole network, and dividing the network is a feasible means. In addition, node attributes

Supported by the Scientific Research Fund of Liaoning Provincial Education Department (L2019048), and Talent Scientific Research Rund of LSHU (2016XJJ-033) of China.

J. Zeng et al. (Eds.): ICPCSEE 2021, CCIS 1452, pp. 69–77, 2021.
https://doi.org/10.1007/978-981-16-5943-0_6

show implicit information that cannot be reflected by meta-path characteristics. Combining node attributes can improve the quality of recommendation.

In this paper, we construct term communities and merge communities with high similarity without destroying the reference relationship. Secondly, we establish the meta path characteristics affected by time effect to define similarity measures and recommend them. The algorithm framework in this paper is shown Fig. 1.

The main contributions of this paper are as follows:

1. We propose a new data structure named the term community, which captures the semantic relationship and similarity of papers between documents. This method can better divide the network structure and reduce the search scope.
2. We use meta path based feature space to represent the association information in the citation network, and add time effect factor to calculate the recommended score in the feature space. Experimental results on real datasets show that we can generate recommendation lists with high efficiency and quality.

The rest of this paper is arranged as follows. The problems studied are described in the second section. The third section introduces the proposed method. In the fourth section, the experiment is carried out with real data set. The fifth section is the summary of this paper.

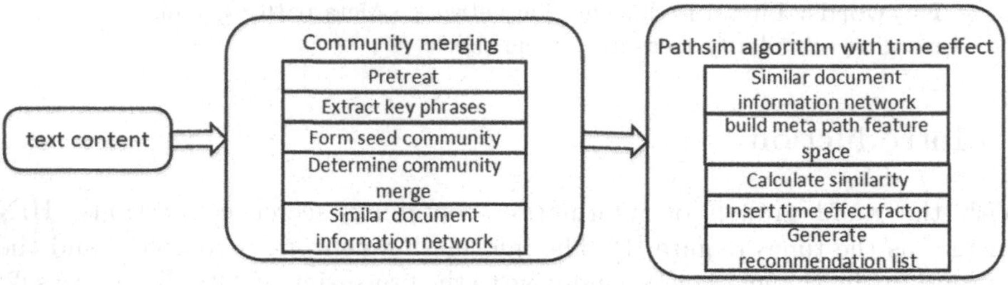

Fig. 1. Algorithm framework

2 Problem Statement

Theorem 1. *Heterogeneous information network is an information network that contains multiple nodes and relationships, which can be represented by* $G = \{V, E, H, R\}$. *In the network* G, *the node set is* V, *the link relation is* E, *the node type set is* H, *and the type set of the link relation is* R. *If and only if* $|H| > 1$ *or* $|R| > 1$, *the network* G *is a heterogeneous information network. If* $|H| > 1$ *and* $|R| > 1$, *then* G *is isomorphic.*

Theorem 2. *Network pattern [6] is usually defined as* $N = (V, E)$, *which is similar to E-R graph in database. Network pattern is a kind of meta description of network.*

Theorem 3. *The expression of meta path [6] is* $A_1 \xrightarrow{R_1} A_2 \xrightarrow{R_2} \ldots \xrightarrow{R_l} A_{l+1}$, *which expresses the complex relationship from* A_1 *to* A_{l+1}.

Theorem 4. *PathSim: A meta path-based similarity measure. Given a symmetric meta path P, PathSim between two objects of the same type x and y is:*

$$s(x,y) = \frac{2 \times |\{p_{x \to y} : p_{x \to y} \in P\}|}{|\{p_{x \to x} : p_{x \to x} \in P\}| + |\{p_{y \to y} : p_{y \to y}\} \in P|} \tag{1}$$

where $p_{x \to y}$ is a path instance between x and y,$p_{x \to x}$ is that between x and x, and $p_{y \to y}$ is that between y and y.

3 Community Merging

The first step of our model is to divide ACM information network. The traditional topic modeling method [13] calculates the similarity by comparing the topic distribution of the paper, but a single word makes the topic too broad, such as "research" and "algorithm". In order to solve the above problems, we propose a community merging method based on multiple key phrases. Community merging is divided into two steps. We extract key phrases from the abstracts of each target paper and its citations to form a community (the target paper is used as the identifier of the community). Key phrases express the topic information or research content of the community. Then, all communities are integrated by merging technology, and each community is merged with the other communities once. The merged large community is the network area successfully divided. The target paper is still used as the identifier of the community. This community is the search space for papers with similar topics as the target paper. A paper can be divided into different communities. In the same community, the citation probability between all citations and the target papers is higher than that of other papers not in the community.

We directly call HanLP [14] key phrase extraction interface. On the premise of extracting key phrases, the target paper p_0 and the papers directly cited by it form a community C_p. We define the community as follows:

$$C_p = \{(p_0, p, T) \mid \exists T, p_0 \in T, p \in T\} \tag{2}$$

Community merging is a measure to judge the similarity between communities. By fusing with other communities, the spatial scope of the community can be expanded. It also shows that there is potential interdependence between the two communities.

We define the community merging measure (CMM) as follows:

$$B_{ij} = \frac{(T_i \cap T_j)}{(T_i \cup T_j)} \tag{3}$$

where T_i and T_j are the key phrases of C_{pi} and C_{pj}.

Community merging needs a pre-defined threshold to limit its value, and the key phrases with high mutual information score should be selected as far as possible. We will discuss this problem in the experiment section of Sect. 5.

4 CMTE-PathSim Model

4.1 Meta Path-Based Feature Space

We need a structural feature to define the structural similarity between papers in ACM information network. Using the meta path and PathSim defined in the Sect. 2, the feature space is defined by the Cartesian product of two sets:

$$\mathrm{F} = \mathrm{P} \times \mathrm{M} \tag{4}$$

where P is the set of possible meta paths, M is the metric, and PathSim algorithm is directly used here. Feature space can be understood from two aspects. On the one hand, P represents the relationship between entities. On the other hand, different metrics can be defined on the same meta path to show the different degree of relationship between entities.

4.2 Citation Probability

After defining community merging and feature space based on meta path, we can comprehensively calculate the similarity of papers in the same community under multiple meta paths. Using the probability calculated by the feature space and the reduced search range, our framework can effectively obtain the recommendation probability of each paper on the premise of determining the target paper. I can define the recommendation probability as follows:

$$Pr\left(p_0, p, T\right) = \frac{e^z}{e^z + 1} \tag{5}$$

where $z = \sum_{f_i \in F} \theta_i \cdot f_i.Pr\left(p_0, p, T\right)$ is the probability that paper p_0 cites paper p. F is the feature space and θ_i is a normalized weight value for feature f_i.

In order to train the recommendation citation model, we decided to use the merged paper community as a high-quality training data set, and design each combined community as a training set containing 0–1 values of citation relationship, which can help the model to capture trivial and detailed information.

The definition of training data set is as follows:

$$\mathcal{T} = \{(p_0, p, label) \mid \exists T_i, p_0 \in T_i, p \in T_i\} \tag{6}$$

where p_0 and p are the target papers and the papers in the same community, respectively, and T_i is the key phrase set corresponding to the community.

In order to train the model, we use L_2 regularized logistic regression to estimate the optimal θ of a given data set $\mathcal{T}$. The L_2 regularization is an effective means to alleviate the overfitting. The regularization term can be added into the cross entropy loss function to correct the parameters in the gradient descent.

$$\hat{\theta} = argmin_\theta \sum_{i=1}^{n} - \log Pr\left(label \mid p_0, p; \theta\right) + \mu \sum_{j=0}^{d} \theta_j^2 \tag{7}$$

With the above objective function, θ_j can be easily estimated by various optimization methods. In the experiment, we use the standard gradient descent method to derive $\hat{\theta}$ to minimize the objective function.

4.3 Time Effect

The publication time of a paper will affect the probability of recommendation. The paper published in 2020 is likely to be more important than the one published in 2010. The shorter the publication time, the more novel the research content of the paper is, which is conducive to the work of researchers. In order to solve the above problems, we define the time effect factor as follows:

$$F_d(t_i, t_j) = \begin{cases} \exp\left(-\ln 2\,(t_i - t_j)/t_d\right) & (if t_i > t_j) \\ 1 & (if t_i = t_j) \\ \exp\left(\ln 2\,(t_j - t_i)/t_d\right) & (if t_i < t_j) \end{cases} \quad (8)$$

where t_i is the publication time of the target document i; t_j is the publication time of the cited article j; t_d is the half-life of the citation time factor.

4.4 Citation Recommendation

With learning the above definition, we can build CMTE-PathSim model and use this model to calculate the reference probability of all papers. CMTE-PathSim is defined as follows:

$$Sc = F_d(t_0, t_i) \cdot Pr(p_0, p_i, T) \quad (p_0 \neq p_i, p_0 \in C_p, p_i \in C_p) \quad (9)$$

With p_0 is the target paper, p_i is all papers in the community excluding p_0, T is the key phrase set of the community, t_0 is the publication time of paper p_0, t_i is the publication time of paper p_i.

Given a query paper p^*, we first find a community whose identifier is paper p^*, and we define the community as C_p^*. All papers in this community are called $P(C_{p*})$, and then the citation probability of $Sc(p_0, p)$ of each paper is calculated. In the following chapters, we conduct a group of experiments to compare CMTE-PathSim model with baseline model, and the results show that our method can recommend papers more effectively.

5 Experiments

5.1 Dataset and Methods Setup

ACM data set is used in the experiment. After removing the literature with incomplete author and abstract information, 9013 papers, 16896 author names and 102625 citation relationships were selected. In order to describe the heterogeneous relationship between entities in ACM information network, we use seven different meta paths, namely$(P-A-P)$, $(P-A-P)^2$, $(P-A-P)^3$, $(P-C-P)$, $(P-T-P)$, $(P-T-P)^2$, $(P-T-P)^3$. We use PathSim algorithm with time effect factor as similarity measurement tool. In order to capture most of the similarity characteristics, we extract three key phrases from each paper abstract, and the threshold of community merging is 0. Using the data structure of community merging, we can make full use of the relationship between the target papers and the papers directly cited by them.

We verify the validity of the time effect factor in the model. Compared with CM-PathSim algorithm without time effect factor, our framework proves that publishing time will affect the content of recommendation list. Then, our framework is compared with P-PageRank and PathSim algorithms. Our method is supervised and needs data sets to train. In order to compare the fair competition of the experiment, we randomly divided all the merged communities into five sets, four of which were used as the training set and the remaining one as the test set.

5.2 Verifying Time Effect Factor

Table 1. Performance using time effect factor.

Methods	Title	Time
CMTE-PathSim	Online help systems: a conspectus	1984
	Rapid prototyping of information management systems	1982
	Rapid prototyping of interactive information systems	1982
	An integral approach to user assistance	1981
	Prefetching in file systems for MIMD multiprocessors	1990
CM-PathSim	An integral approach to user assistance	1981
	Rapid prototyping of information management systems	1982
	Online help systems: a conspectus	1984
	Rapid prototyping of interactive information systems	1982
	Optimization of inverted vector searches	1985

It can be seen from Table 1 that the time effect factor has a good influence on the literature order of recommendation list. In CMTE-PathSim algorithm, the article "Online Help Systems: A Conspectus" ranks first, but in CM-PathSim algorithm, "Online Help Systems: A Conspectus" ranks third. This shows that, regardless of the publication time, the article is not recommended first, but recommends older papers with greater weight. Time effect factor can make the papers published in shorter time be recommended first, increase the score of new papers, and reduce the score of old papers.

5.3 Comparison with Benchmark Algorithm

Table 2. Performance as query processing on ACM network.

Methods	Category 1		Category 2		Category 3	
	recall@20	recall@40	recall@20	recall@40	recall@20	recall@40
P-PageRank	0.1997	0.3022	0.1978	0.2985	0.1965	0.2962
PathSim	0.2311	0.3285	0.2266	0.3209	0.2254	0.3174
CMTE-PathSim	0.2419	0.3373	0.2368	0.3288	0.2242	0.3265

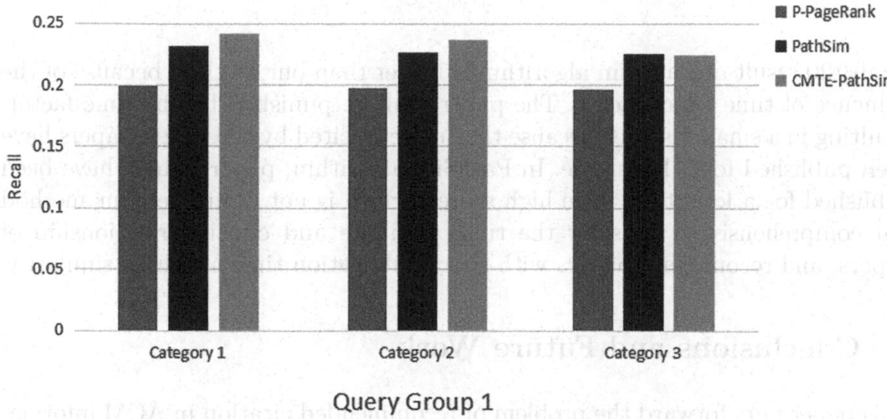

Fig. 2. Recall at 20

We randomly selected 60 papers from the test set and divided them into two groups with 30 papers in each group. Within the same group, there are three categories according to the number of cited relationships. The first category consists of 10 papers, each with no more than 20 references. The second group had 10 papers, each with between 20 and 30 references. The third group had 10 papers, each with more than 30 references. The recommended results are the mean values of the two groups of experiments, as shown in Table 2 and Fig. 2 and 3. We use two query indicators to evaluate the performance of the methods, which are the recall rate of top 20 query results and the recall rate of top 40 query results recall@20 and recall@40. Based on these measurements, we can find that our method can recommend a better list of papers. For example, our method improves recall@20 by 21.1% in category 1 compared with the P-PageRank algorithm. Our method improves recall@20 by 4.6% in category 1 compared with the PathSim algorithm. It is worth noting that in category 3, the

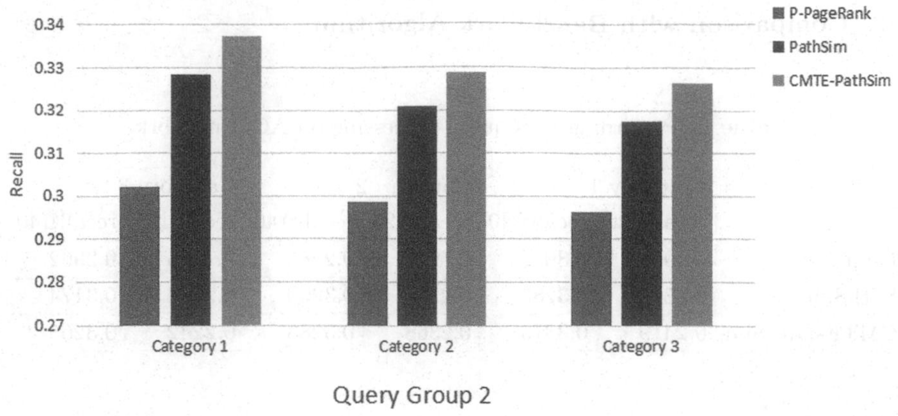

Fig. 3. Recall at 40

recall@20 result of PathSim algorithm is better than our method because of the influence of time effect factor. The papers will be punished by the time factor, resulting in a smaller score. Because they directly cited by the target papers have been published for a long time. In PathSim algorithm, papers which have been published for a long time have high scores, which is not desirable. Our method can comprehensively consider the time attribute and citation relationship of papers, and recommend papers with short publication time and high similarity.

6 Conclusions and Future Work

This paper puts forward the problem of recommended citation in ACM information network. Community merging is a new data structure, which can eliminate irrelevant papers and reduce the search scope. In the new data structure, we extend the classic PathSim algorithm by combining time effect factor. Experiments show that the framework can recommend citation list more effectively. Future work includes the application of community structure in other information network research fields, such as disease analysis information network, which can organize etiologies with similar symptoms into a community for further analysis of etiology.

References

1. Chen, J., Liu, Y., Zhao, S., Zhang, Y.: Citiation recommendation based on weighted heterogeneous information network containing semantic linking. In: 2019 IEEE ICME, Shanghai, China, pp. 31–36. IEEE (2019)
2. Perotti, J.I., Tessone, C.J., Caldarelli, G.: Hierarchical mutual information for the comparison of hierarchical community structures in complex networks. Phys. Rev. E **92**(6), 062825 (2015)

3. West, J.D., Wesley-Smith, I., Bergstrom, C.T.: A recommendation system based on hierarchical clustering of an article-level citation network. IEEE Trans. Big Data **2**(2), 113–123 (2016)
4. Wang, H., Li, W.: Relational collaborative topic regression for recommender systems. IEEE Trans. Knowl. Data Eng. **27**(5), 1343–1355 (2015)
5. Ma, X., Wang, R.: Personalized scientific paper recommendation based on heterogeneous graph representation. IEEE Access **7**, 79887–79894 (2019)
6. Zhang, C.X., Huang, C., Yu, L., Zhang, X.L., Chawla, N.V.: Camel: content-aware and meta-path augmented metric learning for author Identification. In: Proceedings of the 2018 World Wide Web Conference, pp. 709–718. International World Wide Web Conferences Steering Committee, Republic and Canton of Geneva, CHE (2018)
7. Musto, C., Lops, P., Gemmis, M.D., Semeraro, G.: Context-aware graph-based recommendations exploiting Personalized PageRank. Knowl. Based Syst. **216**(3), 106806 (2021)
8. Roul, R.K., Sahoo, J.K.: A novel approach for ranking web documents based on query-optimized personalized pagerank. Int. J. Data Sci. Anal. **11**(1), 37–55 (2020). https://doi.org/10.1007/s41060-020-00232-2
9. Ozsoy, M.G., et al.: MP4Rec: explainable and accurate top-N recommendations in heterogeneous information networks. IEEE Access **8**, 181835–181847 (2020)
10. Do, P., Pham, P.: DW-PathSim: a distributed computing model for topic-driven weighted meta-path-based similarity measure in a large-scale content-based heterogeneous information network. J. Inform. Telecommun. **3**(1), 19–38 (2019)
11. Do, P., Pham, P.: W-PathSim++: the novel approach of topic-driven similarity search in large-scaled heterogeneous network with the support of Spark-based DataLog. In: 2018 10th International Conference on Knowledge and Systems Engineering, Ho Chi Minh City, Vietnam, pp. 102–106. IEEE (2018)
12. Hou, U., L., Yao, K., Mak, H., F.: PathSimExt: revisiting PathSim in heterogeneous information networks. In: Li, F., Li, G., Hwang, S., Yao, B., Zhang, Z. (eds.) WAIM 2014, LNCS, vol. 8485, pp. 38–42. Springer, Cham (2014). https://doi.org/10.1007/97833190801096
13. Wang, W., Gong, Z.G., Ren, J., Xia, F.: Venue topic model–enhanced joint graph modelling for citation recommendation in scholarly big data. ACM Trans. Asian Low Res. Lang. Inform. Process. **20**(1), 1–15 (2020)
14. Yang, Y., X., Ren, G., C.: HanLP-based technology function matrix construction on Chinese process patents. IJMCMC **11**(3), 48–64 (2020)

Sentiment Analysis for MOOC Course Reviews

Tianyi Liu[1(✉)], Wei Hu[1], Fang Liu[2], and Yining Li[1]

[1] Wuhan University of Science and Technology, Wuhan Hubei, China
Lty@wust.edu.cn
[2] Wuhan University, Wuhan Hubei, China

Abstract. Currently, increasing users are using MOOC platforms to choose courses and leave text comments with emotional overtones. The traditional words vector representation method uses static method to extract text information, which ignores the text position information. The traditional convolutional neural network fails to make full use of the semantic features and association information of the text between channels, which will cause inaccurate text sentiment classification. In order to solve the problems, a text classification model based on Albert and Capsule Network and attention mechanism is proposed. The model was verified on the MOOC comment data set, and compared with the traditional user comment sentiment analysis model. The results show the accuracy of the model was improved to a certain extent.

Keywords: MOOC review · ALBERT · Capsule network · Attention mechanism · Sentiment analysis

1 Introduction

MOOC, also known as Massive Open Online Courses (Mooc). The MOOC course platform, as the most well-known kind of education course platform, gathers a large number of online learning users. Learners can leave comments on courses and platforms in the comments section of MOOC. The rapid development of the MOOC platform has brought an explosive growth of text comments. Comments often contain a huge amount of information, which can reflect many issues, such as: whether the quality of the course is guaranteed, whether the technical support of the website is perfect, and what content is the feedback of the students on the teaching content. At present, MOOC platforms at home and abroad also have some shortcomings: for example, the increase in the number of courses leads to uneven course quality, and there are some problems in the platform itself. All of these will seriously affect the quality of the MOOC platform, which makes the study of MOOC course reviews very meaningful. In order to achieve an in-depth understanding of the MOOC course review text, it is necessary to conduct sentiment analysis. Text sentiment analysis is the process of using computer technology to analyze, reason, summarize and reason about subjective texts with emotional colors. There are three kinds of sentiment analysis methods, namely, the method based on the sentiment dictionary, the method based on machine learning and the method based on deep learning.

J. Zeng et al. (Eds.): ICPCSEE 2021, CCIS 1452, pp. 78–87, 2021.
https://doi.org/10.1007/978-981-16-5943-0_7

The traditional sentiment analysis methods are mainly based on the sentiment dictionary method and the method based on machine learning. Literature [1] uses the English sentiment dictionary WordNet to judge the sentiment attitude and sentiment value contained in the English text. Literature [1] uses maximum entropy, naive Bayes and support vector machine to perform text sentiment classification experiments on movie review data, and finds that support vector machine classification can achieve the best results. Literature [2] uses the maximum entropy algorithm to classify electronic product reviews. Literature [3] found that the Naive Bayes model is better in processing short texts, and the support vector machine is better in processing long texts. The words in the sentiment dictionary need to be manually selected. This method largely depends on manual design and prior knowledge. Machine learning methods require a complex feature selection process, also rely on manual design, and have poor generalization capabilities. Therefore, emotion analysis methods based on deep learning have attracted more and more attention. The two mainstream applications in deep learning are Convolutional Neural Networks (CNN) and Recurrent Neural Networks (RNN). With the good performance of the capsule network in the image field, the literature [4] applied the capsule network to the sentiment analysis task for the first time, and the classification performance surpassed CNN.

This paper proposes a text sentiment classification model based on ALBERT that combines BiGRU and capsule network, and adds an attention mechanism to the capsule network layer. Pay attention to the key information of the text through the attention mechanism.

2 Related Work

The prerequisite for the application of deep learning technology in sentiment analysis is to solve the word mapping problem, that is, to convert text into numbers that can be recognized by machines. The commonly used method is to train word vectors for text. Literature [5] proposed the NNLM model in 2003, and then a series of word vector technologies (such as Word2Vec [6], Glove [7], FastTest [8], etc.) provided a numerical representation method for text, but Can't solve the problem of ambiguity. ELMo [9] uses bidirectional Long Short-Term Memory (LSTM) for pre-training. It transforms the word vector from static to dynamic, so that it can be combined with context to give word meaning. BERT [10] used the bidirectional Transformer for the language model for the first time, making the model have a deeper understanding of the context than GPT. Literature [11] proposed a lightweight BERT model ALBERT (A lite BERT). Based on BERT, this model uses a two-way Transformer as a feature extractor to obtain the feature representation of the text, reducing the amount of model parameters and shortening the training time.

In order to accurately determine the emotional tendency of the bullet screen text, the feature classification algorithm is also particularly important. Literature [12] proposes a text convolutional neural network model (TextCNN), which uses convolution kernels of different sizes to train local features of the text to achieve sentence-level classification tasks. Literature [13] proposed Gated Recurrent Unit (GRU), which makes the model structure simpler while maintaining the effect of LSTM. Literature [14] proposed a hierarchical attention model, which shows that the importance of each sentence and word in the model construction text is interpretable. Literature [15] proposed adding a custom capsule model to the recurrent neural network, and introduced three modules: the representation module, the probability module and the reconstruction module, which improved the accuracy of the classification effect.

In summary, the AL-BiGCaps model proposed in this paper combines the ALBERT pre-training language model with BiGRU and Capsule Network, and embeds the attention mechanism in it. This article proves the effectiveness of the AL-BiGCaps model in sentiment analysis of MOOC review texts through comparative experiments with other models.

3 MOOC Text Comment Sentiment Analysis Model

The sentiment analysis of AL-BiGCaps MOOC text comments proposed in this article mainly includes the following steps. (1) Clean and preprocess the comment text data, and label the text data with emotional polarity. (2) Use the ALBERT pre-training language model to obtain the dynamic feature representation of the text. (3) Use a model combining BiGRU and Capsule Network to train text features to obtain deep semantic features. (4) Use the softmax function to classify the deep semantic features of the text, and finally get the emotional polarity of the comment.

3.1 ALBERT Pre-trained Language Model

ALBERT is essentially a lightweight BERT model. BERT (Bidirectional Encoder Representation from Transformers) is a new pre-training model proposed in 2018. It no longer uses the traditional one-way language model or the shallow splicing of two one-way language models for pre-training, but uses a new masked language model (MLM) to generate deep two-way language representations. As shown in Fig. 1, it is mainly composed of input layer, word embedding layer, feature coding layer and output layer.

Compared with BERT, ALBERT has made the following improvements:

Word Embedding Factorization (Factorized Embedding Parameterization):
In BERT and XLNet, the embedding size E is the same as the hidden size H. From the perspective of the model, the embedding layer is used to learn context-independent representations, while the embeddings of the hidden layer are used to learn context-independent representations. The two do not have to keep the same embedding size. The method of matrix decomposition is to project the original embedded layer matrix (size $V \times H$) into a low-dimensional space (dimension E, $E < < H$), and then project from this space to the hidden layer space (dimension H) [16].

The calculation formula of BERT model parameter P_1 is:

$$P_1 = L * H \tag{1}$$

The calculation formula of ALBERT model parameter P_2 is:

$$P_2 = L * V + V * H \tag{2}$$

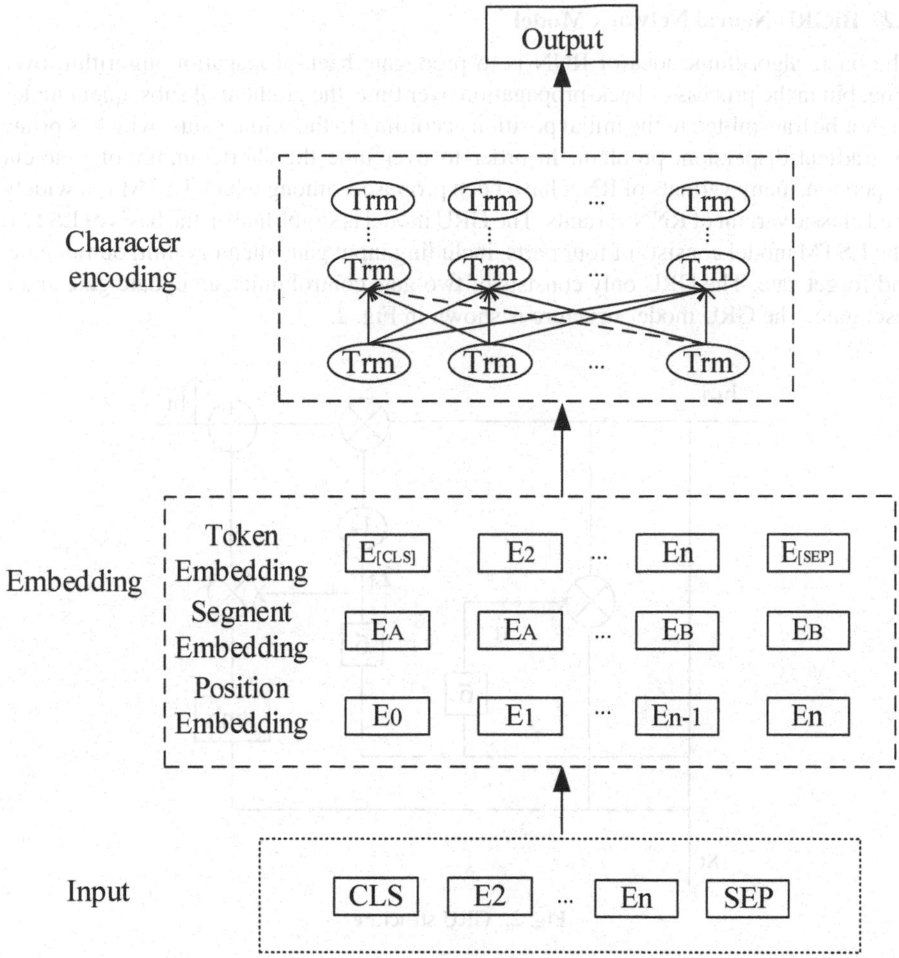

Fig. 1. ALBERT structure

Cross-Layer Parameter Sharing: In the ALBERT model, two parameter sharing methods of Transformer are combined to realize parameter sharing between the fully connected layer and the attention layer. So as to achieve the effect of reducing the amount of parameters and increasing the speed of model training.

Inter-Sentence Coherence Loss: A new training task of SOP (sentence-order prediction) is proposed, allowing the model to recognize the sequence of given two sentences.

Remove Dropout.

3.2 BiGRU Neural Network Model

The basic algorithmic idea of RNN is to propagate back-propagation algorithm over time, but in the process of back-propagation over time, the gradient of subsequent nodes cannot be transmitted to the initial position according to the initial value, which is prone to gradient dispersion. problem. In order to overcome the shortcomings of gradient dispersion, many variants of RNN have been proposed, among which LSTM is a widely used classic variant of RNN variants. The GRU model is simplified on the basis of LSTM. The LSTM model consists of four parts, including input gate, memory unit, output gate, and forget gate. The GRU only consists of two gate control units, an update gate and a reset gate. The GRU model structure is shown in Fig. 2.

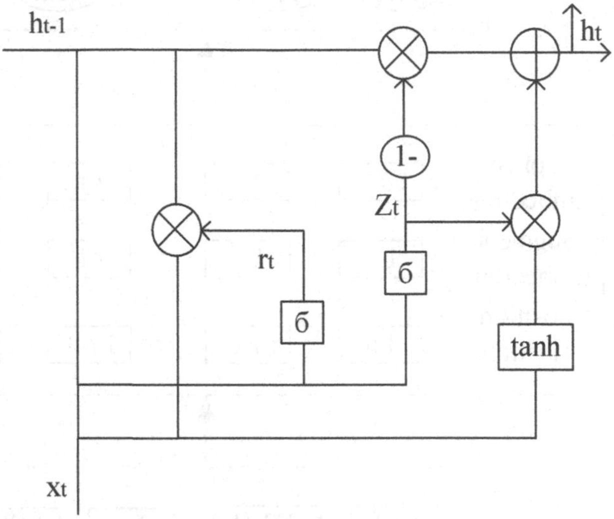

Fig. 2. GRU structure

On the basis of GRU, a layer of GRU model is expanded, and the above information is obtained in a unidirectional neural network by letting the two-layer GRU network flow in opposite directions to process data. The status of a single-layer GRU network is always output from front to back. However, in sentiment classification tasks, if the output at the current moment can be related to the state at the previous moment and the state at the next moment, it will be more conducive to the extraction of deep-level features of the text. BiGRU can establish this connection, so it can more fully extract the information of the text.

The calculation formula in the BiGRU model is as follows. rt represents the reset gate, which is used to indicate the degree of ignoring the state information at the previous moment. The smaller the value of rt, the more ignorance. zt represents the update gate, which is used to indicate the degree to which the state information at the previous moment is substituted into the current state. The value of zt is proportional to the impact of the previous hidden layer output on the current hidden layer.

$$r_t = \sigma\left(W_t \cdot \left[h_{t-1}, x_t\right]\right) \tag{3}$$

$$z_t = \sigma\left(W_z \cdot \left[h_{t-1}, x_t\right]\right) \tag{4}$$

$$\widetilde{h_t} = tanh\left(W_{\widetilde{h_t}} \cdot \left[r_t * h_{t-1}, x_t\right]\right) \tag{5}$$

$$h_t = (1 - z_t) * h_{t-1} + z_t * \widetilde{h_t} \tag{6}$$

$$y_t = \sigma(W_o \cdot h_t) \tag{7}$$

3.3 Capsule Network

In the capsule network, the vector capsule replaces the neurons in the convolutional neural network, dynamic routing replaces the pooling operation, and the Squash function replaces the ReLU activation function. The capsule network in this paper is divided into the initial capsule layer and the main capsule layer.

Initial Capsule Layer. The function of the initial capsule layer is to perform initial feature vectorization on the overall information extracted by the BiGRU layer. The scalar information transmitted by the upper neural unit to this layer memorizes the semantic information between contexts. In this paper, the transformation matrix $W \in R^{N \times d \times d}$ is used to generate the predicted quantity $u_{j|i}$ from the BiGRU layer to the initial capsule layer, and the weight sharing method is adopted. Where d is the dimension of $u_{j|i}$, and N is the number of capsules in the initial capsule layer. The specific calculation formula is:

$$u_{j|i} = W_j g_i + b_{j|i} \tag{8}$$

In the capsule network, dynamic routing is used to replace the pooling operation to reduce the loss of semantic information. Dynamic routing gradually adjusts the distribution of low-level capsules to high-level capsules according to the output of high-level capsules through iteration, and finally achieves an ideal distribution. For each unit of the capsule layer, dynamic routing is used to increase or decrease the connection strength to express the features in the text. It can detect whether the features of the text at different

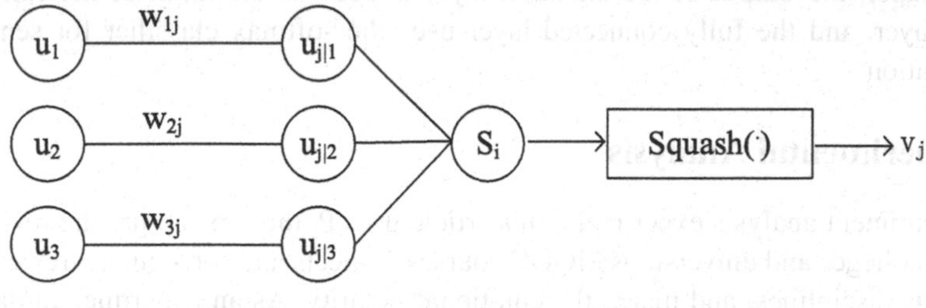

Fig. 3. Capsule network schematic diagram

positions are the same or similar without losing specific spatial information. Figure 3 is Capsule network schematic diagram. The specific formula is as follows:

$$V_j = g(S_j) \tag{9}$$

$$g(s_j) = \frac{\|s_j^2\|}{1 + \|s_j^2\|} \frac{s_j}{\|s_j\|} \tag{10}$$

$$s_j = \sum_i c_{ij} u_{j|i} \tag{11}$$

$$u_{i|j} = W_{ij} u_i \tag{12}$$

$$c_{ij} = Softmax(b_{ij}) \tag{13}$$

Main capsule Layer. Compared with the initial capsule layer using scalar information as input, the main capsule layer takes feature vectors as input. These inputs are recombined between features using dynamic routing algorithms to extract key information that can better characterize text features.

3.4 Attention Mechanism

In text reviews, different words have different degrees of influence on the emotional polarity expressed in the text. Not all words are equally important to the expression of the meaning of the text. Therefore, the introduction of the attention mechanism (Attention Mechanism) to select from a large amount of information is more important for the current goal Information.

$$u_i = tanh(W_w h_i + b_w) \tag{14}$$

$$a_i = softmax(u_i) \tag{15}$$

$$\mathrm{H} = \sum_{\mathrm{i}} \mathrm{a_i h_i} \tag{16}$$

3.5 Fully Connected Layer

In this paper, the output of the attention layer is used as the input of the fully connected layer, and the fully connected layer uses the softmax classifier for sentiment classification.

4 Experimental Analysis

In the sentiment analysis experiment, this article uses Python to capture the reviews of Chinese colleges and universities MOOC courses, collects and sorts out the review texts of various disciplines, and marks the emotional polarity. As an experimental data set, the data set contains three emotion polarities: positive emotion, neutral emotion, and negative emotion, which are represented by 1, 0, and -1, respectively.

4.1 Introduction to the Data Set

In the emotion analysis experiment, Python was used to crawl the course comments on MOOCs of Chinese universities, collect and sort out the comment text of each subject, and mark the emotional polarity. As an experimental data set, the data set contains three emotional polarities: positive emotion, neutral emotion and negative emotion, represented by 1, 0 and −1 respectively. Use 80% of the data for training, 20% of the data is used for testing.

4.2 Evaluation Standard

In order to evaluate the classification effect of the model, a confusion matrix is used to make statistics on the classification results. Using TP to indicate that the actual sample is a positive sample and the prediction is a positive sample, FP indicates that the actual sample is a negative sample but the prediction is a positive sample, TN indicates that the actual sample is a negative sample and the prediction is a negative sample, and FN indicates that the actual sample is a positive sample but the prediction is a negative sample. According to the results of the confusion matrix statistics, the accuracy rate (Ac), the precision rate (P), the recall rate (R), and the harmonic mean value F1 of the precision rate and the recall rate are used to evaluate the effect of the model. The calculation formula is:

$$\mathrm{P} = \frac{TP}{TP + FP}$$

$$\mathrm{R} = \frac{TP}{TP + FN}$$

$$\mathrm{F1} = \frac{2 \times \mathrm{P} \times \mathrm{R}}{P + \mathrm{R}}$$

4.3 Experimental Comparison Model

In order to objectively evaluate the classification effect of the model proposed in this paper, the sentiment classification effect of ALBERT-BiGRU-Caps was studied through comparative analysis with the following models.

In order to verify the effectiveness of the ALBERT-BiGRU-Caps comment text sentiment analysis model, the ALBERT-BiGRU-Caps model was compared with the SVM, CNN, BiGRU, CRNN and ALBERT models. Among them, the SVM, CNN, BiGRU and CRNN models are all based on the Word2Vec model Construct word vectors. The ALBERT-BiGRU and ALBERT-BiGRU-Caps models use the Chinese pre-training model ALBERT released by Google for text feature representation, and this pre-training model is fine-tuned under the data set of this article.

Table 1. Experimental results

Evaluation	Presision	Recall	F1
SVM	81.3	82.1	81.7
CNN	84.4	84.7	84.5
BiGRU	88.1	87.7	87.9
CRNN	89.2	88.8	89.0
BiGRU-Caps	89.5	90.4	89.9
ALBERT-BiGRU	91.2	92.3	91.7
ALBERT-BiGRU-Caps	93.2	94.1	93.6

4.4 Analysis of Experimental Results

In this paper, through experimental comparison, it is found that the BiGRU-based capsule network can extract feature vectors better. At the same time, pre-training the text with the ALBERT model can better get the text feature representation.The ALBERT-BiGRU-Caps model is used to predict the text of the MOOC review, showing the true application function of the model.

5 Conclusion

This paper proposes an Albert based sentiment analysis model for MOOC review text, which combines BiGRU and Capsule Network. By using the Albert pre-training language model to obtain the dynamic feature representation of the text of the bullet screen, the problem that the traditional sentiment analysis method of the bullet screen cannot distinguish the different meanings of the same word in different contexts is solved. The neural network combining BiGRU and Capsule network is used to train the features, which makes full use of the local feature information and contextual semantic association in the text. A comparative experiment was conducted on the MOOC text dataset to prove the effectiveness of the ALBERT-BiGRU-Caps model in the sentiment analysis task of bullet text. As the number of parameters in the ALBERT model is still large in the using process, the training takes a long time. In the next research work, the ALBERT model will be compressed to reduce the complexity of the model as much as possible without great loss of model accuracy, so as to improve the training efficiency of the model.

References

1. Kamps, J., Marx, M., Mokken, R.J., et al.: Words with attitude. Language and Computation, University of Amsterdam, Institute for Logic (2001)
2. Pang, B., Lee, L.: A sentimental education: sentiment analysis using subjectivity summarization based on minimum cuts. In: Proceedings of the 42nd Meeting of the Association for Computational Linguistics (ACL 2004), vol. 2004, pp. 271–278 (2004)

3. Lee, H.Y., Renganathan, H.: Chinese sentiment analysis using maximum entropy. In: Proceedings of the Workshop on Sentiment Analysis Where AI Meets Psychology (SAAIP 2011), pp. 89–93 (2011)
4. Wang, S., Manning, C.D.: Baselines and bigrams: simple, good sentiment and topic classification. In: Proceedings of the 50th Annual Meeting of the Association for Computational Linguistics: Short Papers-volume 2. Association for Computational Linguistics, pp. 90–94 (2012)
5. Zhao, W., Ye, J., Zhang, M., et al.: Investigating capsule networks with dynamic routing for text classification. In: The 2018 Conference on Empirical Methods in Natural Language Processing, Brussels, Belgium, pp. 3110–3119 (2018)
6. Bengio, Y., Ducharme, R., Vincent, P., et al.: A neural probabilistic language model. J. Mach. Learn. Res. **3**, 1137–1155 (2003)
7. Mikolov, T., Chen, K., Corrado, G., et al.: Efficient estimation of word representations in vector space .arXiv: 1301.3781 (2013)
8. Pennington, J., Socher, R., Manning, C.: Glove :global vectors for word representation. In: Proceedings of the 2014 Conference on Empirical Methods in Natural Language Processing (EMNLP), pp. 1532–1543 (2014)
9. Joulin, A., Grave, E., Bojanowski, P., et al.: Bag of tricks for efficient text classification. arXiv: 1607.01 759 (2016)
10. Peters, M.E., Neumann, M., lyyer, M., et al.: Deep contextulized word representations. arXIv : 1802.05365 (2018)
11. Devlin, J., Chang, M.W., Lee, K., et al.: BERT: pretraining of deep bidirectional transformers for language understanding, arXiv:1 810.04805 (2018)
12. Lan, Z., Chen, M., Goodman, S., et al.: Albert: a lite bert for self-supervised learning of language representations . arXiv preprint arXiv:1909.11942 (2019)
13. Kim, Y.: Convolutional neural networks for sentence classification. In: Proceedings of the Conference on Empirical Methods in Natural Language Processing. Stroudsburg: Association for Computational Linguistics, pp. 1746–1751 (2014)
14. Dey, R., Salem, F.M.: Gate-variants of gated recurrent unit (GRU) neural networks. In: Proceedings of the 60th IEEE International Midwest Symposium on Circuits and Systems. IEEE Press, Piscataway, pp. 1597–1600 (2017)
15. Yang, Z.C., Yang, D.Y., Dyer, C., et al.: Hierarchical attention networks for document classification. In: Proceedings of the 2016 Conference of the North American Chapter of the Association for Computational Linguistics: Human Language Technologies. San Diego, California: Association for Computational Linguistics, pp. 1480–1489 (2016)
16. Wang, Y., Sun, A., Han, J., et al.: Sentiment analysis by capsules. In: Proceedings of the 2018 World Wide Web Conference. Republic and Canton of Geneva, CHE: International World Wide Web Conferences Steering Committee, pp. 1165–1174 (2018)

Analyzing Interpretability Semantically via CNN Visualization

Chunqi Qi, Yuechen Zhao, Yue Wang(✉), Yapu Zhao, Qian Hong, Xiuli Wang, and Weiyu Guo

School of Information, Central University of Finance and Economics, Beijing, China

Abstract. Deep convolutional neural networks are widely used in image recognition, but the black box property is always perplexing. In this paper, a method is proposed using visual annotation to interpret the internal structure of CNN from the semantic perspective. First, filters are screened in the high layers of the CNN. For a certain category, the important filters are selected by their activation values, frequencies and classification contribution. Then, deconvolution is used to visualize the filters, and semantic interpretations of the filters are labelled by referring to the visualized activation region in the original image. Thus, the CNN model is interpreted and analyzed through these filters. Finally, the visualization results of some important filters are shown, and the semantic accuracy of filters are verified with reference to the expert feature image sets. In addition, the results verify the semantic consistency of the same important filters under similar categories, which indicates the stability of semantic annotation of these filters.

Keywords: CNN · Deconvolution visualization · Semantic annotations · Interpretability

1 Introduction

With the development of deep learning, convolutional neural network (CNN) has achieved great success in the field of image recognition. However, a complex CNN model is still a black box, and it is difficult to find the basis of its classification reasons. Therefore, the interpretation of the model becomes important, which can effectively improve the transparency of the model and gain the trust of users. In this paper, we use the deconvolution to visualize the activation regions of some filters in the CNN model, excavate the semantic features recognized by the filters, and then analyze the interpretability of the CNN model.

C. Qi, Y. Zhao and Y. Wang—Contributed equally to this paper. This work is supported by: National Defense Science and Tech-nology Innovation Special Zone Project (No. 18-163-11-ZT-002-045-04); Engineering Research Center of State Financial Security, Ministry of Education, Central University of Finance and Economics, Beijing, 102206, China; Program for Innovation Research in Central University of Finance and Economics; National College Students' Innovation and Entrepreneurship Training Program "Research on classification and interpretability of popular goods based on Neural Network".

J. Zeng et al. (Eds.): ICPCSEE 2021, CCIS 1452, pp. 88–102, 2021.
https://doi.org/10.1007/978-981-16-5943-0_8

In CNN, the filters in the low layers usually identify some low-level features such as lines and edges, and the filters in the high layers can identify features from combined edge, some of which having semantic meanings. Therefore, we select filters in the high layers of the CNN model, and screen important filters for the target image category by statistical analysis. Then we visualize these important filters, select the images that the important filters are highly activated, and map their feature maps corresponding to the filters to the activation regions in the original images by deconvolution, so as to excavate the semantic information recognized by the filters. In this way, we can obtain part of the decision basis of CNN model through these important filters. When an important filter is highly activated, it means that there may be semantic features corresponding to this filter in the original image, and it is recognized by the CNN. Then, we can map the feature map to the activation region in the original image, so as to find the semantic feature region in the original image, and we can find some interpretable reasons for the classification decision of CNN.

2 Related Work

Generally, as mentioned by Lipton [1], the approaches used to interpret models fall into two categories. The first the is transparent explanation, which explains how the model works. The second is post-hoc explanations, explaining what the classification model can tell us.

Network Visualization. Model visualization is a kind of post-hoc explanations used to show what the model has learned. Springenberg et al. [2] used a special all convolution net and achieved a great result. Zhou et al. [3] proposed CAM to visualize CNN and Selvaraju et al. [4] improved it. Simonyan et al. [5] proposed saliency maps. Wang et al. [6] proposed Score-CAM to further improve the quality of visualization. Smilkov et al. [7] improved the visualization results by using smooth grad. In addition, Zeiler et al. [8] constructed the deconvolution network. Zeiler et al. [9] used the deconvolution visualization for the first time to explain the CNN model and obtained convincing results.

Semantic Annotation. Semantic annotation can improve the transparency of the model. Zhang et al. [10] proposed a method to modify the traditional CNNs into interpretable CNNs, which can automatically allocate an object part to the filter in a specific convolution layer. Combined with these, Zhang et al. [11] used a decision tree to decode various decision modes hidden in the full connection layer, so as to analyze the prediction principle of CNN. In addition, Zhang et al. [12] similarly proposed a method to extract object parts from CNN and generated interpretable part graphs to understand model behavior hierarchically. Also, in the field of application, Dong Y et al. [13] integrated semantics into the model to obtain interpretable features, which can be effectively used for video subtitle recognition. In the previous papers, while increasing model transparency can clearly show how the model works, they often sacrifice the predictive performance of the model.

Our Contribution. In this paper, we combine two interpretability ideas, model visualization and semantic annotation. We use the deconvolution visualization to make semantic annotation on the selected filters, trying to explain the working mechanism of the classification model from the semantic perspective while not affecting the model. And through the experiments we prove that the semantic annotation method has high accuracy and stability.

3 Interpretation by Filters Screening and Semantic Labeling

We screen filters in the high layers of VGG16, count the filters high activation frequencies in each category, and use sensitivity analysis to screen filters further. After the screening is done, we use the deconvolution network to map the filters from its corresponding feature map to the original image, and the semantic label of a filter is given by human referring to the mapping region in the original image. After completing this, we can use the semantic labels of the filters to explain the corresponding parts of the tested image. These steps are detailed in the following.

3.1 Data Preparation

We use the most commonly used image classification model VGG16[15]. VGG16 model is a traditional CNN model directly stacked by multiple convolutional layers and pooling layers, which has a simpler hierarchical structure compared with Inception Net and ResNet. After fine-tuning training, a good classification result can also be achieved, making it easier for us to construct the corresponding deconvolution network model.

We select 10 categories in different fields from the 1000 classes in the ImageNet [14] for testing. The 10 categories are tank, half-track, chihuahua, Weimaraner, macaw, electric guitar, acoustic guitar, grand piano, kimono and cauliflower. For each category and each test set image, we artificially annotate multiple feature regions and record the coordinates of these regions to form the "expert annotations" test set for the subsequent tests of the explanation model. Some annotation rules are shown in Fig. 1.

3.2 Screening Convolution Features

Some of the filters may not be commonly used to identify this category, only after screening can we find out the representative filters under this category. We select 1300 images for each category from the ImageNet as the training samples for screening. For each category, we input the sample images into the VGG16 model, the feature maps output on the last several convolution layers of these images are obtained, and the maximum activation value in each feature map are taken as representative values. For each convolution layer, we record the serial numbers of the top 5 feature maps with the highest representative values and take them as the high-activation feature maps in the image. Then, we count high-activation frequencies of feature maps in all the images to find the feature map with the highest frequency in this category, and its corresponding filter is the filter with high recognition ability for this category.

After the high-frequency filters are obtained, sensitivity analysis is introduced for further screening. Filters are screened mainly by calculating the contribution of the feature region to the final prediction result. In detail, for an image, the convolution result

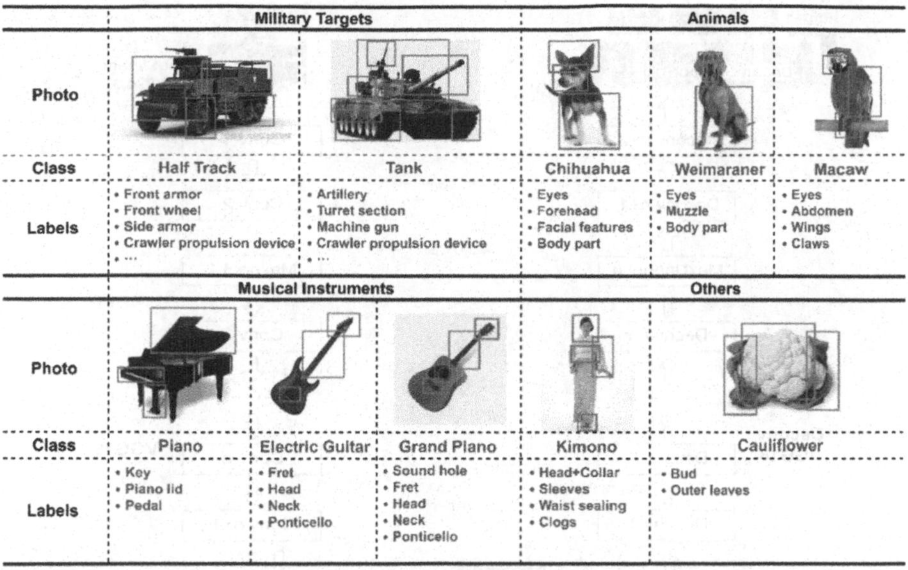
Military Targets
Animals
Photo
Class
Half Track
Tank
Chihuahua
Weimaraner
Macaw
Labels
• Front armor
• Front wheel
• Side armor
• Crawler propulsion device
• ...
• Artillery
• Turret section
• Machine gun
• Crawler propulsion device
• ...
• Eyes
• Forehead
• Facial features
• Body part
• Eyes
• Muzzle
• Body part
• Eyes
• Abdomen
• Wings
• Claws
Musical Instruments
Others
Photo
Class
Piano
Electric Guitar
Grand Piano
Kimono
Cauliflower
Labels
• Key
• Piano lid
• Pedal
• Fret
• Head
• Neck
• Ponticello
• Sound hole
• Fret
• Head
• Neck
• Ponticello
• Head+Collar
• Sleeves
• Waist sealing
• Clogs
• Bud
• Outer leaves

Fig. 1. Expert annotation rules

of a filter can be mapped back to the original image through deconvolution, obtaining the corresponding feature region in the original image. Then we set the pixels of the feature region to 0, i.e., by covering up, while keeping the pixel values of the rest regions unchanged. The processed images are input into the model again to obtain the prediction probability of the correct category after the covering. Compared with the prediction probability before the covering, the change is calculated to be the classification contribution degree. The greater its influence on the classification prediction, the greater the classification contribution of the region is considered. Based on the above two rounds of processing, we can finally obtain the important filters with high recognition ability and high classification contribution for each target category.

3.3 Deconvolutional Visualization

In reference to Zeiler's work on deconvolution [9, 16], we construct a deconvolution model Deconv-VGG16 corresponding to the structure of VGG16 model. Its network structure is exactly opposite to VGG16, and the corresponding convolutional layers in the original VGG model is replaced by the deconvolution layers, and the pooling layers are replaced by unpooling layers, which are constructed in the reverse order of VGG16. The diagram is shown in Fig. 2.

At the same time, we need to establish a corresponding relationship between each layer of the convolutional network and that of the deconvolution network. For example, the feature maps output of the last layer of CNN need to be input from the first layer of the deconvolutional network if we want to map it to the original image through the deconvolution network. The feature maps output of the penultimate layer of the convolutional network can only be mapped to the original image by input into the second

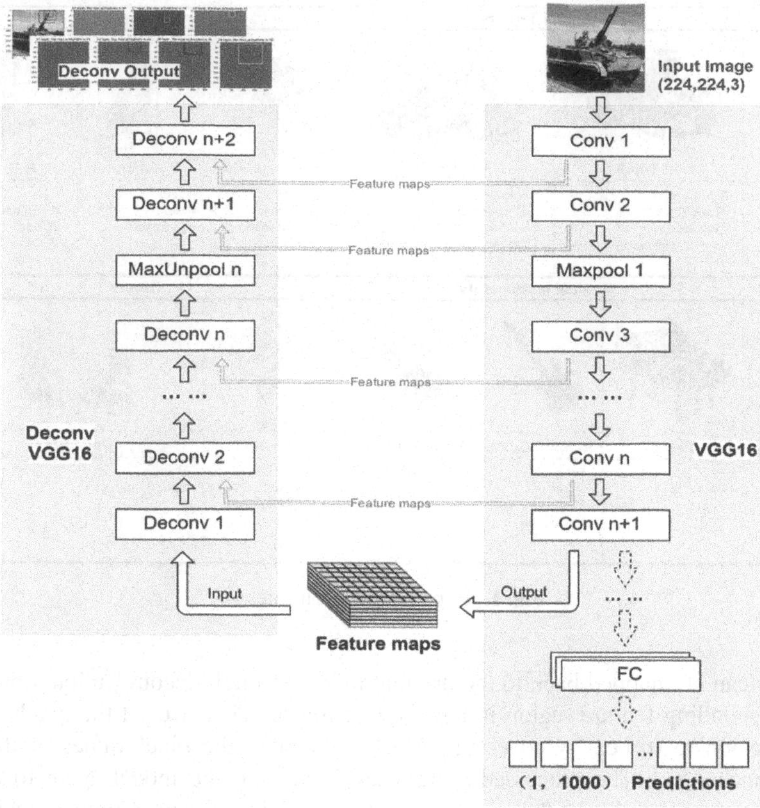

Fig. 2. The Diagrammatic sketch of deconvolution[1] visualization

layer of the deconvolutional network. It can be intuitively understood that the original image has been obtained through several convolution and pooling operations, and its feature maps can only be mapped back to the original image after the same number of deconvolution and uppooling operations. In this way, we only need to input the position information of the maximum in each pooling operation and the feature maps into the deconvolution model, then we can get the results after deconvolution visualization.

After visualization by deconvolution, a feature region is not easy to be directly identified. Therefore, we set a threshold based on the activation value of deconvolution to select the region in the deconvolution result, which was convenient for semantic feature labeling and the subsequent testing. The results are shown in the following Fig. 3.

3.4 Semantic Annotation of Filters

According to the above filters' selection and visualization methods, for a specific category, we can select important filters, and use the deconvolution visualization method to map high-activation feature maps of these filters to the original image region. Since

[1] This type of deconvolution for visualization is also called transpose convolution.

Fig. 3. The deconvolution displays and selects important regions based on some activation threshold

the receptive fields of the convolutional layers vary in size, this kind of flexibility allows us to match these layers to some typical regions of the target class. Based on this, we compared the sizes of the receptive field and the typical feature parts, looked for several convolutional layers with the most fitting sizes and labeled the filters in these layers semantically according to the order of filters' importance.

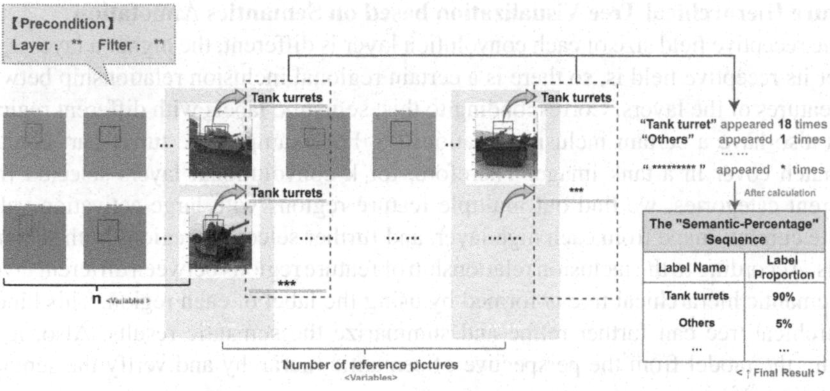

Fig. 4. The diagrammatic sketch of semantic annotation[2]

The general process of semantic annotation is shown in Fig. 4. We find the highly activated images from the image set for a particular filter (we can record these images when we counted the high activation frequency of each filter). The activation region corresponding to the filter on these images can be visualized by deconvolution, and semantic labels are given to the filter according to the region in each image. We can assign a semantic label to each image. After marking the semantics of some target images, we can calculate the proportion of the occurrence times of each semantic label to obtain the "Semantic: Percentage" sequence of the filter. Usually, the image parts corresponding to important filters after screening are mostly consistent, i.e., the highest semantic percentage accounts for the vast majority. But *semantics hybridity* still exists, which is defined as the number of semantics identified by a filter, which is common in

[2] When we annotate a filter, for each image in the set, we can get n regions through deconvolution based on the top n activation values, then give them semantic labels.

VGG16 since the number of filters in a convolution layer is smaller than the number of image categories. When the semantics hybridity is more than 3, it can be considered that the semantics of the filter is somewhat mixed. If necessary, we can set the threshold of semantic hybridity as a screening condition to remove filters of high hybridity.

4 Experiments

4.1 Visualization of Screened Filters

Visualization of Filters with Low Semantics Hybridity
As mentioned above, we can screen and annotate filters for 10 target categories of image data. According to the percentage of semantic tags of each filter after labeling, we select some filters with low semantics hybridity. We visualize them by selecting the feature regions with red boxes. We can select faithful semantic filters by the screening and annotation from the experiments in Fig. 5.

Feature Hierarchical Tree Visualization based on Semantics Annotation
As the receptive field size of each convolution layer is different, the higher a layer is, the larger its receptive field is, so there is a certain regional inclusion relationship between the features of the layers. Corresponding to this, semantic labels with different regional sizes also have a certain inclusion relationship. For example, the turret part contains the hatch cover in a tank image. Therefore, for k convolutional layers selected from different categories, we find out multiple feature regions with large activation values for the current image from each high layer, and further select the regions with semantic labels. According to the inclusion relationship of feature regions between different layers, the semantic hierarchical tree is formed by using the label of each region. This kind of hierarchical tree can further refine and summarize the semantic results. Also, it can explain the model from the perspective of semantic hierarchy and verify the semantic annotation effect.

The hierarchical tree visualization process is roughly shown in Fig. 6. In the example, three convolutional layers (13, 14 and 15) are selected from the test set for feature extraction. The feature regions of the top ten activation values are extracted from each layer, and then the feature regions with semantic labels are further screened out from them (left of the figure below). The regions containing relationship are analyzed between each adjacent two layers, and the semantic labels are used to construct a two-tier hierarchical tree (in the middle of the figure below). The final result is the integration of many two-tier trees of two adjacent layers to create a global tree (the right side of the figure below).

In the construction of the hierarchical tree and the following tests, we refer to the idea of IoU (Intersection over Union) to judge the inclusion relationship among various feature regions. The IoU is the ratio of the Intersection and Union of two sets. However, since the size of an activation region after deconvolution is related to the receptive field and activation value of the feature map, the mapping regions can large or small, so the *inclusion degree* of the two regions may be underestimated by using IoU. Therefore, when calculating the inclusion degree, the ratio between the intersection of two areas and their own areas is used in this paper, and the larger value of both is taken. We set a

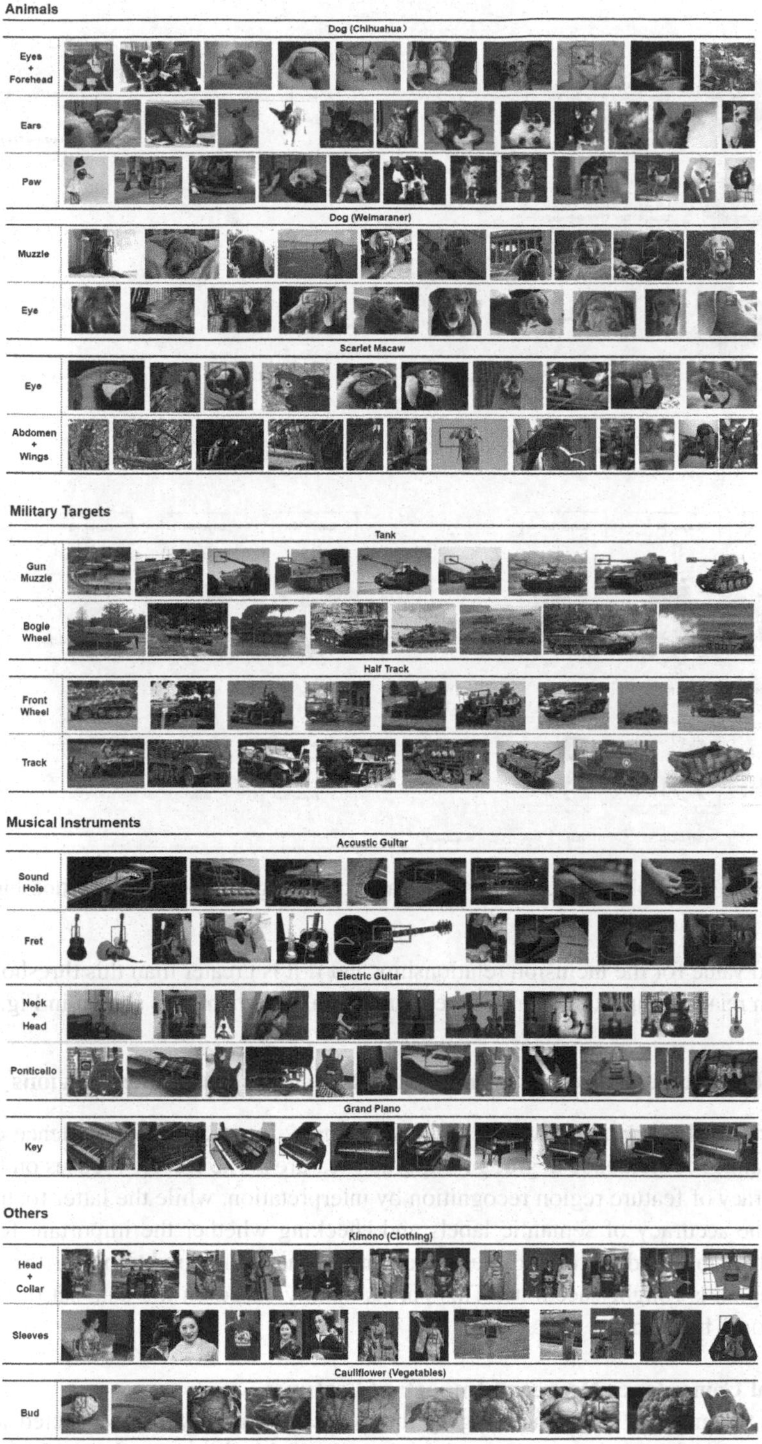

Fig. 5. Filter visualization for filters of low semantic hybridity in 10 categories (each row of images in Fig. 5 is a feature region identified by a same filter)

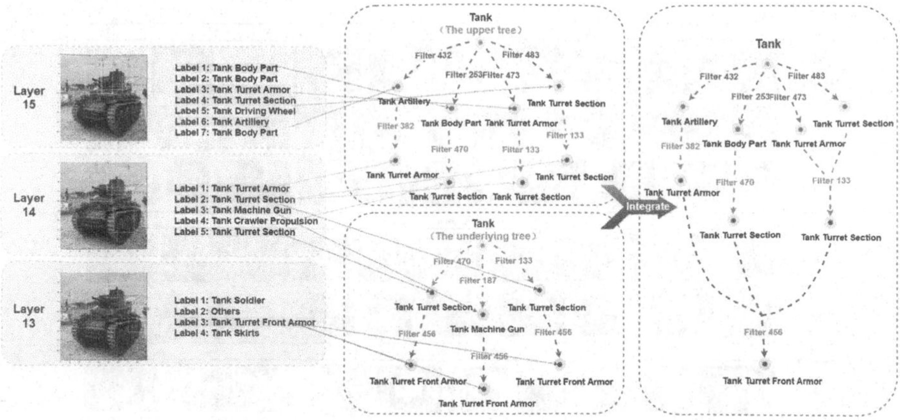

Fig. 6. An example of hierarchical feature tree

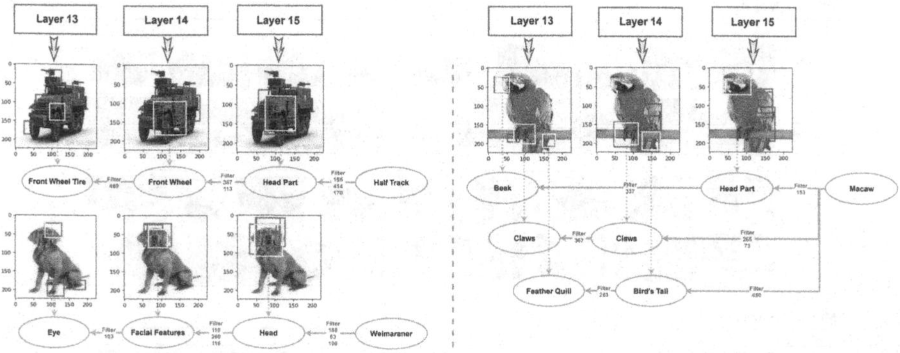

Fig. 7. The display of feature hierarchy based on the regions containing relationship

threshold value for the inclusion relationship, and if it is greater than this threshold, the inclusion relationship exists. Part of the hierarchical tree display is shown in Fig. 7.

4.2 The Coincidence Degree of Interpretable Regions and Expert Regions

This part mainly constructs two test indexes, namely the regional coincidence degree and semantics coincidence degree of important features. The former focuses on testing the accuracy of feature region recognition by interpretation, while the latter focuses on testing the accuracy of semantic labels and checking whether the important features captured by the model are credible by semantics under the condition that the recognition region is roughly accurate. This part is mainly based on the data set of "expert annotations" built above.

Regional Coincidence Degree of Important Features

For each test image of a specific category, three convolutional layers are selected according to the size of the feature region and the receptive field of the convolutional layer. For each layer selected, the filter where the top five activation values located is selected for deconvolution and 15 feature regions are finally obtained. We traverse each feature to check whether there is an inclusion relationship with the feature regions of the current picture in the data set of "expert annotations". If there is, increase the counter by 1, and calculate the percentage of coincidence times in the total as the test result. Among them, the inclusion relation is determined by the parameter *Inclusion Degree* mentioned in 4.1. The regional coincidence degree can test whether the model accurately test the accuracy of the model's recognition of a certain characteristic region identified by human.

Semantics Coincidence Degree of Important Features

This part is divided into the following four steps.

The first step is to screen out filters with low semantics hybridity. For a specific image, five filters with semantic annotation are selected according to activation values from the three convolutional layers selected above. For each filter, calculate the proportion of the 'Others' label (not any characteristic feature label of an image category) in all the labels of the current filter. If the proportion of the "Others" label is greater than the hybridity threshold, which is set to 60% by defaults, it indicates that the current filter is high semantics hybridity. Skip this filter and it will not be included in subsequent calculations.

The second step is to test the regions containing relationship. For each image, traverse each filter after screening to see if there is an inclusion relationship with the features marked by human experts. Whether an inclusion relationship is constituted depends on the *inclusion degree* mentioned in 4.1. If the inclusion relationship holds, the current image is assigned a region containing score a ($0 < a < 1$). If there is an inclusion relationship with an expert feature, and the label of the current filter is consistent with the expert feature (under the hierarchical inclusion relationship), the current picture will be given a semantic label consistent score (1-*a*); Otherwise, the semantic label consistency score will not be given.

The third step is to test the accuracy of semantic labels. Following the previous step, if there is no region inclusion relationship for a tested filter after traversing, but the machine features overlap with the area marked by experts (we use *overlapping degree*, determine whether two regions overlap and the default value is 33%) and the labels are consistent, the filter will be given a score value (1-*a*) that matches semantic labels, without assigning region containing score *a*.

The fourth step is to calculate the final score. After the above three steps, the selected filter of each picture has two parts of the score, that is, the region containing score *a* and the semantics coincidence score (1-*a*). Since the sum of the above two scores is less than or equal to 1, the final test score can be obtained by calculating the ratio between the sum of scores and the number of marked features.

In conclusion, we can test the accuracy of semantics labels in the model with certain region recognition ability. Hybridity determines the strictness of feature filtering. The three parameters, including the weight *a*, inclusion degree and overlapping degree, can dynamically change the influence of regional coincidence on the total score. Parameter *a* determines the proportion of the region containing score. When *a* is 0, only semantic labels are tested. On the contrary, the semantic and regional coincidence degree are integrated to test. Overlapping degree determines the requirement of feature region recognition ability in semantic label testing.

Test Result

Regional Coincidence Degree. The test results based on the above algorithm of regional coincidence degree are shown in the following Table 1.

Table 1. The test results of the regional coincidence degree

Categories	Score (By default)
Tank	82.67%
Half Track	86.00%
Chihuahua	92.53%
Weimaraner	76.67%
Macaw	80.79%
Electric Guitar	62.93%
Acoustic Guitar	74.13%
Grand Piano	62.26%
Kimono	74.80%
Cauliflower	67.13%

According to the results, it can be concluded that most of the important feature regions obtained through model deconvolution have an inclusion relationship with the expert feature regions, indicating that the model in this paper can identify the key regions of human concern and has a certain accuracy. The high consistency between the model and human experience further proves the feasibility and effectiveness of the method of semantic annotation.

We further analyze which key areas identified by the model overlapped with the feature areas labeled by experts. We conduct statistics on the model recognition features that had inclusion relations with the expert feature regions to observe which semantic features are mainly used by the to classify images. Part of the analysis results are shown in the following Table 2.

Table 2. Statistics of semantic importance

Category	Label	Proportion
Tank	Turret Section	19.85%
	Body Part	17.49%
	Artillery	10.73%
	Gun Muzzle	9.27%
	Turret Armor	7.07%
	Crawler Propulsion Device	6.62%

(continued)

Table 2. (*continued*)

Category	Label	Proportion
Half Track	Front Wheel	55.33%
	Front Armor	24.17%
Acoustic Guitar	Sound Hole	40.63%
	Neck	19.26%
	Ponticello	15.57%
Electric Guitar	Fret	49.17%
	Neck	39.59%
Grand Piano	Key	54.48%
	Piano Lid	34.66%
Chihuahua	Facial Feature	44.58%
	Eyes	37.5%
	Forehead	13.92%
Weimaraner	Eye	46.24%
	Muzzle	43.50%
Macaw	Eyes	64.54%
	Abdomen + Wings	22.69%
Cauliflower	Bud	76.39%
Kimono	Collar	82.35%
	Sleeves	7.44%

Semantics Coincidence Degree. Based on the semantics coincidence degree algorithm proposed above, we select two categories (tank and half-track) for detailed annotation and then test them. The test results are shown in the Table 3 below.

Table 3. The test results of the semantics coincidence degree

Inclusion Degree = 0.6; Hybridity = 0.6; Overlapping Degree = 0.30; a = 0.5 [By default]	
Class	Score
Tank	87.79%
Half Track	91.79%

The above results show that the semantic annotation method in this paper has high accuracy and can explain the model from the semantic perspective. Through activation value, sensitivity analysis and semantic annotation, we can get the important and semantically interpretable filters in the current class. These filters can ensure high consistency with experts' labels under high region identification accuracy during testing.

4.3 Semantic Consistency of Filters

Through the previous experiment, we demonstrate and test the semantic annotation results of filters. We deliberately select three groups of similar categories (tank and halftrack, chihuahua and Weimaraner, acoustic guitar and electric guitar) from the experimental data, carry out feature screening and semantic annotation for them, and then compare the common semantics of the two categories to test their semantic consistency.

Take the group of tank and halftrack as an example. First, we compare the important filters of tank and halftrack and select the filters they had in common, i.e., the filters that are important for both categories. Then we select the test semantics for this group. By comparing the expert features, we can find that this group (tanks and halftrack) has some common semantic part, such as artillery, caterpillar, armor, etc. We select the artillery and caterpillar these two relatively obvious characteristics as test semantics, and then the filters with such semantics as tank artillery, tank caterpillar, halftrack artillery and caterpillar are selected from the common filters. By comparing the semantics of these filters, we can find out the filters that have the same semantics on the two categories, and the ratio of these filters is the consistency of the semantics (Table 4).

Table 4. The test results of the semantic consistency[3]

Semantic	Category	Number of important filters	Number of semantic filters	Number of common filters	Consistency of the semantics
Artillery	Tank	155	66	33	50.00%
	Halftrack	137	42		78.57%
Caterpillar	Tank	155	64	40	62.50%
	Halftrack	137	51		78.43%
Eyes or Snout	Chihuahua	127	70	52	74.29%
	Weimaraner	125	58		89.66%
Body Part	Chihuahua	127	41	30	73.17%
	Weimaraner	125	43		69.77%
Fret	Acoustic guitar	128	51	34	66.67%
	Electric guitar	129	42		80.95%

(*continued*)

[3] We select each filter of the 13st layer for experiment.

Table 4. (*continued*)

Semantic	Category	Number of important filters	Number of semantic filters	Number of common filters	Consistency of the semantics
Head	Acoustic guitar	128	41	19	46.34%
	Electric guitar	129	26		73.08%

It can be seen from the above table that, for filters with the same semantics in similar categories, their semantics in different categories are highly consistent, i.e., the same filter can identify the same feature part in similar categories stably. By annotating the semantic of such filters, the prediction results of the CNN model can also be interpreted from the semantic perspective.

5 Conclusion

The semantic annotation method introduced in this paper can help understand the working mechanism of the model from the semantic perspective. We select the filters and use the feature map visualization to make semantic annotation of the important filters. The internal classification basis of CNN model is captured by these semantic annotation results. The above interpretable method is to excavate the decision patterns of the original CNN, so there is no need to modify the model and the classification accuracy of the original model will not be affected.

In the experiments, we further select the filters with obvious semantic features by judging semantics hybridity, and demonstrate results of semantic annotation by semantic visualization, coincidence analysis and semantic consistency. The recognition ability of the filters we finally got is relatively obvious. Its feature activation regions are consistent with expert feature regions well, and its semantic features are also stable.

The annotation method proposed in this paper can explore the latent semantic information of filters. However, in order to obtain more accurate semantic annotation, more reference images are often needed in the annotation process. At the same time, we mainly studied the semantic features of high layer filters in this paper. Later, the same method can be used to explore some attribute features of the low-layer filters and combine the features of the high and low layers to give a more comprehensive explanation of the CNN model.

References

1. Lipton, Z.C.: The Mythos of Model Interpretability. Commun. ACM **61**(10), 1–27 (2018)
2. Springenberg, J.T., Dosovitskiy, A., Brox, T., Riedmiller, M.: Striving for simplicity: the all convolutional net. In: IEEE International Conference on Learning Representations (ICLR), pp. 1–14 (2015)

3. Zhou, B., Khosla, A., Lapedriza, A., et al.: Learning deep features for discriminative localization. In: IEEE Conference on Computer Vision and Pattern Recognition (CVPR), pp. 2921–2929 (2016)
4. Selvaraju, R.R., Cogswell, M., Das, A., et al.: Grad-CAM: visual explanations from deep networks via gradient-based localization. In: IEEE International Conference on Computer Vision (CVPR), pp. 618–626 (2017)
5. Simonyan, K., Vedaldi, A., Zisserman, A.: Deep inside convolutional networks: visualising image classification models and saliency maps. arXiv:1312.6034 (2013)
6. Wang, H., et al.: Score-CAM: score-weighted visual explanations for convolutional neural networks. In: IEEE/CVF Conference on Computer Vision and Pattern Recognition Workshops (CVPRW), pp. 111–119 (2020)
7. Smilkov, D., Thorat, N., Kim, B., Viegas, F., Wattenberg. M.: Smoothgrad: removing noise by adding noise. arXiv:1706.03825 (2017)
8. Zeiler, M D., Taylor, G W., Fergus, R.: Adaptive deconvolutional networks for mid and high level feature learning. In: International Conference on Computer Vision (ICCV), pp. 2018–2025 (2011)
9. Zeiler, M.D., Fergus, R.: Visualizing and understanding convolutional networks. In: Fleet, D., Pajdla, T., Schiele, B., Tuytelaars, T. (eds.) ECCV 2014. LNCS, vol. 8689, pp. 818–833. Springer, Cham (2014). https://doi.org/10.1007/978-3-319-10590-1_53
10. Zhang, Q., Wu, Y.N., Zhu, S.C.: Interpretable convolutional neural networks. In: IEEE/CVF Conference on Computer Vision and Pattern Recognition (CVPR), pp. 8827–8836 (2018)
11. Zhang, Q., Yang, Y., Ma, H., Wu, Y.N.: Interpreting CNNs via decision trees. In: IEEE/CVF Conference on Computer Vision and Pattern Recognition (CVPR), pp. 6254–6263 (2019)
12. Zhang, Q., Cao, R., Wu, Y.N., Zhu, S.C.: Growing interpretable part graphs on ConvNets via multi-shot learning. In: Proceedings of the Thirty-First AAAI Conference on Artificial Intelligence (AAAI 2017), pp. 2898–2906 (2017)
13. Dong, Y., Su, H., Zhu, J., Zhang, B.: Improving interpretability of deep neural networks with semantic information. In: IEEE Conference on Computer Vision and Pattern Recognition (CVPR), pp. 975–983 (2017)
14. Deng, J., Dong, W., Socher, R., Li, L., Li, K., Li, F.-F.: ImageNet: a large-scale hierarchical image database. In: IEEE Conference on Computer Vision and Pattern Recognition (CVPR), pp. 248–255 (2009)
15. Simonyan, K., Zisserman, A.: Very deep convolutional networks for large-scale image recognition. arXiv:1409.1556 (2014)
16. Zeiler, M.D., Krishnan, D., Taylor, G.W., Fergus, R.: Deconvolutional networks. In: IEEE Computer Society Conference on Computer Vision and Pattern Recognition (CVPR), pp. 2528–2535 (2010)

Probing Filters to Interpret CNN Semantic Configurations by Occlusion

Qian Hong, Yue Wang(✉), Huan Li, Yuechen Zhao, Weiyu Guo, and Xiuli Wang

School of Information, Central University of Finance and Economics, Beijing, China

Abstract. Deep neural networks has been widely used in many fields, but there are growing concerns about its black-box nature. Previous interpretability studies provide four types of explanations including logical rules, revealing hidden semantics, sensitivity analysis, and providing examples as prototypes. In this paper, an interpretability method is proposed for revealing semantic representations at hidden layers of CNNs through lightweight annotation by occluding. First, visual semantic configurations are defined for a certain class. Then candidate filters whose activations are related to these specified visual semantics are probed by occluding. Finally, lightweight occlusion annotation and a scoring mechanism is used to screen out the filters that recognize these semantics. The method is applied to the datasets of mechanical equipment, animals and clothing images. The proposed method performs well in the experiments assessing interpretability qualitatively and quantitatively.

Keywords: Image classification · CNN · Interpretability · Image occlusion

1 Introduction

Deep Neural Networks (DNNs) has made great success in computer vision, speech recognition, natural language processing etc. Convolutional neural networks (CNNs) have achieved impressive accuracy in computer vision, especially in image classification. However, in view of the black-box nature of CNNs, we are afraid to apply it to make decisions in the fields with high risk and error costs such as medical imaging diagnosis and military object detection. Due to its black-box nature, we cannot ensure that this model is completely error-free. In fact, there have been many studies about adversarial attack, such as one-pixel attacks [10]. The interpretability of neural networks helps people understand the decision-making processes and enhances the trustworthiness of the network. In addition, as the application range of DNNs continues to expand, it is

Q. Hong and Y. Wang—Contributed equally to this work. This work is supported by: National Defense Science and Technology Innovation Special Zone Project (No. 18-163-11-ZT-002-045-04); Engineering Research Center of State Financial Security, Ministry of Education, Central University of Finance and Economics, Beijing, 102206, China; Program for Innovation Research in Central University of Finance and Economics; National College Students' Innovation and Entrepreneurship Training Program "Research on classification and interpretability of popular goods based on Neural Network".

J. Zeng et al. (Eds.): ICPCSEE 2021, CCIS 1452, pp. 103–115, 2021.
https://doi.org/10.1007/978-981-16-5943-0_9

possible that neural networks and their interpretability can help people explore scientific researches such as genomics and social sciences with complex internal patterns that are still unknown. Therefore, it is imperative that the neural network can be explained.

There have been many studies on interpretability, but the definition of interpretability has not yet been uniformly determined. A review article by Yu Zhang et al. [1] further defines the interpretability of neural networks, and believes that interpretability is based on explanations in understandable terms to a human. Explanations refer to a series of logical rules and understandable terms provide domain knowledge related to specific tasks. For example, computer vision uses super pixels (or image patches) and different levels of visual semantics (colors, materials, textures, parts, objects and scenes) [11] explain the classification of input images. Yu Zhang et al. [1] proposed three classification dimensions for interpretable methods. The first dimension is active/passive interpretation, and the second dimension is the interpretation type, such as providing similar or prototype images, finding the attributes of the object, analyzing semantics of certain latent neurons or layers, providing logical rules. The third dimension is local/global interpretation.

Contributions. In this work, we propose a method to probe semantic representations of latent filters by occluding some specific semantics configurations, providing model-independent, post-hoc and semi-local interpretability.

2 Related Work

Different convolutional layers of the CNNs identify features of different granularities. We can establish connections between a filter at a convolutional layer and abstract semantic concepts through visualization or sensitivity analysis.

CNNs itself is inspired by the human brain visual system. The convolution kernel simulates human's visual neurons -- each convolution kernel recognizes some specific visual features. Therefore, convolutional layers of different depths correspond to different granularity of abstract visual concepts.

Dumitru et al. [12] proposed qualitative interpretability for units in higher layers, finding what the neural is looking for. They looked for the input image which maximize the activations of a hidden filter. Although this proposal could find potential patterns for a specific hidden filter, the visualizing result is too abstract or unrealistic to provide reliable concrete explanations.

Besides detecting the semantic vision patterns implicitly by visualization, Bau et al. [11] managed to find the semantic concepts corresponding to the hidden units explicitly. A pixel-wise labeled with semantic concept dataset was collected to provide the ground truth mask for the receptive field of a hidden unit. The feature map of a unit is converted to a mask with the same size with input images. The alignment between the ground truth mask and the converted mask measures the correlation between the unit and the visual concept. However, the ground truth dataset requires a great quantity of manual annotation, and the hidden semantics rely too much on labeled dataset, so this interpretability method cannot provide explanations for the semantic features specified by users if it has not been labeled. The method proposed in this paper only needs lightweight annotation and the target semantic features can be customized by users.

Although the researches above attempt to bind filters to certain semantic concepts, there are obvious deficiencies. This is because the neural networks need to encode too many features, but there are fewer filters, which means that there is a many-to-many correspondence between filters and semantic concepts. Experiments by Fong et al. [7] proved that a concept requires multiple filters to encode, and there is no clear semantic relationship between one filter and one semantic pattern. Multi-filter encoding has better performance than single-filter encoding. Our method also detects this kind of many-to-many correspondence.

Zeiler and Fergus [13] performed a set of sensitivity analysis by occluding to reveal vital portions of the input images which lead to a significant reduction in the classifier output. Besides, after occluding the same semantic part in many images, the difference vectors (difference of the feature vectors of the original and occluded images) at convolutional layers have lower Hamming distances compared with random occlusion, which means that activations of filters at convolutional layers have the same increase and decrease in each image. It shows that the filters whose activations change together are likely to be related to the semantics of the occluding part. Our method is a generalized application of this sensitivity analysis, probing the hidden sematic representations of filters further.

Zhou et al. [14] replicated each image many times with small random occlusion at different locations in an image, searching for the image patches that cause a large discrepancy in activations. They looked for important units at pooling layers related to each image patch and then displayed the receptive field of each unit in the image dataset. At last, they conducted manual annotation work by identifying the semantic concept of receptive field of each unit and counting the precision of each unit. However, they fed about 5000 occluded images into the network for each original image, which requires a lot of redundant calculations. And this approach still lots of manual annotation.

3 Proposed Method

How does CNNs recognize specific visual concept features? We propose a method that provides a rapid positioning of which parts of the network recognize some specific semantic vision concepts quickly and accurately. Our method is divided into the following three main steps.

Step 1. Choose the class that needs to be studied, define the semantic name of configurations of the object, manually occlude each configuration in the image, replicate each original image with occlusion for each configuration.

Step 2. Obtain feature maps of the original image and the occluded images at convolutional layers in the network. Then compute the difference of the feature maps for the original and occluded images input. If activations of some filters at convolutional layers related to that configuration are significantly reduced after a certain configuration is occluded, these filters will be given a semantic concept of the occluded configuration.

Step 3. After finding the filters corresponding to the semantic configuration, visualize the receptive fields of the filters in original image and evaluate interpretability of our method.

Figure 1 shows our method architecture.

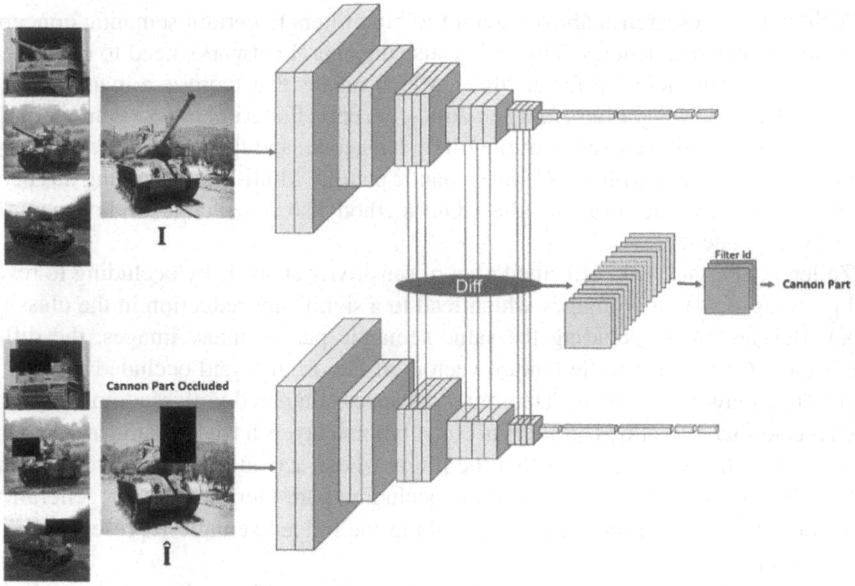

Fig. 1. The method architecture

3.1 Dataset and Semantic Configurations

We select images randomly from ImageNet, mainly focusing on mechanical equipment, animals, clothing, etc. Then we define semantic configurations of these objects as visual semantic concepts that humans can understand.

As convolutional layers are getting higher, visual semantics concepts that they recognize also change from abstract texture features to object-part features with clear contours. Therefore, we add this prior knowledge when defining the semantic configurations of objects. When probing filters at lower convolutional layers, we mainly define some features of texture, color and other granularities for annotation, and when probing filters at higher convolutional layers, we mainly define some object-part level features for annotation.

3.2 Probe Filters to Recognize Specified Visual Semantic Concepts by Occluding

Previously, there have been many studies on the sensitivity analysis of the classification probability of CNNs by occluding, as well as many studies on the impact on activations at convolutional layers by occluding. Zeiler's experiments [13] occluded the same semantic configurations of different images, and activations of some filters at the convolutional layers will undergo similar changes. These commonly changed filters are believed to have a greater correlation with these occluded parts. Our method is the generalized application of this sensitivity analysis, looking for the hidden filters at convolutional layers that recognize the visual concepts specified by the human.

(1) Locate filter whose activations are strongly influenced by specified visual semantic configurations by occluding.

As shown in Fig. 1, we feed the original image I and the occluded image $\hat{I}$ into CNNs at the same time, record the activations of the two inputs at each convolutional layer, and then obtain the *discrepancy map* by computing difference of feature maps for the original and occluded input. For each discrepancy map, we take the maximum activation of the pixels in the map to represent the activation of the discrepancy map. Then the correspondence between the discrepancy map with maximum activation and the semantic configuration related to the occluded image $\hat{I}$ is established.

Specifically, function ϕ. is defined as the process of computing feature map at a certain layer for an input image. We first obtain feature maps of the original image I_0 at convolutional layer L as $\phi\left(I_0^L\right)$, and feature maps of the imageith semantic configuration s occluded I_s at convolutional layer L as $\phi\left(I_s^L\right)$. Then we look for the filter i whose activation decreases mostly after occlusion. The correspondence label $LABEL_s^L$ is established to show the representation of filter i for the semantic configuration s.

$$LABEL_s^L for \underset{\mathrm{i}}{\operatorname{argmax}}\left(\max\left(\phi(I_0)_i^L - \phi(I_s)_i^L\right)\right) \tag{1}$$

where $\phi(I)_i^L$ are the i-th feature map at layer L for input I.

(2) Annotate more occluded images and score the correspondence between filters and semantics to find *reliable candidate filters*.

We call an original image and its copies with different semantic configurations occluded as *a set of images*. After we finish labeling on a set of images, semantics can be found to correspond to the filters. In the practical process, we need to annotate more images for validity and accuracy. Furthermore, a scoring mechanism is proposed to score the degree of correspondence between the semantics and the filters. Therefore, when annotating a certain semantics of a set of images, we not only find the filter i with the maximum discrepancy in activation, we also find the top k filters with maximum activation discrepancy, avoiding missing filters due to individual chance. We assign these top-k filters a descending order of scores, the top 1 filter given the maximum score (e.g. scored k) and the top k-th filter given minimum score (e.g. scored k). Every time we label, the score will be accumulated.

$$SCORE\left(LABEL_s^L, i\right) = \sum_{n=1}^{N} score\left(\left(I_{S,n}\right)^L, i\right), \tag{2}$$

where $score\left(\left(I_{S,n}\right)^L, i\right)$ computes the score assigned to filter i at layer L when annotating the n-th set of images with semantic configuration s occluded. $SCORE\left(LABEL_s^L, i\right)$ is a cumulative score of the degree of correspondence between the semantics s and the filter i.

After annotating all images, we can score the degree of correspondence between the filters and semantics. When probing a specified visual semantic configuration, the higher score a filter gets, the more likely the filter to recognize this visual semantic configuration. At the end, filters with the higher scores will be the candidates for recognizing the corresponding semantics.

(3) Calculate overlaps of filter projection and expert labeled feature regions in the validation set, and select filters with the best representations.
We have found the *reliable candidate filters* that recognize the specified visual semantic concepts. Then we calculate the degree of overlap between the ground truth region and the candidate filters' activation region projected down to the input space.

To build a validation dataset, for each image, we occlude each semantic configuration of the object from the original image and record the coordinates of the occlusion part as the ground truth for each semantics. Then for each semantic configuration of the images, we look up the candidate filters and compute the projected region using the deconvolution method in Sect. 3.3. We compute the overlap between the ground truth and candidate filters' regions projected for each semantic concept. Then we select filters with higher overlap. At this point, we have located the reliable filters that recognize the specified semantic concepts.

3.3 Visualization for Evaluation

In order to evaluate the performance of correspondence between semantics and filters, we need to see how the labeled filters are activated in the input image space. Therefore, we adopt two visualization methods to visualize it. The first is to use the deconvolution network proposed by Zeiler et al. [13] to project the feature map of filters down to the input pixel space. The feature map projected is converted to a mask with the same size of input image by selecting the area where activations are above a certain threshold with a rectangular box, and the rest activations are set to zero. At last, the receptive field is visualized by overlaying the mask on the input image, which helps us to calculate the degree of overlap between each projected box and manually occluded box.

Besides, we can also use a rough projection method. After computing a feature map at a convolutional layer, we directly resize it to the same shape of the input image, and then display the area where activations are above a certain threshold in the original image This method ignores the paddings and so on in the convolution process, but the experimental results show that direct projecting method performed not badly. The advantage of this method is its simplicity and model-independence. Figure 2 shows the two visualization methods, the top row is deconvolution and the bottom row is the rough projecting method.

4 Experiment

4.1 Dataset and Experimental Setup

We mainly use VGG16 [15] and focus on filters in higher convolutional layers (conv5_1, conv5_2 and conv5_3) which are tend to recognize object-part patterns. And we choose tank, chihuahua, egyptian cat, acoustic guitar and T-shirt to annotate. We have defined the configurations for these objects based on expert knowledge. In each round of annotating, we randomly select about 50 images from the ImageNet as the annotation dataset, 50 images as the validation dataset and 100 images as the test dataset for each class.

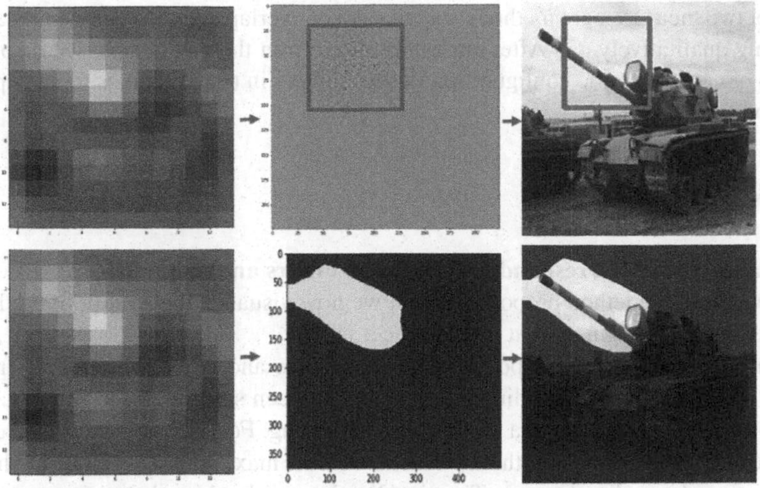

Fig. 2. Visualization Methods for Evaluation (Top row: deconvolution; Bottom row: rough projection)

4.2 Annotation Task

In order to apply the above method, we build a small internal website to complete the labeling task. Our interface is shown in Fig. 3.

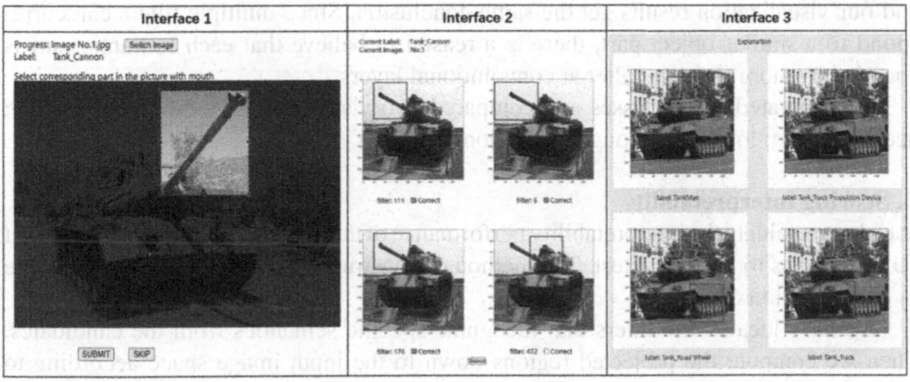

Fig. 3. Annotation Task Interface

The annotation task is divided into three main small phases: (1) In this phase, after defining objects and its configurations, given the CNN network and the layers chosen, interface annotators are asked to use the mouse to select the required semantic configuration of each image. This process needs to be repeated many times to annotate multiple pictures. (2) Interface annotators judge whether the projected area of filters for a certain semantics is accurate by overlapping with human expert annotated feature area. When evaluating the interpretability performance of the annotation results, we adopt two measurement methods: calculating of overlap quantitatively and manually evaluating qualitatively. (3) After choosing filters from the candidates, we display the position of each semantic configuration of these filters in tested images to be explained via deconvolution.

4.3 Experiment Result

A. Visualization of Correspondence Between Filters and Semantics
After applying the method proposed above, we now visualize the feature of the labeled filters down to the original input image space.

Figure 4 shows the correspondence of semantics and labeled filters. After finish labeling, we get the corresponding relationship between semantics and filters $LABEL_s^L$ mentioned in Sect. 3, which is a one-to-many mapping. For each semantic concept, we visualize the activate regions of the top 6 filters with the maximum scores in convolutional layer, displayed in order of scores. The visualization results show that the semantic-filter mapping we got after annotation has a strong relationship. The activation regions of all the top 6 filters can match the semantics very well to a certain extent, so one semantic concept may be encoded by many filters.

Bau et al. [11] once divided the semantic concepts into six levels: color, texture, material, scene, part, and object. The lower convolutional layers of CNNs often recognize low-level semantic features without contours, such as color, textures, materials, scenes, higher convolutional layers often recognize parts and objects with specific shapes. Our method mainly focuses on the higher convolutional layers related to object-part patterns and our visualization results get the same conclusion. Since multiple filters can correspond to a similar object part, there is a reason to believe that each semantic part is encoded by more than one filter at convolutional layers.

We annotated other classes and compared two visualization methods (upper row: deconvolution; lower row: rough projection) in Fig. 5.

B. Scoring Interpretability
In order to evaluate the interpretability performance of the proposed method, we construct 50 test images to conduct a test. The method of constructing the test dataset is the same as the validation dataset.

We have located the filters that recognize specific semantics from the candidates. Then we compute the projected regions down to the input image space according to these filters' activations. The projected rectangular box serves as explanations for the specific semantics. We then compute the degree of overlap between the projected region and the ground truth region for the specified semantics in the test dataset to score interpretability. Table 1 shows the scores of the filters with the maximum score for each semantic configuration.

C. Feature-oriented Image Search
The semantic labels established by this method is in the direction from the semantic features pointing to filters. Based on the experimental results of the above framework, we propose a method for *semantic feature-oriented image search*. This method can locate

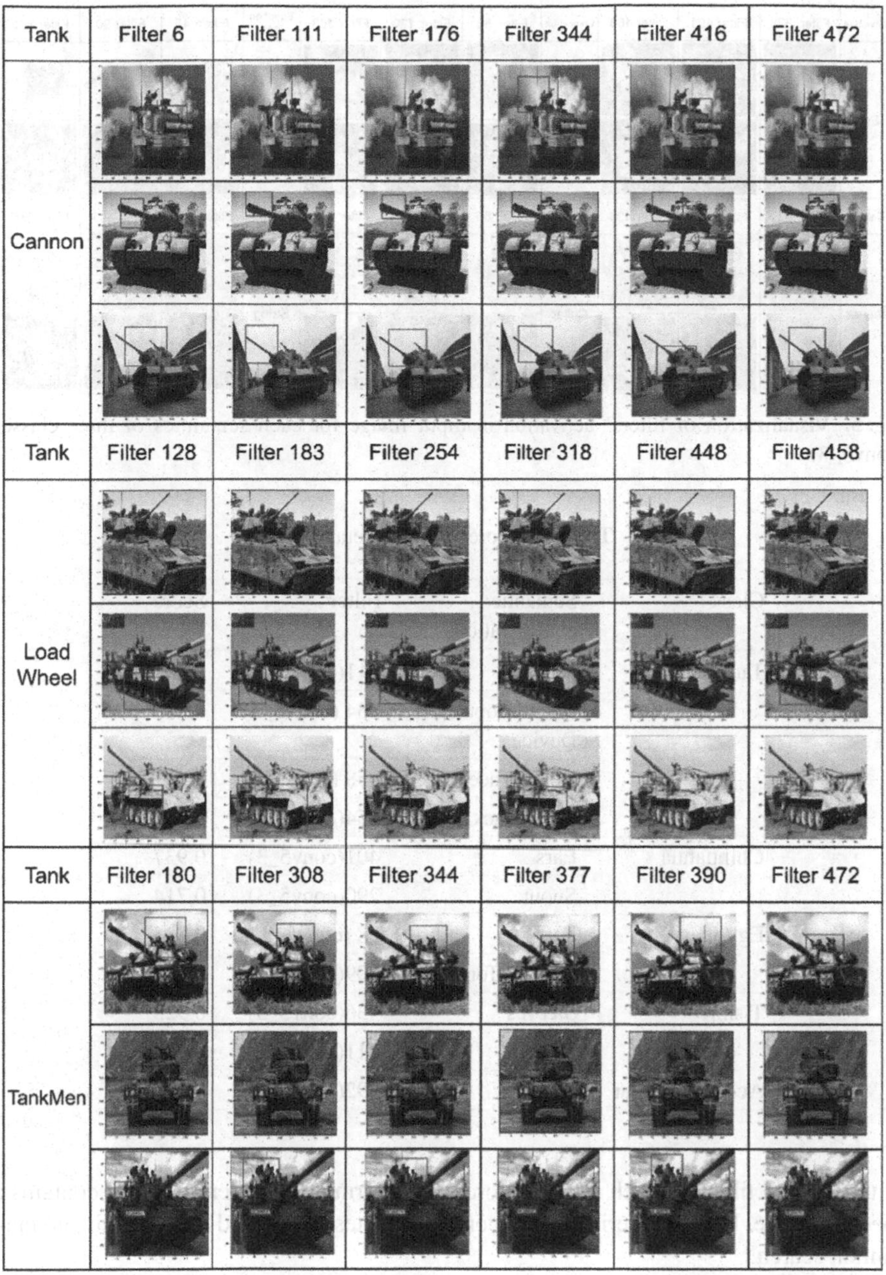

Fig. 4. Visualization of filters' activation in input images for each semantics (conv5_3)

filters that recognize special features by annotating a small number of images with special features. In the subsequent image feature search stage, we rank the correspondence between the image and the special feature by computing ranking index of feature maps

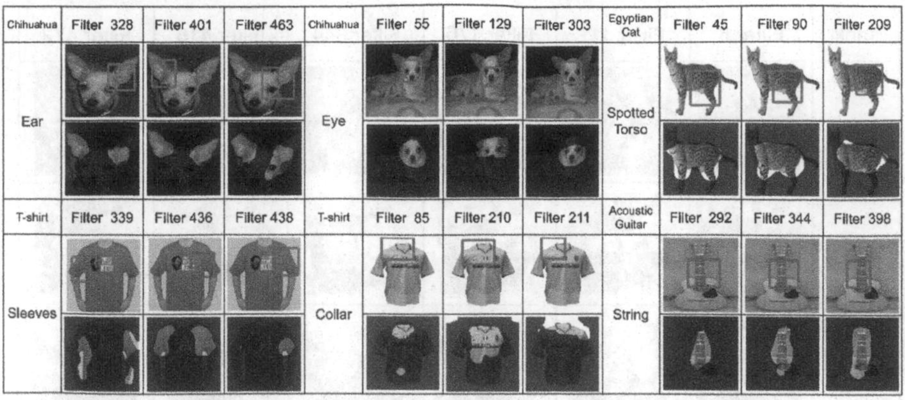

Fig. 5. Visualization of filters' activation in input image for each semantics of three classes (conv5_3)

Table 1. Scores of interpretability

Class	Semantic configurations	Filter	Score
Tank	Cannon	111(conv5_3)	0.869
	Track Propulsion Device	254 (conv5_3)	0.846
	Road Wheel	458 (conv5_3)	0.931
	Tank Track	254(conv5_3)	0.793
Chihuahua	Ears	401(conv5_3)	0.937
	Snout	290(conv5_3)	0.714
Egyptian Cat	Ears	28 (conv5_3)	0.835
	Spotted Torso	209(conv5_3)	0.632
T-shirt	Sleeves	436(conv5_3)	0.923
	Collar	211(conv5_3)	0.709
Acoustic Guitar	String	292(conv5_3)	0.876

of the labeled filter for each image, in order to determine whether an image contains a special feature. In this experiment, we conduct two test cases: red stars search and tank cannon search.

For red stars search, we construct a 30-images set of tanks with red stars. Applying the method proposed in Sect. 3 to annotation, we locate the filters that recognize red stars in the convolutional layer (conv5_3) and the filter with the maximum score is selected. Then we prepare a 100-images test set, of which 10 images are tanks with red stars, and the other 90 images are tanks without red stars. For cannon search, we annotate 30 tank images with cannon. Then we prepare a 100-images test set, of which 10 images are tanks with cannons, and 10 images are other classes including cannon, half-track, train, chihuahua, agaric, acoustic guitar, T-shirt and jeans.

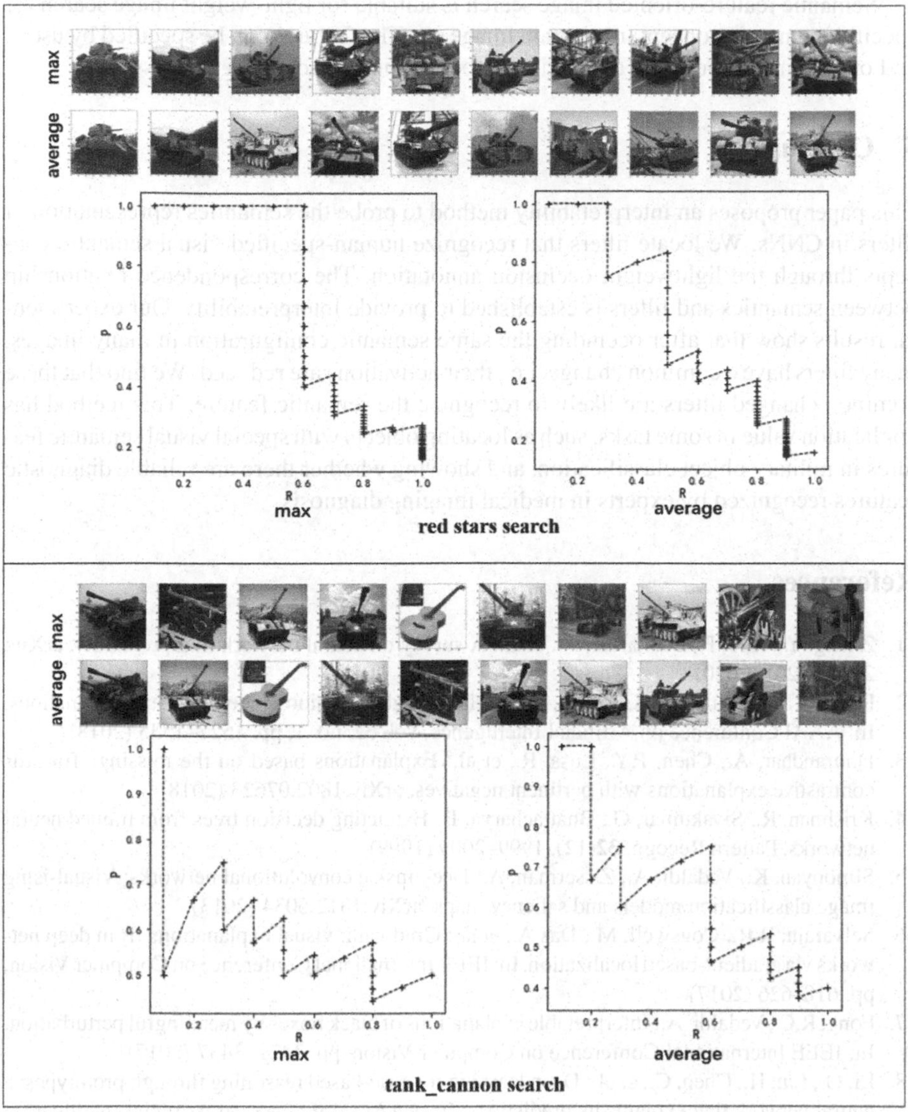

Fig. 6. Precision (P) and recall (R) rates for semantic feature-oriented image search

We conduct the red stars search and tank cannon search in 100 images and record precision and recall rate shown in Fig. 6. We compare the two ranking methods. The first is to rank images according to the average of feature map activations of the labeled filter for each image, and the second is to rank according to the maximum of feature map activations of the labeled filter for each image. We find the second ranking method is better. For example, in the red stars search, the recall rate reaches 1.0 until the 63rd picture is returned using the first method, while the recall rate reaches 1.0 until the 37th image is returned using the second method.

Semantic feature-oriented image search is suitable for light-weight image search for specified semantic concept in a certain image set. The features can be specified by users, and only a small amount of annotation by occluding can achieve good results.

5 Conclusion

This paper proposes an interpretability method to probe the semantics representation of filters in CNNs. We locate filters that recognize human-specified visual semantic concepts through the lightweight occlusion annotation. The correspondence relationship between semantics and filters is established to provide interpretability. Our experimental results show that after occluding the same semantic configuration in many images, many filters have a common change, i.e., their activations are reduced. We find that these common changed filters are likely to recognize the semantic feature. This method has application value in some tasks, such as locating objects with special visual semantic features in military object classification, and showing whether there are reliable diagnostic features recognized by experts in medical imaging diagnosis.

References

1. Zhang, Y., Tiňo, P., Leonardis, A., et al.: A survey on neural network interpretability. arXiv: 2012.14261 (2020)
2. Ribeiro, M.T., Singh, S., Guestrin, C.: Anchors: high-precision model-agnostic explanations. In: AAAI Conference on Artificial Intelligence, vol. 32, no. 1, pp. 1527–1535 (2018)
3. Dhurandhar, A., Chen, P.Y., Luss, R., et al.: Explanations based on the missing: Towards contrastive explanations with pertinent negatives. arXiv:1802.07623 (2018)
4. Krishnan, R., Sivakumar, G., Bhattacharya, P.: Extracting decision trees from trained neural networks. Pattern Recogn. **32**(12), 1999–2009 (1999)
5. Simonyan, K., Vedaldi, A., Zisserman, A.: Deep inside convolutional networks: Visual-ising image classification models and saliency maps. arXiv:1312.6034 (2013)
6. Selvaraju, R.R., Cogswell, M., Das A., et al.: Grad-cam: visual explanations from deep networks via gradient-based localization. In: IEEE International Conference on Computer Vision, pp. 618–626 (2017)
7. Fong, R.C., Vedaldi, A.: Interpretable explanations of black boxes by meaningful perturbation. In: IEEE International Conference on Computer Vision, pp. 3429–3437 (2017)
8. Li, O., Liu, H., Chen, C., et al.: Deep learning for case-based reasoning through prototypes: a neural network that explains its predictions. In: AAAI Conference on Artificial Intelligence, vol. 32, no. 1, pp. 3530–3537 (2018)
9. Chen, C., Li, O., Tao, C., et al.: This looks like that: deep learning for interpretable image recognition. arXiv:1806.10574 (2018)
10. Su, J., Vargas, D.V., Sakurai, K.: One pixel attack for fooling deep neural networks. IEEE Trans. Evol. Comput. **23**(5), 828–841 (2019)
11. Bau, D., Zhou, B., Khosla, A., et al.: Network dissection: quantifying interpretability of deep visual representations. In: IEEE Conference on Computer Vision and Pattern Recognition, pp. 6541–6549 (2017)

12. Erhan, D., Bengio, Y., Courville, A., et al.: Visualizing higher-layer features of a deep network. University of Montreal, Technical Report 1341 (2009)
13. Zeiler, M.D., Fergus, R.: Visualizing and understanding convolutional net-works. In: European Conference on Computer Vision, pp. 818–833 (2014)
14. Zhou, B., Khosla, A., Lapedriza, A., et al.: Object detectors emerge in deep scene CNNs. arXiv:1412.6856 (2014)
15. Simonyan, K., Zisserman, A.: Very deep convolutional networks for large-scale image recognition. arXiv:1409.1556 (2014)

Targeted BERT Pre-training and Fine-Tuning Approach for Entity Relation Extraction

Chao Li and Zhao Qiu(✉)

School of Computer Science and Cyberspace Security, Hainan University, HaiKou, China

Abstract. Entity relation extraction (ERE) is an important task in the field of information extraction. With the wide application of pre-training language model (PLM) in natural language processing (NLP), using PLM has become a brand new research direction of ERE. In this paper, BERT is used to extracting entity-relations, and a separated pipeline architecture is proposed. ERE was decomposed into entity-relation classification sub-task and entity-pair annotation sub-task. Both sub-tasks conduct the pre-training and fine-tuning independently. Combining dynamic and static masking, new Verb-MLM and Entity-MLM BERT pre-training tasks were put forward to enhance the correlation between BERT pre-training and Targeted NLP downstream task-ERE. Inter-layer sharing attention mechanism was added to the model, sharing the attention parameters according to the similarity of the attention matrix. Contrast experiment on the SemEavl 2010 Task8 dataset demonstrates that the new MLM task and inter-layer sharing attention mechanism improve the performance of BERT on the entity relation extraction effectively.

Keywords: Entity relation extraction · BERT · Verb-MLM · Entity-MLM · Inter-layer sharing attention mechanism

1 Introduction

Entity relation extraction plays an important role in construction of knowledge data. The basic goal of ERE is extracting a defined entity-relations from the text data and annotating the position where the entity-pair appears in the text. The traditional ERE method is mainly based on the feature pattern matching [1, 2], these research focus on analyzing the grammatical structure of the text, combing out the syntactic relations between words, and extracting the entity-relations according to the feature templates. In recent years, neural network models [3–9] has gradually become the mainstream research method of ERE, these research focus on constructing word embedded vector and extracting entity-relations by various deep neural networks combined with supervised or semi-supervised training data.

The emergence of pre-training models provides a new solution for ERE. The concepts of upstream and downstream tasks has gradually emerges in NLP. Upstream tasks refer to language models constructed by pre-training models (PLM), and downstream tasks refer to specific NLP tasks, such as NER, ERE, etc. The PLM can be regarded as a very large-scale Transformer group structure [10]. The emergence of BERT triggered a fever

J. Zeng et al. (Eds.): ICPCSEE 2021, CCIS 1452, pp. 116–125, 2021.
https://doi.org/10.1007/978-981-16-5943-0_10

wave of pre-training models in NLP, and it successfully achieved excellent results in the state of the art in 11 NLP tasks [11].

In order to consider the overall performance of various NLP downstream tasks, the model design of original BERT may compromise in many ways. Masked Language Model (MLM) in BERT is undeniably successful, and researchers have also proposed a variety of improvement schemes for MLM. Next Sentence Prediction (NSP) is mainly designed for the sentence level tasks, and it is not helpful for the single sentence annotation task such as ERE. Therefore, there must be space for the performance improvement of the original BERT in ERE task. Enhancing the correlation between the BERT pre-training tasks and ERE will improve the performance furtherly.

The contributions of this paper are as follows:

1) A separated pipeline architecture is proposed to decompose ERE into entity-relation classification sub-task and entity-pair annotation sub-task, both sub-tasks use their own pre-training model and fine-tuning approach independently.
2) We propose Verb-MLM and entity-MLM pre-training tasks, not only mask random words, we also mask predicate verbs and entity-pairs for specific needs in ERE.
3) We add inter-layer sharing attention mechanism to the model. Sharing the attention parameters according to the similarity of the attention weight matrix.

2 Related Work

2.1 Methods with Distant Supervision

Distant supervision is an important research method of ERE. The concept of distant supervision was first put forward by Mike [12]. It is based on a basic assumption: entity-pairs appear repeatedly, so does entity-relations; Daojian proposed the PCNN model to solve the mislabeling problem, and applied the idea of multiple-instance learning in ERE [13]; Yankai added attention mechanism in distant supervision for the first time [6]; A sentence-level attention model (APCNNS) proposed by Guoliang is used to select effective examples and make full use of supervision information in knowledge base [14]; The model proposed by Jinhua introduced the 2-D attention mechanism [15]. The above models are applicable to unsupervised or semi-supervised training.

2.2 Methods with Pre-trained Language Model

The Word2Vec model proposed by Mikolov [16] introduced the concept of Word Embedding, and the word vectors obtained by embedding training can measure the similarity between words well; The GloVe proposed by Jeffrey Pennington distinguished words by the probability ratio of co-occurrence matrix [17]; The representative model of context word embedding is ELMO [18], which enables words to have specific contextual characteristic information; GPT [19] proposed a method combining unsupervised pre-training and supervised fine-tuning by using the transformer structure; ALBERT, as an improved model of BERT, adopts a new sentence order prediction (SOP) pre-training task [20]. All of the above models can be used in ERE.

3 Methodology

3.1 Separated Pipeline Model Architecture

The essence of entity relation extraction is a multi-label annotation task, which needs to label the position of subject-entity and object-entity, and label the entity-relation type of the sentence. In this paper, ERE is decomposed into two sub-tasks: entity-relation classification sub-task and entity-pair annotation sub-task.

Both two sub-tasks can be trained in parallel respectively. In order to enhance the processing ability of BERT to ERE task, we propose a separated pipeline architecture. Considering the characteristics of ERE task, which is different from such tasks as sentence translation and knowledge QA, ERE mainly focuses on single sentence. NSP pre-training task in original BERT is aimed at training the relations between sentences, which will not provide obvious help to ERE and may produce noise interference. Therefore, we removed the NSP pre-training task in our model.

3.2 Verb-mLM Task for Relation Classification Model

In the first sub-task of ERE, the goal is to identify the entity-relation types contained in the sentence, this sub-task can be regarded as short text classification task. In the original BERT. Original BERT adopted static random words masking strategy which enable the language model learning more general linguistic knowledge assuredly. We believe that adjusting the masking strategy properly can improve the performance of the model.

We puts forward the Verb-MLM pre-training task, the basic idea is to use static masking to the predicate verbs and dynamic masking to other words. In the task of entity relation classification, the core predicate verb of a sentence makes more contribution to classification than other words, and the most of relations between subject-entity and object-entity are directly related through verbs. Therefore, we targeted to mask more verbs so that Bert could learn more semantic information about the verb. Considering the generalization ability of the language model generated by BERT, other words will be masked randomly.

Verb-MLM specific training procedure:

1) Selection of masked words: all the verbs in the sentence, 10% of the words remaining in the sentence.
2) Mask rule: 80% probability of all selected words is replaced by [MASK] marks, 10% probability is replaced by random words, and 10% probability remains unchanged.
3) Verb masking: mask all the verbs statically in each iteration with same verbs (static masking).
4) Other word masking: re-select and mask words randomly in each iteration (dynamic masking).

3.3 Entity-MLM Task for Entity-Pair Annotation Model

In the second sub-task of ERE, the goal is to mark the correct position of entity-pairs in the sentence, this sub-task can be regarded as multi-label sequence labeling task. We also design an independent MLM pre-training approach for this sub-task. The original BERT selects the smallest input unit (token) randomly in the sentence to mask. The disadvantages of such processing are as follows: phrase information with strong relevance is divided into independent sentence components forcibly, which is not the optimal training method for entity-pair annotation task.

Inspired by ERNIE [21], We propose the Entity-MLM pre-training task, the basic idea is to do dynamic masking to entities and other words. Considering more entity information is to enable BERT to learn more semantic knowledge about entities. Since some entities are composed of multiple words, we will mask all the words corresponding to the entities.

Entity-MLM specific training procedure:

1) Selection of masked words: 30% of all the entities in the sentence, 10% of the words remaining in the sentence.
2) Mask rule: 80% probability of all selected words is replaced by [MASK] marks, 20% probability is replaced by random words.
3) Entity masking: re-select and mask entities randomly in each iteration (dynamic masking).
4) Other word masking: re-select and mask words randomly in each iteration (dynamic masking).

3.4 Inter-layer Sharing Attention Mechanism

In the original BERT fine-tuning, each transformer layer has its own multi-head self-attention layer with independent attention parameters. The idea of sharing parameters between attention layers comes from Vaswani's exploration and construction process of Transformer architecture [10]. Dehghani also found that sharing parameters between layers can improve the performance of language models, especially in the transformer architecture [22]. In this paper, We propose inter-layer sharing attention mechanism. The basic idea is to do parameter sharing processing for attention layers which has high similarity, and dropout some parameters with certain probability.

Given a sequence of sentence, $sent = \{token_1, ..., token_n\}$ each token is treated with word vector embedding, $sent$ will be transformed into a word embedding matrix $X = \{x_1, ..., x_n\}$. The vector similarity of x_i and x_j can be used to represent the semantic correlation between the words $token_i$ and $token_j$. There is also attention between x_i and x_j. For each x_i, a set of attention vectors is constructed: Query vector q_i, Key vector k_i and Value vector v_i, and the final Value weight matrix W^V are obtained by Query weight matrix W^Q and Key weight matrix W^K. Each Transformer layer has multiple attention-heads, we express the attention calculation method of $layer = l$, $head = h$ as follows:

$$W^V \Big|_{head=h}^{layer=l} = soft\max\left(\frac{W^Q W^{K^T}}{\sqrt{d_k}}\right) \tag{1}$$

After attention calculation, each word x_i will get attention weights v_i between all the words in the sentence, the number of attention weights is equal to the product of the number of transformer layers L and attention-heads H, So we can represent the attention between two words in the statement as:

$$attention^{i,j} = \{v_{i,j}^{1,1}, \ldots, v_{i,j}^{l,h}, \ldots, v_{i,j}^{L,H}\} \quad (2)$$

In the practical application of Inter-layer sharing attention mechanism, if $L \times H$ attention layers are all involved in parameter sharing, it will greatly increase the cost of training calculation of the model. What the model needs to do is to learn which attention layer should be shared. Our sharing approach is based on an assumption: The higher the similarity of the attention weight matrix, the closer the attention layer learned. The similarity calculation method of attention weight matrix as follows:

$$simi(n, m) = \frac{nm}{\|n\|^2 \times \|m\|^2} = \frac{\sum n_i m_i}{\sqrt{\sum n_i^2}\sqrt{\sum m_i^2}} \quad (3)$$

When the similarity is greater than or equal to threshold K, the m and n layers should share the attention weights and dropout part of the weights with a certain probability. The above operation will continue until the end of fine-tuning process.

4 Experiments

The entity relation extraction model we built uses a separated pipeline approach, we will evaluate the performance of both two sub-task models respectively. Firstly, we carry out the experiment of entity-relation classification Model, and mainly investigating the impact of Verb-MLM task on the performance of relation classification. Then, the experiment of entity-pair annotation model was carried out to investigate the effect of entity-MLM task on the performance of entity annotation. We used the original BERT as baseline and compared the effects of inter-layer sharing attention mechanism.

4.1 Datasets

The data sets involved in this paper mainly include FewRel and SemEavl 2010 task8 dataset (SemEavl). Fewrel was produced and released by the Natural Language Processing Laboratory of Tsinghua University [23]. FewRel was built on the basis of the distant supervision dataset. Repeated entity-pairs samples were removed, intercepted samples of 100 entity-relations types with a quantity greater than 1000. A total of 70,000 high-quality labeled sentences with an average length of 24.99, FewRel involving 124,577 different words; SemEavl was derived from Task No.8 of the International Workshop on Semantic Evaluation 2010 [24]. SemEavl contains 9 + 1 entity-relation types, and 8000 high-quality labeled sentences. Because the original BERT uses a large-scale general domain corpus dataset for pre-training, we use the FewRel dataset for BERT pre-training, and the SemEavl dataset for BERT fine-tuning, we tests the entity-relation classification model and entity-pair annotation model on the SemEavl dataset.

4.2 Experiment on Entity Relation Classification

Entity relation classification model uses FewRel dataset to do BERT pre-training. We choose the base version of BERT as the basic model of the experiment, 12 layers of transformer, 768 dimensions of hidden layer, 12 heads of self-attention layer, the feed-forward filter size is set to 4 times the hidden layer dimension, namely 3072. In the pre-training of verb-MLM, 105942 verbs were masked statically. In order to get the Verb information in text, we first made word segmentation and part of speech labeling for the sentences in FewRel, so that the Verb-MLM pre-training task can go smoothly. In this experiment, we set up a total of 4 models for performance comparison:

1) Bert-NoNSP: The original BERT model with the NSP pre-training task removed.
2) Bert- VerbMLM: BERT model with the Verb-MLM pre-training task.
3) Bert-SA: BERT model with the inter-layer sharing attention mechanism.
4) BERT-VerbMLM-SA: BERT model with the Verb-MLM pre-training task and the inter-layer sharing attention mechanism.

Table 1. Entity relation classification experiment results.

Entity relation classification model	SemEavl 2010 task8		
	Precision	Recall	F1-score
(baseline) BERT-NoNSP	0.839	0.790	0.814
(control-group) BERT-VerbMLM	0.876	0.804	0.838
(control-group) BERT-SA	0.844	0.795	0.818
(our approach) BERT-VerbMLM-SA	0.887	0.828	0.856

There are a total of 3500 test samples in this dataset, including 2716 samples in the test set of SemEavl, and additional 584 noise samples we added (without any entity-relation type). Table 1 shows the experimental results of entity-relation classification. The experimental results show that The classification performance of the model was significantly improved by the Verb-MLM pre-training task. The addition of Verb-MLM improved the classification accuracy by 4%, which also confirmed that masking verbs statically in the sentence is beneficial to the ERE task. Relatively, the inter-layer sharing attention mechanism has little impact on the entity-relation classification model, but it can also improve the classification accuracy slightly. Under the combined effect of the Verb-MLM pre-training task and the inter-layer sharing attention mechanism, the classification accuracy of the model reached the highest value of 88.7%.

4.3 Experiment on Entity-Pair Annotation

The entity-pair annotation model uses the FewRel dataset for BERT pre-training. We also choose the base version of BERT as the basic model of the experiment, FewRel is designed to train ERE task, so we can obtain fine-labeled entity-pair information artificially, which will provide great convenience to Entity-MLM pre-training. In the pre-training of Entity-MLM, 187336 entities were dynamically masked, including 59105 pairs of paired entities. In this experiment, we set up 4 models for performance comparison:

1) Bert-NoNSP: The original BERT model with the NSP pre-training task removed.
2) Bert- EntMLM: BERT model with the Entity -MLM pre-training task.
3) Bert-SA: BERT model with the inter-layer sharing attention mechanism.
4) BERT-EntMLM-SA: BERT model with the Entity -MLM pre-training task and the inter-layer sharing attention mechanism.

Table 2. Entity-pair annotation experiment results.

Entity-pair annotation model	SemEavl 2010 task8		
	Precision	Recall	F1-score
(baseline) BERT-NoNSP	0.810	0.776	0.793
(control-group) BERT-EntMLM	0.868	0.808	0.837
(control-group) BERT-SA	0.837	0.792	0.814
(our approach) BERT-EntMLM-SA	0.891	0.818	0.853

This experiment also used a total of 3,500 test samples. Table 2 shows the experimental results of the entity-pair annotation. It can be found that the Entity-MLM pre-training task is also beneficial to the ERE task. After adding the dynamic masking for the entities, the accuracy of the model annotation is improved. The inter-layer sharing attention mechanism also has a greater impact on the model. In the control model which only added the attention mechanism, the annotation accuracy of the model was also improved by nearly 3%. Under the combined effect of the Entity-MLM pre-training task and the inter-layer sharing attention mechanism, the accuracy of annotation has been increased by 8%.

4.4 Results Analysis

Verb-MLM and Entity-MLM enable BERT to learn more semantic knowledge which is beneficial to ERE. We will observe their impact on the model from a test sample in SemEavl:

{"The < e1 > singer < /e1 > arrived to the < e2 > outdoor stage < /e2 > for rehearsal.", Entity-Destination(e1, e2)}.

This sentence contains two entities e1: singer, e2: outdoor stage. Their entity-relation is Entity-Destination. The Fig. 1 shows the attention map of the sample in the model prediction process. BERT-EntMLM-SA (our approach), BERT-NoNSP (original BERT model). BERT-NoNSP use a completely random masking approach, the boundaries of entities cannot be clearly seen in the attention map, and the attention between entities is relatively blurred; BERT-EntMLM-SA use dynamic entity masking approach, we can clearly see the boundary of the entity, and the attention between the two entities has been further enhanced. That is to say, BERT has learned more semantic knowledge about entity-pairs through Entity-MLM pre-training, which is beneficial for ERE doubtlessly.

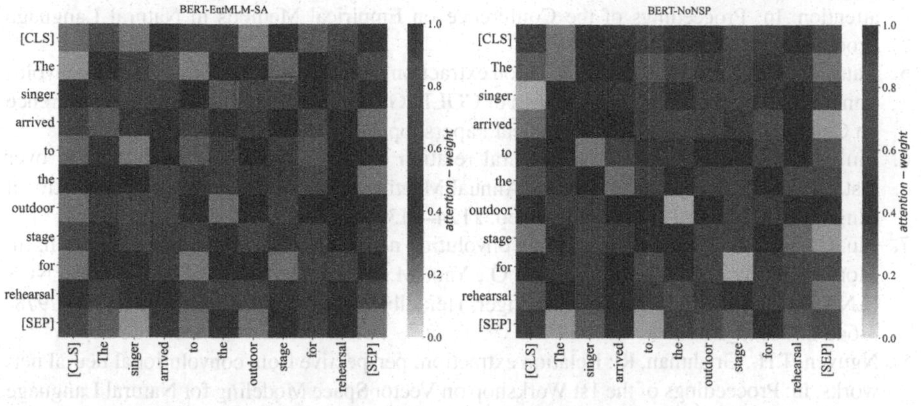

Fig. 1. Visual comparison of attention maps.

5 Conclusion

We proposes a BERT pre-training and fine-tuning approach for ERE by using a separated pipeline architecture, and decompose ERE into entity-relation classification sub-task and entity-pair annotation sub-task; Combining dynamic and static masking, we propose new Verb-MLM and Entity-MLM BERT pre-training tasks to enhance the correlation between BERT pre-training and Targeted NLP downstream task-ERE; We adding inter-layer sharing attention mechanism to the model, sharing the attention parameters according to the similarity of the attention matrix. The experimental results on the SemEavl 2010 dataset show that the addition of Verb-MLM and Entity-MLM greatly improves the performance of BERT on ERE, and attention parameters sharing also improves the performance of entity-relation classification and entity-pair annotation.

Acknowledgment. The writing of this paper has received support and help from many people. Thanks to Professor Zhao Qiu, the corresponding author of the paper. This paper is supported by Hainan Province High level talent project of basic and applied basic research plan (Natural Science Field) in 2019 (No. 2019RC100), Haikou City Key Science and Technology Plan Project (2020–049), Hainan Province Key Research and Development Project (ZDYF2020018).

References

1. Rink, B., Harabagiu, S.: Utd: classifying semantic relations by combining lexical and semantic resources. In: Proceedings of the 5th International Workshop on Semantic Evaluation, pp. 256–259 (2010)

2. Zelenko, D., Aone, C., Richardella, A.: Kernel methods for relation extraction. J. Mach. Learn. Res. **3**, 1083–1106 (2003)
3. Fu, T.J., Li, P.H., Ma, W.Y.: GraphRel: modeling text as relational graphs for joint entity and relation extraction. In: Proceedings of the 57th Annual Meeting of the Association for Computational Linguistics, pp.1409–1418 (2019)
4. Han, X., Yu, P., Liu, Z., et al.: Hierarchical relation extraction with coarse-to-fine grained attention. In: Proceedings of the Conference on Empirical Methods in Natural Language Processing, pp. 2236–2245 (2018)
5. Jiang, X., Wang, Q., Li, P., et al.: Relation extraction with multi-instance multi-label convolutional neural networks. In: Proceedings of COLING 2016, the 26th International Conference on Computational Linguistics: Technical Papers, pp. 1471–1480 (2016)
6. Lin, Y., Shen, S., Liu, Z., et al.: Neural relation extraction with selective attention over instances. In: Proceedings of the 54th Annual Meeting of the Association for Computational Linguistics (Volume 1: Long Papers), pp. 2124–2133 (2016)
7. Liu, C., Sun, W., Chao, W., Che, W.: Convolution neural network for relation extraction. In: Motoda, H., Wu, Z., Cao, L., Zaiane, O., Yao, M., Wang, W. (eds.) ADMA 2013. LNCS (LNAI), vol. 8347, pp. 231–242. Springer, Heidelberg (2013). https://doi.org/10.1007/978-3-642-53917-6_21
8. Nguyen, T.H., Grishman, R.: Relation extraction: perspective from convolutional neural networks. In: Proceedings of the 1st Workshop on Vector Space Modeling for Natural Language Processing, pp. 39–48 (2015)
9. Zeng, D., Liu, K., Lai, S., et al.: Relation classification via convolutional deep neural network. In: Proceedings of COLING 2014, the 25th International Conference on Computational Linguistics: Technical Papers, pp. 2335–2344 (2014)
10. Vaswani, A., Shazeer, N., Parmar, N., et al.: Attention is all you need. arXiv preprint arXiv: 1706.03762 (2017)
11. Devlin, J., Chang, M.W., Lee, K., et al.: Bert: pre-training of deep bidirectional transformers for language understanding. arXiv preprint arXiv:1810.04805 (2018)
12. Mintz, M., Bills, S., Snow, R., et al.: Distant supervision for relation extraction without labeled data. In: Proceedings of the Joint Conference of the 47th Annual Meeting of the ACL and the 4th International Joint Conference on Natural Language Processing of the AFNLP, pp. 1003–1011 (2009)
13. Zeng, D., Liu, K., Chen, Y., et al.: Distant supervision for relation extraction via piecewise convolutional neural networks. In: Proceedings of the 2015 Conference on Empirical Methods in Natural Language Processing, pp. 1753–1762 (2015)
14. Ji, G., Liu, K., He, S., et al.: Distant supervision for relation extraction with sentence-level attention and entity descriptions. In: Proceedings of the AAAI Conference on Artificial Intelligence, vol. 31, no. 1 (2017)
15. Du, J., Han, J., Way, A., et al.: Multi-level structured self-attentions for distantly supervised relation extraction. arXiv preprint arXiv:1809.00699 (2018)
16. Mikolov, T., Chen, K., Corrado, G., et al.: Efficient estimation of word representations in vector space. arXiv preprint arXiv:1301.3781 (2013)
17. Pennington, J., Socher, R., Manning, C.D.: Glove: global vectors for word representation. In: Proceedings of the 2014 conference on empirical methods in natural language processing (EMNLP), pp. 1532–1543 (2014)
18. Peters, M.E., Neumann, M., Iyyer, M., et al.: Deep contextualized word representations. arXiv preprint arXiv:1802.05365 (2018)
19. Radford, A., Narasimhan, K., Salimans, T., et al.: Improving language understanding by generative pre-training (2018)
20. Lan, Z., Chen, M., Goodman, S., et al.: Albert: A lite bert for self-supervised learning of language representations. arXiv preprint arXiv:1909.11942 (2019)

21. Zhang, Z., Han, X., Liu, Z., et al.: ERNIE: enhanced language representation with informative entities. arXiv preprint arXiv:1905.07129 (2019)
22. Dehghani, M., Gouws, S., Vinyals, O., et al.: Universal transformers. arXiv preprint arXiv: 1807.03819 (2018)
23. Han, X., Zhu, H., Yu, P., et al.: Fewrel: a large-scale supervised few-shot relation classification dataset with state-of-the-art evaluation. arXiv preprint arXiv:1810.10147 (2018). Iris Hendrickx
24. Hendrickx, I., Kim, S.N., Kozareva, Z., et al.: Semeval-2010 task 8: Multi-way classification of semantic relations between pairs of nominals. arXiv preprint arXiv:1911.10422 (2019)

Data Security and Privacy

Security-as-a-Service with Cyberspace Mimic Defense Technologies in Cloud

Junchao Wang[1(✉)], Jianmin Pang[1], and Jin Wei[2,3]

[1] State Key Laboratory of Mathematical Engineering and Advanced Computing, ZhengZhou, China
[2] School of Computer Science, Fudan University, Shanghai, China
[3] Shanghai Key Laboratory of Data Science, Fudan University, Shanghai, China
jwei17@fudan.edu.cn

Abstract. Users usually focus on the application-level requirements which are quite friendly and direct to them. However, there are no existing tools automating the application-level requirements to infrastructure provisioning and application deployment. Although some security issues have been solved during the development phase, the undiscovered vulnerabilities remain hidden threats to the application's security. Cyberspace mimic defense (CMD) technologies can help to enhance the application's security despite the existence of the vulnerability. In this paper, the concept of SECurity-as-a-Service (SECaaS) is proposed with CMD technologies in cloud environments. The experiment on it was implemented. It is found that the application's security is greatly improved to meet the user's security and performance requirements within budgets through SECaaS. The experimental results show that SECaaS can help the users to focus on application-level requirements (monetary costs, required security level, etc.) and automate the process of application orchestration.

Keywords: Cyberspace mimic defense · Software diversity · Security-as-a-Service · Multi-compiler · Application deployment

1 Introduction

In cloud environments, infrastructures, platforms, services, etc. are delivered as services to end users which are widely known as Infrastructure-as-a-Service (IaaS), Platform-as-a-Service (PaaS) and Software-as-a-Service (SaaS). A user can easily specify the service level he wants and uses it in a pay-as-you-go manner, which brings great convenience to users. In this paper, we propose the

This research is supported by National Key Research and Development Program of China (2017YFB0803202), Major Scientific Research Project of Zhejiang Lab (No. 2018FD0ZX01), National Core Electronic Devices, High-end Generic Chips and Basic Software Major Projects (2017ZX01030301)and the National Natural Science Foundation of China (No. 61309020) and the National Natural Science Fund for Creative Research Groups Project (No. 61521003).

J. Zeng et al. (Eds.): ICPCSEE 2021, CCIS 1452, pp. 129–138, 2021.
https://doi.org/10.1007/978-981-16-5943-0_11

concept of SECurity-as-a-Service (SECaaS) to guarantee the security of the end user's software with cyberspace mimic defense technology. SECaaS can also help to automatically provision and execute the software in cloud environments.

Cyberspace mimic defense technology (CMD) is an innovative cyberspace security defense mechanism which can enhance a system's security through three characteristics intrinsic in CMD: Dynamic, Heterogeneous, Redundancy[1]. Similar defense mechanisms include Moving-Target-Defense (MTD) technology [2] and security through software diversity [3]. Classical defense technologies usually collect existing approaches and remedy the software by fixing vulnerabilities. The general steps of an attack to software include target identification, vulnerability detection and exploitation and eventually finish an attack. It is believed that vulnerabilities exist in almost all software. To improve a software's security, the software vendors usually need to update patches to their software once a vulnerability is found. But before that, the vulnerability may be exploited by attackers. Before fixing the vulnerability, unpredictable damages can happen during this period. A typical example of such attack is the famous zero-day attack. Therefore, almost all software is suffered from these potential threats which can cause enormous consequences.

To defend against unknown attacks, the CMD technology is proposed to prevent software from damages caused by zero-day attacks. The CMD technology identifies the original issues in software security. In CMD's theory, the attack and defense are unbalanced in existing defense architectures. A successful attack can be conducted through trials and errors. An attacker can try many times by dynamically debugging a software with dynamic debugging tools like OllyDbg [4] and statically analyzing a software with static analyzing tools such as IDA [5], etc. However, during the process, the software remains static without changing. Cyberspace mimic defense technologies try to increase the software's security by introducing uncertainty and dynamic properties on the software side. Such an approach is similar to the security enhancement with a dynamic password. Compared with a static password, a dynamic password can greatly increase its security, making the decryption of the attacker more difficult.

Transformations to software can make the software exhibit different behaviors to an attack. An attacker always takes advantage of exact memory positions (sometimes the position is relative) or instructions to exploit an attack. Any changes to the position can lead to the failure of the attack [6]. Memory errors or unexpected outputs can occur during the attack which can be easily detected.

Although the CMD technology can help to enhance software security, there is no existing approach on how to help the consumer to orchestrate their application in the cloud meeting their security requirements. To meet this end, we propose the concept of SECaaS and automate the process from application-level requirements to software orchestration and execution.

2 Background

An attacker may utilize different attacking methods to achieve his malicious objective. Classical defense methods usually work in a make-up-afterwards manner.

Once an attack is detected, it means that damages have already been done by attackers. Software vendors have to take a quick reaction to fix the vulnerability and remedy the damages. On such occasions, software defenders are comparatively passive to attackers.

Software diversity has been introduced to increase the attacker's difficulty. By introducing randomization in software, an attacker cannot explicitly locate the precise address he wants to use. Therefore, regular attacks will fail in attacking the randomized software. Once the attacker knows how the defending works, he can devise more sophisticated attacking methods to successfully finish the attack. One typical example of such attacking and defending scenario is the Address Space Layout Randomization (ASLR) and Return-Oriented Programming (ROP) attack. ASLR can introduce randomization during the execution of software. For instance, ASLR can randomize the stack base address or heap address so that attackers cannot directly locate the address of his shellcode [7]. To defeat ASLR, the ROP attack is proposed by combining small gadgets together to finish an attack [8].

CMD technology is a proactive defense mechanism that can make the user end more diversified by combining and optimizing the existing diversification technologies together. Moreover, it can also introduce dynamism during the execution of software. At randomized checkpoints, the executed software executable binaries can be changed or substituted by another functional-equal binary. In this paper, we also use "variant" to refer to a software executable binary that is functionally equal to the original software but with different outside characteristics. By using redundancy, key points of the software will be checked whether the output is generated as expected. This is based on an assumption that an attacked software will exhibit abnormal behavior compared with the normal software. Due to CMD's DHR property, CMD technology has been proven to be an effective approach for increasing software's diversity.

3 SECaaS

The overall architecture of SECaaS is shown as below. On the user side, users only need to focus on and specify the application-level requirements. A Service Level Agreement (SLA) will be negotiated between the end user and SECaaS service provider. Other processes will be automated by SECaaS. The Proactive Decision module can help the user determine the number of variants he needed and the virtual infrastructures he needs to use. A multi-compiler can help to generate the diversified variants with the assistance of a heterogeneity evaluation module and a security evaluation module. Then the Cyberspace Mimic Environment will help to load the diversified variants in a dynamic environment. During the execution of the application, the Cyberspace Mimic Environment will change the variants randomly and monitor the status of the executed variants. Once an attack is detected, the Cyberspace Mimic Environment will substitute the malicious variant.

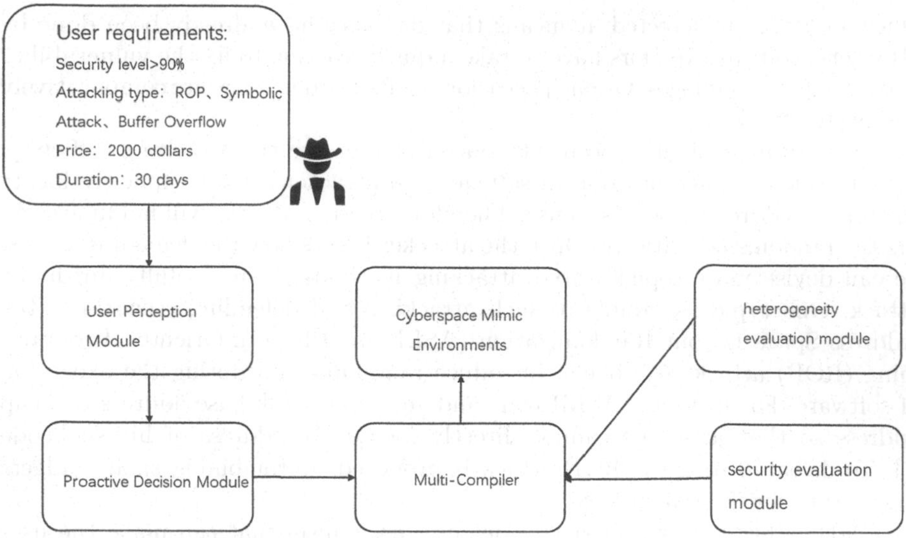

Fig. 1. The overall architecture of SECaaS.

3.1 User's Perception Module

This module is used to negotiate with users on service level agreements. The users can make their tradeoff on the monetary cost they want to pay for their security service and the security level they want to get.

Before consolidating their service, Service-Level-Agreements (SLAs) should be instantiated. The consumers need to prepare a ready-to-run application. They also have their security requirements and budget constraints. For example, a company needs a software running in the cloud continually for 30 days. The maximum budget is 2000 dollars. They want their software to be secured for more than 99% against Return-Oriented Programming (ROP) and symbolic attacks. They also want their software to guarantee certain level of Quality-of-Service (QoS) to end users. For example, the throughput should be higher than 1000 transactions per second. These parameters are quite straightforward to the application providers.

3.2 Decision Module

With users' specified application-level requirements and constraints, the proactive decision module should be able to help determine the right type of Virtual Machines (VMs) and the security strategies with cyberspace mimic defense technologies. Once the parameters are specified, the decision module should translate the users' specified requirements to infrastructure-level parameters.

In this paper, we use the number of variants as the decision variable. The more variants mean more cost from the compiler, storage and MVEE perspective.

The compiling process consumes quite a lot of CPUs and memories. Moreover, the variants should be stored in a certain place of a disk. During the execution of the MVEE, the more variants may incur more resource utilization. Thus, the objective of the decision module is to minimize the number of variants.

The ultimate binaries generated by a multi-compiler (to be discussed in Sect. 3.3) are compiled by different transformation techniques. They may use a single transformation technique or a combination of several strategies. The transformation may also incur extra costs. Different transformation techniques can make the performance of the same software different and make the software's resistance to a certain type of attack different. For instance, encryption of function names can make the debugging process of the attacker more difficult, but cannot resist buffer overflow attacks. Thus, we need to quantify to what extent a transformation can improve the software's security under an attack. Machine learning techniques have been proposed to quantify different transformation techniques under symbolic attack [9].

As is known to all, the more diversity of an ecosystem can be more resistant to outside attacks. Similarly, we believe that the more diversity of the generated binaries can make the whole system more secure. Thus, we also need to measure the diversity of the generated binaries. The decision module needs to take all the above factors into consideration and generate the binaries making trade-offs between these factors.

3.3 Multi-compiler

A multi-compiler can produce binaries with different sizes, control flows but exactly the same function given the same source code. A multi-compiler can take full advantage of the existing works on code obfuscation.

Code obfuscation has long been used to protect a software from reverse engineering issues. This technique has also been used by attackers to conceal their attacking codes. For instance, OLLVM is widely used to protect the attacker's code from detection by anti-virus detection.

Apart from code obfuscation, other techniques that can introduce software diversity such as ASLR can also be applied.

We should note that the function of a multi-compiler is quite different from a code obfuscator. The goal of a multi-compiler is to change the external characteristics of a software. For instance, insertion of NOP instructions can change the size of software, but may not make the software harder to reverse engineered.

When an attacker launches an attack on a software, the attack may be successful for the originally generated binaries. If we insert NOPs to the binary, the stack or heap's address may change, which can cause the attack failing. The multi-compiler can guarantee the heterogeneous property of a system.

Security Evaluation Module. The generated binaries can be used in cyberspace mimic defense environments in a COTS manner. The attackers may use different ways to attack software. Thus, we should evaluate the resistance of software to a certain attack.

For instance, Banescu et al. [9] propose a machine learning-based approach to evaluate the robustness of a software to reverse engineering attack.

Diversity Evaluation Module. Cyberspace mimic defense technology is inspired by the bio-diversity in nature. The more diversity of a population means more tolerance to unexpected disasters. To make the system more secure and robust, we expect the multi-compiler to generate binaries with more diversity. Thus, a diversity evaluation module is required.

Intuitively, NOP insertion can generate an unlimited number of executables. However, the attacker can easily make minor revisions to their attacking codes, finally making the attack successful for all the NOP insertion-generated executables. The diversity evaluation module can help to guide the multi-compiler on generating the executables.

3.4 Cyberspace Mimic Environments

As described above, the cyberspace mimic defense technology has three basic characteristics: dynamic, heterogeneous, redundancy.

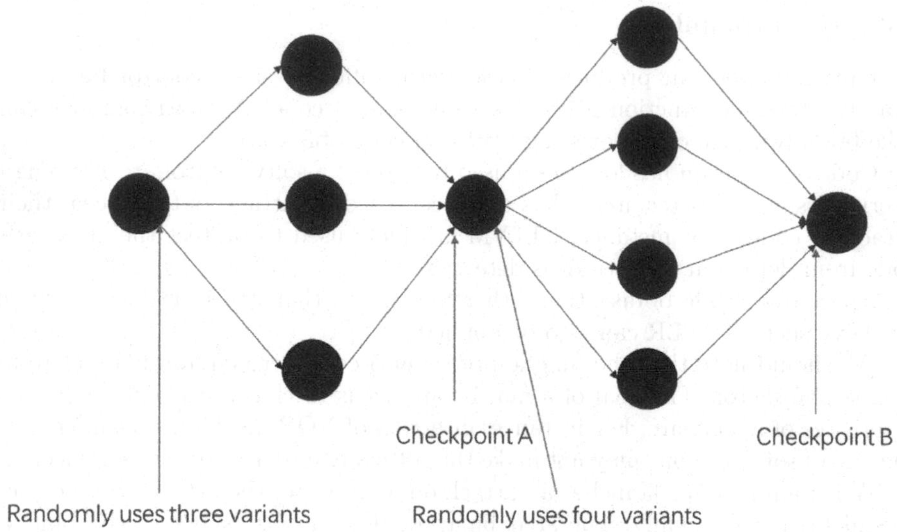

Fig. 2. The overall architecture of CMD.

As is shown in Fig. 2, the executables can change during the execution of a system. At time A, there are four executables in total; in time B, there are three executables. The executables generated by the multi-compiler can also change.

A monitor will detect whether these executables generate the expected output. Sometimes the output of the variant is related to the operating system state or affected by calling some randomized functions. For instance, gettimeofday is one common system call in the Linux operating system. Each output can be different. Despite this, we assume no attack happens. Thus, we should setup rules for some specific function calls to determine on what occasion does an attack happen. When the output of a certain binary is different from the others, we assume an attack may happen to the executable. The executable is no longer trusted. Thus, we remove the executable. When the output of the system is not generated, the software may crash.

For a software, sometimes it is extremely difficult to restore to its expected state. Thus, we use the process duplicate tools in the operating system to make duplications of the crashed binary. When the monitor detects the software's diversity no longer satisfies our threshold, we will notify the user to make back ups.

4 Experiments

We mainly focus on two indexes of our proposed SECaaS—security aspect and performance aspect. In the security aspect, we use cases that can test the assumption we use in our paper that two variants will perform differently under attacks. In the performance aspect, we evaluate the performance degradation with multi-compiler.

4.1 Security Experiments

We use the CTF2017 Babyheap as the case to evaluate the effectiveness of our proposed solution. There is a "fastbin" overflow vulnerability in the file. As fastbin only has one single list, the double free may leak the base address of libc. When only one small/large chunk is released, small/large chunk's fd and bk will point to the address of main_arena. Then fastbin attack can have writing privileges to certain addresses. The shell can be got by using the "malloc_hook" function. A Return-Oriented Programming (ROP) can be further implemented and call the system's shell. Executing the software without using our framework can execute the software normally but start a hidden shell. However, with our system, we can detect abnormal system call when the attacker tries to start the shellcode.

4.2 Performance Evaluation

In this paper we use SPEC 2006 to benchmark the performance of SECaaS. We set the number of variants from 1 to 3 and evaluate their performance compared with the SPEC 2006 without SECaaS. We set the checkpoint as system calls defined in [10]. From the experimental results we can see that the performance is barely not heavily affected by SECaaS. From experimental results shown in Fig. 3, with three variants executed in parallel, we have only 45% percent increase to the performance at most.

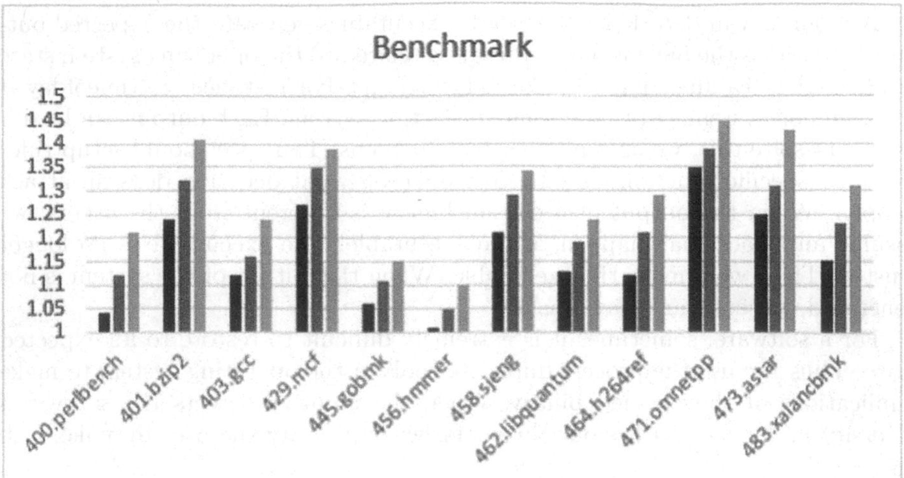

Fig. 3. The performance of SECaaS.

5 Related Works

Sun et al. [11] study the security of microservices which can be composed as a loud application. Although the microservices bring flexibility, the complexity of an application can lead to difficulties in the management of these microservices. The trust issues among distributed microservices are involved. They add a new API primitive FlowTap to monitor and change the policies of network traffic. Although the "Security-as-a-Service" is mentioned in the paper, it has different meanings in their paper. We summarize existing works which can serve as supporting techniques to CMD or is similar to the idea of CMD.

5.1 Multi-compiler

The runtime loading can introduce the diversity of the simultaneous copies. However, this is not enough for attacks like ROP. Thus, we turn to compiler level software diversity. Compiler level software diversity has long been applied by the attackers to conceal their malware behavior. Trojans are usually packaged with an outside packer. Some sensitive instructions are encrypted to avoid the detection of Anti-Virus software. These transformation approaches are called "obfuscation" techniques, which are widely applied for such scenarios.

In this paper, we try to use an "eye to eye, teeth to teeth" approach by using obfuscation techniques to introduce software diversity in the system's protection. In the scope of this paper, we focus on compiler level transformation and propose the concept of multi-compiler. Given the source codes of a software, multi-compiler can help to generate executables with equal function but different sizes, control flows, different instructions, etc., namely variants. The multi-compiler can take full advantage of existing obfuscation techniques. Banescu et al. summarize the obfuscation transformation techniques at different levels [12]. For instance, opaque predicates are used to extend loop condition in a software. Variables reordering can transform the location or the name of variables in the program code by permutation or substitution [12]. All these obfuscation techniques can be applied to CMD to generate diversified variants.

Hyungjoon et al. [13] proposed a compiler-assisted code randomization (CCR) approach to transform a software executable binary with metadata. Basic block reordering is applied to generated different variants by extending the LLVM compiler toolchain.

The NOP insertion is examined in [2]. Different NOP insertion algorithms' performance is also evaluated.

The multi-compiler can generate different binaries with the same source code. However, these compiled binaries cannot solely improve the security of software. They need to be executed and monitored by a MVEE environment.

5.2 MVEE

The idea of using duplications to avoid the failure originates from the research of the system's fault tolerance [14]. Once a software's executable binary fails, another one can quickly serve as a substitute. Multi-Variant Execution Environment (MVEE) duplicates several copies of a software and executes these copies in parallel. The copies of a software have an equal function but may exhibit different behaviors on attacks. A monitor is used to detect the key-points in a software such as system call. Once a difference is detected, the monitor will notify the user that an attack is detected. The idea of MVEE is similar to the Heterogeneous and Redundancy property in MDT. However, MVEEs only make minor changes to a copy to introduce heterogeneity. For instance, GHUMVEE uses Disjoint Code Layouts (DCL) to make two software different [15]. The synchronization is indeed both a problem in MVEE which might incur performance degradation. Thus, the GHUMVEE proposed several synchronization strategies to enhance the performance.

The security of the monitor is another issue in MVEE. An attacker may not directly target the software but the monitor. To solve such a problem, MvArmor put the software at user privilege rather than kernel privilege [10]. This makes sense because the attacker needs to use kernel-level privilege to call sensitive system-level APIs to finish his malicious objective.

Existing MVEEs don't introduce the dynamism during the execution of a software. In our previous work [16], we introduce an extensive framework to introduce the dynamic properties which can substitutes several copies of software during execution at random time slices. This can introduce extra efforts for the attacker to analyze the software and exploit a vulnerability.

6 Conclusion and Future Works

In this paper, we propose a SECaaS architecture to help the user enhance their software's security through cyberspace mimic defense technology. Our architecture can help the user to make a tradeoff between the monetary cost and the security level they want to get. It can also help the user to automatically instantiate their software in cyberspace mimic defense environments.

Data synchronization during the execution of the software is extremely difficult. Moreover, in our future works, we will focus on the performance and security tradeoff.

References

1. Chen, Z., Cui, G., Zhang, L., et al.: Optimal strategy for cyberspace mimic defense based on game theory. IEEE Access PP(99), 1 (2021)
2. Jajodia, S., Ghosh, A.K., Swarup, V., et al.: Moving Target Defense. Springer, New York (2011). https://doi.org/10.1007/978-1-4614-0977-9
3. Voulimeneas, A., Song, D., Larsen, P., Franz, M., Volckaert, S.: dMVX: secure and efficient multi-variant execution in a distributed setting. In: 14th European Workshop on Systems Security (EuroSec 2021), Edinburgh, Scotland, April 2021
4. OllyDbg. http://www.ollydbg.de/ http://www.ollydbg.de/
5. IDA Pro. https://www.hex-rays.com/products/ida/ https://www.hex-rays.com/products/ida/
6. Voulimeneas, A., et al.: Distributed heterogeneous N-variant execution. In: 17th International Conference on Detection of Intrusions and Malware, and Vulnerability Assessment (DIMVA 2020), Lisbon, Portugal, June 2020
7. Evtyushkin, D., Ponomarev, D., Abu-Ghazaleh, N.: Jump over ASLR: attacking branch predictors to bypass ASLR. In: IEEE/ACM International Symposium on Microarchitecture ACM (2016)
8. Borrello, P., Coppa, E., D'Elia, D.C.: Hiding in the particles: when return-oriented programming meets program obfuscation. In: 51st Annual IEEE/IFIP International Conference on Dependable Systems and Networks (DSN 2021). IEEE (2021)
9. Banescu, S., Collberg, C., Pretschner, A.: Predicting the resilience of obfuscated code against symbolic execution attacks via machine learning. In: 26th USENIX Security Symposium (USENIX Security 17), pp. 661–678 (2017)
10. Koschel, J., Giuffrida, C., Bos, H., Razavi, K.: TagBleed: breaking KASLR on the isolated kernel address space using tagged TLBs. In: EuroS&P, September 2020
11. Sun, Y., Nanda, S., Jaeger, T.: Security-as-a-service for microservices-based cloud applications. In: 2015 IEEE 7th International Conference on Cloud Computing Technology and Science (CloudCom), pp. 50–57. IEEE (2015)
12. Banescu, S., Pretschner, A.: A tutorial on software obfuscation. Adv. Comput. **108**, 283–353 (2018)
13. Koo, H., Chen, Y., Lu, L., Kemerlis, V.P., Polychronakis, M.: Compiler-assisted code randomization. In: IEEE Symposium on Security and Privacy (SP), San Francisco, CA, vol. 2018, pp. 461–477 (2018). https://doi.org/10.1109/SP.2018.00029
14. Cox, B., Evans, D., Filipi, A., et al.: N-variant systems: a secretless framework for security through diversity. In: USENIX Security Symposium, pp. 105–120 (2006)
15. Volckaert, Stijn, De Sutter, Bjorn, De Baets, Tim, De Bosschere, Koen: GHUMVEE: efficient, effective, and flexible replication. In: Garcia-Alfaro, Joaquin, Cuppens, Frédéric., Cuppens-Boulahia, Nora, Miri, Ali, Tawbi, Nadia (eds.) FPS 2012. LNCS, vol. 7743, pp. 261–277. Springer, Heidelberg (2013). https://doi.org/10.1007/978-3-642-37119-6_17
16. Junchao, W., et al.: A framework for multi-variant execution environment. J. Phys. Conf. Ser. **1325**(1), 012005 (2019)

Searchable Encryption System for Big Data Storage

Yuxiang Chen[1,2,3](✉), Yao Hao[1,2], Zhongqiang Yi[1,2], Kaijun Wu[1,2], Qi Zhao[1,2], and Xue Wang[1,2]

[1] Science and Technology on Communication Security Laboratory, Chengdu 610041, China
[2] No. 30 Inst, China Electronics Technology Group Corporation, Chengdu 610041, China
[3] School of Computer Science and Engineering, University of Electronic Science and Technology of China, Chengdu 611731, China

Abstract. Big data cloud platforms provide users with on-demand configurable computing, storage resources to users, thus involving a large amount of user data. However, most of the data is processed and stored in plaintext, resulting in data leakage. At the same time, simple encrypted storage ensures the confidentiality of the cloud data, but has the following problems: if the encrypted data is downloaded to the client and then decrypted, the search efficiency will be low. If the encrypted data is decrypted and searched on the server side, the security will be reduced. Data availability is finally reduced, and indiscriminate protection measures make the risk of data leakage uncontrollable. To solve the problems, based on searchable encryption and key derivation, a cipher search system is designed in this paper considering both data security and availability, and the use of a search encryption algorithm that supports dynamic update is listed. Moreover, the system structure has the advantage of adapting different searchable encryption algorithm. In particular, a user-centered key derivation mechanism is designed to realize file-level fine-grained encryption. Finally, extensive experiment and analysis show that the scheme greatly improves the data security of big data platform.

Keywords: Big data platform · Searchable encryption · Fine-grained · Secure storage

1 Introduction

Currently, big data storage platforms store user's data in plain text, which brings the risk of data leakage, and it is difficult for users to protect their stored data from being stolen. At the same time encrypted data storage faces the following problems: if the encrypted data is downloaded to the user's terminal and then decrypted and searched, the search efficiency will be inefficient. Decrypting and searching the encrypted data on the server will reduce the security [1–4].

At the same time, when the distributed file system is applied to provide storage services for users, users usually upload and store plaintext data directly, and it is difficult for users to control whether their stored data is leaked or stolen. In addition, the file

J. Zeng et al. (Eds.): ICPCSEE 2021, CCIS 1452, pp. 139–150, 2021.
https://doi.org/10.1007/978-981-16-5943-0_12

storage service provider may monitor and analyze the user's file retrieval behavior, perform correlation analysis on the user's file content through the keywords retrieved by the user, and then focus on cracking and stealing user's data [5–9]. Besides, users store plaintext data in third-party organizations, losing the initiative to control data and failing to know whether their data has been stolen or leaked [10–13].

For example, in June 2017, the analysis report of the Shodan Internet device search engine showed that the Hadoop server was exposed due to insecure configuration and plaintext storage. It involves nearly 4500 servers using the Hadoop Distributed File System (HDFS), with a data volume of up to 5120 TB. The requirements corresponding to the above risks are:

1. File system data encryption protection.
2. The data of the file system can realize the search query of the ciphertext to obtain the corresponding file content.

A feasible solusion is to use the ciphertext computing algorithm such as searchable encryption, homomorphic encryption, etc. They can efficiently perform retrieval operations in ciphertext and also have diverse scenarios, such as medical data, financial data and government data, etc. [13–16].

In response to the above risks and requirements, we developed our system based on searchable encryption algorithm and hierarchical key management to realize three major functions, including data encryption protection, encrypted search of encrypted data on the storages side, and data security sharing. By classifying users, distributing corresponding master keys, the keys of file contents are derived from users' master keys and file unique identifiers. Hierarchical and fine-grained control is realized through hierarchical keys, thus, users of corresponding levels can complete file encryption, upload and sharing only by holding encryption keys of corresponding levels, which is simple and convenient.

To sum up, our system can realize ciphertext storage, cipher retrieval and file access control based on user identity level and master key, effectively solve the content security problem of traditional distributed file system, and prevent file content leakage caused by malicious administrators and network attacks.

The rest of the paper is organized as follows: In Sect. 2, we first familiarize with the scheme's functional interaction model and system architecture. In Sect. 3, we list the usage of our algorithm and deployment of our system. Section 4 present the system analysis and feasibility. Section 5 concludes the paper.

2 Functional Interaction Model and System Architecture

2.1 Functional Interaction Model

Distributed file system supporting searchable encryption technology mainly provides users with file secure search and storage, Fig. 1 shows the functional model of distributed file system supporting searchable encryption technology proposed for this project. The core parts of functional interaction model mainly includes 3 parts: file data encryption protection, ciphertext search, download and decryption of the file data at the storage end and file secure sharing.

1. Data encryption protection: File data encryption protection means that users generate the encryption keys of files based on file meta data and their own master keys, then encrypt files, generate indexes and encrypt them at the same time, then upload ciphertext files and cipher text index to the cloud server.
2. Ciphertext search: Cipher search of encrypted file data at the storage end means that users generate ciphertext retrieval key value through data keywords and keys, then send the key value to the server, search the ciphertext index of the file through the retrieval algorithm to obtain the corresponding file, finally download and decrypt the retrieved file to obtain the plain text.
3. Data secure sharing: Security sharing of file data means that user A can directly share file ciphertext to user B by sharing the derived key and index, and user B updates its own ciphertext retrieval set, as long as user B's security level meets the security level of the file, the file can be obtained through ciphertext retrieval, download and decryption of file data.

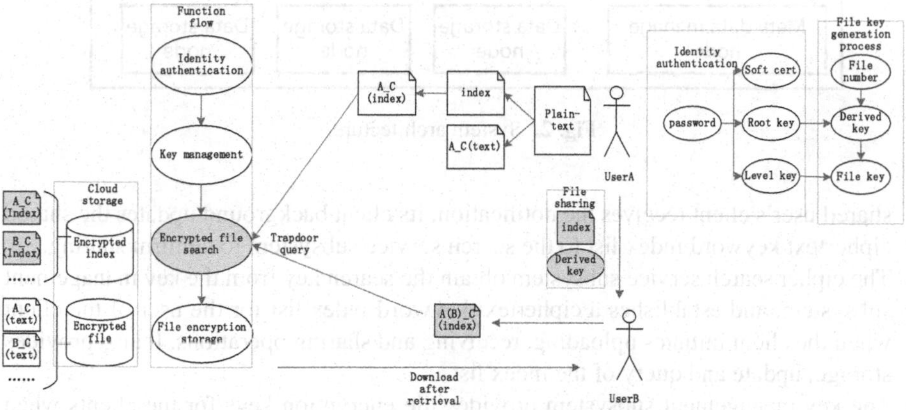

Fig. 1. Functional interaction model.

2.2 System Architecture

As is shown in Fig. 2, our system architecture mainly includes cipher search system client, cipher search service subsystem, key management subsystem and the existed big data platform it relies on (take HDFS as an example in the figure).

- The cipher search client uploads and downloads files to the HDFS storage system, performs encryption and decryption operations in the background of the client during uploading and downloading. When the client uploads the ciphertext file to the HDFS storage system, it also establishes a cipher keyword index list in the cipher search service subsystem for locating the ciphertext file.
  When the client initiates the sharing operation, it distributes the file key to the shared user's client through secure channel and notifies the shared user's client. After the

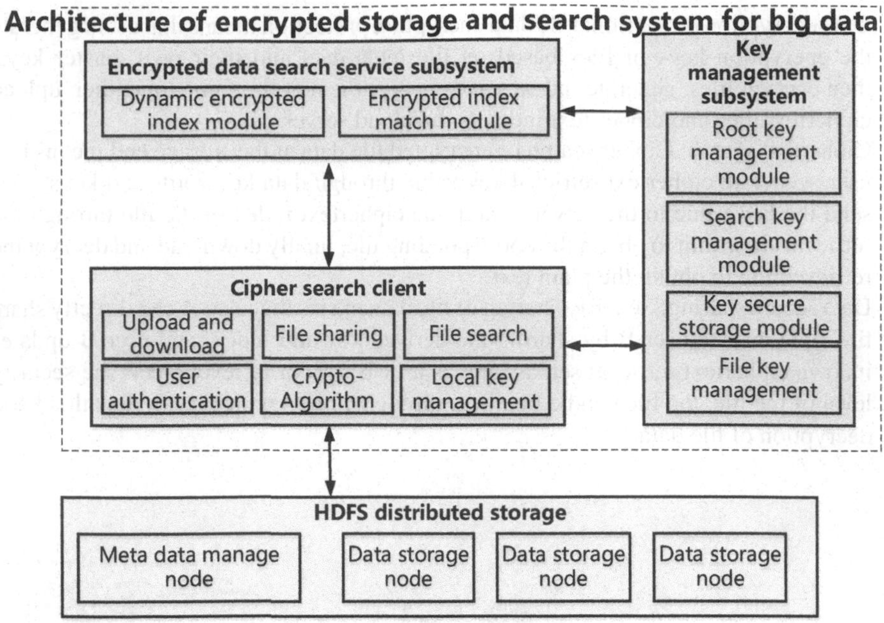

Fig. 2. System architecture.

shared user's client receives the notification, its client background updates the shared ciphertext keyword index list to the search service subsystem to confirm sharing.

- The cipher search service subsystem obtain the search key from the key management subsystem, and establishes a ciphertext keyword index list for the user of the client when the client initiates uploading, receiving and sharing operations. It also provides storage, update and query of the index list.
- The key management subsystem provides file encryption keys for the clients when they perform upload and download operations, provides file encryption keys for shared user clients when sharing, provides search keys for clients when searching. At the same time, key management system provides search keys for search service system to establish, update and search the cipher keyword index.

3 Main Cryptographic Algorithm

3.1 Key Management Algorithm

Suppose SKE = (Setup, Enc, Dec) is a secure traditional symmetric encryption algorithm, $Hash$:$\{0, 1\}^* \rightarrow \{0, 1\}^*$ is an anti-collision hash function, key management scheme includes following steps:

1. System boot: Key management center initiate, generate system public parameter Param and master key MK.

2. User register: User U_i use his identity and attributions to initiate registration, key management centre compute user's root key RK from its master key MK (using IBE scheme to generate secret key), then issue a certificate PK for user as the public key corresponding to RK. Meanwhile, issue hierarchical key LK_u to users according to their secure level, of which $LK_u \in \{LK_1, LK_2, LK_3\}$. Hierarchical keys are divided into several levels, users of same level have the same classification key, let 1st level key be LK_u, which is a pseudo-random number generated by using pseudo-random function Rand. The 2nd level key is $LK_2 = Hash(LK_1|2)$, 3rd level key is $LK_3 = Hash(LK_2|3)$, of which | means string concatenation.
3. Encryption key derivation, encryption and management algorithm: A file may have different security levels, in order to improve security levels, users have to encrypt different levels of files with different keys.

A user first generates a file key FK (where $FK = Hash(RK|Filename|\mathrm{LK}_u)$) by using his root key RK, filename and file security level, of which LK_u corresponds to user's authorization file security level. Besides, the user randomly choose a string nonce to compute secret key SK for searchable encryption, of which SK = Hash(RK|nonce). The user can use SK to compute cipher index of the file, which is used to identify the owner of the file and facilitate the query operation of other users.

Finally, the user set a password to encrypt the file key FK (that is, C_K = Encrypt(FK, hash(password))).

3.2 File Classification Encryption/Decryption Algorithm

For file encryption, we adopt the method of digital envelope, that is, use symmetric encryption algorithm to encrypt the file, use searchable encryption to generate the cipher index of the file, finally encrypt the file key. Further, we adopt national algorithm of China SM2 in the encryption/decryption procedure.

For example, when it comes to a file with security level A, the user first calculates its file key by using steps in key derivation, that is, $FK_{A_u} = Hash(RK|filename|LK_u)$. Then use file key to $FK = FK_{A_u}$ to encrypt file A to get its ciphertext C_F. ($C_F = Enc(FileA, FK_{A_u})$).

Finally, the user use his password key K to encrypt file key, get $C_K = Enc(FK_{A_u})$, stores the final cipher result $C = (C_F, C_K)$.

When it comes to decryption, user first decrypt file key $FK = FK_{A_u} = Dec(C_K, K)$, of which K = hash(password). Then restore his original file key FK_{A_u} according to his authorization. Finally, restore the file by computing $File = Dec(C_F, FK_{A_u})$.

3.3 Symmetric Searchable Encryption Algorithm

Suppose user has file collection $D = (d_1, d_2, ..., d_n)$ corresponding unique identifiers $(\overline{d}_1, ...\overline{d}_n)$, all of them has keywords collection $W = (w_1, w_2, ..., w_n)$, suppose SKE_1 and SKE_2 are two traditional secure symmetric encryption algorithms, $f : \{0, 1\}^k \times \{0, 1\}^l \rightarrow \{0, 1\}^{k+log_2 s}$ is a pseudo-random function. $\pi : \{0, 1\}^k \times \{0, 1\}^l \rightarrow \{0, 1\}^l$ and $\psi : \{0, 1\}^k \times \{0, 1\}^{log_2 s} \rightarrow \{0, 1\}^{log_2 s}$ are two pseudorandom permutation functions. The main steps listed as follows:

1. Key generation: generate random key K_1, K_2, K_3, separately used for pseudorandom permutation function ψ, pseudorandom function f and pseudorandom permutation function π, generate key K_4 for SKE_2.
2. Keyword matching: scan the file collection to get corresponding keywords collection $\delta(D)$, for each $w \in \delta(D)$, there exist a document subset $D(w)$ corresponding to this keyword, of which $D(w) \subseteq D$, we set a global counter $ctr = 1$ in the traversal.
3. Construction of file index array A: for $1 \leq i \leq |\delta(D)|$, construct linked list L_i corresponding to i-th keyword. Each L_i's nodes (expressed as $N_{i,j}$) order is random and the length $|D(w)|$ may be different. Suppose $id(D_{i,j})$ is j-th unique identifier in$D(w)$, generate each node $N_{i,j} \leq id(D_{i,j})||K_{i,j}||\psi_{K_1}(ctr + 1) >$, encrypt node $N_{i,j}$ with key and write the cipher in position $\psi_{K_1}(ctr)$, of which $K_{i,j}$ is used for encrypting next node $N_{i,j+1}$ while $\psi_{K1}(ctr + 1)$ is the pointer of next node. Every time the encryption storage cycle is completed, the counter is incremented by 1 (ctr $++$). In addition, the pointer and key of the last node in the linked list are empty, that is, $N_{i,|D(w_i)|} \leq id(D_{i,|D(w_i)|})||0^k||NULL >$.
   Let $s' = \sum_{w_i \in \delta(D)} |D(w_i)|$ \$, it means adding up the number of documents to which each keyword belongs. if $s' < s$, fill the remaining positions with random strings, s is the upper limit of array A's element capacity.
4. Construction of head node index T: the size of T is $\{0, 1\}^l \times \{0, 1\}^{k+log_2 s} \times ||)$, $|\Delta|$ is the total number of dictionary set keywords. For all the $w_i \in \delta(D)$, set the index entry of the first node as $T[\pi_{K_3}(w_i)] = (addr_A(N_{i,1})||K_{i,0}) \oplus f_{K_2}(w_i)$. Fill the remaining items $|\Delta| - |\delta(D)|$ with random numbers.
5. Encrypt original data: for each document $d \in D$, compute $c \leftarrow SKE_2.Enc(K_4, d)$, get the final result $I = (A, T), C = (c_1, ..., c_n)$.
6. Generate query token: compute $t = (\pi_{K_3}(w), f_{K_2}(w))$ and send it to cloud server.
7. Cloud search: Parse the query token t as (γ, η), where $\gamma = \pi_{K_3}(w_i)$, $\eta = f_{K_2}(w_i)$, find out if there is a result θ at position γ in the head node index T (that is, $\theta \leftarrow T(\gamma)$). If θ exists, then compute:

$$\begin{aligned}&\theta+\eta\\&=(addr_A(N_{i,1})||K_{i,0}) \oplus f_{K_2}(w_i) \oplus f_{K_2}(w_i)\\&=(addr_A(N_{i,1})||K_{i,0})\end{aligned} \tag{1}$$

   Use key K' to decrypt the node in position $addr_A(N_{i,1})$, finally output all the identifiers in list L_i one by one.
8. For each encrypted document, compute $d \leftarrow SKE_2.Dec(K_4, c)$, of which K_4 is derived as is shown in file classification encryption/decryption algorithm.

3.4 Dynamic Update of Cipher Index

Due to the cipher state exposed to cloud, dynamic update has become an extremely important issue, which is related to security.

When user uploads a new file d_k, which includes keyword w_n and w_m, of which w_n is a keyword that has been established in the cloud's ciphertext index, while w_m is a new keyword that hasn't been established.

- For w_n, user has already constructed head node in the cipher index server, that is, $T[\pi_{K_3}(w_n)] = (addr_A(N_{n,1})||K_{n,0}) \oplus f_{K_2}(w_n)$, the file uploading process is equivalent to resubmitting the trapdoor ofw_n, which means that the server need to parse out all the identifiers containingw_n.

The original linked list is:

$$L_n = N_{n,1}||\ldots||N_{n,j-1}||N_{n,j}||N_{n,j+1}||\ldots||N_{n,|D(w_n)|} \tag{2}$$

where $N_{n,j} \leq id(D_{n,j})||K_{n,j}||\psi_{K_1}(ctr{+}1)>, N_{n,|D(w_n)|} \leq id(D_{n,|D(w_n)|})||0^k||NULL)>$

The server will reconstruct the linked list L_n after parses out all the identifiers, that is reconstruct the tail node and attach a new node, which has the minimum computational overhead.

Generate random key $K'_{n,|D(w_n)|}$, then use it to encrypt the new tail node $N_{n,|D(w_n)|+1}$ through symmetric encryption algorithm SKE_1. The tail node $N_{n,|D(w_n)|}$ of original linked list L_n is updated to

$$N_{n,|D(w_n)|} \leq id(D_{n,|D(w_n)|})||K_{n,|D(w_n)|}||\psi_{K_1}(ctr_{|N_{ij}|}+1))>$$

and then attach a new node $N_{n,|D(w_n)|+1} \leq id(D_{n,|D(w_n)|+1})||0^k||NULL\ >$, of which $id(D_{n,|D(w_n)|+1}) = id(d_k)$, at this time the new linked list is:

$$L'_n = N_{n,1}||\cdots||N_{n,j-1}||N_{n,j}||N_{n,j+1}||\cdots||N_{n,|D(w_n)|-1}||N'_{n,|D(w_n)|}||N_{n,|D(w_n)|+1} \tag{3}$$

So when the client updates the server array, it only needs to submit the server:

$$A\left[\psi_{K_1}\left(ctr_{|D(w_n)|}\right)\right] \leftarrow SKE_1.Enc(K_{n,|D(w_n)|-1}, N'_{n,|D(w_n)|}) \tag{4}$$

$$A\left[\psi_{K_1}\left(ctr_{|N_{ij}|}+1\right)\right] \leftarrow SKE_1.Enc(K'_{n,|D(w_n)|}, N_{n,|D(w_n)|+1}) \tag{5}$$

Equation (4) overwrite the encrypted node in the original position with the new node. Equation (5) overwrite the random string of position $A\left[\psi_{K_1}\left(ctr_{|D(w_n)|}\right)\right]$ with the new tail node.

- For w_m, the user has not established the cipher index of w_m before, thus need to construct new linked list L_m, at this time, the linked list only needs to construct one node $N_{m,1} \leq id(D_{m,1})||0^k||NULL>$, where $id(D_{m,1}) = id(D_{n,|D(w_n)|+1}) = id(d_k)$, that is, the same identifier corresponds to different keywords.

Under this condition, the head node cipher index needs to be newly constructed:$T\left[\pi_{K_3}(w_m)\right] = (addr_A(N_{m,1})||K_{m,0}) \oplus f_{K_2}(w_m)$, the node storage position is calculated by client:$addr_A\left(N_{m,1}\right) = \psi_{K_1}(ctr_{|N_{ij}|}+1)$, of which $ctr_{|N_{ij}|}$ is the number of filled nodes in array A, besides, the remaining $s - s'$ positions still fill with random strings, $s' = ctr_{|N_{ij}|}$.

So when the client updates keyword w_m in the server array, it only needs to submit the server:

$$A\left[\psi_{K_1}\left(ctr_{|N_{ij}|}+1\right)\right] \leftarrow SKE_1.Enc(K_{m,0}, N_{m,1}) \tag{6}$$

$$T[\pi_{K_3}(w_m)] \leftarrow (addr_A(N_{m,1})||K_{m,0}) \oplus f_{K_2}(w_m) \tag{7}$$

File content encryption only needs to refer to step 5) in Sect. 3.

From the new uploaded items in index update Eq. (4), (5), (6), (7), we can see that every time a new document with k keywords is uploaded, whether it is an existing keyword or a new keyword, it is equivalent to the client re-executing the search process for each keyword, and updating the trapdoor of the keywords at 2k index positions in array A, without downloading the whole cipher index $I = (A, T)$ locally and updating it totally, thus reducing the computational overhead and bandwidth consumption of ciphertext update.

4 Deployment and System Test

4.1 Deployment

Figure 3 shows the application deployment and network connectivity of different modules of searchable encryption system. The client software is deployed on the user side, which provides encryption and decryption of user files, file indexes and search requests, sends encrypted ciphertext files to the distributed storage system, and sends encrypted search requests to the cipher search service subsystem.

Service side is constructed based on existed big data platform(take HDFS as an example), so users need to provide HDFS storage system and read-write interface for clients to call. Clients directly store ciphertext files in HDFS, and locate ciphertext files stored in HDFS based on ciphertext keyword index list stored in cipher search subsystem.

Cipher search service subsystem is deployed on a separate server, which is thought to be "honest but curious". It establishes a ciphertext keyword index list for locating ciphertext files stored in HDFS. It responds to the client search request, searches the ciphertext fole location and returns it to the client.

The key management subsystem provides search keys and file encryption keys for client and provides search keys for search service subsystem.

4.2 System Test

Based on the deployment, we can further construct the environment. We adopted Huawei's big data platform FusionInsight HDFS system, which is mainly used to store encrypted files and cipher index uploaded by users. The performance indexes of the test equipment are shown in Table 1.

Test of Encryption

In terms of encryption efficiency, we perform 10 encryption tests on files of 100 Mb size, record the time consumption of each file encryption and calculate the average time consumption. As is Shown in Fig. 4. We can see that the encryption efficiency is close to 300 Mbps.

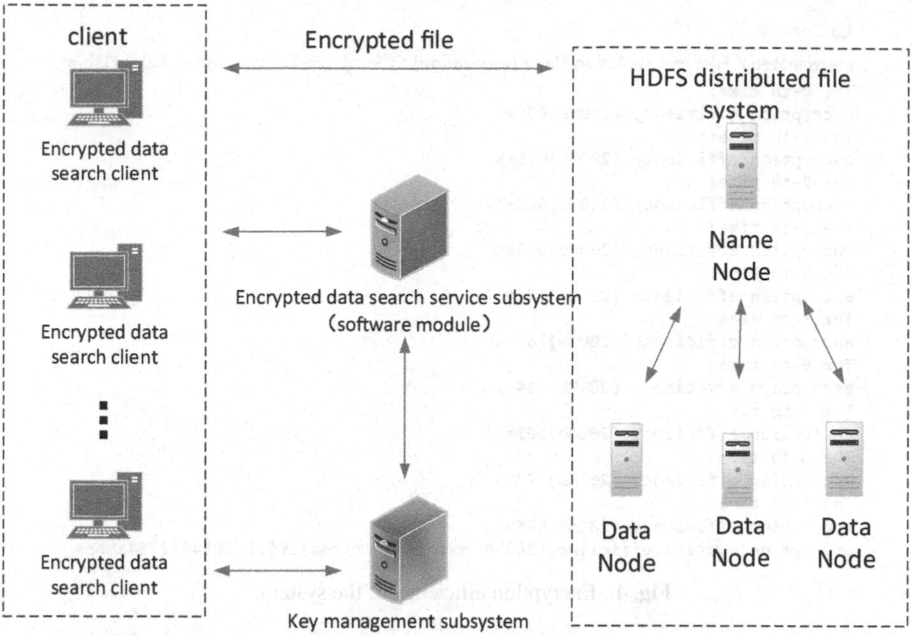

Fig. 3. Deployment of cipher search system.

Table 1. System configuration and experimental environment.

Hardware requirements	Operating system	Usage
2 servers, 8 core, Intel-i7, 32 GB memory	Ubuntu 16.04	Deployment of cipher search and key management software
Desktops ≥1, Intel-i7, 16 GB memory	win7/win10	Deployment of client software
Servers ≥3, Intel-i7, 8 core, 32 GB memory	Ubuntu 16.04	Deployment of HDFS

Test of Efficiency

When it comes to search efficiency, considering the index size of keywords is much smallerthat the size of file. We constructed an index of 100 Mb size for retrieval tests, recorded time consumption of search efficiency, and calculated the average time consumption. As is shown in Fig. 5. We can see that the search efficiency is more than 4000 Mbps, All in all, whether it is encryption or search efficiency, even if the network transmission delay is included, the efficiency will be on the order of milliseconds, which will not affect user experience and gives consideration to security and availability.

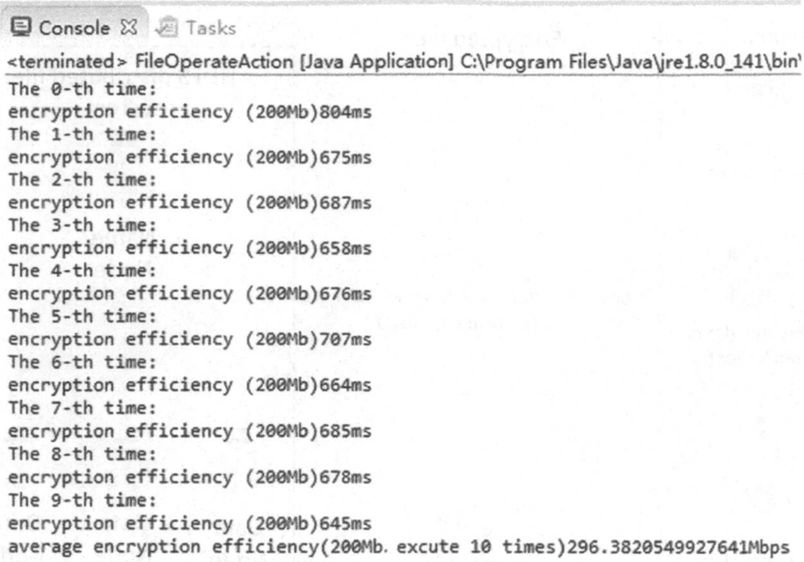

```
Console ⊠  Tasks
<terminated> FileOperateAction [Java Application] C:\Program Files\Java\jre1.8.0_141\bin'
The 0-th time:
encryption efficiency (200Mb)804ms
The 1-th time:
encryption efficiency (200Mb)675ms
The 2-th time:
encryption efficiency (200Mb)687ms
The 3-th time:
encryption efficiency (200Mb)658ms
The 4-th time:
encryption efficiency (200Mb)676ms
The 5-th time:
encryption efficiency (200Mb)707ms
The 6-th time:
encryption efficiency (200Mb)664ms
The 7-th time:
encryption efficiency (200Mb)685ms
The 8-th time:
encryption efficiency (200Mb)678ms
The 9-th time:
encryption efficiency (200Mb)645ms
average encryption efficiency(200Mb. excute 10 times)296.3820549927641Mbps
```

Fig. 4. Encryption efficiency of the system.

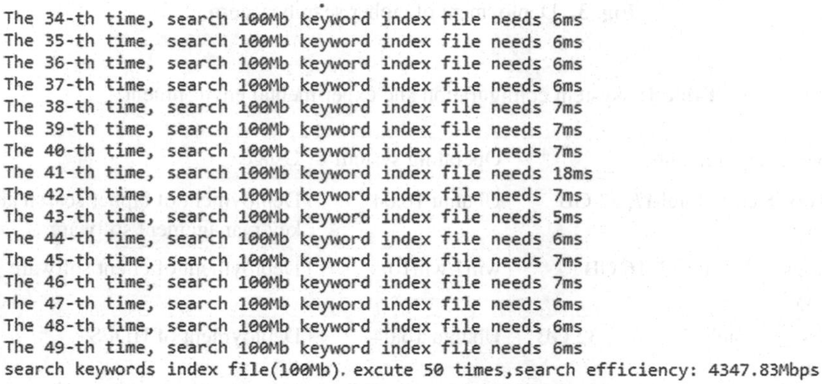

```
The 34-th time, search 100Mb keyword index file needs 6ms
The 35-th time, search 100Mb keyword index file needs 6ms
The 36-th time, search 100Mb keyword index file needs 6ms
The 37-th time, search 100Mb keyword index file needs 6ms
The 38-th time, search 100Mb keyword index file needs 7ms
The 39-th time, search 100Mb keyword index file needs 7ms
The 40-th time, search 100Mb keyword index file needs 7ms
The 41-th time, search 100Mb keyword index file needs 18ms
The 42-th time, search 100Mb keyword index file needs 7ms
The 43-th time, search 100Mb keyword index file needs 5ms
The 44-th time, search 100Mb keyword index file needs 6ms
The 45-th time, search 100Mb keyword index file needs 7ms
The 46-th time, search 100Mb keyword index file needs 7ms
The 47-th time, search 100Mb keyword index file needs 6ms
The 48-th time, search 100Mb keyword index file needs 6ms
The 49-th time, search 100Mb keyword index file needs 6ms
search keywords index file(100Mb). excute 50 times,search efficiency: 4347.83Mbps
```

Fig. 5. Search efficiency of the system.

4.3 Advantage Analysis

Our system supports the storage of ciphertext data and the direct query of the stored ciphertext data. In the query process, the storage end can not know what the user's query content is, nor can it know the user's file data.

The system supports the data owner to actively share the ciphertext data stored by himself with other users. In the process of sharing, the storage side cannot know the sharing behavior of users, nor can it know the plaintext of sharex file data. Meanwhile, encrypting the search request ensures that the server can't perform correlation analysis on the user's search behavior.

Users can completely control their own data through private keys and the storage end cannot know, steal or disclose users' plaintext data.

The cipher search system is loosely coupled with the big data platform, which means it can be deployed quickly only by providing the file data read-write interface of the big data platform, which is more practicability than other solutions.

5 Conclusion

In this paper, we have proposed a cipher search system for big data platform on the basis of searchable encryption algorithm. Especially, we have constructed a scheme.

that takes into account both the security and efficient use of the data. Meanwhile, we designed user-centric key management and file level fine-grained encryption and decryption, effectively preventing the risk of data leakage from getting out of control, which greatly improve the security of encrypted data storage and utilization. In the future work, we will further extend data protection to other ciphertext calculate algorithms. For instance, we will perform fully homomorphic encryption, order-preserving encryption and secure multi-Party computation in a big data fashion.

Acknowledgements. This work is supported by the Sichuan Science and Technology Program (2021JDRC0077), the Sichuan Province's Key Research and Development Plan.

"Distributed Secure Storage Technology for Massive Sensitive Data" Project (2020YFG0298), and Applied Basic Research Project of Sichuan Province (No. 2018JY0370).

References

1. Li, H., Yang, Y., Dai, Y., Yu, S., Xiang, Y.: Achieving secure and efficient dynamic searchable symmetric encryption over medical cloud data. IEEE Trans. Cloud Comput. **8**(2), 484–494 (2020). https://doi.org/10.1109/TCC.2017.2769645
2. He, K., Chen, J., Zhou, Q., Du, R., Xiang, Y.: Secure dynamic searchable symmetric encryption with constant client storage cost. IEEE Trans. Inf. Forensics Secur. **16**, 1538–1549 (2021). https://doi.org/10.1109/TIFS.2020.3033412
3. Shen, J., Wang, C., Wang, A., Ji, S., Zhang, Y.: A searchable and verifiable data protection scheme for scholarly big data. IEEE Trans. Emerg. Topics Comput. **9**(1), 216–225 (2021). https://doi.org/10.1109/TETC.2018.2830368
4. Chen, G., et al.: Differentially private access patterns for searchable symmetric encryption. In: IEEE Conference on Computer Communications, Honolulu, USA, pp. 810–818 (2018)
5. Song, Q., et al.: SAP-SSE: protecting search patterns and access patterns in searchable symmetric encryption. IEEE Trans. Inf. Forensics Secur. **16**, 1795–1809 (2021). https://doi.org/10.1109/TIFS.2020.3042058
6. Mishra, P., et al.: Oblix: an efficient oblivious search index. In: IEEE Symposium on Security and Privacy San Francisco, USA, pp. 279–296 (2018)
7. Liu, X., Yang, G., Mu, Y., Deng, R.H.: Multi-user verifiable searchable symmetric encryption for cloud storage. IEEE Trans. Dependable Secure Comput. **17**(6), 1322–1332 (2020). https://doi.org/10.1109/TDSC.2018.2876831
8. Wang, Y., et al.: Towards multi-user searchable encryption supporting Boolean query and fast decryption. J. Univ. Comput. Sci. **25**(3), 222–244 (2019)

9. Pang, H., Zhang, J., Mouratidis, K.: Scalable verification for outsourced dynamic databases. VLDB Endowment **2**(1), 802–813 (2019)
10. Belguith, S., et al.: Phoabe: securely outsourcing multi-authority attribute based encryption with policy hidden for cloud assisted IOT. Comput. Netw. **133**, 141–156 (2018)
11. Liu, X., et al.: Privacy-preserving multi-keyword searchable encryption for distributed systems. IEEE Trans. Parallel Distrib. Syst. **32**(3), 561–574 (2021). https://doi.org/10.1109/TPDS.2020.3027003
12. Zhang, K., et al.: Lightweight searchable encryption protocol for industrial Internet of Things. IEEE Trans. Industr. Inf. **17**(6), 4248–4259 (2021). https://doi.org/10.1109/TII.2020.3014168
13. Ge, X., et al.: Towards achieving keyword search over dynamic encrypted cloud data with symmetric-key based verification. IEEE Trans. Dependable Secure Comput. **18**(1), 490–504 (2021). https://doi.org/10.1109/TDSC.2019.2896258
14. Wang, H., et al.: Encrypted data retrieval and sharing scheme in space-air-ground integrated vehicular networks. IEEE Internet of Things J. https://doi.org/10.1109/JIOT.2021.3062626
15. Sultan, N.H., Laurent, M., Varadharajan, V.: Securing organization's data: a role-based authorized keyword search scheme with efficient decryption. IEEE Trans. Cloud Comput. https://doi.org/10.1109/TCC.2021.3071304
16. Mante, R.V., Bajad, N.R.: A study of searchable and auditable attribute based encryption in cloud. In: 2020 5th International Conference on Communication and Electronics Systems (ICCES), pp. 1411–1415 (2020). https://doi.org/10.1109/ICCES48766.2020.9137860

Research on Plaintext Resilient Encrypted Communication Based on Generative Adversarial Network

Xiaowei Duan[1,2], Yiliang Han[1,2](✉), Chao Wang[1,2], and Xinliang Bi[1,2]

[1] College of Cryptographic Engineering, Engineering University of PAP, Xi'an 710086, China
[2] Key Laboratory of PAP for Cryptology and Information Security, Xi'an 710086, China

Abstract. Encrypted communication using artificial intelligence is a new challenging research direction. Google Brain first proposed the generation of encrypted communication based on Generative Adversarial Networks, resulting in related discussions and research. In the encrypted communication model, when part of the plaintext is leaked to the attacker, it will cause slow decryption or even being unable to decrypt the communication party, and the high success rate of the attacker decryption, making the encrypted communication no longer secure. In the case of 16-bit plaintext symmetrical encrypted communication, the neural network model used in the original encrypted communication was optimized. The optimized communication party can complete the decryption within 1000 steps, and the error rate of the attacker is increased to more than 0.9 without affecting the decryption of the communicator, which reduces the loss rate of the entire communication to less than 0.05. The optimized neural network can ensure secure encrypted communication of information.

Keywords: Generative adversarial networks · Artificial intelligence · Convolutional neural network · Plaintext leakage

1 Introduction

Information leakage in the communication process will cause the message to be stolen by an attacker, making the communication insecure. With the development of various side channel attack [1–3], the ways of information leakage in encrypted communication have become diversified, and the attack efficiency of side channel attack is much higher than that of traditional cryptanalysis attacks. Akavia [4], Naor and Segev [5], Alwen [6] and other scholars have all conducted researches against leaked communication encryption. How to reduce the impact of information leakage on encrypted communication and ensure the security of communication is one of the issues that need to be considered in modern cryptography.

Foundation Items: The National Natural Science Foundation of China (No. 61572521), Engineering University of PAP Innovation Team Science Foundation (No. KYTD201805), Natural Science Basic Research Plan in Shaanxi Province of China (2021JM252).

J. Zeng et al. (Eds.): ICPCSEE 2021, CCIS 1452, pp. 151–165, 2021.
https://doi.org/10.1007/978-981-16-5943-0_13

Generative Adversarial Networks (GAN) was proposed by Lan Goodfellow [7] in 2014. The network structure is mainly composed of two parts: a generator and a discriminator. The goal of the generator is to generate sufficiently realistic data to deceive the discriminator, the purpose of the discriminator is to correctly judge the real data and the fake data generated by the generator, and the two compete against each other to improve their ability, and finally reach a balanced state. GAN has been developing and widely used since its birth, such as computer vision [8], image processing [9], image generation and segmentation [10], etc.

In 2016, Abadi et al. [11] designed an encrypted communication system based on the idea of generative adversarial. The communication parties and the attacker were divided into two opposite parts. The two parties in communication use neural networks to encrypt and decrypt the plaintext, and the attacker uses the neural network to crack the ciphertext when he only knows the ciphertext. The communication parties continue to fight to improve their own capabilities. The communication parties increase the complexity of encryption and maintain communication security under the premise of successful communication, attackers continue to learn to improve their own cracking success rate and decrypt the plaintext. Through the training of the neural network, a secure encrypted communication model is finally generated under the premise of ensuring the success of the communication. Since then, many experts and scholars have conducted research on this basis. In 2017, Tirumala [12] explored the ability of deep neural network encryption and decryption. In 2018, Coutinho et al. [13] used a selected plaintext attack to improve and analyze the work of Abadi et al. Zhu et al. [14] used SUNA to realize the encryption and decryption under the adversarial network. In 2019, Hitaj et al. [15] designed PASSGAN, which uses generative adversarial networks to learn the real password distribution from information leakage and generate high-quality password guesses. Zhou et al. [16] conducted a security analysis on the communication model of Abadi et al. LI et al. [17] designed a fuzzy key communication scheme under the communication model of generative adversarial networks.

This paper first briefly introduces the encrypted communication based on the generative adversarial networks, including the principle of the generative adversarial networks, and the main functions used in encrypted communication. Then, analyzed the security of 16 bits plaintext and key plaintext leakage communication, which mainly analyzes the convergence speed of secret party, the error rate of adversary and the loss rate of communication. Aiming at the security defects of communication, the structure of neural network is optimized. Aiming at the security defects of communication, the structure of neural network is optimized. The optimized communication party can decrypt in less than 1000 steps in the case of partial plaintext leakage, which can improve the error rate of attacker cracking to more than 0.9 without affecting the communication party's decryption, so that the loss rate of communication is less than 0.05, which ensures the security of encrypted communication.

This paper is organized as follows: in Sect. 2, we mainly introduce the basic principles and related functions of encrypted communication in Generative adversarial network. In Sect. 3, we mainly analyze the encrypted communication in the case of partial plaintext leakage, and finds out its defects. In Sect. 4, the original communication model is optimized to improve the decryption stability and security, and the experimental test is carried out. The fifth chapter summarizes this paper and prospects the next work.

2 Adversarial Neural Cryptography

2.1 GANs Principle

Generative adversarial network is a competitive network based on two antagonistic model generators and discriminators. The two competitors compete with each other and reach a Nash equilibrium. The generator continuously learns from real data samples $\{x^1, x^2, ..., x^m\}$, generating data G(z) very close to real samples. For the discriminator, the purpose is to correctly judge the input data. When the input is real sample data, the target output of the discriminator is 1. When the input of the discriminator is G(z), the target output of the discriminator is 0. For the generator, the optimized formula is as follows

$$\mathrm{V} = \frac{1}{m}\sum\nolimits_{i=1}^{m} \log(\mathrm{D}(\mathrm{G}(z^i))) \tag{1}$$

Where m is the number of data, G(z) is the output of generator and D(x) is the output of discriminator. The purpose of generator optimization is to make the output of discriminator closer to 1, that is to say, the larger the value of V, the better. For the discriminator, the optimization formula is as follows

$$\min_G \max_D \mathrm{V}(D, G) = E_{x-p_{data}(x)}[\log \mathrm{D}(x)] + E_{x-p_z(z)}[\log(1 - \mathrm{D}(\mathrm{G}(z)))] \tag{2}$$

$p_z(z)$ and $p_{data}(x)$ represents the probability distribution in the hidden space and the distribution of real samples, $\mathrm{V}(G, D)$ represents the difference between true and false data. The whole optimization formula represents the generator with the smallest difference between true and false data under the condition that the discriminator remains unchanged.

2.2 Adversarial Neural Cryptography Model

Abadi et al. designed an encrypted communication system based on generative adversarial network in reference [11]. In this paper, the communication system is divided into three parts: the encrypting party, the decrypting party and the attacking party, which are represented by Alice, Bob and Eve respectively. The system realizes secure communication under the monitoring of the attacker through countermeasure promotion. The communication model is a classic symmetric encryption scenario, as shown in Fig. 1. Alice uses key K to encrypt plaintext P to get ciphertext C and send it to Bob. After receiving ciphertext C, Bob uses key K to decrypt, and the decrypted message is P_{Bob}. The attacker Eve can listen to the ciphertext C transmitted by both sides of the communication, but does not know the key K. the output of the attacker after cracking is P_{Eve}.

In the communication model, the encryption, decryption and attack are all completed by neural network, and no specific encryption and decryption and attack algorithm is input into neural network, all are completed by neural network autonomous learning. Finally, through the generation of GANs training, both sides of the communication can complete the communication, and the attacker can't crack the plaintext.

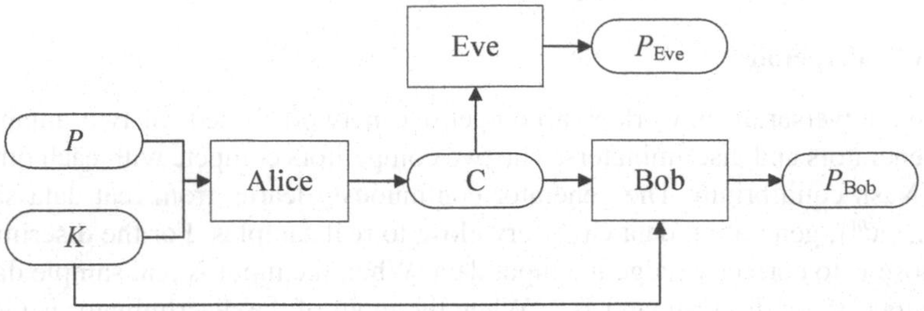

Fig. 1. Encrypted communication model based on generative adversarial network

2.3 Function Design

In neural networks, an important criterion for optimization is the difference between the generated data and the real data. In reference [11], L1 distance is used to calculate the difference between the two. The calculation function is as follows, in which the length of key, plaintext and ciphertext are N.

$$d(P, P') = \frac{1}{N}\sum \left|P_i - P'_i\right| \tag{3}$$

The encryption function is set to $E_A(\theta_A, P, K)$, θ_A represents the parameters when Alice encrypts. The decryption function is set to $D_B(\theta_B, C, K)$, θ_B represent the relevant parameters when Bob decrypts. Eve's decryption function is set to $D_E(\theta_E, C)$, θ_E represent the parameters of Eve attack.

Define the loss function of Bob:

$$L_B(\theta_A, \theta_B, P, K) = d(P, D_B(\theta_B, E_A(\theta_A, P, K), K)) \tag{4}$$

The formula represents the distance difference between Bob's decrypted and the original plaintext. By setting an expected value, it is extended to the corresponding plaintext and key loss function. The expected value is set as follows:

$$L_B(\theta_A, \theta_B, P, K) = E_{P,K}(d(P, D_E(\theta_E, E_A(\theta_A, P, K)), K) = E_{P,K}[L_B(\theta_A, \theta_B, P, K)] \tag{5}$$

Similarly, we can define the loss function and expectation of Eve attack.

$$L_E(\theta_A, \theta_E, P, K) = d(P, D_E(\theta_E, E_A(\theta_A, P, K))) \tag{6}$$

$$L_E(\theta_A, \theta_E) = E_{P,K}[L_E(\theta_A, \theta_E, P, K)] \tag{7}$$

The optimal Eve is optimized by finding the minimum loss. The function of Eve optimization is as follows:

$$O_E(\theta_A) = \text{argmin}_{\theta_E}(L_E(\theta_A, \theta_E)) \tag{8}$$

According to the loss function of Bob and Eve, the loss function of the communication party can be defined for optimization.

$$L_{AB}(\theta_A, \theta_B) = L_B(\theta_A, \theta_B) - L_E(\theta_A, O_E(\theta_A)) \tag{9}$$

By minimizing the loss of the decryptor and maximizing the loss of the attacker to find the optimal encrypted communication, so we need to find the minimum $L_{AB}(\theta_A, \theta_B)$.

$$(O_A, O_B) = \text{argmin}_{(\theta_A, \theta_B)}(L_{AB}(\theta_A, \theta_B)) \tag{10}$$

3 Plaintext Leakage in the Original Model

3.1 Neural Network Design in the Original Model

In the encrypted communication based on generation adversarial network, the structure diagram of neural network is shown in Fig. 2. If the attacker obtains part of the plaintext through illegal means, the attacker obtains part of the advantages compared with the decryptor, which will affect the generation of the network and cause the decryption instability of the decryptor, such as decryption failure and very slow decryption speed.

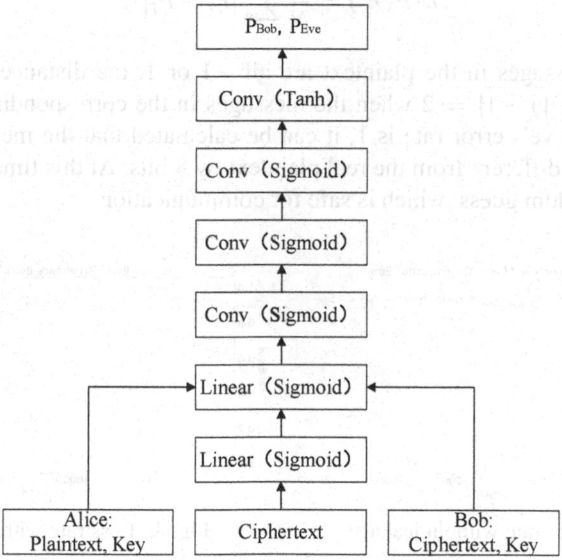

Fig. 2. Original model neural network structure

As shown in Fig. 2, the plaintext and key input by Alice and the ciphertext and key input by Bob only need to use the full connected layer once, and then the output can be completed through four layers of convolution layer. The activation function used in the first three layers of convolution layer is sigmoid function, and the activation function used in the last layer is tanh function. Since Eve has only ciphertext input, it uses two convolution layers to make the input length of convolution layer the same as Alice and Bob.

When Eve acquires some plaintext illegally, it will impact the balance of the network. Causes the decryptor can not to decrypt or decrypt slowly and so on. At this time, Eve has some plaintext and ciphertext, which will have certain advantages for the neural network to crack out all plaintext. Bob only knows the ciphertext and key, so he has less or no advantages in the competition with Eve, causing unstable final decryption. The structure of the neural network should be optimized to solve the unstable problem of Bob's decryption.

3.2 Leakage Experiments Under the Original Model

Abadi et al. used the neural network model described in Sect. 3.1 to establish communication. Plaintext and keys are randomly generated when encrypted and are 16-bit strings of -1 or 1. Adam Optimizer was used to optimize the training, and the learning rate was set to 0.0008.

The effect of communication is shown in Fig. 3 and Fig. 4. The final error rate of Bob is close to 0, which indicates that the plaintext can be decrypted successfully. The error rate of Eve is close to 1, according to the formula

$$d(P, P') = \frac{1}{N} \sum |P_i - P'_i| \tag{11}$$

Since the messages in the plaintext are all -1 or 1, the distance is $|P_i - P'_i| = |1 - (-1)| = |(-1) - 1| = 2$ when the messages in the corresponding positions are different. When Eve's error rate is 1, it can be calculated that the message decrypted by the attacker is different from the real plaintext by 8 bits. At this time, the attacker is equivalent to random guess, which is safe for communication.

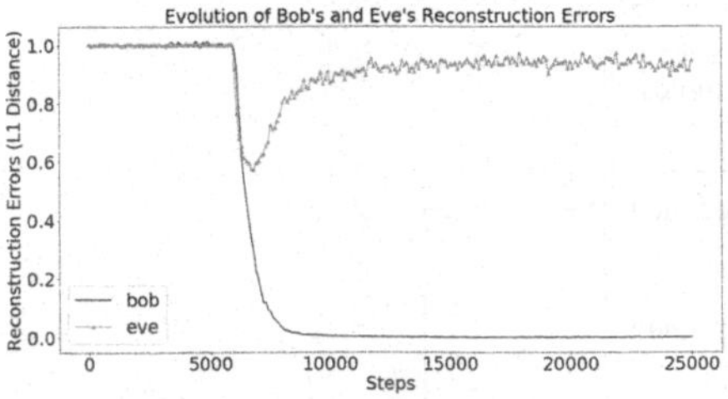

Fig. 3. Error rate without leakage

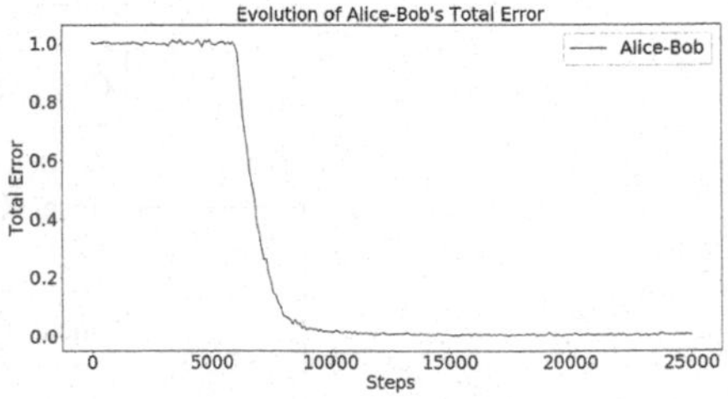

Fig. 4. Loss rate without leakage

Under the original neural network model, four bits of plaintext are leaked to Eve. At this time, Eve's input is $4 + 16 = 20$ bits, and Bob's input is 16-bit ciphertext and 16-bit key. Under the condition of 15000 steps of training, the implementation results are shown in Fig. 5.

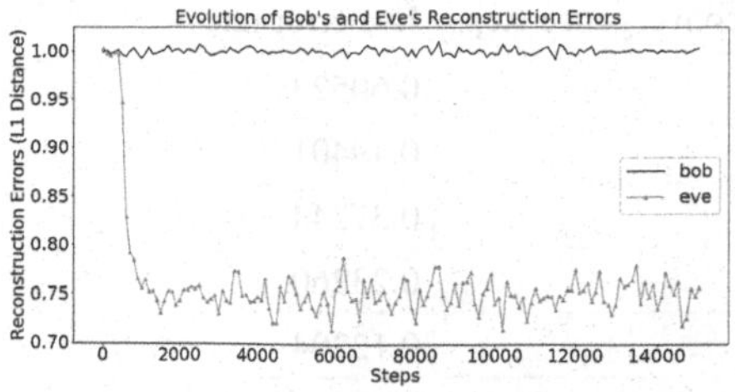

Fig. 5. Error rate of 4 bits plaintext leakage

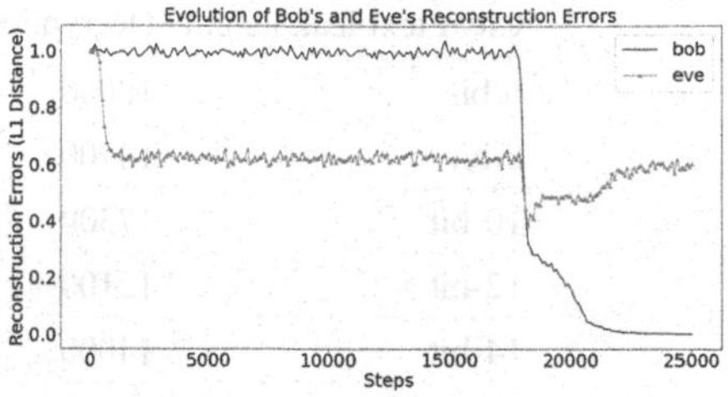

Fig. 6. Error rate of 6 bits plaintext leakage

It can be seen that in 15000 steps, Bob can't achieve convergence. Under the interference of Eve, Bob can't obtain the competitive advantage, decrypt the correct plaintext, and the loss rate of communication is also very high.

On the basis of the original model, the training steps were enlarged to 25,000 steps, and 6 bits of plaintext were leaked to Eve. As shown in Fig. 6, the error rate of Bob began to decrease at 18,000 steps, and converged at 21,000 steps. Eve's error rate finally reached about 0.6.

In the case of disclosing 12 bits of plaintext to Eve, the results are shown in Figs. 7 and 8.

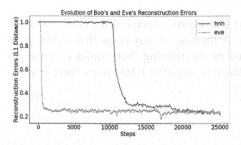

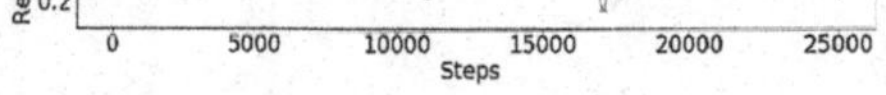

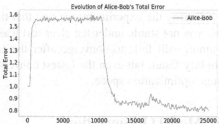

Fig. 7. Error rate of 12-bit plaintext leakage

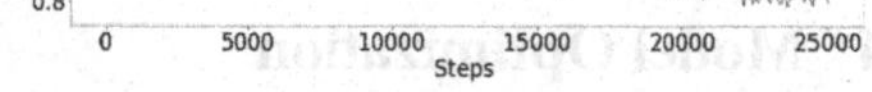

Fig. 8. Loss rate of 12 bits plaintext leakage

When 12 bits of plaintext are leaked to Eve and 25,000 steps are trained, Eve's final error rate through learning and training is only 0.25 due to leaking more plaintext. Bob's error rate starts to decrease until about 10,400 steps, 10,000 steps slower than Eve, and the final error rate is also around 0.25, at which time the communication loss rate is as high as 0.82. Bob has no advantage over Eve in communication.

The goal of improving the neural network under the original model is to make Bob's decryption efficiency stable through structural improvement and detail optimization, make Bob's decryption speed faster, increase Eve's error rate when Eve's error rate is low, reduce Eve's advantage, and make communication more secure. Due to the limitation of the length of the article, the experimental data under some explicit leaks are shown in Table 1 and Table 2.

Table 1. Decryption speed and error rate for partial clear text disclosure in original model

Clear text leak length	Decryption convergence steps	Eve error rate
6-bit	18000	0.60523
8-bit	14700	0.49401
10-bit	17300	0.37344
12-bit	12100	0.23260
14-bit	14000	0.12204

Table 2. Communication loss rate in partial clear text leakage in original model

Clear text leak length	Communication Loss Rate
6-bit	0.22466
8-bit	0.36536
10-bit	0.44576
12-bit	0.79715
14-bit	0.80310

During the experiment, we find that the convergence speed of the communication party is not stable under the clear text leak. Sometimes it may occur that 25,000 steps training still fails to converge after the end of the training. Sometimes it converges slightly faster, but even the fastest convergence is still after 6,000 steps, there is still a huge optimization space.

4 Model Optimization

In the model designed by Abadi et al. to generate adversarial network encrypted communication, Alice, Bob and Eve use the same neural network. The only difference is that Eve uses two full connected layers, one layer more than Alice and Bob, because the inputs are half less than Alice and Bob.

In the optimization, the main idea is to separate the neural network of both sides of communication and Eve, optimize the neural network structure of both sides of communication, enhance the ability of decryption, improve the speed of decryption, make Bob obviously superior to Eve in the competition, and protect the security of communication.

4.1 Increase Decryption Speed

Neural Network Design. To Solve the Problem that Bob Decrypts Slowly, the Communication party's Neural Network is Optimized. This Paper Uses Batch Normalization [18] to Process the Neural Network. Batch Normalization Can Make the Solution Space of the Optimization Problem Smoother and Ensure the Accuracy and Stability of Gradient Prediction. The Main Purpose is to Speed up the Training of the Neural Network, Map the Data Values to a Specific Range, and Make the Training of the Model More Stable and Insensitive to the Initialization of Parameters.

In the training of the neural network in this paper, because there are many iterations of parameters, when data transfer is done between two layers of neural networks, the parameters in the network will affect the distribution of data after the training of the previous layer of neural network. When data is input to the next layer of neural network, the learning of the next layer of neural network will become difficult, so the speed of decryption will be slower. After batch normalization, each input layer of the deep neural network keeps the same distribution, which accelerates the learning speed and makes Bob's error rate lower and convergence faster when decrypting.

This paper modifies the communication party's neural network, uses batch normalization after the full connected layer, and then inputs the processed data into the convolution layer for learning. Eve's neural network does not make any modifications. After batch normalization optimization, the neural network structure of both sides of the communication is shown in Fig. 9.

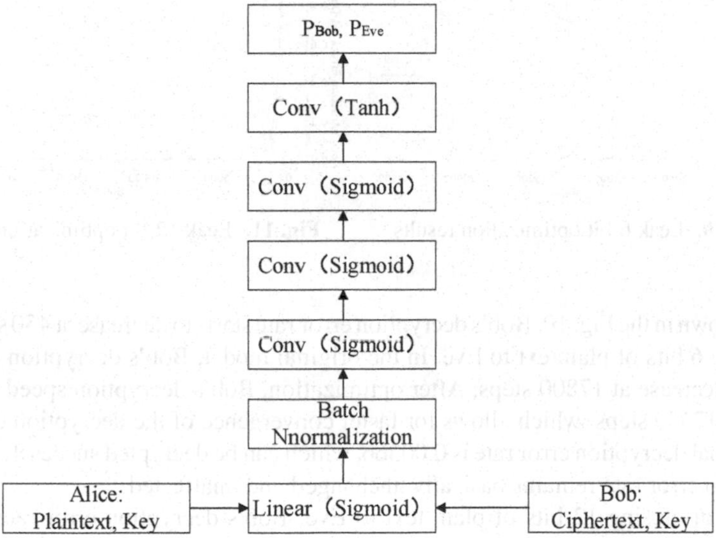

Fig. 9. Communication party neural network structure after optimizing decryption speed

As shown in the Fig. 9, Bob's network is built as follows after optimization.

(1): The ciphertext and the key are spliced as Bob's input, and the input length is 32 bits.
(2): Transmit Bob's input to the fully connected layer, and the activation function used by the fully connected layer is the sigmoid function.
(3): Batch normalize the output of the fully connected layer.
(4): Input the batch normalized data into the first convolutional layer, the activation function is the sigmoid function, the output depth is 2, the step size is 1, and the convolution kernel size is 4.
(5): The activation function of the second convolutional layer is the sigmoid function, the output depth is 4, the step size is 2, and the convolution kernel size is 2.
(6): The activation function of the third convolutional layer is the sigmoid function, the output depth is 4, the step size is 1, and the convolution kernel size is 1.
(7): The activation function of the third convolutional layer is the sigmoid function, the output depth is 4, the step size is 1, and the convolution kernel size is 1.

Experiment Analysis. When leaking 6-bit and 12-bit plaintext to Eve, set the number of training steps to 25,000, and use the optimized neural network for learning and training. The training error rate effect is shown in Fig. 10 and Fig. 11:

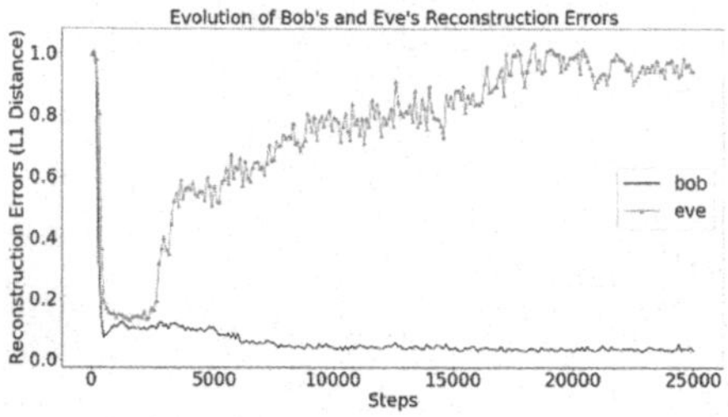

Fig. 10. Leak 6-bit optimization results

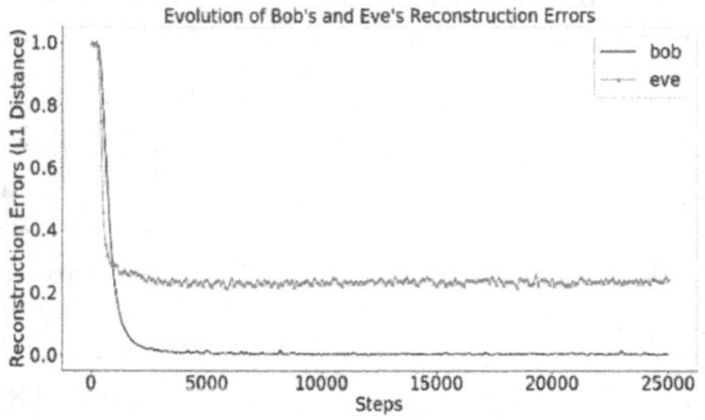

Fig. 11. Leak 12-bit optimization results

As shown in the Fig. 10, Bob's decryption error rate starts to decrease at 450 steps after disclosing 6 bits of plaintext to Eve. In the original model, Bob's decryption error rate starts to decrease at 17800 steps. After optimization, Bob's decryption speed increases by about 17350 steps, which allows for faster convergence of the decryption error rate, and the final decryption error rate is 0.00366, which can be decrypted successfully. Eve's decryption error rate remains basically unchanged and unaffected.

After disclosing 12 bits of plain text to Eve, Bob's decryption error rate starts to decrease at 150 steps. In the original model, it starts to decrease at 10400 steps, and the optimized model decryption convergence rate increases by about 10250 steps. Finally, Bob's decryption error rate is 0.15408, which is basically successful. Eve's decryption error rate is basically unchanged from the original model.

Due to the limitation of article size, Bob's decryption performance when optimized to leak different plaintext to Eve is shown in the following Table 3.

Table 3. Decryption speed comparison before and after optimization

Clear text leak length	Optimized decryption convergence steps	Primary convergence step
6-bit	800 steps	18000 steps
8-bit	780 steps	14700 steps
10-bit	910 steps	17300 steps
12-bit	780 steps	12100 steps
14-bit	960 steps	14000 steps

According to the Table 3, the optimized model can be decrypted at least 10,000 steps faster than the original model, and the decryption speed can be greatly improved.

4.2 Decrease the Correct Rate of Attacker's Cracking

The idea of optimizing the network model is similar to that in Sect. 4.1. Because in the original model, the communicator and the adversary use the same neural network model, when more plaintext is leaked to Eve, the communicator does not have too many advantages to compete to surpass Eve, which makes the error rate of Eve lower. In the optimization, the neural network models of the communication party and the adversary are distinguished, and the neural network of the communication party for encryption and decryption is optimized.

In this paper, we change the activation function in the convolution layer of the communication neural network, change the sigmoid function used in the convolution layer of the original model to the ELU function, and keep the rest unchanged. Eve's neural network structure is consistent with the original model.

The exponential linear unit (ELU) function belongs to the modified activation function, which can still have non-zero output when the input is negative. Because it includes negative exponential term, it can prevent the appearance of silent neurons, so that the derivative converges to 0. The formula of ELU function is as follows:

$$\mathrm{ELU}(x) = \max(0, x) + \min(0, \alpha(e^{x} - 1) \qquad (12)$$

The model structure of the optimized communication party neural network is shown in Fig. 12.

Experimental Analysis. When the 6-bit Plaintext is Leaked to Eve, the Error Rate of Communication After Optimizing the Neural Network is Shown in Fig. 13. The Error Rate of Communication When the Original Model Leaks 6-bit Plaintext to Eve is Shown in Fig. 14. According to the Error Rate in the Figure, It Can Be Seen that Bob's Decryption also Shows an Advantage in the Competition with Eve on the Basis of Improving the Speed. Finally, the Error Rate of Eve is also Increased from 0.6 to 0.97, Which Indicates that the attacker's Guess of Plaintext is Similar to Random Guess, and can't Gain Any Advantage in the Case of Leaking Plaintext. At This Time, the Encrypted Communication of the Whole Communication Model is Secure.

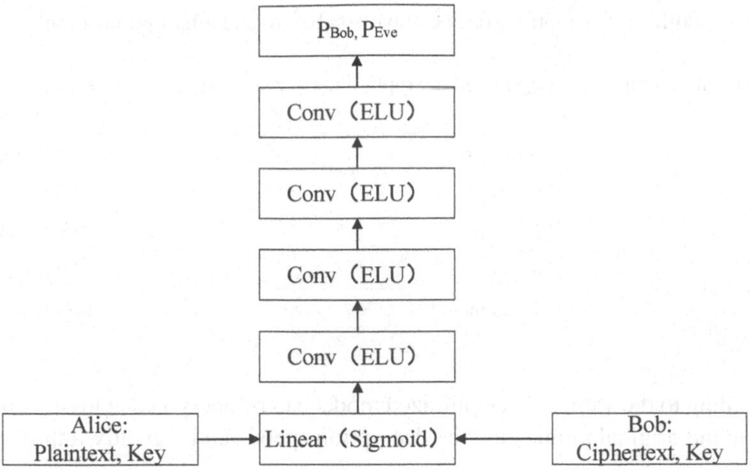

Fig. 12. Neural network structure after optimizing the activation function of the communication party

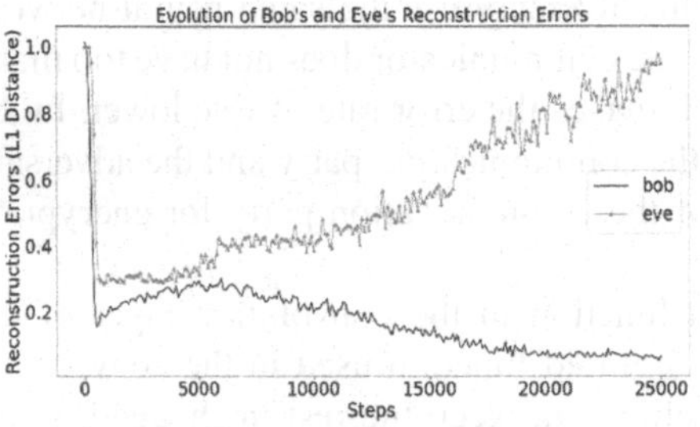

Fig. 13. Leak 6-bit optimization results

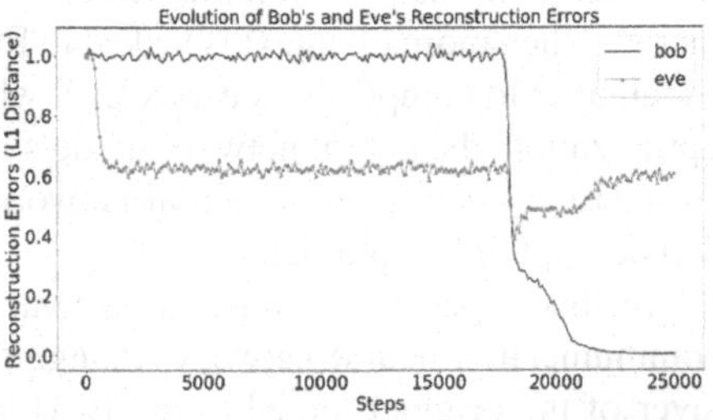

Fig. 14. Leak 6-bit in original model

When 12-bit plaintext is leaked to Eve, the improvement effect on Eve error rate is more significant. The optimized communication error rate is shown in Fig. 15. The communication error rate when 12-bit plaintext is leaked under the original model is shown in Fig. 16. As shown in the Fig. 16, in the original model, the error rate of Eve after 25000 steps of training is only 0.23, which indicates that Eve can crack most plaintext. After the model is optimized, the error rate of Eve is lower It can be increased to about 0.88, an increase of about 65%. At this time, it can basically meet the requirements of secure communication. Compared with the original Eve, it can guess most of the plaintext. At this time, the communication is based on the distance formula in Sect. 2.3.

$$d(P, P') = \frac{1}{N} \sum |P_i - P'_i| \tag{13}$$

It can be calculated that Eve guessed 7-bit plaintext wrong. Compared with 8-bit plaintext under arbitrary guess, Eve only lost 1-bit advantage, and the security of communication was greatly improved.

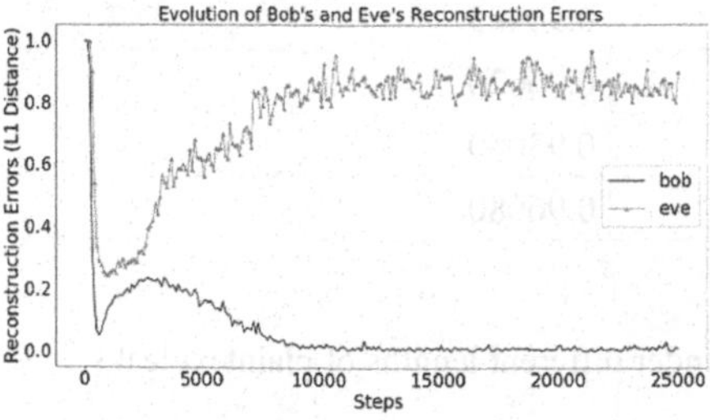

Fig. 15. Leak 12-bit optimization results

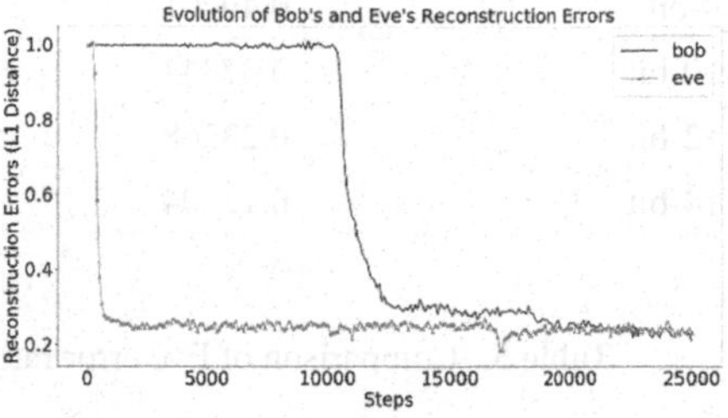

Fig. 16. Leak 12-bit in original model

At the same time, the optimized communication loss rate is compared with that of the original model, as shown in Fig. 17 and Fig. 18.

The abscissa represents the number of training steps, and the ordinate represents the loss rate of the communication party.

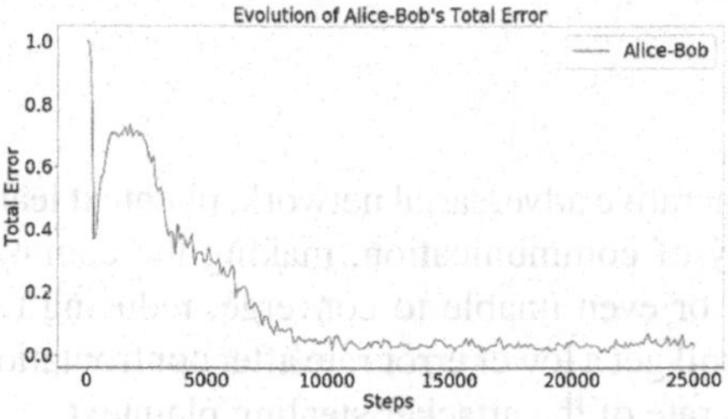

Fig. 17. Leak 12-bit optimization loss rate

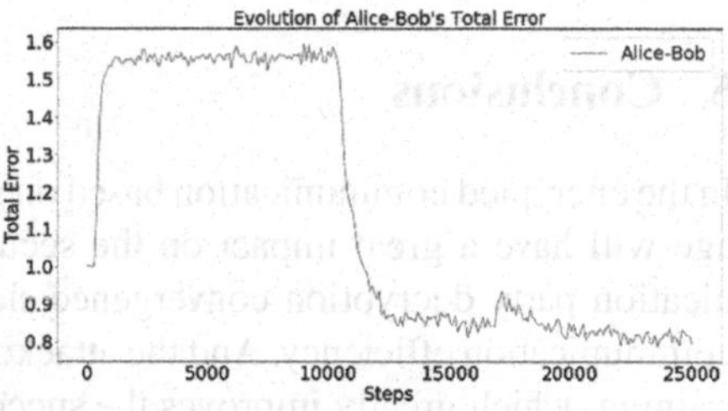

Fig. 18. Leak 12-bit in original model loss rate

It can be seen from the figure of communication rate that when the original model leaks 12-bit plaintext, the loss rate of communication is about 0.82, which is very high. After the optimization of the model, the loss rate of communication is only about 0.02, which is reduced by about 80%, so the optimization of the model also greatly reduces the loss rate of communication.

In this paper, the communication error rate and loss rate of Eve with 6, 8, 10, 12 and 14-bit plaintext leakage are tested. Due to the limitation of the article, the experimental data are listed in the Table 4 and Table 5 below. By comparing the error rate and loss rate before and after optimization, it can be seen that the optimized communication model can greatly improve the error rate of Eve, reduce the loss rate of communication, and ensure the security of communication.

Table 4. Comparison of Eve error rates under different lengths of plaintext leaks

Clear text leak length	Original Eve error rate	Error rate of Eve after optimization
6-bit	0.60523	0.97407
8-bit	0.49401	0.90483
10-bit	0.37344	0.96620
12-bit	0.23268	0.93080
14-bit	0.12204	0.96680

Table 5. Comparison of Eve error rates under different lengths of plaintext leaks

Clear text leak length	Original loss rate	Loss rate after optimization
6-bit	0.22466	0.00485
8-bit	0.36536	0.00751
10-bit	0.44576	0.04503
12-bit	0.79715	0.01662
14-bit	0.80310	0.03009

5 Conclusions

In the encrypted communication based on generative adversarial network, plaintext leakage will have a great impact on the security of communication, making the communication party decryption convergence slow or even unable to converge, reducing the communication efficiency. And the attacker will get a lower error rate after confrontation training, which greatly improves the success rate of the attacker stealing plaintext.

In view of the high error rate of attacker decryption, this paper modifies the activation function of the communication party's neural network, so that the communication party can obtain more advantages in the confrontation training with the attacker. After optimization, the error rate of the adversary decrypting plaintext is basically above 0.9, and the loss rate of communication is basically reduced below 0.05. The optimized communication is safe communication.

Although this paper optimizes the encrypted communication based on generative adversarial network under plaintext leakage, the error rate of the final attacker after training is still slightly floating, but it is basically above 0.9. In the next step, we continue to optimize it to make the final training effect smoother. In addition, we need to analyze the security and continue to optimize when both plaintext and key are leaked.

References

1. Kocher, P.C.: Timing attacks on implementations of Diffie-Hellman, RSA, DSS, and other systems. In: Annual International Cryptology Conference, pp. 104–113. Springer, Berlin, Heidelberg (1996)

2. Biham, E., Shamir, A.: Differential fault analysis of secret key cryptosystems. In: Kaliski, B.S. (ed.) CRYPTO 1997. LNCS, vol. 1294, pp. 513–525. Springer, Heidelberg (1997). https://doi.org/10.1007/BFb0052259
3. Kocher, P., Jaffe, J., Jun, B.: Differential power analysis. In: Annual International Cryptology Conference, pp. 388–397. Springer, Berlin, Heidelberg (1999)
4. Akavia, A., Goldwasser, S., Vaikuntanathan, V.: Simultaneous hardcore bits and cryptography against memory attacks. In: Reingold, O. (ed.) TCC 2009. LNCS, vol. 5444, pp. 474–495. Springer, Heidelberg (2009). https://doi.org/10.1007/978-3-642-00457-5_28
5. Naor, M., Segev, G.: Public-key cryptosystems resilient to key leakage. In: Advances in Cryptology: 29th Annual International Cryptology Conference, pp. 18–35. Springer, Berlin (2009)
6. Alwen, J., et al.: Public-key encryption in the bounded-retrieval model. In: Annual International Conference on the Theory and Applications of Cryptographic Techniques, pp. 113–134. Springer, Berlin, Heidelberg (2010)
7. Goodfellow, I., et al.: Generative adversarial nets. Adv. Neural. Inf. Process. Syst. **27**, 2672–2680 (2014)
8. Hinton, G.E., Salakhutdinov, R.R.: Reducing the dimensionality of data with neural networks. Science **313**(5786), 504–507 (2006)
9. Sonka, M., Hlavac, V., Boyle, R.: Segmentation. In: Image Processing, Analysis and Machine Vision, pp. 112–191. Springer, Boston (1993)
10. Long, J., Shelhamer, E., Darrell, T.: Fully convolutional networks for semantic segmentation. In: Proceedings of the IEEE Conference on Computer Vision and Pattern Recognition, pp. 3431–3440 (2015)
11. Abadi, M., Andersen, D.G.: Learning to protect communications with adversarial neural cryptography. arXiv:1610.06918 (2016)
12. Tirumala, S.S., Narayanan, A.: Transpositional neuro cryptography using deep learning. In: Proceedings of the 2017 International Conference on Information Technology, pp. 330–334 (2017)
13. Coutinho, M., et al: Learning perfectly secure cryptography to protect communications with adversarial neural cryptography. Sensors **18**(5), 1306 (2018)
14. Zhu, Y., Vargas, D.V., Sakurai, K.: Neural cryptography based on the topology evolving neural networks. In: 2018 Sixth International Symposium on Computing and Networking Workshops (CANDARW), pp. 472–478. IEEE (2018)
15. Hitaj, B., et al.: PassGan: a deep learning approach for password guessing. In: International Conference on Applied Cryptography and Network Security, pp. 217–237. Springer, Cham (2019)
16. Zhou, L., et al.: Security analysis and new models on the intelligent symmetric key encryption. Comput. Secur. **80**, 14–24 (2019)
17. Li, X.M., et al.: Study on fuzzy key encryption based on GAN. Appl. Res. Comput. **6**(37), 1779–1781 (2020)
18. Ioffe, S., Szegedy, C.: Batch normalization: accelerating deep network training by reducing internal covariate shift. In: International Conference on Machine Learning, PMLR, pp. 448–456 (2015)

An Improved LeNet-5 Model Based on Encrypted Data

Huanhuan Ni[1,2(✉)], Yiliang Han[1,2], Xiaowei Duan[1,2], and Guohui Yang[1,2]

[1] College of Cryptographic Engineering, Engineering University of PAP, Xi'an 710086, China
[2] Key Laboratory of PAP for Cryptology and Information Security, Xi'an 710086, China

Abstract. In recent years, the problem of privacy leakage has attracted increasing attentions. Therefore, machine learning privacy protection becomes crucial research topic. In this paper, the Paillier homomorphic encryption algorithm is proposed to protect the privacy data. The original LeNet-5 convolutional neural network model was first improved. Then the activation function was modified and the C5 layer was removed to reduce the number of model parameters and improve the operation efficiency. Finally, by mapping the operation of each layer in the convolutional neural network from the plaintext domain to the ciphertext domain, an improved LeNet-5 model that can run on encrypted data was constructed. The purpose of using machine learning algorithm was realized and privacy was ensured at the same time. The analysis shows that the model is feasible and the efficiency is improved.

Keywords: Paillier homomorphic encryption · LeNet-5 model · Convolutional neural network · Privacy protection

1 Introduction

With the rapid development of information technology, thanks to the two "boosters" of large data and high-performance computers, Machine Learning (ML) has made breakthrough progress, and is widely used in computer vision, natural language processing, intelligent assisted driving and other fields. Machine learning provides a more convenient service to life, but at the same time, the problem of privacy data leakage can't be ignored [1]. Efficient and accurate machine learning models are built on a large amount of high-quality data. In recent years, with the wide application of Machine Learning as a Service (MLaaS) mode, more and more data holders choose to upload their data to the cloud for prediction or training models, which undoubtedly increases the risk of data leakage. Privacy data contains sensitive information, which will cause huge losses if it is leaked. Therefore, it is urgent to study the machine learning privacy protection.

Foundation Items: The National Natural Science Foundation of China (No.61572521), Engineering University of PAP Innovation Team Science Foundation (No. KYTD201805), Natural Science Basic Research Plan in Shaanxi Province of China (2021JM252).

J. Zeng et al. (Eds.): ICPCSEE 2021, CCIS 1452, pp. 166–178, 2021.
https://doi.org/10.1007/978-981-16-5943-0_14

Homomorphic encryption [2] and differential privacy [3] are two main types of privacy protection technologies currently used for machine learning. Homomorphic encryption allows users to perform operations directly on ciphertext, and the results obtained after decryption are consistent with the result of the corresponding operation on plaintext. Therefore, homomorphic encryption is currently the most direct and effective technology to protect users' privacy. Wang et al. [4] designed a privacy protection neural network based on homomorphic encryption, used the homomorphic encryption algorithm to encrypt data, and rewritten the operation process of the neural network with the properties of homomorphic addition and multiplication. The computability of the data is preserved while the privacy of the data is guaranteed. Compared with previous privacy protection neural networks, it can be better applied to complex neural networks and has higher security. Based on Paillier homomorphic encryption algorithm, Zhu et al. [5] proposed a privacy-protected deep neural network model 2P-DNN (Privacy-Preserving Deep Neural Network), which can be applied to machine learning in cryptographic domains. Zhang et al. [6] added the homomorphic encryption algorithm into federal learning, proposed a federated deep neural network model PFDNN (Privacy-preserving Federated Deep Neural Network), which supports data privacy protection. This model ensures data privacy by homomorphic encryption of weight parameters, and greatly reduces the amount of encryption and decryption computation during training. Shokri and Shamtikov [7] historically proposed a scheme to protect privacy in deep learning in 2015, and ensured its security by adding differential privacy. On this basis, Le Trieu Phong et al. [8] proposed a deep learning privacy protection scheme based on additional homomorphic encryption, which can ensure accuracy without leaking information to the server. Nathan Dowlin et al. [9] creatively combined homomorphic encryption with neural networks and tested them on MNIST datasets to achieve high performance, high accuracy, and privacy prediction.

Based on Paillier homomorphic encryption algorithm [10] and thc improved LeNet-5 convolution neural network model, the improved LeNet-5 model can run on encrypted data by mapping the model's operations in the plaintext domain to the ciphertext domain, thus achieving the purpose of privacy protection. Through analysis, the correctness of the improved LeNet-5 model in plaintext and ciphertext domains is not much different. In ciphertext domain, the improved LeNet-5 model has some advantages over the original LeNet-5 model in computational efficiency. The organization of this paper is as follows: The first section mainly introduces the necessity of researching the machine learning privacy protection and the related development process. The second section mainly introduces the relevant theoretical knowledge in this paper, including homomorphic encryption, convolution neural network and LeNet-5 model. The third section explains the scheme in theory, including the improvement of LeNet-5 model and the mapping of operations in convolution neural network from plaintext domain to ciphertext domain. The fourth section analyses the scheme and illustrates its advantages of feasibility and efficiency through comparison. The fifth section summarizes this paper and prospects the next work.

2 Preliminaries

2.1 Homomorphic Encryption

The concept of Homomorphic Encryption (HE) was first introduced by Rivest et al. [2] in the 1970s. It is a cryptographic technique based on the computational complexity theory of mathematical problems. Compared with general encryption algorithms, homomorphic encryption not only implements basic encryption operations, but also implements a variety of computing functions between ciphertext, that is, it processes and decrypts the homomorphically encrypted ciphertext data, and the result is consistent with processing plaintext data in the corresponding way. The formula is as follows:

$$\mathrm{D}(\mathrm{E}(m_1) \oplus \mathrm{E}(m_2)) = m_1 \otimes m_2 \tag{1}$$

E denotes encryption, D denotes decryption, m_1 and m_2 denote plaintext, and $\oplus$ and $\otimes$ denote two operations corresponding to plaintext space and ciphertext space respectively. This feature is of great significance for data privacy protection. Using homomorphic encryption technology can directly operate on the data in the ciphertext state, thereby effectively ensuring the privacy of the data. Because of the advantages of homomorphic encryption technology in communication security, more and more scholars are devoted to the exploration of its theories and applications [11–13].

Homomorphic encryption usually consists of four components: key generation, encryption algorithm, evaluation algorithm, and decryption algorithm. Homomorphic encryption schemes can generally be divided into two categories: one is partial homomorphic encryption scheme, which can only perform a limited number of additive or multiplicative homomorphic operations, such as RSA multiplicative homomorphic encryption scheme, ElGamal multiplicative homomorphic encryption scheme [14], Paillier additive homomorphic encryption scheme [10]. The other is fully homomorphic encryption scheme, which can perform any number of additive and multiplicative homomorphic operations on ciphertext. Gentry [15] proposed the first fully homomorphic encryption scheme based on ideal lattice in 2009. Since then, scholars have proposed continuously optimized fully homomorphic encryption scheme [16]. However, due to the low computational efficiency of the fully homomorphic encryption scheme, it cannot be put into use on a large scale at present.

This paper uses the Paillier additive homomorphic encryption scheme, which has four steps as follows:

(1): KeyGen() $\longrightarrow (pk, sk)$: Randomly select two large prime numbers p and q of the same length to satisfy

$$\gcd(pq, (p-1)(q-1)) = 1 \tag{2}$$

Then, calculate

$$N = pq \tag{3}$$

$$\lambda = \mathrm{lcm}(p-1, q-1) \tag{4}$$

Randomly select $g \in Z_{N^2}^*$, public key is $pk = (N, g)$ and private key is $sk = (\lambda)$.

(2): Encryption$(pk, m) \longrightarrow c$: Input public key pk and plaintext m, randomly select $r \in Z_N^*$, and calculate ciphertext as follows:

$$c = g^m r^N \left(\text{mod } N^2\right) \tag{5}$$

(3): Decryption$(sk, c) \longrightarrow m$: Input private key sk and ciphertext c, where:

$$\text{L}(x) = \frac{x-1}{N} \tag{6}$$

calculates plaintext

$$m = \frac{\text{L}(c^\lambda(\text{mod } N^2))}{\text{L}(g^\lambda(\text{mod } N^2))} \text{ mod } N \tag{7}$$

(4): Evaluation algorithm:

1. Additive homomorphism:

$$\begin{aligned} & E(m_1) \times E(m_2) \\ & = \left(g^{m_1} r_1^N \left(\text{mod} N^2\right)\right) \times \left(g^{m_2} r_2^N \left(\text{mod} N^2\right)\right) \\ & = g^{m_1+m_2}(r_1 \times r_2)^N \left(\text{mod} N^2\right) \\ & = E(m_1 + m_2) \end{aligned} \tag{8}$$

2. Multiplication homomorphism:

$$\begin{aligned} & (E(m))^k \\ & = (g^m r^{\text{N}})^k \text{ mod N}^2 \\ & = g^{k \times m}(r^k)^{\text{N}} \text{ mod N}^2 \\ & = E(k \times m) \end{aligned} \tag{9}$$

2.2 Convolutional Neural Network

Convolutional Neural Network (CNN) [17], as one of the important algorithms in machine learning, has excellent performance in many scientific research fields. It is a kind of feed-forward neural network with deep structure and convolution calculation, which has the ability of representational learning. CNN usually consist of an input layer, convolution layers, pooling layers, activation layers, fully connected layers and an output layer. Figure 1 shows the structure of a simple CNN.

The operation of convolution layer can be regarded as sliding window of convolution core on upper input layer and doing inner product operation, which can reduce the dimension of input data and automatically extract the core features of original image. Its formula is expressed as:

$$y_{i,j} = \sum_m \sum_n w_{m,n} \times x_{i+m,j+n} + b \tag{10}$$

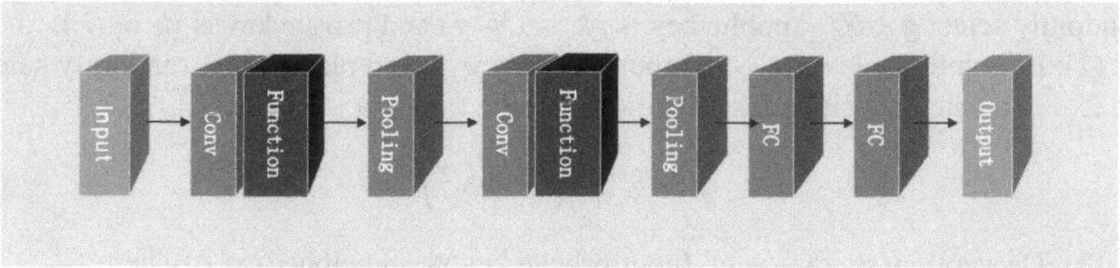

Fig. 1. Diagram of a simple CNN structure

Where $x_{i+m,j+n}$ is the input value for the convolution operation, $y_{i,j}$ is the output value for the convolution operation, $w_{m,n}$ is the weight parameter for the convolution kernel, and b is the bias parameter in the convolution calculation.

The role of the pooling layer is to reduce the parameters and computations of the next layer while preserving the main features, while also preventing overfitting. Pooling generally involves AvgPooling and MaxPooling.

The activation layer is used to increase the non-linear mapping. The commonly used activation functions include Sigmoid, Tanh, and ReLU, as shown in Fig. 2.

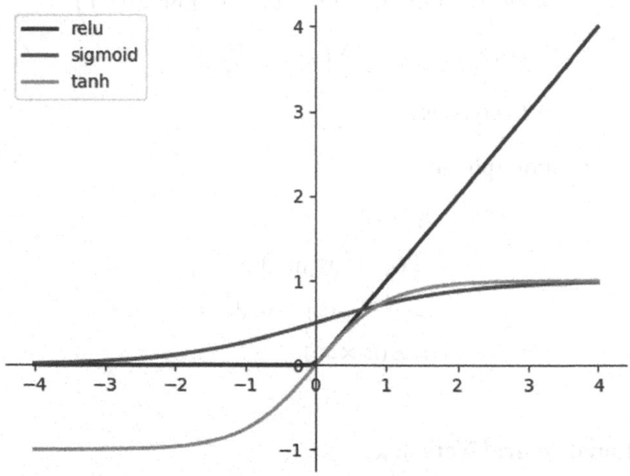

Fig. 2. Graphs of three activation functions

Now most convolution neural network models use ReLU function as the activation function. Its advantage is that it can retain the data characteristics to the maximum, converge faster and will not cause gradient disappearance problems. The expression of ReLU function is:

$$f(x) = \max(0, x) = \begin{cases} 0, x \leq 0 \\ x, x > 0 \end{cases} \tag{11}$$

The fully connected layer acts as a classifier in a convolution neural network, where each node is connected to all the nodes in the upper layer to synthesize the features extracted from the front. Because of its fully connected features, the parameters of the fully connected layer are generally the most. Each output of a fully connected layer can be viewed as each node of the upper layer multiplied by a weight factor w, and finally obtained by adding a bias value b. In the full connected layer shown in Fig. 3 x_i and y_j represent the input and output, $w_{i,j}$ represents the weight of the connection between input node i and output node j, and b_j represents the bias value of output node j.

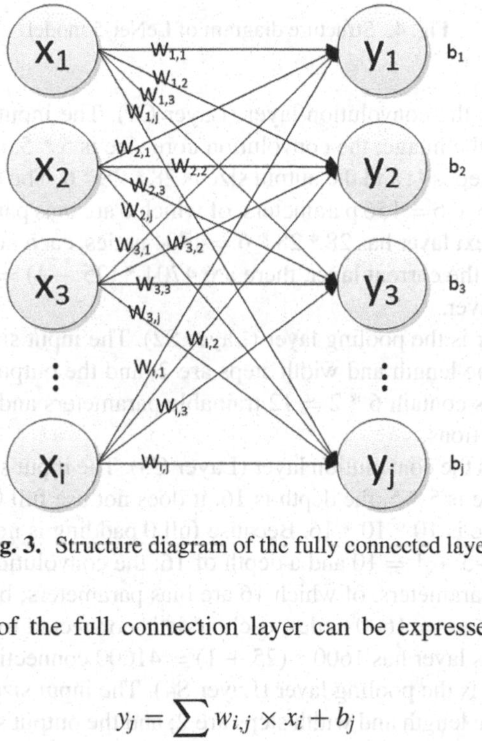

Fig. 3. Structure diagram of the fully connected layer

The calculation of the full connection layer can be expressed by the following formula:

$$y_j = \sum_i w_{i,j} \times x_i + b_j \tag{12}$$

2.3 LeNet-5 Model

The LeNet-5 model, proposed by Professor Yann LeCun [18] in 1998, uses a gradient-based inverse propagation algorithm to supervise the training of the network. By alternately connect convolution layers and pooling layers, the original image is transformed into a series of feature maps, which are transferred to a fully connected network to classify the image. LeNet-5 is a multiple iteration model of LeNet and is widely used in handwritten number recognition. The LeNet-5 model has seven layers, and its architecture is shown in Fig. 4:

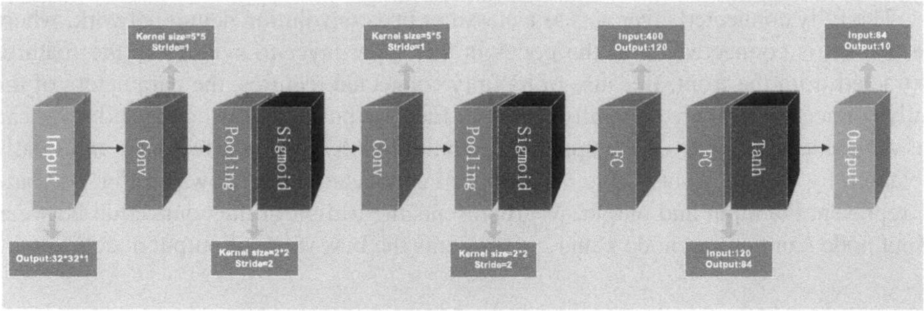

Fig. 4. Structure diagram of LeNet-5 model

The first layer is the convolution layer (Layer C1). The input size is 32 * 32 * 1, which is the size of the image; the convolution core size is 5 * 5, depth is 6, it does not use full 0 padding, step is 1; and the output size is 28 * 28 * 6. The convolution layer has a total of 5 * 5 * 1 * 6 + 6 = 156 parameters, of which 6 are bias parameters; because the node matrix of the next layer has 28 * 28 * 6 = 4704 nodes, each node is connected with 5 * 5 = 25 nodes of the current layer, there are 4704 * (25 + 1) = 122304 connections in the convolution layer.

The second layer is the pooling layer (Layer S2). The input size is 28 * 28 * 6; the filter size is 2 * 2, the length and width steps are 2; and the output size is 14 * 14 * 6. The six feature maps contain 6 * 2 = 12 trainable parameters and 14 * 14 * 6 * (2 * 2 + 1) = 5880 connections.

The third layer is the convolution layer (Layer C3). The input size is 14 * 14 * 6; the convolution core size is 5 * 5, the depth is 16, it does not use full 0 padding, the step is 1; and the output size is 10 * 10 * 16. Because full 0 padding is not used, this layer has an output size of 14–5 + 1 = 10 and a depth of 16; the convolution layer has 5 * 5 * 6 * 16 + 16 = 2416 parameters, of which 16 are bias parameters; because the next node matrix has 10 * 10 * 16 = 1600 nodes, each node is connected to 5 * 5 = 25 nodes in the current layer, this layer has 1600 * (25 + 1) = 41600 connections.

The fourth layer is the pooling layer (Layer S4). The input size is 10 * 10 * 16; the filter size is 2 * 2, the length and width steps are 2; and the output size is 5 * 5 * 16. The 16 feature maps contain 16 * 2 = 32 trainable parameters and 5 * 5 * 16 * (2 * 2 + 1) = 2000 connections.

The fifth layer is the full connected layer (Layer C5). The number of input nodes is 5 * 5 * 16 = 400, the number of parameters is 5 * 5 * 16 * 120 + 120 = 48120, and the number of output nodes is 120.

The sixth layer is the full connected layer (Layer F6). The number of input nodes is 120, the number of parameters is 120 * 84 + 84 = 10164, and the number of output nodes is 84.

The seventh layer is the output layer (Layer F7). The number of input nodes is 84, the number of parameters is 84 * 10 + 10 = 850, and the number of output nodes is 10.

3 Improved LeNet-5 Model in Ciphertext Domain

3.1 Mapping from the Plaintext Domain to the Ciphertext Domain

(1) Convolutional layer: According to the properties of addition and multiplication homomorphism of Paillier encryption algorithm introduced in Sect. 2.1 and convolution operation in convolutional neural network, convolution operation can be mapped from plaintext domain to ciphertext domain. The specific calculation method can be expressed as follows:

$$\begin{aligned} E(y_{i,j}) &= E(\sum_m \sum_n w_{m,n} \times x_{i+m,j+n} + b) \\ &= E(b) \times \prod_m \prod_n E(x_{i+m,j+n})^{w_{m,n}} \end{aligned} \tag{13}$$

$x_{i+m,j+n}$ and $y_{i,j}$ represent the input and output of the convolution, $w_{m,n}$ and b represent the weight and bias parameters of the convolution, respectively. As can be seen from the above formula, using Paillier algorithm for homomorphic encryption of data, encrypted output can be obtained by computing encrypted input and encrypted bias in ciphertext domain, which effect is consistent with the convolution operation in the plaintext domain. In plaintext domain, it can fill the dimensions of convolution output with Padding, and in ciphertext domain, filling 0 directly as in the plaintext domain will achieve the same effect.

(2) Pooling layer: From the introduction in Sect. 2.2, we can see that the pooling methods usually have MaxPooling and AvgPooling, which do not differ much in performance, but because it is difficult to compare the sizes of the two values in ciphertext domain, AvgPooling is chosen as the pooling method for this model. An enlarged version of AvgPooling [19], which calculates the sum of values without dividing by the number of values, is used to AvgPooling by addition only without affecting the depth of the algorithm. Specific calculation can be expressed in the following formula:

$$\begin{aligned} E(y) &= E(\sum_i x_i) \\ &= \prod_i E(x_i) \end{aligned} \tag{14}$$

x_i and y represent the input and output of the pooling window, respectively. In the plaintext domain, the input of the pooling layer can be symmetrical by Padding. Similar to plaintext domain, filling 0 into ciphertext domain achieves the same effect.

(3) Activation Layer: According to Sect. 2.2, ReLU function is the most commonly used and has better performance, so the activation layer of this model uses ReLU function as the activation function. The curve of the ReLU function is relatively simple, when the input is less than or equal to 0, the function's output value is 0; when the input is greater than 0, the function's output value is equal to the input value. Therefore, it only needs to judge the positive or negative of the input value of the activation layer, and then the output result through the ReLU function can be calculated. Based on this feature,

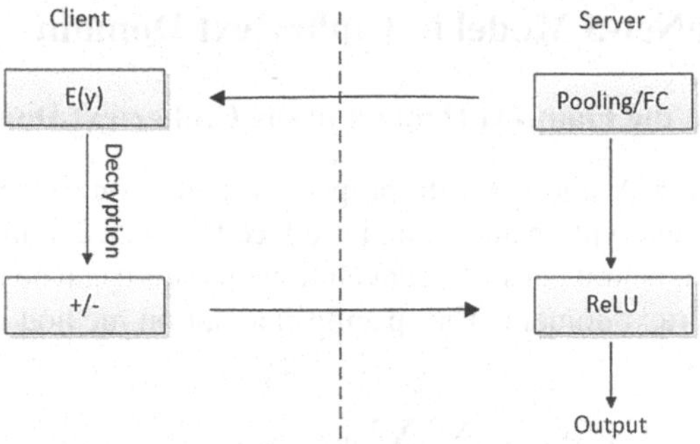

Fig. 5. Interactive Activation Layer Flowchart

interaction can be used to realize the calculation of the ReLU function in the ciphertext domain. The process of the method is shown in Fig. 5:

Before the activation calculation of the plaintext domain, the server side that holds the model sends the ciphertext value of the upper output to the client side. The user uses the private key to decrypt the data homomorphically encrypted by Paillier and feeds back the positive and negative values to the server side (positive or negative feedback does not affect the result when the value is 0). If positive, the output of the active layer is equal to the input; if negative, the output of the active layer is 0. By feeding the positive and negative values directly to the server side instead of returning the results calculated by the ReLU function to the server side, users can effectively reduce the computing burden on the client side. In ciphertext domain, LeNet-5 model uses this interactive method to compute the ReLU function in activation layer.

(4) Full Connected Layer: Through the properties of addition and multiplication homomorphism of Paillier encryption algorithm described in Sect. 2.1, and the operation of full connected layer in convolutional neural network, the calculation of linear transformation of full connected layer in ciphertext domain can be deduced. The operation process is as follows:

$$\begin{aligned} E(y_j) &= E(\sum_i w_{i,j} \times x_i + b_j) \\ &= E(b_j) \times \prod_i E(x_i)^{w_{i,j}} \end{aligned} \tag{15}$$

Among them, x_i and y_j represent the input and output of the fully connected layer, $w_{i,j}$ and b_j represent the weights and bias in the linear transformation, respectively. From the above formula, it can be seen that in ciphertext domain the encrypted output can be calculated by the encrypted input and encrypted bias in full connected layer, which achieve the same effect as the fully connected layer in the plaintext domain.

3.2 Improved LeNet-5 Model

The original LeNet-5 convolutional neural network model is introduced in Sect. 2.3. Now the model in the plaintext domain is improved, mainly including the following two aspects:

1. Change the activation function of LeNet-5 model from Sigmoid and Tanh to ReLU. Compared with Sigmoid and Tanh, the image of ReLU function is simpler, and its performance is better. It can make the network sparse, train faster, and solve the problem of gradient vanishing. In the ciphertext domain, after the activation function is changed to ReLU, the activation layer uses the interactive method in Sect. 3.1 to realize the operation of the activation function.

2. Cancel C5 layer and connect S4 layer and F6 layer in a full connected way. According to the introduction of the original LeNet-5 convolutional neural network model in Sect. 2.3, the number of training parameters of C5 layer is 48120, accounting for 78% of the total number of training parameters. After the data is encrypted by Paillier homomorphic algorithm, the running speed of convolutional neural network will be slowed down. If C5 layer in LeNet-5 model is removed and replaced by F5 full connected layer, the total number of training parameters of the model can be greatly reduced, so as to effectively improve the operation efficiency of the model. The back propagation (BP) neural network with single hidden layer can be formed in the last three layers S4, F5 and output layer of the improved LeNet-5 model, which can be used for classification output.

The improved LeNet-5 model includes six layers, which are convolution layer C1, pooling layer S2, convolution layer C3, pooling layer S4, full connection layer F5 and output layer F6, as shown in Fig. 6.

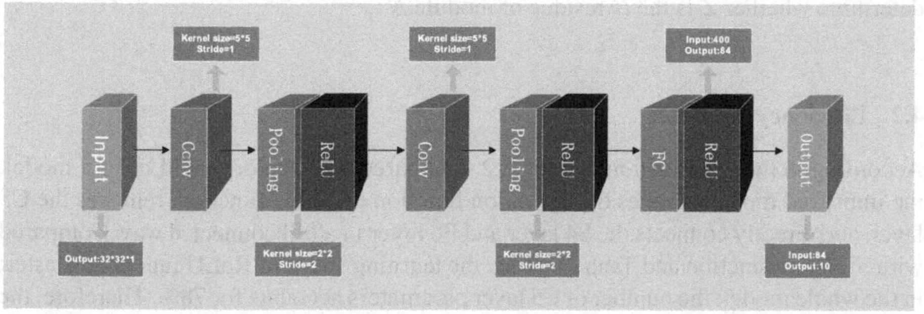

Fig. 6. Structure diagram of improved LeNet-5 model

The first four layers are consistent with the original network structure of LeNet-5 model.

The fifth layer is layer F5. The number of input nodes is $5 * 5 * 16 = 400$, the number of parameters is $400 * 84 + 84 = 33684$, and the number of output nodes is 84.

The sixth layer is the output layer (layer F6). The number of input nodes is 84, the number of parameters is $84 * 10 + 10 = 850$, and the number of output nodes is 10.

4 Model Analysis

4.1 Feasibility Analysis

Correctness Analysis. Section 3.1 describes the mapping of convolutional neural network from plaintext domain to ciphertext domain. In the convolution layer, using the properties of additive homomorphism and multiplicative homomorphism of the Paillier encryption algorithm, by calculating the encrypted input and encrypted bias can get the encrypted output. In the pooling layer, the enlarged version of AvgPooling is selected, that is, only the numerical value is added. By using the additive homomorphism property of Paillier encryption algorithm, the encrypted output can be obtained by adding the encrypted input. In the activation layer, ReLU function is used to get the encrypted output through interaction. In the fully connected layer, using the properties of additive homomorphism and multiplicative homomorphism of the Paillier encryption algorithm, the encrypted output can be obtained from the encrypted input and encrypted bias. The correctness of the model operation depends on the correctness of the decryption of the Paillier homomorphic encryption algorithm, the properties of the additive homomorphism, the multiplication homomorphism, and the correctness of the interaction.

Safety Analysis. The improved LeNet-5 designed in this paper can run on encrypted data, which can protect data privacy. Users who hold data use the Paillier homomorphic encryption algorithm to encrypt the data. In the activation layer of the model, the user needs to interact with the server that holds model. Therefore, the security of the model is based on the difficulty of the decryption of the Paillier algorithm and the security of the interaction. Paillier encryption algorithm is a public key encryption algorithm, which is based on the difficult problem of compound remaining classes. When $N = p \times q$, p and q are large primes, given $g \in Z_{N^2}^*$ randomly, so that $Z = g^N \bmod N^2$, it is difficult to determine whether Z is the N residue of module N^2.

4.2 Efficiency Analysis

According to the introduction in Sect. 3.2, compared with the original LeNet-5 model, the improved model changes the activation function to ReLU function, removes the C5 layer, and directly connects the S4 layer and F6 layer in a full connected way. Compared with Sigmoid function and Tanh function, the learning speed of ReLU function is faster; in the whole model, the number of C5 layer parameters accounts for 78%. Therefore, the efficiency of the improved LeNet-5 model will be improved. The comparison between the original LeNet-5 model and the improved LeNet-5 model is shown in Table 1.

It can be seen from Table 1 that the improved LeNet-5 model has fewer layers in the structure, which helps to improve the overall operation speed. The comparison of the number of parameters between the original LeNet-5 model and the improved LeNet-5 model is shown in Table 2.

It can be seen from Table 2 that although the improved LeNet-5 model adds 23720 parameters in F6 layer, it cancels C5 layer and directly reduces 48120 parameters. Therefore, for the whole model, the number of parameters is reduced by 24400, which can effectively improve the operation speed of the model.

Table 1. Comparison of feature maps

Model	Original LeNet-5	Improved LeNet-5
Layer C1	32 * 32 * 1	32 * 32 * 1
Layer S2	28 * 28 * 6	28 * 28 * 6
Layer C3	14 * 14 * 6	14 * 14 * 6
Layer S4	10 * 10 * 6	10 * 10 * 16
Layer C5	5 * 5 * 16	–
Layer F6	84 neurons	84 neurons
Layer F7	10 neurons	10 neurons

Table 2. Comparison of parameter quantity

Model	Original LeNet-5	Improved LeNet-5
Layer C1	156	156
Layer S2	12	12
Layer C3	2416	2416
Layer S4	32	32
Layer C5	48120	–
Layer F6	10164	33884
Layer F7	850	850

5 Conclusions

With the rapid development of machine learning, it is inevitable to bring a lot of privacy leakage problems, so how to use machine learning algorithm to ensure that privacy is not leaked is a crucial research topic. This paper uses Paillier homomorphic encryption algorithm to encrypt data to achieve privacy protection, and improves the LeNet-5 convolutional neural network model. By mapping the operations of each layer of the convolutional neural network in the plaintext domain to the ciphertext domain, the improved LeNet-5 model is implemented to run on encrypted data. The comparative analysis shows that the improved LeNet-5 model is feasible and the efficiency is improved.

In the next step of the work, there are still the following aspects to be further studied. One is to choose and improve the homomorphic encryption algorithm to further improve the operation efficiency and save the time cost; the other is to improve the mapping method from plaintext domain to ciphertext domain, such as replacing the activation function with low degree polynomial, so that the encrypted data can be directly calculated in the activation layer, and can be correctly decrypted by using the homomorphic property, while reducing the communication cost; the third is to improve the structure of convolutional neural network to improve the efficiency.

References

1. Ji, S.L., Du, T.Y., Li, J.F., Shen, C., Li, B.: Security and privacy of machine learning models: a survey. J. Softw. **32**(1), 41–67 (2021)
2. Rivest, R.L., Adleman, L., Dertouzos, M.L.: On data banks and privacy homomorphisms. Found. Secure Comput. **4**(11), 169–180 (1978)
3. Abadi, M., et al.: Deep learning with differential privacy. In: ACM SIGSAC Conference on Computer and Communications Security, pp. 308–318 (2016)
4. Wang, Q.Z.: GAO L: neural network for processing privacy-protected data. J. Cryptologic Res. **6**(2), 258–268 (2019)
5. Zhu, Q., Lv, X.: 2P-DNN: privacy-preserving deep neural networks based on homomorphic cryptosystem (2018). arXiv:1807.08459
6. Zhang, Z.H., Fu, Y., Gao, T.G.: Research on federated deep neural network model for data privacy protection. Acta Automatica Sinica (2020). https://doi.org/10.16383/j.aas.c200236
7. Shokri, R., Shmatikov, V.: Privacy-preserving deep learning. In: ACM SIGSAC Conference on Computer and Communications Security, pp. 1310–1321 (2015)
8. Phong, L.T., et al.: Privacy-preserving deep learning via additively homomorphic encryption. IEEE Trans. Inf. Forensics Secur. **13**(5), 1333–1345 (2018)
9. Dowlin, N., et al.: CryptoNets: applying neural networks to encrypted data with high throughput and accuracy. In: International Conference on Machine Learning, pp. 201–210 (2016)
10. Paillier, P.: Public-key cryptosystems based on composite degree residuosity classes. In: Stern, J. (ed.) EUROCRYPT 1999. LNCS, vol. 1592, pp. 223–238. Springer, Heidelberg (1999). https://doi.org/10.1007/3-540-48910-X_16
11. Arita, S., Nakasato, S.: Fully homomorphic encryption for classification in machine learning. In: IEEE International Conference on Smart Computing, pp. 1–4 (2017)
12. Sun, X., et al.: Private machine learning classification based on fully homomorphic encryption. IEEE Trans. Emerg. Top. Comput. **8**(2), 352–364 (2018)
13. Li, J., et al.: Privacy preservation for machine learning training and classification based on homomorphic encryption schemes. Inf. Sci. **526**, 166–179 (2020)
14. Eigamal, T.: A public key cryptosystem and a signature scheme based on discrete logarithms. IEEE Trans. Inf. Theory **31**(4), 469–472 (1985)
15. Gentry, C.: Fully homomorphic encryption using ideal lattices. In: ACM Symposium on Theory of Computing, ACM, pp.169–178 (2009)
16. Brakerski, Z., Gentry, C., Vaikuntanathan, V.: (Leveled) fully homomorphic encryption without bootstrapping. ACM Trans. Comput. Theory **6**(3), 1–36 (2014)
17. LeCun, Y., et al.: Backpropagation applied to handwritten zip code recognition. Neural Comput. **1**(4), 541–551 (1989)
18. LeCun, Y., et al.: Gradient-based learning applied to document recognition. Proc. IEEE **86**(11), 2278–2324 (1998)
19. Hesamifard, E., Takabi, H., Ghasemi, M.: CryptoDL: Deep neural networks over encrypted data (2017). arXiv:1711.05189

Design and Implementation of Intrusion Detection System Based on Neural Network

Zengyu Cai, Jingchao Wang, Jianwei Zhang(✉), and Xi Chen

Zhengzhou University of Light Industry, Zhengzhou 450000, China

Abstract. With the continuous emergence of cyber-attacks, traditional intrusion detection methods become increasingly limited. In the field of network security, new intrusion detection methods are needed to ensure network security. To solve the problems, relevant knowledge involved was first introduced. Then, an intrusion detection system based on neural network was designed according to the general intrusion detection framework, and the design of the event collector and analyzer in the system was described in detail. Experiments were conducted with the weight initialization method of the neural network model, the selection of the activation function, and the selection of the optimizer. Finally, the most suitable hyperparameters were determined and the optimal neural network model was trained. The test results show that the application of neural network to the intrusion detection system can greatly improve the accuracy of intrusion detection, thereby improving the security of computer networks.

Keywords: Intrusion detection system · Intrusion detection technology · Neural network

1 Introduction

Intrusion detection system is one of the commonly used equipment to prevent cyber-attacks. The intrusion detection system is a system used to detect the security status of a computer network. It can detect malicious intrusions and attacks that cannot be detected by the firewall [1]. At present, the intrusion detection system is not only used in the internal network of enterprises and institutions, but also widely used in various fields such as smart home systems, smart grid, and smart cities [2–5]. However, traditional intrusion detection systems require a large number of normal or abnormal behavior characteristics to detect whether intrusions have occurred. Its detection ability largely depends on the size of the rule base, and the rule base requires a lot of manpower and material resources to maintain, and the detection effect of emerging and unknown cyber-attacks is not satisfactory.

In recent years, artificial intelligence technology has been born under the mutual influence of other science and technology, and has been applied to various fields, and network security defense has also been greatly affected. W. Alhakami [6] stated that integrating machine learning into an intrusion detection system can greatly improve the

J. Zeng et al. (Eds.): ICPCSEE 2021, CCIS 1452, pp. 179–189, 2021.
https://doi.org/10.1007/978-981-16-5943-0_15

efficiency of intrusion detection. X. K. Li and A. Khraisat [7, 8] proposed random forest or decision tree algorithm in machine learning algorithm as an intrusion detection algorithm. Its detection speed is fast, but the detection accuracy is low. I. Ahmad, W. L. Al-Yaseen and J. X. Wei [9–11] proposed support vector machines as an intrusion detection algorithm, it has better performance than the former, but the accuracy of classification is still not ideal. G. J. Liu and H. Wang [12, 13] designed an intrusion detection system based on convolutional neural network. Convolutional neural network has great advantages in image processing and also shows good results in the field of intrusion detection. In summary, a variety of machine learning algorithms have been applied to the field of intrusion detection, but different machine learning algorithms have shown great differences in practical applications, and the selection and processing of data sets also have a great impact on detection effects. In this paper, the neural network with better effect is selected as the core detection algorithm of intrusion detection, and the NSLKDD data set which has been improved on the KDDCUP99 data set is selected, and the data set is processed by one-hot encoding. Finally, after adjusting the hyperparameters, the expected intrusion detection classifier is obtained and the whole system is realized.

2 Related Technologies

2.1 Intrusion Detection System

Intrusion Detection System (IDS) is a network security device that monitors the data flowing in the network in real time, and issues an alarm or takes active defense actions when suspicious behavior is found. It monitors the operating status of the network and computer in real time, continuously collects information and analyzes the collected information, thereby discovering signs of system attacks or intrusions. When the system detects suspicious behavior or signs of abnormality, it will generate a detection log and issue an alarm, and take active defense measures when necessary [14, 15].

2.2 BP Neural Network

BP (Back Propagation) neural network algorithm is a kind of machine learning algorithm. The main idea is to perform back propagation through error calculation and adjust the parameters of the neural network model to achieve a more accurate neural network model [16, 17]. The BP neural network is composed of a stack of multi-layer neuron models. After multiple iterations of training and weight updates, a model that can be used for prediction or classification is finally obtained. Each iteration consists of three steps: First, forward calculation, send the training input to the neural network to get the model output; second, back propagation, calculate the difference between the model output and the corresponding real output to obtain the output error of the model. Finally, the weight is updated. The partial derivative of each weight is obtained for the output error, the gradient of the output error is obtained, and the gradient is multiplied by a ratio and the inverse is added to the current weight. At this point, one iteration is completed. After many iterations in the neural network training process, the difference between the model output and the real output is getting smaller and smaller, which means that the model output is getting closer and closer to the real output, and the accuracy of the model is getting higher and higher.

3 Design of Intrusion Detection System Based on Neural Network

3.1 System Framework

The design of the intrusion detection system based on neural network is based on the general intrusion detection framework, which mainly includes four modules, event collector, event analyzer, system monitoring platform and database system, as shown in Fig. 1.

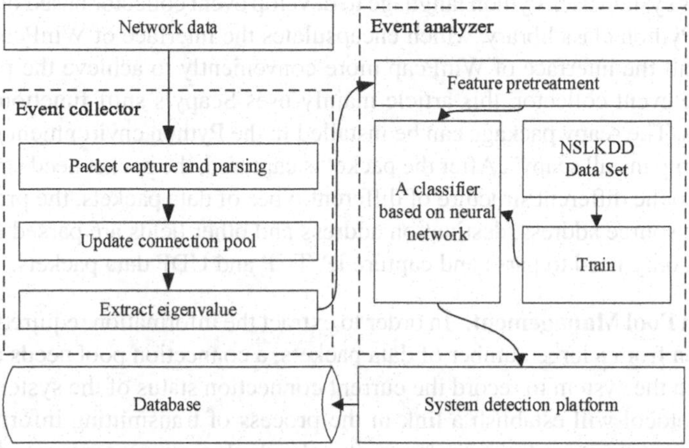

Fig. 1. System diagram of Intrusion Detection System based on Neural Network

Event collector: The information exchange on the network is carried out in the form of data packets, so a large number of data packets can be captured from the network through a specific technology. However, the captured data packets are in binary form, which requires further analysis of the captured data packets, and converts the binary value into the meaning it expresses according to the structure of the data packet. However, it is impossible to judge whether the current network status is safe by only relying on the captured data packets. The data packets need to be further sorted to extract the characteristic information required by the system, such as the number of simultaneous TCP connections and whether the data packets contain sensitive permission information, etc.

Event analyzer: The event analyzer first needs to load the trained model, and then receives the event feature values transmitted from the event collector. After preprocessing the event feature values, input a series of feature values into the neural network-based Classifier for classification. According to the classification results, determine whether there is intrusion behavior in the network environment where the system is located, and then send the analysis results to the system monitoring platform.

System monitoring platform: The system monitoring platform accepts messages from the event analyzer, organizes the messages and displays them to users and stores them in the database. If the system receives a report of intrusion behavior sent by the event analysis platform, it will send an alert and intrusion behavior information to the system administrator.

Database: The database can be used to store the results of system analysis and captured packet information, so that the administrator can use the stored information to further analyze the security status of the system.

3.2 Event Collector

Packet Capture and Parsing. The capture and analysis of data packets are the most basic part of the entire system, and network data packets are the information source of the system. This system uses Python language to develop event collector based on PyCharm. Scapy is a Python class library, which encapsulates the interface of WinPcap, allowing python to call the interface of WinPcap more conveniently to achieve the purpose you want. In the event collector, this article mainly uses Scapy's sniff function to capture data packets. The scapy package can be installed in the Python environment through the command "pip install scapy". After the packet is captured, the packet needs to be parsed. According to the different structure of different types of data packets, the protocol type, service type source address, destination address and other fields are parsed from it. For this system, only need to parse and capture IP, TCP and UDP data packets.

Connection Pool Management. In order to extract the information required for system classification from a large number of data packets, a connection pool needs to be established within the system to record the current connection status of the system. But only the TCP protocol will establish a link in the process of transmitting information, and UDP is a connectionless transmission protocol, which requires corresponding adjustments to the definition of the link in the connection pool. The information to be recorded for each link in the connection pool is shown in Table 1.

Table 1. Variables in the connection pool

Name	Explain	Name	Explain
src_ip	Source IP	flag	Connect flag
dst_ip	Destination IP	src_port	Source port
protocol	Protocol type	dst_port	Dextination port
time	Connect update time	is_fin	Connect status
s_time	Connect start time	num_urg	Emergency packets count
src_bytes	Source traffic count	num_wro	Error packets count
dst_bytes	Destination traffic count		

For UDP data packets, each UDP data packet will be regarded as a connection, the protocol is set to UDP, time and s_time are set to the time when the data packet is received, and the flag field is set to the normal state by default.

For TCP, the connection status can be obtained through TCP data packets. For example, when the SYN flag bit in the TCP data packet is 1, it means that a host sends a request to create a link to another host. At this time, a new connection information is created in the connection pool, the protocol is set to TCP, and the s_time and time fields are set to the current time and flag to S0, which means that only the connection establishment request is received, but the link is not established. Set the service field according to the port number. When a data packet with the same source address and port and destination address and port is received, update the connection information, update the time field and the connection information is placed at the top of the entire connection pool, and at the same time according to the data packet content updates flag, num_wro, num_urg, is_fin, src_bytes, dst_bytes and other fields.

Every time the connection pool is updated, the system will scan the entire connection pool. When the number of connections in the connection pool exceeds 100, the connection information that ends in more than two seconds and is after the first 100 connections will be deleted to ensure a low cost of system operation.

Extract Characteristic Values. After the necessary conditions for extracting feature values are prepared, the feature values required by the intrusion detection classifier can be counted from the connection pool. Because the classifier of this system is trained by using the NSLKDD data set, it is necessary to count the feature value types in NSLKDD when counting the feature values. Each sample in the data set has 43 values, including 42 feature values and 1 label value. The process of extracting feature values is to extract and count the captured data packet information and the information in the connection pool.

Extraction of Basic Features. There are 9 types of eigenvalues in this part. Duration represents the duration of the connection, which can be obtained by subtracting s_time from the time in the connection information. The protocol_type can be obtained from the protocol field of the connection information. The service can be obtained according to the destination port number. The flag, src_bytes, dst_bytes, wrong_fragmenta, urgent can be obtained directly from the connection information, but due to the lack of detailed descriptions of some states and feature extraction methods in the official documents of the NSLKDD data set, it is temporarily defined as 4 states, and SF means normal Status, S0 means that only SYN packets are received without SYN/ACK packets, that is, a SYN error occurs, an RST flag is a REJ error, and OTH means other connection error conditions. The land feature value indicates whether the address and port of the source host and the address and port of the destination host are the same, which can be obtained by comparing the corresponding fields in the connection information.

Extraction of Content Features. There are 13 types of feature values in this part, namely hot, num_failed_logins, logged_in, num_compromised, root_shell, su_attempted, num_root, num_file_creations, num_shells, num_access_files, num_outbound_cmds, is_hot_login, is_guest_login. The extraction of the feature value in this part requires analysis of the load of the data packet, but there is less information currently implemented, which is difficult to achieve in this article, so it is basically filled with 0, so these features can be ignored in the feature extraction stage.

Extraction of Statistical Characteristics of Network Traffic Based on Time and Host. This is part of a total of 19 kinds of characteristic values, respectively, count, srv_count, serror_rate, srv_serror_rate, rerror_rate, srv_rerror_rate, same_srv_rate, diff_srv_rate, srv_diff_host_rate,dst_host_count, dst_host_srv_count, dst_host_same_srv_rate, dst_host_diff_srv_rate, dst_host_same_src_port_rate, dst_host_srv_diff_host_rate, dst_host_serror_rate,dst_host_srv_serror_rate, dst_host_rerror_rate, dst_host_srv_rerror_rate. The characteristic value of this part needs to be statistically obtained on the connection information. In the connection pool, the characteristic value of this part can be obtained through the statistics of the connection information of the corresponding condition.

3.3 Event Analyzer

Character Feature Numerical Processing. Among the 41 feature values extracted, three feature values are expressed in the form of strings, which requires processing them and converting them into numbers. The three characteristic values are the protocol type, the service type of the target host, and the normal or error status of the connection. There are two ways to digitize. As far as the protocol type is concerned, you can replace TCP with 1, UDP with 2, and ICMP with 3. However, this method will not avoid the size relationship of the number itself, but for the protocol type, there is no size relationship between them, so it is inappropriate to use this method. Therefore, this system adopts one-hot encoding method, which represents a one-dimensional vector of bits, such as TCP-[1,0,0], UDP-[0,1,0], ICMP-[0,0,1]. In this way, the size relationship among the numbers themselves is avoided, making the feature value more scientific. After numerical processing, the feature value of each event has changed from 41 to 122.

Classifier Design Based on Neural Network. This system uses a neural network containing two hidden layers and an output layer as the classifier for intrusion detection, and uses the NSLKDD data set as the training set and test set of the neural network model. In the NSLKDD data set, each piece of data consists of 42 numbers, including 41 feature values and a label value. The feature values in the training set also need to be pre-processed to change the 41-dimensional feature values into 122-dimensional feature values. For the label values in the training set, one-hot encoding is also required. There are two types of label values. One is normal, which means that no intrusion has occurred on the network, and the other is abnormal, which means that intrusion has occurred. After processing with one-hot encoding, the label value becomes a two-dimensional vector.

The classifier model is divided into three layers. Input the processed feature values into the first layer to get the result, and then use the output of the first layer as the input of the second layer, and then use the output of the second layer as the input of the output layer. Finally, the output layer outputs a two-dimensional vector as the result. Tensorflow commonly used activation functions include relu, sigmoid, tanh, etc. The commonly used weight initialization methods include initializing weights with random numbers generated by uniform distribution, initializing weights with random numbers generated by normal distribution, and initializing weights with random constants. Commonly used types of optimizers include SGD, Adam, Adamax, etc.

The choice of the three hyperparameters needs to be further determined by analyzing the following experimental results. Since the final classification result is a two-dimensional vector, the output layer must be composed of two neurons. However, for different event eigenvalue models, the output values may differ greatly. Therefore, the softmax activation function needs to be used for the output layer to map the sum of the output two-dimensional vectors to 1. When the first value in the two-dimensional vector is greater than 0.5, the second value must be less than 0.5, and vice versa. According to the one-hot encoding rule, for events with the second digit value greater than or equal to 0.5, the system determines that there is an intrusion in the network and issues an alarm.

Classifier Training Based on Neural Network. In the training process of the neural network model, it is necessary to use the same training set for multiple iteration training until the minimum loss value is obtained. At the beginning of training, the weights in the model need to be initialized, and then the data set is sent to the neural network model, and the model calculates the corresponding output value according to the received data set. After obtaining the model output value, it is necessary to calculate the degree of difference between the output value and the true value.

This article chooses to use the Euclidean distance to measure the degree of difference between the two, and then calculate the gradient of the loss function, and according to the given learning Rate to update the weight and bias of the model, and then send it

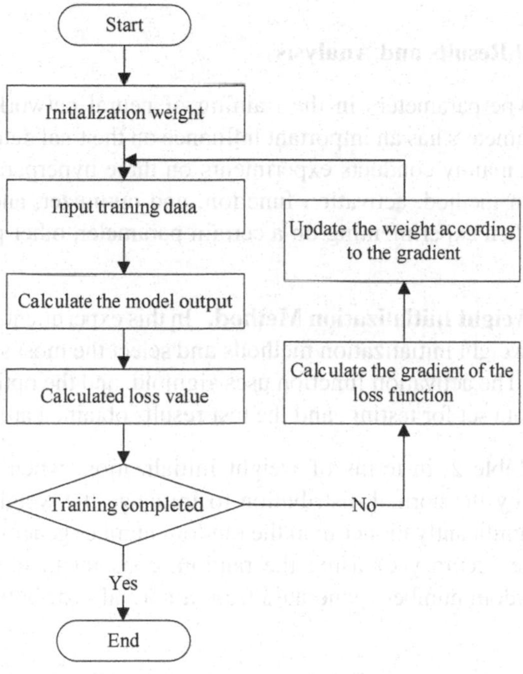

Fig. 2. Classifier training process

to the training set again and repeat the above process, until the size of the loss value no longer drops or the decline is small, you can stop the training, at this time a trained neural network model is obtained which is also called a classifier. The flow of the entire training process is shown in Fig. 2.

4 Intrusion Detection Experiment Based on Neural Network

4.1 Experimental Environment

This article mainly uses the Python programming language to implement the intrusion detection system. The database used is SQL Server 2019. The Python packages used are Scapy 2.4.3, Scikit-learn 0.23.1, Tensorflow 2.1.0, Numpy 1.18.5, and WinPcap 4.1.2 is required for packet capture.

The main workload in the subject includes: using Scapy to call the WinPcap interface for packet capture and analysis; using Numpy to read and process the original data set; using Scikit-learn, Tensorflow to train and test the neural network model of the event analyzer; optimize the neural network model for indicators such as accuracy rate, false alarm rate, and false alarm rate.

The neural network model used in this article is a three-layer fully connected neural network. The first and second layers have 128 neurons respectively. The third layer, the output layer, has 2 neurons. BATCH_SIZE (the number of samples selected for one training) is set to 100, EPOCHS (the number of cycles of the entire data set) is set to 100, and LEARN_RATE (learning rate) is set to 0.01.

4.2 Experimental Results and Analysis

There are many hyperparameters in the training of neural network models, and the choice of hyperparameters has an important influence on the final detection effect of the model. This article mainly conducts experiments on three hyperparameters, including weight initialization method, activation function, and optimizer, and selects the optimal parameters. When experimenting on a certain parameter, other parameters remain unchanged.

Neural Network Weight Initialization Method. In this experiment, the main task is to compare the three weight initialization methods and select the most suitable weight initialization method. The activation function uses sigmoid, and the optimizer uses Adam. Use the prepared data set for testing, and the test results obtained are shown in Table 2.

According to Table 2, in terms of weight initialization, when using the random number generated by the normal distribution to initialize the weight, the accuracy is 0.9128, which is significantly higher than the random number generated by the uniform distribution and the accuracy of using the random constant to initialize the weight. This paper uses random numbers generated from a normal distribution to initialize the weights.

Table 2. Comparison of experimental results with different weight initialization methods

	Random normal intializer	Random uniform initializer	Constant initializer
Accuracy	0.9128	0.4654	0.8594
Non-response rates	0.0273	0	0.0158
False positives rates	0.0597	0.5345	0.1246

Activation Function of Neural Network. In this experiment, the main task is to compare the three activation functions such as relu, sigmoid, and tanh to select the most appropriate activation function. The weight initialization method has been determined by the first experiment to initialize the random number generated by the normal distribution. For weight, the optimizer uses Adam. Use the prepared data set for testing, and the test results obtained are shown in Table 3.

Table 3. Comparison of experimental results with different activation functions

	sigmiod	relu	tanh
Accuracy	0.9128	0.4654	0.9166
Non-response rates	0.0273	0.0	0.0265
False positives rates	0.0597	0.5345	0.0567

According to Table 3, among the three activation functions, the tanh activation function performs the best among the three activation functions. Although the effect of using the sigmoid activation function is close to that of using the tanh activation function, it has the accuracy rate, false negative rate and the false alarm rate in three aspects is worse than using the tanh activation function, so this article finally uses the tanh activation function to apply to the neural network.

Neural Network Optimizer. In this experiment, the main task is to compare three optimizers such as SGD, Adam, and Adamax to select the most suitable optimizer. The weight initialization method has been determined by the first experiment to initialize the random number generated by the normal distribution. The activation function is determined by the second experiment and use the relu function as the activation function of the neural network model. Use the prepared data set for testing, and the test results obtained are shown in Table 4.

It can be seen from Table 4 that for the optimizer, the use of the Adam optimizer and the use of the Adamax optimizer have similar effects, but the effect of using the Adam optimizer in the false negative rate is obviously better than using the Adamax optimizer, so the final decision to use the Adam optimizer is for the training of neural networks.

Table 4. Comparison of experimental results using different optimizers

	Adam optimizer	Adamax optimizer	SGD optimizer
Accuracy	0.9166	0.9134	0.9094
Non-response rates	0.0265	0.0324	0.0210
False positives rates	0.0567	0.0541	0.0694

In summary, after multiple rounds of testing, the optimized neural network model on the NSL-KDD dataset is obtained: the weights of the neural network are initialized with random numbers generated by the normal distribution, the activation function uses the tanh function, and the optimizer uses Adam optimizer.

5 Conclusions

This paper takes the current significant problems in the field of network security as the demand, and uses the current popular neural network algorithm in the field of artificial intelligence to evaluate the engineering application value of the field of network security. The neural network model is used as the core of intrusion detection algorithm based on network data packets to identify intrusions in the network. Multiple hyperparameters of the neural network are tuned and built on the basis of Tensorflow open source machine learning architecture. Good results are achieved in actual tests.

Acknowledgements. This work is supported by National Natural Science Foundation of China (62072416), Henan Province Science and Technology Research Project (202102210176), Zhongyuan Science and Technology Innovation Leader of Zhongyuan Talent Project (214200510026), Henan Province Science and Technology Project (212102210429), The fourth batch of innovative leading talents of Zhihui Zhengzhou 1125 talent gathering plan (ZhengZheng[2019] No. 21).

References

1. Yin, C., Xia, L., Zhang, S., Sun, R., Wang, J.: Improved clustering algorithm based on high-speed network data stream. Soft. Comput. **22**(13), 4185–4195 (2017). https://doi.org/10.1007/s00500-017-2708-2
2. Zhang, R., et al.: An intrusion detection scheme based on repeated game in smart home. Mob. Inf. Syst. **2020**, 9 (2020). Article ID. 8844116
3. Zou, X., Cao, J.H., Guo, Q., Wen, T.: A novel network security algorithm based on improved support vector machine from smart city perspective. Comput. Electr. Eng. **65**, 67–78 (2018)
4. Jow, J., Xiao, Y., Han, W.L.: A survey of intrusion detection systems in smart grid. Int. J. Sensor Netw. **23**(3), 170–186 (2017)
5. Zuo, X., Chen, Z., Dong, L., Chang, J., Hou, B.: Power information network intrusion detection based on data mining algorithm. J. Supercomput. **76**(7), 5521–5539 (2019). https://doi.org/10.1007/s11227-019-02899-2

6. Alhakami, W.: Alerts clustering for intrusion detection systems: overview and machine learning perspectives. Int. J. Adv. Comput. Sci. Appl. **10**(5), 1–10 (2019)
7. Li, X.K., Chen, W., Zhang, Q.R., Wu, L.F.: Building auto-encoder intrusion detection system based on random forest feature selection. Comput. Secur. **95** (2020). Article ID. 101851
8. Khraisat, A., Gondal, I., Vamplew, P., Kamruzzaman, J., Alazaab, A.: Hybrid intrusion detection system based on the stacking ensemble of C5 decision tree classifier and one class support vector machine. Electronics **9**, 173 (2020)
9. Ahmad, I., Basheri, M., Iqbal, M.J., Rahim, A.: Performance comparison of support vector machine, random forest, and extreme learning machine for intrusion detection. IEEE Access **6**, 33789–33795 (2018)
10. Al-Yaseen, W.L., Othman, Z.A., Nazri, M.Z.A.: Multi-level hybrid support vector machine and extreme learning machine based on modified K-means for intrusion detection system. Expert Syst. Appl. **67**, 296–303 (2017)
11. Wei, J.X., Long, C., Li, J.W., Zhao, J.: An intrusion detection algorithm based on bag representation with ensemble support vector machine in cloud computing. Concurrency Comput. Pract. Exp. **32**(24), e5922 (2020)
12. Liu, G.J., Zhang, J.B.: CNID: research of network intrusion detection based on convolutional neural network. Discrete Dyn. Nat. Soc. **2020**, 11 (2020)
13. Wang, H., Cao, Z.J., Hong, B.: A network intrusion detection system based on convolutional neural network. J. Intell. Fuzzy Syst. **38**(6), 7623–7637 (2020)
14. Li, P., et al.: Intrusion detection methods based on incomplete RFID traces. Chin. J. Electron. **26**(4), 675–680 (2017)
15. Keegan, N., Ji, S.-Y., Chaudhary, A., Concolato, C., Yu, B., Jeong, D.H.: A survey of cloud-based network intrusion detection analysis. HCIS **6**(1), 1–16 (2016). https://doi.org/10.1186/s13673-016-0076-z
16. Khan, R., Al-Shehri, M.: A futuristic analysis approach of neural network for intrusion detection system. Dilemas Contemporaneos-Educacion Politica Y Valores **6**(3), 1–16 (2019)
17. Liu, Y., Zhu, L.: A new intrusion detection and alarm correlation technology based on neural network. EURASIP J. Wirel. Commun. Netw. **2019**(1), 1 (2019). https://doi.org/10.1186/s13638-019-1419-z

IoT Security Situational Awareness Based on Q-Learning and Bayesian Game

Yang Li[1,2](✉), Tianying Liu[1], Jianming Zhu[1,2], and Xiuli Wang[1,2]

[1] School of Information, Central University of Finance and Economics, Beijing 100081, China
liyang@cufe.edu.cn

[2] Engineering Research Center of State Financial Security, Ministry of Education, Central University of Finance and Economics, Beijing 102206, China

Abstract. IoT security is very crucial to IoT applications, and security situational awareness can assess the overall security status of the IoT. Traditional situational awareness methods only consider the unilateral impact of attack or defense, but lackconsideration of joint actions by both parties. Applying game theory to security situational awareness can measure the impact of the opposition and interdependence of the offensive and defensive parties. This paper proposes an IoT security situational awareness method based on Q-Learning and Bayesian game. Through Q-Learning update, the long-term benefits of action strategies in specific states were obtained, and static Bayesian game methods were used to solve the Bayesian Nash Equilibrium of participants of different types. The proposed method comprehensively considers offensive and defensive actions, obtains optimal defense decisions in multi-state and multi-type situations, and evaluates security situation. Experimental results prove the effectiveness of this method.

Keywords: IoT security · Q-Learning · Bayesian game

1 Introduction

The concept of Internet of Things (IoT) appeared in 1995. With the development of sensor network technology, IoT was formally proposed at an international conference in 2005. Today, the Internet of Things has become a widespread expansion of the Internet, not only connecting things with things, but also improving the quality of industrial production, agricultural production and daily life through various application technologies. With the rapid development and application of the IoT, its security issues have gradually become the focus of attention. As the devices and sensors of IoT are widely distributed, they are vulnerable to a large number of attacks from physical domains and virtual domains. The network is composed of many heterogeneous networks with complex structures, which is difficult to implement security protection, and has insufficient defense capabilities. At the same time, data collection by IoT applications involves commercial secrets or personal privacy information. If those data leaks, it may cause serious consequences. The vulnerability of IoT devices in the face of attacks makes data leakage incidents more likely to occur. In daily life, if a large number of devices that implement smart homes are

J. Zeng et al. (Eds.): ICPCSEE 2021, CCIS 1452, pp. 190–203, 2021.
https://doi.org/10.1007/978-981-16-5943-0_16

attacked, it will not only affect the security of cyberspace, but also pose a serious threat to the security of real life. In the field of industrial control, the Stuxnet virus once caused great harm in large-scale facility systems around the world. In order to improve the security of the Internet of Things, researchers have proposed many security solutions in the perception layer, transmission layer, processing layer and application layer of IoT, such as lightweight encryption technology, cloud computing security technology and user privacy protection technology. However, these technologies are based on a specific security problem and then propose solutions without considering the awareness and assessment of the overall security status. The security of IoT is an entirety, and its analysis should also be comprehensive. It is necessary to comprehensively analyze the threat level and security situation of IoT network. IoT security situation awareness is a risk assessment method that can help administrators or users analyze the current situation in a complex security environment, and can meet the needs of comprehensive analysis.

In the overall security situational awareness and defense, the Internet of Things is also different from traditional networks. Many devices are limited by resources and only have simple security measures such as password verification, and cannot be configured with large-scale security analysis systems such as intrusion detection that can be applied to traditional networks. If a security system that requires lots of computing resources is adopted, it will affect the normal operation of the system when analyzing data flow, malware, and user behavior, and it is difficult to apply to small devices such as sensors. Therefore, the IoT security situation awareness solution requires low energy consumption with the ability of helping security defenders make optimal decisions in a complex dynamic environment and evaluating the overall security situation.

2 Related Works

Network Security Situation Awareness (NSSA) is a research field that perceives the overall network security status. It analyzes and judges the security situation by fusing various information about network security, and provides help for network administrators and participants in decision-making process. In 1988, Endsley proposed the concept of situation awareness [1]. Many researchers later elaborated on the definition and connotation of NSSA from different aspects, but most of them involved "safety", "whole" and "decision-making helper". Perceiving, analyzing and evaluating the security situation of the Internet of Things can grasp the overall network security status, predict future trends, and make decisions that are most conducive to security defense. Aiming at IoT security and security situational awareness, literature [2] constructs a situation awareness solution from the perspective of fog computing, transforming security analysis from a vertical framework to a horizontal structure suitable for distributed computing, which can be applied to complex security environments. Literature [3] proposed a risk assessment framework for IoT devices, which measures threats and risks from the perspective of information leakage, and predicts the future security situation to make reasonable decisions through perception of the environment and risks. Literature [4] introduced security assessment methods for IoT applications, and gave suggestions on enhancing security in IoT security areas such as local devices, data transmission, and data storage. The above-mentioned literature analyzes the risks and threats in the network, and perceives the security situation from the perspective of the entire IoT network, but did not consider the mutual impact between malicious attackers and defenders in the IoT network.

Game theory is a mathematical theory for studying countermeasures. In a security game, the attacker and the defender confront each other, and strategy and benefits are interdependent. It can consider the impact of the actions of both offensive and defensive parties at the same time, instead of only focusing on the behavior of attacking or defending party, which meets the need for overall security situation awareness, and is suitable for IoT security research. Literature [5] uses SCPN to calculate the attack path, builds a dynamic game model to conduct situational awareness in IoT environment, and verifies the model through typical attack scenarios. Literature [6] combines stochastic Petri nets and game theory to build a stochastic game net model, which makes the game method suitable for complex IoT networks. The administrator node can detect and prevent attacks in a dynamic and scalable environment, and make a reasonable action plan. Literature [7] proposed a security situation awareness method based on Markov game. Threats are used as the unit that affects network security. A game model integrating attackers, administrators and users is established, and the security situation in the network system is dynamically evaluated. The above-mentioned literature applies game theory to the analysis of the Internet of Things and network security, which can integrate the influence of both offensive and defensive actions on the security status, but lacks specific analysis of the game process and cannot use the game matrix to make security decisions.

In the application of matrix game for attack path prediction and optimal defense strategy selection, literature [8] combines evolutionary game theory and Markov decision process to solve Markov evolutionary game equilibrium under multi-state conditions, taking into account the imperfect rationality of the decision maker. Literature [9] constructed a Markov game model for moving target defense, comprehensively quantified defense benefits and defense costs, and proposed an optimal strategy selection algorithm. Literature [10] dynamically analyzes the network attack and defense process and determines the optimal pure strategy through the results of the Bayesian game, which is more practical in use than the mixed strategy. Those literature can be applied to a variety of situations such as imperfect rationality and multi-state security situation, but they mainly focus on predicting specific attack and defense behaviors, which lacks analysis of the overall security situation.

This paper combines Q-Learning algorithm with static Bayesian game to analyze the security status of IoT, and conduct security situation awareness under multi-state and incomplete information conditions. Through solving the optimal defense strategy and quantitative analysis of the overall network security situation, the method can help network administrators to take better defense actions and predict future security status.

3 Q-Learning Algorithm

In the process of IoT security attack and defense game, as the attacker attacks, the defender takes measures to defend. There are transitions and jumps between multiple states. It is necessary to consider the current and future benefits at the same time to make defensive decisions that are more in line with long-term interests. Therefore, the idea of multi-state stochastic game can be introduced into the attack and defense process. Traditional stochastic games mostly use the model-based Markov decision process to describe, but there is a problem that the state transition probability is difficult to determine. Using the model-free Q-Learning algorithm does not need to obtain the state transition probability in advance, but through exploring and interacting with the environment to autonomously obtain unknown environmental information. Considering these characteristics, this paper constructs a security situation awareness model combined with the Q-Learning algorithm to update state and behavior information through exploration in the unknown environment.

Q-Learning is a model-free reinforcement learning algorithm. It is based on the Markov decision process. By learning the action-value function, it can estimate the best action-value function q* independently of future strategies. Its task is to choose each step under the condition of finite state and finite action set, and realize the Markov process that the agent can control.

3.1 Markov Decision Process

In the Markov decision process, the state-value function of the Bellman equation used to select the optimal decision needs to satisfy the relationship:

$$\mathrm{v}_\pi(\mathrm{s}) = \sum\nolimits_{a\in A} \pi(a|s)(R_s^a + \gamma \sum\nolimits_{s'\in S} P_{ss'}^a v_\pi(s')) \tag{1}$$

Equation (1) represents the sum of the expected discounted reward of the state, which is a function value that can represent the current and future returns, while taking into account the decay of future returns. Since the future benefit occurs after a period of time, its current value is smaller than the immediate benefit that can be obtained immediately. In order to measure this feature, the attenuation value γ $(0 < \gamma < 1)$ is introduced to more accurately describe the time value of reward. The algorithmic goal of the Markov decision process is to select the strategy that maximizes the total expected discounted reward as the optimal strategy.

Equation (1) represents the state value, that is, the value calculated by considering all actions in a state and their selection probabilities. The value Q of the selected action in a certain state can be expressed by the following formula. Assuming that the agent chooses the strategy π, the expected discounted income of taking action a in state s can be expressed as:

$$Q^\pi(s, a) = R_s(a) + \gamma \sum_{s'} P_{ss'}[\pi(s)]\mathrm{V}_\pi(\mathrm{s}') \tag{2}$$

$P_{ss'}$ in formulas (1) and (2) represents the transition probability between states. Under the assumption that both the instantaneous return $R_s(a)$ and the state transition probability $P_{ss'}$ are known, the dynamic programming method can be used to calculate the maximum expected state value V, the maximum expected action value Q and the optimal strategy π.

However, in the actual IoT attack and defense process, the transition probability between different security states can only be determined with historical records and expert experience, and it is difficult to obtain accurate values. At the same time, as the IoT security environment changes, the state transition probability may also change accordingly. Therefore, an algorithm that can perform reinforcement learning under the condition of unknown state transition probability is needed.

3.2 Q-Learning Method

In the Q-Learning method, solving the optimal strategy does not need to understand and initialize the state transition probability. Participants interact with the environment, and constantly update the Q value in multiple steps of exploration, and finally achieve the purpose of selecting the optimal strategy. A step of Q-Learning can be defined as:

$$\mathrm{Q}(S_t, A_t) \leftarrow \mathrm{Q}(S_t, A_t) + \alpha\Big[R_{t+1} + \gamma \max_a Q(S_{t+1}, a)\Big) - \mathrm{Q}(S_t, A_t)] \tag{3}$$

Where $\mathrm{Q}(S_t, A_t)$ is the current Q value, α represents the learning rate of the difference between the two calculations of the state value, and γ is an attenuation value, which represents the extent of how future rewards affect the present. In each step, the agent is in a certain state S_t, and the corresponding action A_t can be selected according to S_t. Each state-action pair (S_t, A_t) determines a current benefit R_{t+1} as the direct benefit of making a certain choice in a specific state. At the same time, in the entire state space, the transition from the current state to the next state is also determined by the action selected, and different actions may lead to different next states.

The tasks that the agent needs to complete in a step of updating the Q value are: observe the current state S_t, select the action A_t, determine the next state S_{t+1}, get the current reward R_{t+1}, and recalculates Q to obtain the updated Q value according to the learning rate α and attenuation value γ. Through Q-Learning, agent in the environment continue to explore and obtain current and future revenue information of different actions, taking into account the transfer and change between multiple states.

Since the security game involves both parties, in the strategy of updating the Q value, the behavior and influence of multiple agents need to be considered at the same time. If the attacker takes action A_t and the defender takes action D_t, an update of one step can be defined as:

$$\mathrm{Q}(S_t, A_t, D_t) \leftarrow \mathrm{Q}(S_t, A_t, D_t) + \alpha\left[R_{t+1} + \gamma \max_{a,d} Q(S_{t+1}, a, d)\right) - \mathrm{Q}(S_t, A_t, D_t)] \tag{4}$$

4 IoT Security Situational Awareness Model Based on Q-Learning and Bayesian Game

4.1 Characteristics of IoT Security

IoT security and its situational awareness have many characteristics different from traditional network security. First, most of the IoT network have wireless network structure, and there is no central node to monitor the data transmitted in the network. Therefore, distributed nodes are required to cooperate with each other to ensure security, and the existence of malicious nodes among many nodes must be considered. Besides, the bandwidth of the IoT is limited. If the traditional security perception system is adopted, the large amount of information flow generated by it may cause network congestion and affect the normal operation of the network system. Therefore, the data flow transmitted by the security perception system should be reduced as much as possible. In addition, computing power and storage resources are very limited in IoT devices. Security perception and prediction are performed under the situation where resources are limited, so traditional security methods need to be optimized.

Traditional network security situational awareness and risk assessment analyze a large number of data streams. When applied to the Internet of Things, it will consume excessive power and computing resources, making it difficult to conduct situation monitoring for a long time, resulting in a decline in security assurance capabilities and real-time response capabilities to the attack. Besides, only considering the attacker or the defender unilaterally makes it impossible to accurately measure the factors affecting each other on the offense and defense, nor does it take into account the interconnection between nodes in the Internet of Things. Using game theory to analyze the security of the IoT can solve these problems to a certain extent.

4.2 Model Selection and Definition

The research of applying game process to network security analysis can adopt static game model or dynamic game model. In a static game, usually the game has only a single stage, which cannot reflect the changes between multiple states in IoT security. In a multi-stage game, it is also assumed that the strategies and rewards of the participants are fixed and will not change over time. These models cannot be applied to the dynamic IoT security environment. In a dynamic environment, there are multiple states. Participants adopt different strategies in different states. The influence of strategies is also long-term. Therefore, the Q-Learning algorithm is used to measure the benefits of offensive and defensive participants in a dynamic IoT environment with state transitions.

Whether there is existence of complete information is a prerequisite for constructing game models and obtaining optimal strategies. Many security game models are based on the assumption of complete information, that is, for all participants, it is assumed that they have all the information about the game in advance. Such as the strength of the attacker, the success rate of the attack, and the amount of attack resources. For the defender, that are the defense capability, success rate of the defense, and resources that can be used for defense such as the remaining power of the node. However, in the process of IoT security attack and defense, it is difficult for both sides to obtain complete information on the strength of each other's attack and defense capabilities. They can only make general judgments, which is a situation of incomplete information. Bayesian game is a game process between participants in a state of incomplete information. In a static Bayesian game, participants act at the same time. Each participant may have different types. The strategies and benefits of different types of participants may also be different. Compared with the complete information game, the Bayesian game expands the participants from a single type to several types, which is more in line with the different abilities of the participants in reality.

Therefore, the security situational awareness model of the Internet of Things is defined as a nine-tuple (N, Θ, P, M, S, U, W, π*, *NS*), where:

(1) N = (N_A, N_D) is the participants of the game. N_A is the attacker (malicious node) in the IoT network, and N_D is the defender (normal node) in the IoT network.
(2) $\Theta = (\Theta_A, \Theta_D) = \left(\theta_A^1, \ldots, \theta_A^x; \theta_D^1, \ldots, \theta_D^y\right)$, which is the type set of attacker and defender.
(3) P = (P_A, P_D) is the set of types of the other participants inferred by the attacker and the defender when they know their own type.
(4) M = $(M_A, M_D) = (a_1, \ldots, a_m; d_1, \ldots, d_n)$, which is the action set of offensive and defensive parties. The types of participants are different, and the set of strategies under those types may also be different. The strategy set of each type is a subset of the action set.
(5) S = $(s_1, \ldots, s_k)$, which is the security state space of the IoT network, and there are transfers between security states.
(6) U = (U_A, U_D) is the reward space of the participants. The reward obtained by the participants is determined by state, type, strategy adopted and the strategy selected by the other party.
(7) W = (w, A_w, D_w), which is the importance of different nodes and their distribution.
(8) $\pi^* = \left(\pi^*\left(\theta_A^1\right), \ldots, \pi^*\left(\theta_A^x\right); \pi^*\left(\theta_D^1\right), \ldots, \pi^*\left(\theta_D^y\right)\right)$, is a collection of optimal mixed strategies for different types of attackers and defenders.
(9) NS = $\left(NS^1, \ldots, NS^k\right)$, which is a collection of security situation values in different states.

In order to simplify the model, in the game process, all malicious nodes in the IoT network are collectively regarded as an attacker and a series of strategies are adopted to attack the network. All normal nodes in the IoT network are collectively regarded as a defender and a series of defense strategies are adopted to maintain network security.

4.3 Selection of the Optimal Defense Strategy

In the state of incomplete information, solving the static Bayesian Nash equilibrium can obtain the optimal strategies for different types of attackers and defenders.

The standard expression of a Bayesian game is $G = \left\{A_1, \ldots, A_n; T_1, \ldots, T_n; p_1, \ldots, p_n; u_1, \ldots, u_n\right\}$, where A is the action set of the participants and T is the type Set. $p_i(t_{-i}|t_i)$ is the uncertainty of other participant's type when its own type is t_i. The reward function is related to the type t_i. The reward function of participant i is $u_i(a_1, \ldots, a_n; t_i)$, which represents the income obtained by the action in the case of the type t_i. When offensive and defensive parties act together, use $U_n\left(a_i, d_j, \theta_n^k\right)$ to indicate participant n's income when the attacker's action is a_i, the defender's action is d_j, and the type of participant n is θ_n^k.

Since any finite game has a mixed strategy Nash equilibrium, there is an optimal mixed strategy for the attacker and the defender in each state in the model, that is, the mixed strategy Bayesian Nash equilibrium [12].

Given:

$$((N_A, N_D), (\Theta_A, \Theta_D), (M_A, M_D), (P_A, P_D), (U_A, U_D)) \tag{5}$$

The mixed strategy for offense and defense is:

$$\begin{aligned} \pi(\theta_A^i) &= \{f_1^A(\theta_A^i), f_2^A(\theta_A^i), \ldots, f_{n1}^A(\theta_A^i)\}, \\ \pi(\theta_D^k) &= \{f_1^D(\theta_D^k), f_2^D(\theta_D^k), \ldots, f_{n2}^D(\theta_D^k)\} \end{aligned} \tag{6}$$

If the following conditions are met:

$$\sum_{\theta_D^k \in \Theta_D} P_A(\theta_D^k|\theta_A^i) U_A\left(\pi^*\left(\theta_A^i\right), \pi^*\left(\theta_D^k\right), \theta_A^i\right) \geq \sum_{\theta_D^k \in \Theta_D} P_A(\theta_D^k|\theta_A^i) U_A\left(\pi\left(\theta_A^i\right), \pi^*\left(\theta_D^k\right), \theta_A^i\right), \tag{7}$$

$$\sum_{\theta_A^i \in \Theta_A} P_D(\theta_A^i|\theta_D^k) U_D\left(\pi^*\left(\theta_A^i\right), \pi^*\left(\theta_D^k\right), \theta_D^k\right) \geq \sum_{\theta_A^i \in \Theta_A} P_D(\theta_A^i|\theta_D^k) U_D\left(\pi^*\left(\theta_A^i\right), \pi\left(\theta_D^k\right), \theta_D^k\right), \tag{8}$$

Then the mixed strategy $(\pi^*(\theta_A^i), \pi^*(\theta_D^k))$ is a Bayesian Nash equilibrium with offensive and defensive types of θ_A^i and θ_D^k in a state. By solving equilibrium, the optimal strategy of both offense and defense is obtained, which provides reference for the defender to make specific defense decisions.

4.4 IoT Security Situational Assessment

The security game of the Internet of Things is a constantly changing process with the different actions of offensive and defensive parties and the transfer of the overall network system states. Its security situational assessment also needs to take into account the specific actions taken by malicious nodes and normal nodes, and the current and future state changes of the network. Bayesian equilibrium can help attackers and defenders predict each other's strategy, but when making their own decisions, they can only choose one specific action at a time.

Assuming that in the state s_o, the attacker type is θ_A^i and the action taken is a_x, then the threat value of the attacker is:

$$\mathrm{NS}_A^{s_o} = \sum_{\theta_D^k} P_A(\theta_D^k|\theta_A^i) \sum_y U_a\left(a_x, d_y, \theta_D^k\right) \pi^*\left(\theta_D^k\right) \tag{9}$$

Assuming that in the state s_o, the defender type is θ_D^k and the action taken is d_y, then the defender's benefit value is:

$$\mathrm{NS}_D^{s_o} = \sum_{\theta_A^i} P_D(\theta_A^i|\theta_D^k) \sum_x U_d\left(a_x, d_y, \theta_A^i\right) \pi^*\left(\theta_A^i\right) \tag{10}$$

In the IoT network, factors such as the number of malicious nodes and normal nodes and the importance of nodes have a certain impact on the overall network security situation. Suppose node importance from high to low is $\mathrm{w} = \{w_H, w_M, w_L\}$, the distribution of the three types of nodes in malicious nodes is $A_w = \{A_H, A_M, A_L\}$, and the distribution in normal nodes is $D_w = \{D_H, D_M, D_L\}$, then the security situation value of the IoT is:

$$\mathrm{NS}^{s_o} = \mathrm{NS}_D^{s_o} \cdot \frac{F(D)}{\sum_k A_k + \sum_k D_k} - \mathrm{NS}_A^{s_o} \cdot \frac{F(A)}{\sum_k A_k + \sum_k D_k} \tag{11}$$

Where $\mathrm{F(A)} = \sum_k w_k \cdot A_k$, $\mathrm{F(D)} = \sum_k w_k \cdot D_k$, $\mathrm{k} \in (\mathrm{H, M, L})$.

4.5 Model Algorithm

The algorithm for solving the optimal defense strategy and network security situation is as follows:

Model algorithm

Input: state set S, action set M, type space Θ, priori inference P, reward space U, state of node W

Output: optimal defense strategy π^*, network security situation NS

Begin

For $s_o \in S$

For $\theta_A^i, \theta_D^k \in \Theta$, $n \in N$

$$Q^n(\theta_A^i, \theta_D^k) = \sum_x \sum_y U_n(a_x, d_y)\,\pi^*(\theta_A^i, \theta_D^k)$$

For $\theta_A^i, \theta_D^k \in \Theta$, $n \in N$

Repeat

$$Q_{x+1}^n(S_t, A_t, D_t) = Q_x^n(S_t, A_t, D_t) + \alpha\left[R^n{}_{t+1} + \gamma \max_{a,d} Q^n(S_{t+1}, a, d) - Q_x^n(S_t, A_t, D_t)\right]$$

Until

$$|Q_{x+1}^n(S_t, A_t, D_t) - Q_x^n(S_t, A_t, D_t)| < \varepsilon$$

Solve

$$\pi(\theta_A^i) = \{f_1^A(\theta_A^i), f_2^A(\theta_A^i), \ldots, f_{n1}^A(\theta_A^i)\},$$
$$\pi(\theta_D^k) = \{f_1^D(\theta_D^k), f_2^D(\theta_D^k), \ldots, f_{n2}^D(\theta_D^k)\}$$

Calculate

$$NS_A^{s_o} = \sum_{\theta_D^k} P_A(\theta_D^k|\theta_A^i) \sum_y U_a(a_x, d_y, \theta_D^k)\,\pi^*(\theta_D^k),$$
$$NS_D^{s_o} = \sum_{\theta_A^i} P_D(\theta_A^i|\theta_D^k) \sum_x U_d(a_x, d_y, \theta_A^i)\,\pi^*(\theta_A^i)$$

Output

$$\pi^*, NS^{s_o} = NS_D^{s_o} \cdot \frac{F(D)}{\sum_k A_k + \sum_k D_k} - NS_A^{s_o} \cdot \frac{F(A)}{\sum_k A_k + \sum_k D_k}$$

This algorithm combines Q-Learning with Bayesian game, and is suitable for the situation of multi-state and multiple types. It can solve the optimal strategy, and evaluate the security situation value according to the distribution of normal nodes and malicious nodes in the network. Compared with traditional methods, this algorithm takes actions of both offensive and defensive parties as security influence factors, and the security situational assessment is more accurate.

5 Experimental Simulation

5.1 Experimental Setup

We build an IoT network as shown in Fig. 1. There are 1,100 IoT nodes distributed in the network, and the devices are connected to the network through routers. Among them, there are 100 malicious nodes, 1,000 normal nodes, and one administrator node among the normal nodes.

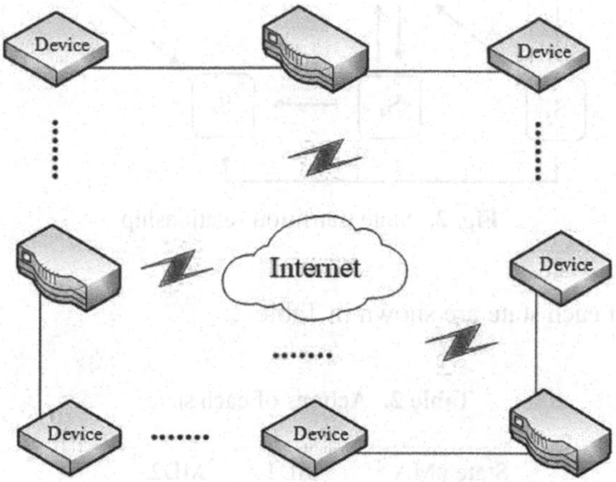

Fig. 1. Simulated IoT network

Refering to reference [6], the strategies of offense and defense are shown in Table 1:

Table 1. Offensive and defensive strategies

Number	Offensive strategies	Number	Defensive strategies
a1	Dos attack	d1	Restore device
a2	Dictionary attack	d2	Remove compromised account
a3	Man in the middle attack	d3	Locking account
a4	Continue Dos attack	d4	Delay response
a5	Server spoofing attack	d5	Use digital signatures
a6	Replay to other devices	d6	Reconfigure DNS setting
a7	Shutdown network	d7	Keep data encrypted
a8	Stop attacking	d8	Multiple authentication

There are 7 simplified network security states, which are: s_0 = {normal operation}, s_1 = {device Dos}, s_2 = {device attacked}, s_3 = {injects malicious data}, s_4 = {device direct to malicious portal}, s_5 = {device get malicious data}, s_6 = {network shutdown}. The state transition relationship is shown in Fig. 2:

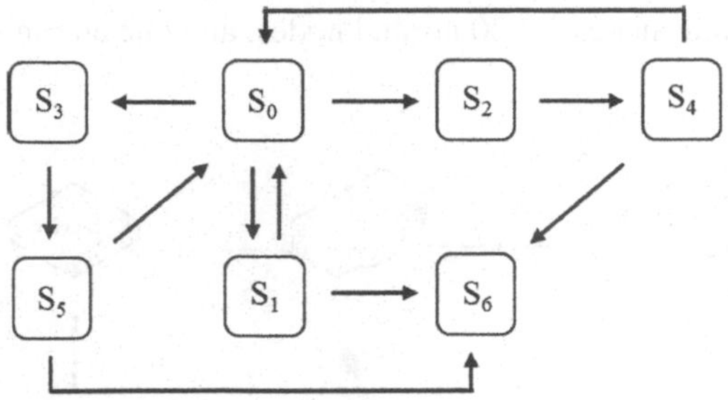

Fig. 2. State transition relationship

The actions of each state are shown in Table 2:

Table 2. Actions of each state

State	MA	MD1	MD2
s0	a1,a2,a3	d1,d2,d3	d2,d3,d4
s1	a4,a8	d1,d3	d2,d3
s2	a5,a8	d2,d5	d6,d5
s3	a6,a8	d2,d6	d5,d6
s4	a7,a6	d1,d7	d2,d7
s5	a7,a4	d1,d4	d2,d4
s6	Terminal	Terminal	Terminal

Since the defense measures selected by normal nodes will be determined according to network situations and their own conditions, the defense of normal nodes is divided into two types: high-level defense and low-level defense. Assuming that the malicious nodes have only one attacker type: high-level attack, the prior belief of the malicious nodes for the two defense types is (0.7, 0.3). Taking (high-level attack, high-level defense) as an example, its returns matrix in state s_0 is shown in Table 3:

Table 3. Returns matrix of (high-level attack, high-level defense) in state s0

s0	d1	d2	d3
a1	30,30	40,20	45,15
a2	30,10	20,15	10,20
a3	10,10	15,15	5,10

5.2 Analysis of Results

Since the algorithm uses Q-Learning to update the state-action value, calculating strategy benefits does not need to obtain the transition probability between states in advance, avoiding possible errors, and its operation process can be adjusted independently according to the actual situation. When changes occur in IoT network, new environmental information can be obtained through exploration again. At the same time, the introduction of Bayesian game allows offensive and defensive participants to make reasonable guesses about the actions of another participant without knowledge of complete information, which improves the practical value of the algorithm.

In the stage of optimal defense strategy selection, according to the algorithm in this paper, the optimal defense strategy can be obtained as: $\pi_{s_0} = \{(0,0.25,0.75),(0,1,0)\}$, $\pi_{s_1} = \{(0,1),(0.91,0.09)\}$, $\pi_{s_2} = \{(1,0),(1,0)\}$, $\pi_{s_3} = \{(1,0),(1,0)\}$, $\pi_{s_4} = \{(1,0),(1,0)\}$, $\pi_{s_5} = \{(1,0),(1,0)\}$. It can be seen from the results that in the initial state, both high-level defenders and low-level defenders should prefer to choose strategy d3. As the network attack continues, different types of defenders have different optimal strategy choices.

In the stage of network security situational assessment, based on the actions taken by both offensive and defensive parties and the prediction of the other's behaviors, the security threat value of the attacker and the security benefit value of the defender are obtained, and the network security situation is comprehensively evaluated. If the attacker and the defender choose an optimal strategy in each state, the network security situation value of the IoT is shown in Fig. 3:

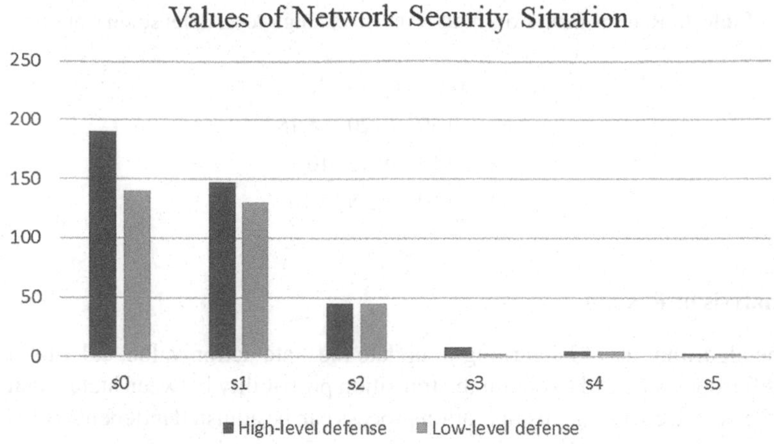

Fig. 3. Changes in the network security situation

It can be seen that as the states changes and the degree of attack becomes stronger, the network security situation value decreases. In addition, in the initial stage of the network being attacked, the gap between the high-level defense and the low-level defense type is large. In the later stages of the attack, the network faces very serious threats. The gap between the security status of the two types of defenders is narrowed, and the security situation is in a low value. This shows that even a defender with strong defensive capabilities needs to stop the attack in time to prevent the network from facing more serious threats.

6 Conclusion

There are multiple security states in the Internet of Things, and there may also be multiple types of attack and defense nodes. Based on Q-Learning and Bayesian game, this paper proposes an IoT security situational awareness method that can be applied to the mutual conversion of different security states and fully considers different types and capabilities of attacker and defender. Experiments provethat the method proposed is effective, and can realize reasonable and efficient IoT security situational awareness while making optimal decisions.

Acknowledgment. This work is supported by the National Key Research and Development Program of China (No. 2017YFB1400700).

References

1. Shi, L., et al.: Survey of research on network security situation awareness. Comput. Eng. Appl. **55**(24), 1–9 (2019)

2. Rapuzzi, R., Repetto, M.: Building situational awareness for network threats in fog/edge computing: emerging paradigms beyond the security perimeter model. Futur. Gener. Comput. Syst. **85**, 235–249 (2018)
3. Park, M., Oh, H., Lee, K.: Security risk measurement for information leakage in IoT-based smart homes from a situational awareness perspective. Sensors **19**(9), 2148 (2019)
4. Chouhan, P.K., McClean, S., Shackleton, M.: Situation assessment to secure IoT applications. In: 2018 Fifth International Conference on Internet of Things: Systems, Management and Security, pp. 70–77. IEEE (2018)
5. He, F., et al.: SCPN-based game model for security situational awareness in the intenet of things. In: 2018 IEEE Conference on Communications and Network Security (CNS), pp. 1–5. IEEE (2018)
6. Kaur, R., Kaur, N., Sood, S.K.: Security in IoT network based on stochastic game net model. Int. J. Netw. Manag. **27**(4), e1975 (2017)
7. Zhang, Y., Tan, X.B., Cui, X.L., Xi, H.S.: Network security situation awareness approach based on Markov game model. J. Softw. **22**(3), 495–508 (2011)
8. Huang, J., Zhang, H., Wang, J.: Markov evolutionary games for network defense strategy selection. IEEE Access **5**, 19505–19516 (2017)
9. Lei, C., Ma, D.H., Zhang, H.Q.: Optimal strategy selection for moving target defense based on Markov game. IEEE Access **5**, 156–169 (2017)
10. Wang, Z., et al.: Optimal network defense strategy selection based on Markov Bayesian game. KSII Trans. Internet Inform. Syst. **13**(11), 5631–5652 (2019)
11. Sutton, R.S., Barto, A.G.: Reinforcement Learning: An Introduction. MIT Press (2018)
12. Wang, J., Yu, D., Zhang, H., Wang, N.: Active defense strategy selection based on the static Bayesian game. J. Xidian Univ. **43**(01), 144–150 (2016)
13. Hu, J., Wellman, M.P.: Nash Q-learning for general-sum stochastic games. J. Mach. Learn. Res. **4**, 1039–1069 (2003)
14. Gibbons, R.: A Primer in Game Theory (1992)
15. Subba, B., Biswas, S., Karmakar, S.: Intrusion detection in mobile ad-hoc networks: Bayesian game formulation. Eng. Sci. Technol. **19**(2), 782–799 (2016)
16. Wang, Y., et al.: A survey of game theoretic methods for cyber security. In: 2016 IEEE First International Conference on Data Science in Cyberspace (DSC), pp. 631–636. IEEE (2016)
17. Watkins, C.J.C.H., Dayan, P.: Q-learning. Mach. Learn. **8**(3–4), 279–292 (1992)
18. Shamshirband, S., et al.: Cooperative game theoretic approach using fuzzy Q-learning for detecting and preventing intrusions in wireless sensor networks. Eng. Appl. Artif. Intell. **32**, 228–241 (2014)

Protecting Web Application Code and Sensitive Data with Symmetric and Identity-Based Cryptosystems

Jinchao Ni, Ziyan Liu(✉), Ning Li, Cheng Zhang, Bo Cui, and Hanzhang Kong

Information and Telecommunication Company, State Grid Shandong Electric Power Company, Jinan 250000, China
ziyanliu2021@126.com

Abstract. How to protect the security of web application code and sensitive data has become one of the primary concerns in web services. In this paper, symmetric cryptosystem combined with identity-based public key cryptosystem is proposed to protect web application programs and sensitive data. The key generation center generates the private and public key pairs for the web server and users, which are used to implement identity authentication and data integrity. When web application code and sensitive data are transmitted between the web server and the user's browser, a random session key is generated for encrypting the web application code and sensitive data. Meanwhile, a digital signature is generated and added to the encrypted program code and sensitive data. The security analysis shows that the proposed security scheme can ensure the confidentiality, integrity and authentication of web application code and sensitive data.

Keywords: Data security · Web security · Symmetric cryptosystem · Public key cryptosystem · Identity-based cryptosystem

1 Introduction

With the evolution of various emerging technologies such as cloud computing [1], big data [2], 5G telecommunication [3], mobile Internet and Internet of things [4], data security [5] has become a critical problem in network information systems. As one type of popular client-server architectures, web applications offer the network service that a browser can access a web server over the internet. However, web-based applications are facing lots of security threats and vulnerabilities. M. Awad, M. Ali, M. Takruri and S. Ismail [6] gave a detailed discussion on several web attacks including database SQL-injection attacks, cookie poisoning attacks, cross-site scripting attacks (XSS for short), and buffer overflow attacks. They also proposed some methods to detect and prevent from these web attacks. In [7], A. Razzaq et al. discussed the drawbacks of traditional signature-based web security mechanisms including scanners, intrusion detection systems (IDS), and web application firewalls. Furthermore, they proposed an ontology-based method [7] to detect and classify various sophisticated web attacks by

J. Zeng et al. (Eds.): ICPCSEE 2021, CCIS 1452, pp. 204–216, 2021.
https://doi.org/10.1007/978-981-16-5943-0_17

using semantic rules. In [8], A. Futoransky, E. Gutesman and A. Waissbein presented a dynamic security and privacy protection technique for web applications, which can prevent web services from some well-known attacks that try to tamper with databases or perform other exploitations. In [9], D. Akhawe, A. Barth, P. E. Lam, J. Mitchell and D. Song tried to construct a formal foundation for web security and presented a formal model to analyze the security of some web applications. In [10], M. Jensen, N. Gruschka and R. Herkenhoener gave a detailed survey of security vulnerabilities in web services and the corresponding countermeasures for preventing these attacks.

The goals of web application security is to protect information assets of websites and web users against different types of attacks. One of the most critical objectives is to ensure the confidentiality, integrity, authenticity and availability of program codes and sensitive data transmitted between the websites and the users' browsers. In [11], G. Wassermann and Z. Su proposed a static detection method for finding cross-site scripting vulnerabilities. In [12], H. Shahriar and M. Zulkernine presented mutation-based testing technique to generate a large test data sets for detecting Cross Site Scripting (XSS) vulnerabilities. In [13], H. Shahriar and M. Zulkernine proposed a server-side JavaScript code injection detection method by inserting comment statements containing random tokens among the JavaScript code blocks. E. Stark [14] designed a set of security tools, called CryptFrame, to ensure the user's data private against the compromised web server. X. Dong et al. [15] presented a browser primitive CRYPTON to protect sensitive data in the user's browser. S. Tople et al. [16] developed a tool, called AUTOCRYPT, to protect sensitive web data in untrusted servers by using partially-homomorphic encryption schemes. In [17–22], lots of JavaScript program libraries and other API are designed and implemented. Lots of cryptographic algorithms [23–28] are widely used in web applications.

In [29], P. H. Phung, H. D. Pham, J. Armentrout, P. N. Hiremath, and Q. Tran-Minh proposed a method to monitor the behaviors of JavaScript code within a web origin based on the source of the code to detect and prevent malicious actions that would compromise users' privacy. In [30], H. Shimamoto, N. Yanai, S. Okamura, J. P. Cruz, and T. Okubo designed an implementation and evaluation of temporal logic in Alloy to express time series and parallel computation for web security analysis and showed that their proposed syntax can analyze state-of-the-art attacks. In [31], A. Figueiredo, T. Lide, D. Matos and M. Correia proposed an approach that aims to improve security of web applications by identifying vulnerabilities in code written in different languages. In [32] F. Caturano, G. Perrone, and S. P. Romano developed a black-box testing methodology consisting of sending a sequence of strings to a web application and observing the responses. In [33], M. Mohammadi, B. Chu, and H. R. Lipford proposed a new approach that can automatically fix this common type of Cross Site Scripting (XSS) vulnerability in many situations.

In this paper, we propose a security scheme to protect web application code and sensitive data in web services. By combining SM4 symmetric cryptographic algorithm with SM9 identity-based public key cryptosystem, we can achieve data confidentiality, integrity and authentication. In SM9 identity-based cryptosystem, the key generation center generates the private and public key pairs according to the web server and the users' identity information. In the key generation phase, a trusted Key Generation Center is constructed to generate users' private keys and secure communication channels are used to distribute the keys. Before web application code and sensitive data are transmitted, a random session key is generated for encrypting the plaintext data. Meanwhile, digital signature of the secret key is wrapped by a digital envelope with the secret key together, which is applied to implement data integrity and authentication. The security analysis shows that the proposed security scheme can ensure the confidentiality, integrity and authentication of web application code and sensitive data.

The rest of this paper is organized as follows. In Sect. 2, we review symmetric cryptosystems and public-key cryptosystems. In Sect. 3, we propose the security scheme for protecting web sensitive data and web program code. Moreover, we give a security analysis. Finally, Sect. 4 presents the conclusion.

2 Symmetric and Public Key Cryptographic Algorithms

In this section, we give a brief review of several widely-used cryptographic algorithms including AES [23], SM4 [24], SM9 [27] and so on.

2.1 Symmetric Cryptographic Algorithms

Symmetric cryptosystems use the same secret key for data encryption and decryption process. AES and SM4 block algorithms are two of the most popular cryptographic algorithms. The lengths of the plaintext/ciphertext block and secret key are 128 bits in SM4 and AES-128. Now we descript the SM4 symmetric encryption algorithm briefly.

Let the plaintext block be $(X_0, X_1, X_2, X_3) \in (Z_2^{32})^4$, the ciphertext block be $(Y_0, Y_1, Y_2, Y_3) \in (Z_2^{32})^4$, the round keys be $rk_i \in Z_2^{32}$. $(i = 0, 1, \ldots, 31)$. The encryption process of SM4 block encryption algorithm [24] is described as follows:

Algorithm 1: SM4 symmetric encryption algorithm [24]

Input: The plaintext block is $(X_0, X_1, X_2, X_3) \in (Z_2^{32})^4$, and the round keys are $rk_i \in Z_2^{32}$. $(i = 0, 1, \ldots, 31)$.

Output: The ciphertext block be $(Y_0, Y_1, Y_2, Y_3) \in (Z_2^{32})^4$.

Encryption:

1. For i =0 to 31 , perform the following 32-round comutations:
   $X_{i+4} = F(X_i, X_{i+1}, X_{i+2}, X_{i+3}, rk_i) = X_i \oplus T(X_{i+1} \oplus X_{i+2} \oplus X_{i+3} \oplus rk_i)$,

where the composed permutation is $T(\cdot) = L(\tau(\cdot))$. The nonlinear transformation τ consists of 4 parallel S-boxes. The linear transformation $L(B) = B \oplus (B <<< 2) \oplus (B <<< 10) \oplus (B <<< 18) \oplus (B <<< 24)$;

2. Finally, the ciphertext block $(Y_0, Y_1, Y_2, Y_3) = R(X_{32}, X_{33}, X_{34}, X_{35}) = (X_{35}, X_{34}, X_{33}, X_{32})$.

The decryption process of SM4 cryptosystem has a similar computation as the encryption process except for using 32 round keys in reverse order.

2.2 Public-Key Cryptographic Algorithms

Public-key cryptosystems can be used to implement identity authentication, data integrity, non-repudiation and symmetric key distribution. In recent years, elliptic curve cryptosystems [25, 26] received lots of attention due to their stronger security and shorter key length than RSA cryptosystem. As a rapid emerging type of cryptosystem, identity-based cryptosystem has much simpler key management and greater flexibility than traditional Public key infrastructure (PKI). In an identity-based cryptosystem, an entity's IP address, device ID number, e-mail address, telephone number and so on can be considered as its public key.

SM9 identity-based cryptosystem [27] is one of the identity-based cryptographic algorithm standards. SM9 cryptosystem consists of digital signature algorithm, key encapsulation mechanism, public key encryption algorithm and key exchange protocol. In the proposed scheme, we apply SM9 identity-based public key cryptosystem.

3 Web Security Scheme for Protecting Sensitive Data and Program Code

In this section, we propose the security scheme for protecting web sensitive data and web program code, which are transmitted between web server and user's browser.

3.1 The Idea and General Architecture

Cryptographic algorithms can provide confidentiality, integrity, authentication and non-repudiation by encryption, message authentication code, digital signature and other schemes.

One of the main drawbacks in traditional JavaScript encryption schemes are the risk of leaking the cryptographic secret key, which is stored directly in the web program code. Moreover, a lot of JavaScript programs implement only symmetric encryption and lack identity authentication and data integrity protection on the secret key and sensitive data.

In [17–22], lots of JavaScript program libraries and other API are designed and implemented. Some cryptographic algorithms [23–28] are widely used in web applications. For JavaScript cryptography programs, the cryptographic secret key can be obtained directly from the program code. White-box cryptography [34] hides the secret key within the program code by combing encryption process with key obfuscation. However, as long as the attacker gets the JavaScript program code, he/she can encrypt or decrypt data. Therefore, we apply public key cryptosystem to protect the session symmetric key.

Our security scheme combines symmetric encryption algorithm with public-key cryptographic algorithm to insure data confidentiality, integrity and identity authentication. The proposed security scheme has the following benefits:

- The symmetric secret key isn't leaked in web JavaScript program. The secret key is wrapped by a digital envelope, in which SM9 public key encryption algorithm is applied to encrypt the symmetric secret key.
- The digital signature on the symmetric secret key is added to ensure the key's integrity and authenticate the identity of the web server or user sending the web application code or sensitive data.

- The web user's public key is bound with the user's device ID, IP address, user name and other identity information. This approach can enhance the security strength of the overall information system.

3.2 The Security Scheme

In this section, we will present the scheme for protecting web application code and sensitive data. In this scheme, a trusted authority, called Key Generation Center (KGC), is constructed to generate users' private keys according to their public keys (users' identities). The key generation method is described in the following key generation phase.

The security scheme consists of three phases: key generation phase, web application code retrieving phase and sensitive data sending phase. Now we give the detailed description of three phases as follows.

For key generation phase, we use a similar notations as SM9 public key cryptosystem in [27]. Let F_q be a finite field with q elements and $E(F_q)$ be the set of rational points on an elliptic curve E. Let G_1 and G_2 be two additive groups of order prime N and G_T be a multiplicative group of order prime N. Let P_1 denote a generator of group G_1 and P_2 denote a generator of group G_2. Let e: $G_1 \times G_2 \rightarrow G_T$ be a bilinear pairing. Let *Hash*() denote a secure cryptographic Hash function, for example SM3 Hash algorithm [28].

In key generation phase, the KGC generates its master key pair and web server or users' key pairs, which is described as follows.

Phase 1: Key generation phase	
1	Generate the master key pair:
	a) The KGC generates a random number $ks \in [1, N-1]$ as the master private key
	b) The KGC computes $P_{pub\text{-}s} = [ks]P_2$ as the master public key
	c) The KGC's master key pair is (ks, $P_{pub\text{-}s}$), where the private key ks is kept secret and the public key $P_{pub\text{-}s}$ is published to all
2	Generate the key pair for web server:
	a) Suppose the identity of web server be the following:
	ID_Server = Server Name‖IP address‖Domain Name‖ other ID Information, where "‖" denote bit string contatennation
	b) Compute $t_1 = H_1(ID\_Server, Time\_Stamp, N) + ks$, where *Time_Stamp* is the current date and N is the order. The viriable ks is the master private key of KGC
	c) Compute $t_2 = ks \bullet t_1^{-1}$ and $Pri\_Key\_Server = [t_2]P_1$, where *Pri_Key_Server* is the private key of web server
3	Generate the key pair for a web user:
	a) Suppose the identity of a web user be the following:
	ID_User = User Name‖IP address‖Device ID‖ other ID Information, where "‖" denote bit string contatennation
	b) Compute $t_1 = H_1(ID\_User, Time\_Stamp, N) + ks$, where *Time_Stamp* is the current date and N is the order. The viriable ks is the master private key of KGC

(*continued*)

(*continued*)

	c) Compute $t_2 = ks \bullet t_1^{-1}$ and $Pri\_Key\_User = [t_2]P_1$, where *Pri_Key_User* is the private key of the web user

Once the private/public key pairs of web server and users is generated, these key pairs must be sent to web server and all the users in secure communication channel. Then the key pair will be loaded into web server and the users' browsers for use in the future web communication.

When a user types the URL (Uniform Resource Locator) into the address bar in a web browser, the web server will return the corresponding HTML web page, in which JavaScript code maybe be embedded. In the proposed scheme, encryption and digital signature techniques are used to ensure the confidentiality, integrity and data origin authentication of web application code. We give the description of web application code retrieving phase in phase 2 and illustrate it in Fig. 1.

Phase 2: Web application code retrieving phase	
1	The user's browser sends a web request to web server
2	Once receiving the web request, web server performs the following processes:
	a) Pick a random session key *Session_Key_Server* and use it to encrypt the important web application code *Plain_Program_Code*. We apply the SM4 symmetric algorithm with GCM encryption mode. The ciphertext is *Cipher_Program_Code*
	b) Compute a digital signature on the session key denoted by $Sig_{\_Server}$(*Session_Key_Server*) using the web server's private key *Pri_Key_Server*
	c) Generate a digital envelope denoted by *Pub_Enc*(*Session_Key_Server*, $Sig_{\_Server}$(*Session_Key_Server*)) using the web user's public key
	d) Send the encrypted *Cipher_Program_Code* and *Pub_Enc*(*Session_Key_Server*, $Sig_{\_Server}$(*Session_Key_Server*)) to the web user
3	After receiving the encrypted web application code, the user do the following procedure:
	a) Using the private key *Pri_Key_User*, the user decrypts the received digital envelope and gets the *Session_Key_Server* and its signature $Sig_{\_Server}$(*Session_Key_Server*)
	b) Using the public key *ID_Server*, the user verifies the signature $Sig_{\_Server}$(*Session_Key_Server*). If the verification isn't passed, the web application ends with an error
	c) The web user's browser decrypts *Cipher_Program_Code*. And verifies the integrity of web application code
	d) If Step a), b) and c) are passed, the web user's browser runs the decrypted web application code

Some important data including user passwords, bank account number, social security numbers and so on may be transmitted between web server and the web users. Cryptographic algorithms are required to encrypt these sensitive data. However, traditional JavaScript encryption will reveal the symmetric key. Therefore, we adopt identity-based cryptosystem to protect the session secret key.

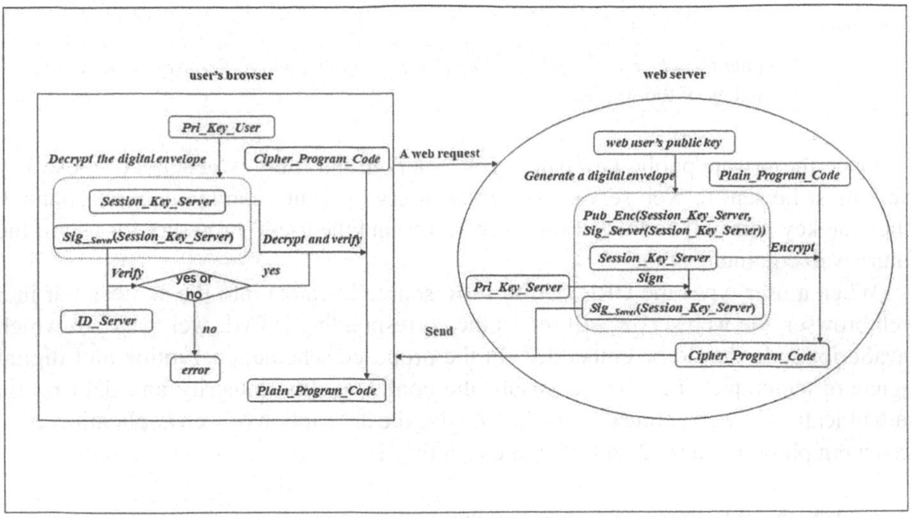

Fig. 1. Web application code retrieving phase.

The transmitted sensitive data may be from on party of web server and a web user to the other party. As the applied security mechanisms are very similar, we describe the security scheme for protecting sensitive data transmitted from the user browser to web server. The scheme of sensitive data sending phase is described in Phase 3 and illustrated in Fig. 2.

Phase 3: Sensitive data sending phase	
1	The web user's browser fills sensitive data *Plain_data* such as password, bank card number and so on
2	The web user's browser performs the following processes:
	a) Pick a random session key *Session_Key_User* and use SM4 with GCM mode to encrypt the sensitive data *Plain_Data*. The ciphertext is *Cipher_Data*
	b) Compute a digital signature on the session key denoted by Sig_{User}(*Session_Key_User*) using the web user's private key *Pri_Key_User*
	c) Encrypt the session key and its signature denoted by *Pub_Enc*(*Session_Key_User*, Sig_{User}(*Session_Key_User*)) using the web server's public key *ID_Server*
	d) Send the ciphertext *Cipher_Data* and the digital envelope of the session key to the web server
3	After receiving the encrypted sensitive data, the web server do the following precedure:
	a) Using its private key *Pri_Key_Server*, the web server decrypts the received digital envelope and obtains the key *Session_Key_User* and Sig_{User}(*Session_Key_User*)
	b) Using the public key *ID_User*, the web server verifies the signature Sig_{User}(*Session_Key_User*). If the verification isn't passed, the program ends with an error

(*continued*)

(*continued*)

	c) The web server uses the session key *Session_Key_User* to decrypt *Cipher_Data* and verifies the integrity of sensitive data
	d) If Step a), b) and c) are passed, the web server can get the plain sensitive data

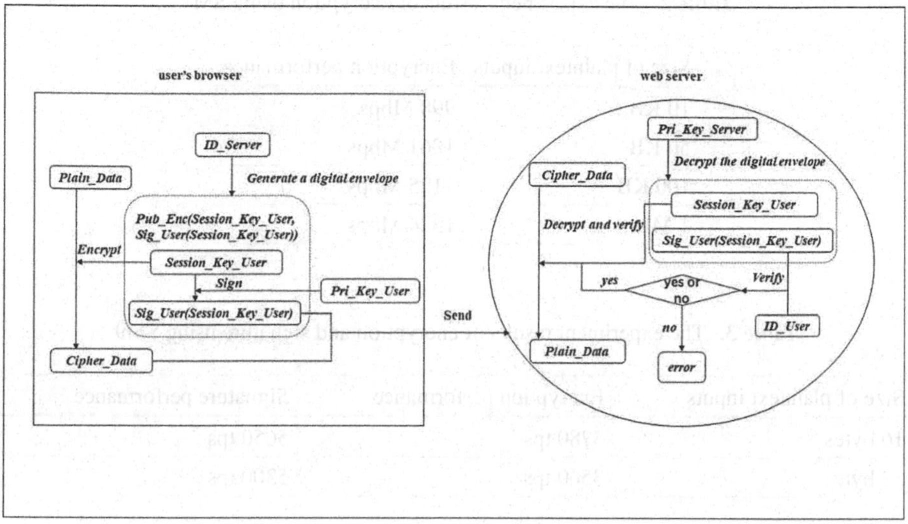

Fig. 2. Sensitive data sending phase.

3.3 Experimental Results and Performance

Based on the above schemes, we implemented SM4 symmetric cryptographic algorithm and SM9 identity-based cryptosystem for performance evaluation. The performances of the cryptosystems have been tested by adopting SM4 algorithm and SM9 algorithm. In order to evaluate the performance of the implementation, we choose a PC server to execute the encryption and signature process. SM4 algorithm and SM9 algorithm are implemented in C++, by using library GmSSL. Table 1 summarizes the experimental environment.

Table 1. The configurations of experimental environment

CPU	Intel(R) Xeon(R) 64-bit CPU E5-2620 v3 @ 2.40 GHz
Memory	64G
OS	Ubuntu 16.0.4
Library	GmSSL 2.5.4

We evaluate the performance of SM4 implementation and SM9 implementation in encryption and signature efficiency under different sizes of plaintext inputs. The test results show each of them can protect the web application code and sensitive data, and the performance of data encryption and signature has practical value. The experimental results are shown in Table 2 and Table 3 respectively.

Table 2. The experiment results of encryption using SM4

Size of plaintext inputs	Encryption performance
10 KB	998 Mbps
50 KB	1061 Mbps
100 KB	1125 Mbps
1 MB	1324 Mbps

Table 3. The experiment results of encryption and signature using SM9

Size of plaintext inputs	Encryption performance	Signature performance
16 bytes	3780 tps	5650 tps
32 bytes	3560 tps	5300 tps

When encrypting sensitive data using SM4 cryptographic algorithm, for example, encrypting 10 KB plaintext inputs on the PC server got the speed of 998 Mbps. SM9 identity-based cryptographic algorithms are only used to encrypt or sign short cryptographic keys such as 16 bytes SM4 key. It can be seen in Table 3 that SM9 has a good computational performance. Thus the proposed method can efficiently protect the process of web communication.

3.4 Security Analysis

Now we will give a security analysis of the proposed scheme, which includes data confidentiality, integrity, authentication, and key management system.

Data Confidentiality and Integrity Analysis In the proposed scheme, we apply SM4 symmetric encryption algorithm to ensure the confidentiality of web application code and sensitive data. SM4 symmetric cryptographic algorithm has a 128-bit key length and a 128-bit plaintext/ciphertext block size. Its computational process consists of 32-round operations and each round operation deals with four 32-bit words.

Let the plaintext block be $(X_0, X_1, X_2, X_3) \in (Z_2^{32})^4$, and the ciphertext block be $(Y_0, Y_1, Y_2, Y_3) \in (Z_2^{32})^4$. The round keys are $rk_i \in Z_2^{32}$, $(i = 0, 1, \ldots, 31)$. The SM4 encryption process is described as follows.

$X_{i+4} = F(X_i, X_{i+1}, X_{i+2}, X_{i+3}, rk_i) = X_i \oplus T(X_{i+1} \oplus X_{i+2} \oplus X_{i+3} \oplus rk_i)$, $i = 0, 1, \ldots, 31$.

The function $T(\cdot) = L(\tau(\cdot))$ and the non-linear transformation τ consists of four parallel S-boxes. The linear transformation $L(B) = B \oplus (B <<< 2) \oplus (B <<< 10) \oplus (B <<< 18) \oplus (B <<< 24)$.

Finally, the ciphertext block $(Y_0, Y_1, Y_2, Y_3) = R(X_{32}, X_{33}, X_{34}, X_{35}) = (X_{35}, X_{34}, X_{33}, X_{32})$.

There are no practical attacks on SM4 symmetric cryptographic algorithm up to now. The brute force attack on SM4 algorithm requires the time complexity 2^{128} operations. SM4 symmetric cryptographic algorithm can resist against some known cryptanalysis techniques such as differential cryptanalysis, linear cryptanalysis, algebraic attack and other attack methods. While SM4 consists of 32 rounds operations, no known attack techniques can attack successfully beyond 24 rounds of SM4 algorithm. Thus we can achieve a tightly high security by using SM4 algorithm.

Furthermore, the GCM (Galois Counter Mode of Operation) mode, which combines encryption technique with message authentication code, is used to implement data integrity. GCM authentication tag is established based on polynomial computation over finite field $GF(2^{128})$. In GCM mode, the encryption and authentication operation is defined as $E:\{0, 1\}^k \times \{0, 1\}^{128} \rightarrow \{0, 1\}^{128}$ and a authentication tag to protect data integrity. Assuming the underlying block cryptographic algorithm achieves random permutations, GCM mode can realize provable security of privacy and integrity. Thus the confidentiality and integrity of web program code and sensitive data can be ensured.

Encryption Protection of the Session Symmetric Key. Different with traditional JavaScript encryption programs, the session symmetric key is encrypted with SM9 identity-based cryptographic algorithms and stored in our proposed scheme. SM9 cryptographic algorithms are established on bilincar pairing over elliptic curves. A bilinear pairing is defined as a bilinear map e: $G_1 \times G_2 \rightarrow G_T$, where G_1, G_2 are two additional groups and G_T is a multiplicative group. SM9 encryption algorithm achieves the ID-IND-CCA2 security (Identity Indistinguishability under Adaptively Chosen Ciphertext Attack) and thus can protect the session symmetric key well. When the browser receives the encrypted program, it will firstly decrypt the encrypted session key by using its private key with SM9 encryption algorithm. Thus the session symmetric key can be protected against the malicious attacker's stealing the session key.

Entity Authentication and Tamper-Resistant Mechanism. While lots of JavaScript encryption programs use only encryption techniques, the proposed scheme applies SM9 digital signature technique. Digital signature technique can be applied to implement identity authentication between web server and users. Once the digital envelope is decrypted, one party can verify the source of the received session key. Or not, without digital signature, we cannot avoid impersonating or modifying the data.

Binding Web Server/Users' Identities With Their Key Pairs. In PKI (Public Key Infrastructure) architecture, users' identities is certified with a certificate issued by the trusted CA (Certification Authority). In identity-based cryptography applied in the proposed scheme, users' public keys are their identity information, which binds their public keys with identities. Thus a higher security can be achieved due to the access authorization restricted to specific devices and IP addresses.

In the proposed scheme, we apply SM4 symmetric cryptographic algorithm, SM9 identity-based public key algorithm and SM3 hash algorithm. These cryptographic algorithms received adequate security evaluations in the past years and had been considered among the outstanding cryptographic algorithms.

4 Conclusion

While web technology has been one of the most important network and Internet technologies, its security mechanism receives lots of attention. In this paper, we proposed a security scheme for protecting web application code and sensitive data transmitted between web server and users. By combining symmetric encryption algorithm with identity-based cryptosystem, we can protect the session key in a digital envelope encrypted by SM9 public key encryption algorithm. Furthermore, digital signature is applied to defeat masquerading and data modification. According to a security analysis, the proposed scheme can be used to enhance the security of various web services.

Acknowledgments. This work was supported by project of State Grid Shandong Electric Power Company (No.520627200001).

References

1. Tabrizchi, H., Rafsanjani, M.K.: A survey on security challenges in cloud computing: issues, threats, and solutions. J. Supercomput. **76**(12), 9493–9532 (2020). https://doi.org/10.1007/s11227-020-03213-1
2. Tawalbeh, L.A., Saldamli, G.: Reconsidering big data security and privacy in cloud and mobile cloud systems. J. King Saud Univ. – Comput. Inform. Sci. (in press). Available online 29 May 2019
3. Khan, R., Kumar, P., Jayakody, D.N.K., Liyanage, M.: A survey on security and privacy of 5G technologies: potential solutions, recent advancements, and future directions. IEEE Commun. Surv. Tutorials **22**(1), 196–248 (2020)
4. Mena, D.M., Papapanagiotou, I., Yang, B.: Internet of things: survey on security. Inform. Security J.: Global Perspect. **27**(3), 162–182 (2018)
5. Toch, E., et al.: The privacy implications of cyber security systems: a technological survey. ACM Comput. Surv **51**(2), 1–27 (2018)
6. Awad, M., Ali, M., Takruri, M., Ismail, S.: Security vulnerabilities related to web-based data. Telkomnika Telecommun. Comput. Electron. Control **17**(2), 852–856 (2019)
7. Razzaq, A., Latif, K., Ahmad, H.F., Hur, A., Anwar, Z., Bloodsworth, P.C.: Semantic security against web application attacks. Inform. Sci. **254**(3), 19–38 (2014)
8. Futoransky, A., Gutesman, E., Waissbein, A.: A dynamic technique for enhancing the security and privacy of web applications. In: Proc. of Black Hat USA, Las Vegas (2007)
9. Akhawe, D., Barth, A., Lam, P.E., Mitchell, J., Song, D.: Towards a formal foundation of web security. In: 2010 23rd IEEE Computer Security Foundations Symposium, Edinburgh, UK, pp. 290–304 (2010)

10. Jensen, M., Gruschka, N., Herkenhoener, R.: A survey of attacks on web services. Comput. Sci. Res. Dev. **24**(4), 185–197 (2009)
11. Wassermann, G., Su, Z: Static detection of cross-site scripting vulnerabilities. In: ICSE-ACM/IEEE International Conference on Software Engineering, Germany, pp. 171–180 (2008)
12. Shahriar, H., Zulkernine, M.: MUTEC: mutation-based testing of cross site scripting. In: Proc. of the 5th ICSE Workshop SESS, Vancouver, Canada, pp. 47–53 (2009)
13. Shahriar, H., Zulkernine, M.: Injecting comments to detect JavaScript code injection attacks. In: COMPSACW-IEEE 35th Annual Computer Software & Applications Conference Workshops, IEEE, pp. 104–109 (2011)
14. Stark, E.: From Client-side Encryption to Secure Web Applications. Thesis. Massachusetts Institute of Technology (2013)
15. Dong, X., Chen, Z., Siadati, H., Tople, S., Saxena, P., Liang, Z.: Protecting sensitive web content from client-side vulnerabilities with cryptons. In: CCS 2013: Proceedings of the 2013 ACM SIGSAC Conference on Computer & Communications Security, pp. 1311–1324 (2013)
16. Tople, S., Shinde, S., Chen, Z., Saxena, P.: AUTOCRYPT: enabling homomorphic computation on servers to protect sensitive web content. In: CCS 2013: Proceedings of the 2013 ACM SIGSAC Conference on Computer & Communications Security, pp. 1297–1310 (2013)
17. Wei, R., Zheng, F.Y., Lin, J.Q.: Implementation of a general-purpose cryptography library supporting domestic algorithm with JavaScript. J. Cryptologic Res **7**(5), 595–604 (2020)
18. Cairns, K., Halpin, H., Steel, G.: Security Analysis of the W3C Web Cryptography API. In: Chen, L., McGrew, D., Mitchell, C. (eds.) SSR 2016. LNCS, vol. 10074, pp. 112–140. Springer, Cham (2016). https://doi.org/10.1007/978-3-319-49100-4_5
19. Sleevi, R., Watson, M.: Web cryptography API. Candidate recommendation, IETF (2014)
20. Halpin, H.: The W3C web cryptography API: motivation and overview. In Proceedings of the Companion Publication of the 23rd International Conference on World Wide Web Companion-WWWCompanion 2014, Switzerland, pp. 959–964 (2014)
21. Stark, E., Hamburg, M., Boneh, D.: Symmetric cryptography in Javascript. In: Proceedings of the 2009 Annual Computer Security Applications Conference-ACSAC 2009, Washington, DC, USA, pp. 373–381 (2009)
22. Matasano Security: Javascript cryptography considered harmful. http://www.matasano.com/articles/javascript-cryptography/
23. Daemen, J., Rijmen, V.: AES Proposal: Rijndael. NIST AES Algorithm Submission (1999)
24. GM/T 0002-2012: SM4 block cipher algorithm. Chinese Cryptography Standard (2012)
25. GM/T 0003.1-0003.5-2012: Public key cryptographic algorithm SM2 based on elliptic curves. Chinese Cryptography Standard (2012)
26. National Institute of Standards and Technology: FIPS PUB 186-4: Digital Signature Standard (DSS) (2013)
27. GM/T 0044.1-2016: Identity-based cryptographic algorithms SM9. Chinese Cryptography Standard (2012)
28. GM/T 0004-2012: SM3 cryptographic hash algorithm. Chinese Cryptography Standard (2012)
29. Phung, P., Pham, Huu-Danh., Armentrout, J., Hiremath, P., Tran-Minh, Q.: A user-oriented approach and tool for security and privacy protection on the web. SN Comput. Sci. **1**(4), 1–16 (2020). https://doi.org/10.1007/s42979-020-00237-5
30. Shimamoto, H., Yanai, N., Okamura, S., Cruz, J.P., Okubo, T.: Towards further formal foundation of web security: expression of temporal logic in alloy and its application to a security model with cache. IEEE Access **7**, 74941–74960 (2019)
31. Figueiredo, A., Lide, T., Matos, D., Correia, M.: MERLIN: multi-language web vulnerability detection. In: 2020 IEEE 19th International Symposium on Network Computing and Applications (NCA), IEEE, pp. 1–9 (2020)

32. Caturano, F., Perrone, G., Romano, S.P.: Discovering reflected cross-site scripting vulnerabilities using a multiobjective reinforcement learning environment. Comput. Security 103 (2021)
33. Mohammadi, M., Chu, B., Lipford, H.R.: Automated repair of cross-site scripting vulnerabilities through unit testing. In: 2019 IEEE International Symposium on Software Reliability Engineering Workshops (ISSREW), IEEE (2019)
34. Chow, S., Eisen, P., Johnson, H., Van Oorschot, P.: White-box cryptography and an AES implementation. In: Nyberg, K., Heys, H. (eds.) SAC 2002. LNCS, vol. 2595, pp. 250–270. Springer, Heidelberg (2003). https://doi.org/10.1007/3-540-36492-7_17

IoT Honeypot Scanning and Detection System Based on Authorization Mechanism

Ning Li, Bo Cui, Ziyan Liu(✉), Jinchao Ni, Cheng Zhang, and Hanzhang Kong

Information and Telecommunication Company, State Grid Shandong Electric Power Company, Jinan 250000, China

ziyanliu2021@126.com

Abstract. In this paper, an Internet of Things (IoT) honeypot scanning and detection system is proposed based on an authorization mechanism. For the functional characteristics of different devices existing in the IoT environment, an authorization and authentication system was designed based on device MAC and randomly generated key for requesting permissions from devices with asset management and traffic monitoring. Subsequently, an authorized access network model was constructed between devices and the authorization system, which inveigles the scanning requests from unauthorized devices into the IoT honeypot based on the authorized authentication algorithm. Specifically, an IoT honeypot system was built and a data collection module, a data preprocessing module, and a scan detection module were installed in it to perform detection and output feedback on the traffic redirected to the honeypot. The experimental results show that our designed system can efficiently identify whether the device is authorized or not in the IoT system and successfully detect the illegal scanning requests from non-authorized devices.

Keywords: Authorization · Honeypot · Internet of Things · Scan detection

1 Introduction

In recent years, the rapid development of IoT technology has brought convenience to society. However, it has been accompanied by increasingly serious network security incidents, which seriously threaten the privacy of users and even life and property security. According to the statistics from the Statista website, the total production volume of IoT devices today has exceeded 30 billion [1]. However, many IoT devices ignore product protection when manufacturing and lack necessary security detection mechanisms, which causes many IoT devices are exposed to the open network. Making use of the device vulnerabilities, the hackers could launch large-scale network attacks on the IoT devices, which seriously threaten the security of IoT devices and cause great losses. Currently, with the growing scale of the Internet of Things, hacking techniques are becoming more sophisticated. Therefore, many security attacks, such as malicious code insertion, viruses, and botnet, are endless and cannot be efficiently prevented.

In 2016, attackers used the Mirai [2] virus to launch a massive DDOS botnet attack on the West Coast of the United States, causing almost half of the U.S. network system to be

J. Zeng et al. (Eds.): ICPCSEE 2021, CCIS 1452, pp. 217–228, 2021.
https://doi.org/10.1007/978-981-16-5943-0_18

paralyzed. The main target of this attack was the exposed IoT device routers and cameras that had security vulnerabilities [3]. Hackers manipulated the Mirai virus to frequently change the control nodes, making it difficult for security researchers to control the spread of the virus [4]. Since then, the Mirai virus has spread more rapidly, infecting more different types of devices. It has evolved several times into the more complex variant of the virus, and the number of devices controlled by the Mirai botnet has grown rapidly from 213,000 to 493,000 [5]. Attacks for IoT have seriously affected global cybersecurity and brought a great impact to the world cybersecurity order. In the medical and health fields, IoT devices also have attack vulnerabilities that can be exploited by hackers. For example, medical devices such as insulin pumps and pacemakers, which are currently used on a large scale, are not effectively secured [6]. If IoT attacks against medical devices are launched, the consequences will be unimaginable. In addition, reports of IoT security incidents are rising year by year, with cameras invaded to spy on privacy, smart cars controlled by attackers and thus losing direction and brakes, etc. Overall the security risks of IoT devices have posed a significant threat to users' life and property safety.

The first step of the cyber-attack is to scan the devices in cyberspace to discover ports, services, and possible vulnerabilities in cyberspace and take further cyber-attacks. The user who initiated the scan request can be a normal user, a network administrator or a hacker. Different users scan ports for different purposes. Normal users scan the port to find out whether the target machine has opened specific services they need, so as to carry out legal access operations; The network administrator scans the ports to find the ports, which are not commonly used or unsafe, and then closes these ports to ensure the security of the device; The attackers scan the port to obtain the information of hosts and services, so as to carry out targeted vulnerability scanning and network attacks.

To combat this, this paper designs a scanning detection system based on authorization mechanism. The system opens the scanning request authority to the devices with asset management and traffic monitoring in the IoT network, while other devices only have normal access rights. In this way, the network scanning request is strictly limited to ensure the security of cyberspace and reduce the exposure risk of equipment and services. Besides, we build an IoT honeypot system to perform detection and output feedback on the traffic redirected to the honeypot. Specifically, if an unauthorized device is trapped and attempts to initiate a scanning request, we can redirect this scanning request initiated by the unauthorized device to the honeypot system. In this way, we can locate the device with security risks, and trace the vulnerability of the device according to the attack behavior in the honeypot. While ensuring that other devices are not exposed, the security defense capability of the Internet of things system can be guaranteed by timely repairing the loopholes of the trapped devices.

2 Related Work

2.1 Network Scanning and Detection

Network scanning enables network system managers to identify security problems in computer networks and keep track of the network operation status. In addition, port scanning technology can help network administrators make a scientific evaluation of the risk level of the network.

Li and Chen et al. [7, 8] studied port scanning techniques and models. According to the difference in the active detection of packets sent, they divided port scanning techniques into various categories: TCP scanning, UDP scanning, FTP bounce scanning, etc. Besides, they analyzed the response message content and keywords to identify the corresponding port-specific service and host information. Their research provides theoretical support for the scanning of IoT device hosts [9, 10]. For example, we can classify the endpoint devices by analyzing traffic characteristics. Moore et al. [11] classified IoT devices based on port traffic. However, the accuracy of this method is low. Haffner et al. [12] parsed inter-host interaction messages and extracted keyword segments in the message payload. Then, they identified the application service protocol used by the current session. This detection method hatched a payload-based deep detection technique, which has better detection accuracy. Liu et al. [13] studied the characteristics of mobile internet in-depth and proposed a real-time classification method for mobile traffic using a combination of lightweight flow table and deep packet inspection (DPI) technique. This method can be applied to temporal flow extended by network flow according to the time interval relationship and accurately classifies the traffic by DPI temporal flow feature packets. However, this method requires payload data of deep detection, which violates the user's privacy and is not available to current network traffic classification. Roughan et al. [14] used the K-nearest neighbor to classify network traffic. Radhakrishnan et al. [15] used IAT as a feature to distinguish the devices. This method can be extended to individual classification. Miettinen et al. [16] used the random forest to classify 24 different IoT devices. The experimental results show that this method can achieve an accuracy of more than 90%. An in-depth analysis of the traffic can go further for traffic anomaly detection, and the network state can be expressed by the information entropy characteristics of the messages [17, 18]. When a device encounters a DDOS attack, duplicate IPs will appear in large numbers. Then, the information entropy decreases accordingly. So, the information entropy of IP addresses can reflect the randomness of IP address distribution of current network messages and determine whether the current network is under attack. Based on the CUSUM control chart depiction theory [19, 20], the cumulative sum of deviations between the network characteristic values and the target is calculated, and the state of the network is judged accordingly. However, this method requires many simulation experiments in the early stage, and its applicability and accuracy are slightly inferior.

Based on the existing researches, this paper designs a traffic detection model, including data collection module, data preprocessing module and scanning detection module. The proposed model can characterize the network and time attributes in the traffic, and then judge the anomaly based on the statistical characteristics.

2.2 IoT Honeypot

When the attacked devices appear in the intranet of IoT, hackers will take this as a springboard to carry out the intranet penetration attack. To avoid secondary damage to the system, we need to trace the source of the attacked devices and locate specific vulnerabilities. The honeypot technology of IoT can realize the security monitoring and security investigation for the IoT efficiently. Specifically, the honeypot can be disguised as an device to identify the trend of abnormal traffic in the intranet. Overall, researches on Honeypot Technology of IoT plays an important role in protecting intranet security.

Internet of Things (IoT) technology is developing rapidly, and IoT devices such as webcams, network printers, and smart wearable devices are being used on a large scale. However, due to the generally poor security of IoT devices and the lack of security awareness among enterprises and developers, there are a lot of security vulnerabilities in IoT devices. Accordingly, various types of honeypots have emerged for specific intrusion interfaces. For example, Poeplau et al. [21] implemented USB-based honeypots and verified the capability in detecting and capturing USB viruses. Podhradsky et al. [22] proposed the Bluepot, a honeypot tool for Bluetooth attacks, which can monitor Bluetooth network intrusions and support various intrusion detection such as Blue Bugging, Blue Snarfing, etc. Dowling et al. [23] built a honeypot simulating the ZigBee gateway to discover ZigBee-specific attacks in the network by forging sensitive medical traffic. Erdem et al. [24] designed a home gateway honeypot-HoneyThing and implanted three CVE vulnerabilities in this honeypot to enhance the exploitability of the honeypot. Besides, there are many honeypots for specific communication protocols. For instance, Chakkaravarthy [25] proposed a robust Intrusion Detection Honeypot (IDH) to address the issues that existing security systems like Intrusion Detection and Prevention System (IDPS) and Anti-virus (AV) as a single monitoring agent are complicated and time-consuming, thus fails in ransomware detection. Guarnizo et al. [26] designed the SIPHON, an IoT honeypot. Administrators can combine it with a cloud agent for centralized deployment to solve the problem of deployment and maintenance of the environment. Zhang et al. [27] implement a medium-high interaction honeypot that can simulate a specific series of router UPnP services. It has functions, such as service simulation, log recording, malicious sample download, and service self-check. Saputro et al. [28] created a medium interaction honeypot using Cowrie, which is used to maintain the Internet of Things device from malware attacks or even attack patterns and collect information about the attacker's machine. From the result analysis, the honeypot can record all trials and attack activities, with CPU loads averagely below 6.3%. Ziaie et al. [29] presented a multi-faceted and multi-phased approach to building an IoT honeypot ecosystem. An evolving honeypot ecosystem can attract more interesting attacks that could yield higher utility for research and operation, compared to one that is built once and deployed for a short period of time (e.g., a couple of months). Sedlar et al. [30] presented a prototype implementation of an iteratively improving low-interaction IoT honeypot, based on serving responses of real IoT devices obtained through IoT search engines, as well as devices and services under our own control.

This paper proposes a honeypot scanning detection system based on authorization mechanism. The system redirects traffic according to whether the device has the authority to scan. At the same time, the scanning detection system in the IoT honeypot can detect the scanning request traffic.

3 IoT Honeypot Scanning and Detection System Based on Authorization Mechanism

3.1 Authorization System

The authorization mechanism-based IoT honeypot scanning and detection system proposed in this paper mainly aims at security protection and scanning detection of IoT devices. Compared with ordinary PC, IoT devices are equipped with a miniaturized and simplified operating system with single operation command and function. However, building a unified IoT management control system is difficult because there are various IoT device types and communication protocols, including public protocols, private protocols, etc. Based on the above premise, this paper proposes an IoT scanning control system based on an authorization mechanism, which restricts the scanning operation of specific devices by an authorized way and forwards the scanning requests of unauthorized devices to the honeypot system for specific analysis.

Table 1. Pre-processing modules of the authorization system.

Pre-processing modules of the authorization system
1) Get the MAC address of device A, noted as MAC(A);
2) Generate the key Key(A) using a pseudo-random number generator;
3) Compute the authorization code AK(A) for device A using Hash: $AK(A) = Md5(MAC(A) : Md5(Key(A)))$.
4) Issuing Key(A) to Device A;
5) The authorization system stores pairwise authorization information for device A: $Device(A) = \{MAC(A): AK(A)\}$.

With a certain network scale and number of terminals, IoT terminal devices combine terminals with synergistic, identical, or similar functions into a network, such as a video surveillance system composed of wireless routers, network cameras, network video recorders, etc. Send probe messages to all terminals in the system to determine the normal working status of the device, assess the operational security status of the terminal, and control the network access capability of the terminal device based on the data packets returned through the terminal device. Firstly, the device with asset management and scanning monitoring functions is selected as a pre-authorized device in the intranet system, and the authorization and verification algorithm is combined to ensure the normal operation of this system.

As shown in Table 1, the pre-processing module of the authorization system is executed first to establish the connection between device A and the authorization system. The Key (A) information required for authorization is stored in device A, and the MAC address and authorization code information of device A are also stored in the authorization system.

After that, a scan request operation is initiated by device A to the authorization system, and the execution process is shown in Fig. 1.

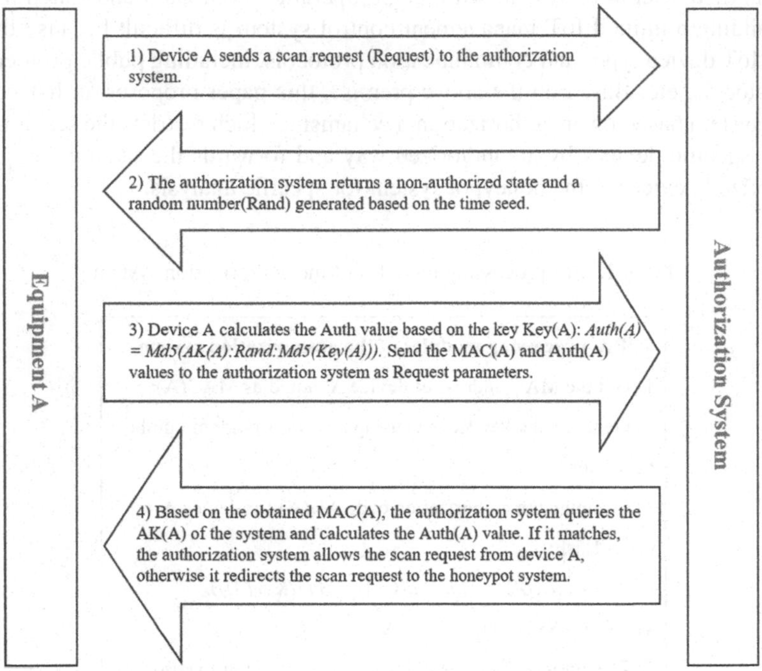
Equipment A
Authorization System
1) Device A sends a scan request (Request) to the authorization system.
2) The authorization system returns a non-authorized state and a random number(Rand) generated based on the time seed.
3) Device A calculates the Auth value based on the key Key(A): Auth(A) = Md5(AK(A):Rand:Md5(Key(A))). Send the MAC(A) and Auth(A) values to the authorization system as Request parameters.
4) Based on the obtained MAC(A), the authorization system queries the AK(A) of the system and calculates the Auth(A) value. If it matches, the authorization system allows the scan request from device A, otherwise it redirects the scan request to the honeypot system.

Fig. 1. Request process of authorization system.

3.2 Honeypot System

If attackers have taken control of an IoT endpoint device in the system, they will scan the network to collect surviving hosts and discover the next node that may be vulnerable. To ensure the concealment and efficiency of such a process, the attacker will use some more complex and stealthy attack strategies, such as scanning space, scanning speed, and jump interval. The scanning space represents the IP address space to be probed. It is usually the address of this network segment or the access network segment, which can be set by the IP address prefix of the network segment. The scanning speed represents the time interval for initiating a scanning request for each target. Each scanning process by the attacker costs a certain amount of money and increases the exposure risk of the attack. The attacker may use more advanced and intelligent scanning methods to dynamically and adaptively update the time interval of scanning based on the scanning results. The jump interval represents the distance between the address currently being probed and the next address to be scanned.

To prevent attackers from obtaining valid information about active hosts and devices in the IoT, the distribution of devices in the IoT is changed by setting up IoT honeypot fake nodes. By this way, we deceive the attacker to attack the fake target and capture the attacker's attack operations and commands, thus increasing the difficulty of the attacker's attack hits and preventing the scanning behavior from exposing the actual system's ports and application services.

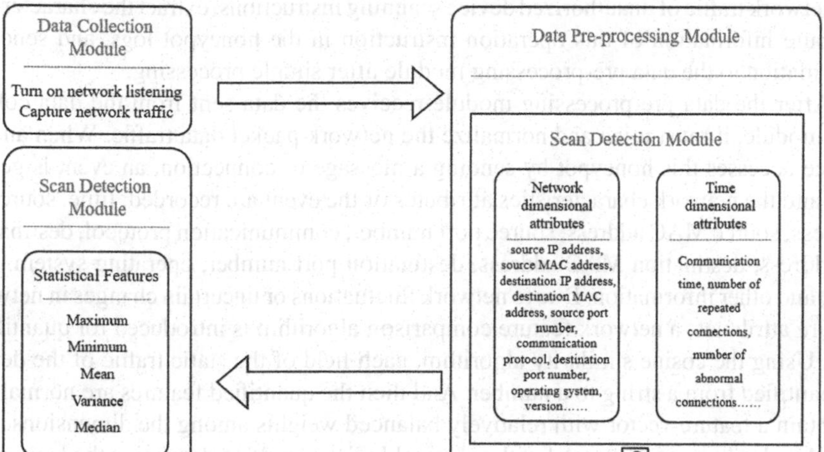

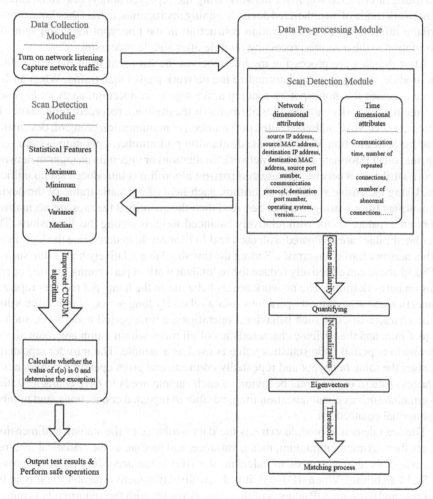

Fig. 2. The modules of the honeypot system.

Combined with the authorization mechanism and based on the real IoT environment architecture, the honeypot system lays virtual honeypot nodes according to the functional characteristics of the IoT devices of the system and simulates the normal network interaction of the IoT devices. According to the results obtained by the authorization system, when the scanning request is initiated by a non-authorized device, the request is redirected to the honeypot system, and the subsequent operation is recorded and analyzed in detail in the honeypot system, and other real IoT devices in the network environment are protected and disconnected from the operation.

The modules and architecture of the honeypot system are shown in Fig. 2. The honeypot system includes a data collection module, a data pre-processing module, and a scan detection module.

The main purpose of the data collection module is to obtain a large amount of network data traffic information within the network using the deployed honeypot system, capture the network traffic of unauthorized device scanning instructions, extract the characteristic attribute information of this operation instruction in the honeypot logs, and send the information to the data pre-processing module after simple processing.

After the data pre-processing module receives the data sent from the data collection module, it must mine and normalize the network packet data traffic. When an IoT device accesses this honeypot by sending a message or connection, an event is generated and the network characteristics attributes of the event are recorded: time, source IP address, source MAC address, source port number, communication protocol, destination IP address, destination MAC address, destination port number, operating system, version, and other information. Due to network fluctuations or uncertain changes in network feature attributes, a network feature comparison algorithm is introduced for quantification. Using the cosine similarity algorithm, each field of the static traffic of the device is quantified from a string to a number. And then the quantified features are normalized to obtain a feature vector with relatively balanced weights among the dimensions. The calculated values are compared with each real IoT threshold to determine the legitimacy of this network traffic fingerprint. Setting the threshold to a relatively high value such as 90% and above can effectively reduce the redundant work of performing filtering operations on network traffic. The network access behavior in the honeypot requires repeated connection and scan request operations over a relatively long period of time. According to the characteristics of such behavioral operations, a time period node is set, such as set to 1 min, and the attribute characteristics of all times within 1 min are counted, and this one time period node statistics value is used as a sample. The intruder repeatedly accesses the same honeypot and repeatedly connects and gives operation commands is an access pattern of abnormal behavior, so each sample needs to have characterization information such as communication time, number of repeated connections, and number of abnormal connections.

The scan detection module extracts the data attributes of the statistical dimension, selects the maximum, minimum, mean, variance, and median as the statistical features, and uses 5 feature sequences to calculate the r(O) value according to the improved CUSUM algorithm. When r(O) = 0, it is determined that there is an abnormality in the network, and the source IP address of the network packet with the abnormality is output. And further locate the IoT device that initiates the network scan request and cut off other IoT terminals connected to this device. Based on the command operation request of this IoT device in the honeypot system, security traceability is performed to find the source of security hazards of this device.

In general, any interaction with a honeypot is likely to be unauthorized or malicious, and any request to establish a connection with a honeypot is most likely to be a probe, attack, or post-penetration attack. Any traffic passing through the honeypot is highly suspicious, but not all traffic passing through the honeypot is a request to launch an attack, such as an unintentional request to the honeypot in a casual situation. In this paper, when the honeypot system is designed, a network scanning access authorization mechanism is added, and according to the characteristics of IoT terminal network connection, the source IP address is directed to the honeypot system for network request connections from unauthorized IoT terminals to prevent the occurrence of attack misses.

4 Experiments

In this paper, we designed the network topology shown in Fig. 3 for testing, where device A has obtained authorization, and devices B and C are unauthorized devices in the same network segment.

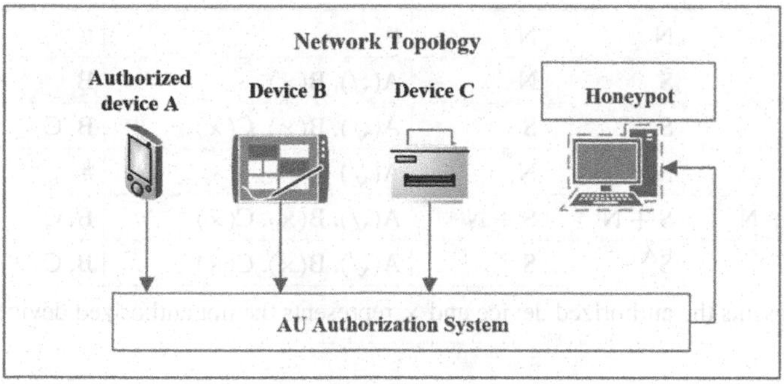

Fig. 3. Net topology.

First, we write two script files Script_Normal and Script_Scan, where the first one sends normal network requests and the second one sends network scan requests. We start the authorization system and honeypot system, and then monitor and capture the traffic of devices A, B and C.

We designed six groups of experiments to test the system. According to the requirements of each group of experiments, the running time of each group is about 20 min. The detailed experiments is as follows:

① EXP_1: Device A, device B, and device C all send normal network requests, but no scanning requests;

② EXP_2: Device A and B send scanning request, and device C send normal network requests;

③ EXP_3: Device A, device B, and device C all send scanning requests, but do not send normal network requests;

④ EXP_4: Device A sends scanning request, device B and device C send normal network requests;

⑤ EXP_5: Device A, device B, and device C send normal network requests and scanning requests;

⑥ EXP_6: Device B disguises as the MAC address of device A, while device A, device B, and device C all send scanning requests.

During the experiment, the device number of the authorization scanning request of the authorization system and the device number of the abnormal traffic captured by the IoT honeypot system are recorded.

The experimental results are shown in Table 2. N represents that Script_Normal sends normal requests; S represents that Script_Scan sends scanning requests; S + N represents that all of Script_Scan and Script_Normal send normal and scanning requests in the same time. S^A represents that Script_Scan sends scanning requests through masquerading as the MAC address of device A.

Table 2. Results of experiment.

	Device A	Device B	Device C	Is the device authorized?	Device number of abnormal scanning
EXP_1	N	N	N	#	#
EXP_2	S	S	N	A(√), B(×)	B
EXP_3	S	S	S	A(√), B(×), C(×)	B, C
EXP_4	S	N	N	A(√)	#
EXP_5	S + N	S + N	S + N	A(√), B(×), C(×)	B, C
EXP_6	S	S^A	S	A(√), B(×), C(×)	B, C

Note: √ represents the authorized device and × represents the not authorized device

The experimental results show that our system can effectively and accurately identify the authorization of the device sending scanning requests. As long as the key of the device is not disclosed, it is difficult for the attacked device to successfully bypass the authorization system for illegal scanning request even if forging the MAC address of other authorized devices. In this experiment, we design six different contract combinations to simulate the complex network request. Overall, the experimental results indicate that the network feature extraction and detection algorithm used in this paper can effectively and accurately detect network anomalies and scanning attacks.

5 Conclusion

The IoT honeypot scanning and detection system based on the authorization mechanism proposed in this paper designs an authorization mechanism based on the network connection and access characteristics of IoT devices, and constructs a set of authorization authentication system based on the MAC address of the authorized device and the random key generated by the authorization system for specifically authorized devices within the network, which guarantees the access rights of different operations of the devices.

We design six different contract combinations to simulate the complex network request to evaluate the IoT honeypots deployed in the IoT network. Network feature extraction and detection algorithms are used to efficiently detect network anomalies and can detect scanning attacks and make abnormal responses relatively well. However, due to the limitation of experimental conditions, we lack enough experimental samples. The next step will be to construct more perfect experimental conditions and experimental samples, design more accurate detection algorithms, and improve and perfect the experiments.

Acknowledgments. This work was supported by project of State Grid Shandong Electric Power Company (No.520627200001).

References

1. Statista Inc.: Internet of Things (IoT) connected devices installed base worldwide from 2015 to 2025 (in billions) [EB/OL] [2017-05-30]. https://www.statista.com/statistics/471264/iot-number-of-connected-devices-worldwide/
2. Antonakakis, M., et al.: Understanding the Mirai Botnet. In: Proceedings of the USENIX Security Symposium, pp. 4–18, August 2017
3. Xu, Y., Koide, H., Vargas, D.V., et al.: Tracing MIRAI Malware in Networked System. In: 2018 Sixth International Symposium on Computing and Networking Workshops (CANDARW). IEEE Computer Society, pp. 18–34 (2018)
4. Marzano, A., et al.: The Evolution of Bashlite and Mirai IoT Botnets. In: 2018 IEEE Symposium on Computers and Communications (ISCC). IEEE, pp. 20–24 (2018)
5. Kolias, C., et al.: DDoS in the IoT: Mirai and other Botnets. Computer **50**(7), 80–84 (2017)
6. Goodman, M.: Hacking the Human Heart[EB/OL]. [2017-04-24]. http://bigthink.com/future-crimes/hacking-the-human-heart
7. Ruimin, L.: Network Scanning Technology Reveals: Principle, Practice and Implementation of Network Scanner. Mechanical Industry Press, Beijing (2012)
8. Nanping, C.: Network scanner and the design principle of the model study. Softw. Guide J. **9**(11), 134–136 (2010)
9. Kohno, T., Broido, A., Claffy, K.C.: Remote physical device fingerprinting. In: Proceedings of 2005 IEEE Symposium on Security and Privacy. IEEE, pp. 211–225 (2005)
10. Polcak, L., Jirasek, J., Matousek, P.: Comment on "remote physical device fingerprinting." IEEE Trans. Depend. Secure Comput. **11**(5), 494–496 (2014)
11. Moore, A.W., Papagiannaki, K.: Toward the accurate identification of network applications. Lect. Notes Comput. Sci. **3431**, 41–54 (2005)
12. Haffner, P., et al.: ACAS: automated construction of application signatures. In: Proceedings of ACM Workshop on Mining Network Data, Minenet 2005. Philadelphia, USA. ACM, pp. 197–202 (2005)
13. Yi, L., Tian, S., Lejian, L.: A real-time mobile traffic classification approach based on timing sequence flow. Trans. Beijing Inst. Technol. **38**(5), 537–544 (2018)
14. Roughan, M., et al.: Class-of-service mapping for QoS: a statistical signature-based approach to IP traffic classification. In: Proceedings of the 4th ACM SIGCOMM Conference on Internet Measurement. ACM, pp. 135–148 (2004)
15. Beyah, R., et al.: GTID: a technique for physical device and device type fingerprinting. IEEE Trans. Depend. Secure Comput. **22**(7), 112–120 (2015)

16. Miettinen, M., et al.: IoT sentinel demo: automated device-type identification for security enforcement in IoT. In: Proceedings of International Conference on Distributed Computing Systems. IEEE, pp. 2177–2184 (2017)
17. Giotis, K., et al.: Combining OpenFlow and sFlow for an effective and scalable anomaly detection and mitigation mechanism on SDN environments. Comput. Netw. **62**(62), 122–136 (2014)
18. Mousavi, S.M., St-Hilaire, M.: Early detection of DDoS attacks against SDN controllers. In: Proceedings of 2015 International Conference on Computing, Networking and Communications (ICNC). IEEE (2015)
19. Conti, M., Gangwal, A., Gaur, M.S.: A comprehensive and effective mechanism for DDoS detection in SDN. In: Proceedings of the 13th IEEE International Conference on Wireless and Mobile Computing, Networking and Communications. IEEE (2017)
20. Xiulei, W., et al.: Defending DDoS attacks in software-defined networking based on legitimate source and destination IP address database. IEICE Trans. Inf. Syst. **99**(4), 850–859 (2016)
21. Poeplau, S., Gassen, J.: A honeypot for arbitrary malware on usb storage devices. In: 2012 7th International Conference on Risks and Security of Internet and Systems (CRiSIS'12), pp. 1–8 (2012)
22. Podhradsky, A., Casey, C., Ceretti, P.: The bluetooth honeypot project: measuring and managing bluetooth risks in the workplace. Int. J. Interdisciplinary Telecommun. Network. **4**(3), 1–22 (2012)
23. Dowling, S., Schukat, M., Melvin, H.: A ZigBee honeypot to assess IoT cyberattack behavior. In: 28th Irish Signals and Systems Conference (ISSC'17), pp. 1–6 (2017)
24. Kara, M., İkinci, A.: HoneyThing: Nesnelerinİnterneti icin Tuzak Sistem. In: 8th International Conference on Information Security and Cryptology (ISCTurkey'15), pp. 258–264 (2015)
25. Chakkaravarthy, S.S., et al.: Design of intrusion detection honeypot using social leopard algorithm to detect IoT ransomware attacks. IEEE Access **8**, 169944–169956 (2020)
26. Guarnizo, J.D., et al.: SIPHON: Towards scalable high-interaction physical honeypots. In: 3rd ACM Workshop on Cyber-Physical System Security (CPSS'17), pp. 456–462 (2017)
27. Zhang, W., et al.: An IoT honeynet based on multiport honeypots for capturing IoT attacks. IEEE Internet Things J. **7**(5), 3991–3999 (2019)
28. Saputro, E.D., Purwanto, Y., Ruriawan, M.F.: Medium interaction honeypot infrastructure on the internet of things. In: 2020 IEEE International Conference on Internet of Things and Intelligence System (IoTaIS). IEEE, pp. 98–102 (2021)
29. Ziaie Tabari, A., Ou, X.: A multi-phased multi-faceted IoT honeypot ecosystem. In: Proceedings of the 2020 ACM SIGSAC Conference on Computer and Communications Security, pp. 2121–2123 (2020)
30. Sedlar, U., Južnič, L.Š., Volk, M.: An iteratively-improving internet-of-things honeypot experiment. In: 2020 International Conference on Broadband Communications for Next Generation Networks and Multimedia Applications (CoBCom). IEEE, pp. 1–6 (2020)

Privacy Protection Model for Blockchain Data Sharing Based on zk-SNARK

Yang Li[1,2], Guangzong Zhang[1(✉)], Jianming Zhu[1,2], and Xiuli Wang[1,2]

[1] School of Information, Central University of Finance and Economics, Beijing 100081, China
[2] Engineering Research Center of State Financial Security, Ministry of Education, Central University of Finance and Economics, Beijing 102206, China

Abstract. In the era of big data, data sharing and communication play a crucial role. The blockchain data sharing model based on ciphertext policy attributed-based encryption (CP-ABE) is an existing solution to data sharing. However, it puts the access policy and attributes directly on the blockchain, so every one in the blockchain can access these access policies and attributes, which will cause privacy leakage. To solve this problem, a privacy protection model based on zk-SNARK is proposed in this paper. The blockchain double-chain structure was applied in this model, and the fine-grained access control of data sharing was realized based on CP-ABE scheme. At the same time, zk-SNARK technologies were adopted to protect sensitive access policies and sensitive attributes from disclosure, effectively defending the privacy of users when data sharing in blockchain.

Keywords: Blockchain · Data sharing · CP-ABE · zk-SNARK

1 Introduction

With the advent of the era of massive data, various industries have proposed different blockchain data sharing solutions based on business scenarios. Many schemes are encrypted by the CP-ABE scheme to meet fine-grained access control. However, this may cause some of the sensitive information leaks. For example, when the access policy involves the income attributes of the company, the attribute of participation of the project, or the expense attributes, etc. If the competitor on the same blockchain gets this information, it may have a negative impact on the company. Therefore, during data sharing in blockchain, how to meet the fine-grained access control requirements while protecting the privacy of user nodes is worth studying.

Zero-knowledge proof and attribute-based encryption technology have a wide range of applications in the field of data sharing privacy protection. Zero-knowledge proof is a technology that has developed very rapidly in recent years. This technology has been widely used in blockchain such as data privacy protection of the blockchain, data compression on the blockchain, and offline expansion of the blockchain. Attribute-based encryption also has a wide range of applications in privacy protection. Especially ciphertext policy attributed-based encryption scheme, CP-ABE scheme uses an access structure

J. Zeng et al. (Eds.): ICPCSEE 2021, CCIS 1452, pp. 229–239, 2021.
https://doi.org/10.1007/978-981-16-5943-0_19

to encrypt ciphertext, and an attribute set to generate user keys, so that only users with attributes specified by the access policy can access.

At present, there are few researches on the privacy protection of data sharing on the blockchain. The zero-knowledge proof technology zk-SNARK has the characteristics of proof-shortness and zero-knowledge. So it can be used for sensitive access policy and sensitive attributes hiding in data sharing. This paper proposes a privacy protection model for blockchain data sharing that applies zero-knowledge technology zk-SNARK to the field of ciphertext policy attribute-based encryption. Using this scheme can effectively protect the privacy of the sensitive attributes and sensitive access policy of the CP-ABE scheme and realize the fine-grained access control.

2 Related Works

Data Sharing between company is an important application scenario in the era of big data, some researchers have proposed appropriate scheme in this scenario. Wang Xiuli et al. [20] have studied the application of attribute-based encryption in combination with the blockchain system, they applied this model to data sharing between enterprises. However, their model directly put access policies and attributes on the blockchain, which may cause privacy risks since some access policies are very sensitive. If privacy protection is not carried out, this may leak the privacy of users who share data [21].

Zero-knowledge proof is an effective technology to solve the privacy leakage problem. The current technologies that realize zero-knowledge proof mainly include zk-SNARK [17], zk-STARK [2], Bullet proof [5], etc. Among which zk-SNARK is due to its short proof, it is easy to verify and widely used. The current typical technical blochchain solution based on zk-SNARK technology is Zcash [3]. In addition, PLONK [10] and Sonic [19] schemes are also based on zk-SNARK technology. These two schemes achieve complete simplicity and linearity in terms of circuit and proof size. Agrawal et al. [1] studied the design of NIZKs, which is used to combine compound sentences of algebra and arithmetic sentences in any way, and provided a framework to prove sentences composed of functions composed of and, ors, algebra and algorithms. When using this method to generate anonymous credentials, Which effectively reduces the amount of calculation of the prover, but increases the size of the proof. Groth et al. [12] proposed an updatable CRS model, so that the public reference string does not need to be set multiple times.

There are also a lot of researches using the zk-SNARK technology to solve privacy problem. Khalil et al. [15] used zk-SNARK technology to provide a commitment chain that can prove correct operation design to improve the general blockchain model, and better improve the scalability of the blockchain. Guan Zhangshuang [13] designed a blockchain trading system based on account model based on zk-SNARK zero-knowledge proof technology, which can effectively protect the transaction privacy of users. Wei Xiaosong [22] designed an anonymous electronic survey system based on NSE-NIZK technology, which can realize the anonymity of questionnaire fillers. Li Gongliang [16] proposed a block chain privacy protection method based on zero knowledge proof, which satisfied zero knowledge when verifying the block chain amount. Fu Yonggui [9] studied the information sharing mechanism and management mode of block chain based on the scene of supply chain. Zhang Siliang [23] designed a traceable blockchain privacy protection scheme based on zero-knowledge proof Bulletproof technology and ring signature.

Attribute-based encryption is widely used in data sharing scheme since it can realize fine-grained access control. Sahai et al. [18] first proposed a cryptosystem based on identity encryption, which is the basis for subsequent derivative attribute-based encryption. Brent waters [4] proposed an attribute-based encryption method based on a ciphertext strategy, which uses an access strategy to encrypt the plaintext and uses an attribute set to decrypt. Goyal et al. [11] proposed a fine-grained access control scheme, and many subsequent attribute-based encryption schemes are based on this scheme. Chen et al. [7] proposed an Inner Product Predicate Encryption (IPE) scheme with adaptive security and complete attribute hiding. Katz et al. [14] studied the application of predicate encryption in attribute-based encryption. Many current strategy hiding and attribute hiding are based on this technology. Camenisch et al. [6] proposed a hidden access control strategy scheme. Users can query the database to find out whether they meet the access control strategy, and the database cannot understand the ciphertext that the user decrypts. Cui et al. [8] proposed a CP-ABE scheme that partially hides the access strategy. This scheme has better access efficiency than the scheme that completely hides the access strategy.

In general, scholars in recent years have focused on applying zk-SNARK to different business scenarios and improving the efficiency of zero-knowledge proof technology. Besides, these researches are optimizing the size of the proof, expanding the scope of proof, and applying the zero-knowledge proof technology to appropriate scenarios, and researching attribute-based encryption to solve blockchain privacy issues and so on.

However, there are few researches and models that solve the problem of exposing the sensitive attributes and access polices directly on the blockchain. Most of them are the application of attribute-based encryption combined with the blockchain model. The zero-knowledge proof technology zk-SNARK is increasingly used in the construction of blockchain models due to its good proof performance. Combining the zero-knowledge proof technology with the ciphertext policy attribute-based encryption scheme can effectively protect sensitive access policies and attributes on blockchain, while meeting the requirements of fine-grained access control. In this paper, we proposed a privacy protection model to solve the privacy problem of sensitive attributes and access policies on blockchain based on zk-SNARK and CP-ABE.

3 Preliminaries

The privacy model in this paper is mainly based on zk-SNARK and CP-ABE to realize privacy protection in the process of blockchain data sharing.

3.1 zk-SNARK

zk-SNARK is a technology of zero-knowledge proof, which has the characteristics of proof-shortness and zero-knowledge. The full name of zk-SNARK technology is Zero Knowledge Succinct Non-interactive ARgument of Knowledge. Generally speaking, the basic structure of zk-SNARK includes knowledge, NP problems, Rank 1 Constraint System, Quadratic Arithmetic Problem, arithmetic circuit verification, polynomial verification, polynomial constraints and so on. zk-SNARK first needs to generate a public parameter, that is, a public character reference string, and then send the parameters to the prover and the verifier. The prover uses the public parameters to generate the proof, and then sends the proof to the verifier. The verifier can use the public parameters and the sent proof to verify whether the prover has knowledge. But in this verification process, the verifier does not get any information about the knowledge proven, which is zero knowledge. The proof generated by the prover that needs to be verified is very concise, and the verifier can verify it in a very short time to determine whether it is correct.

zk-SNARK can be described to four steps. First, set up the public parameter pp_z. Second, The knowledge required to prove is converted into a circuit C, and then the circuit C is converted into a proof of common parameters,. Third, Use the PK, circuit C, the knowledge K to generate the proof π. At last, Verify the zero-knowledge proof. If result equals to 1, this proof is legal, if not, it is illegal. In this process, circuit C, pk_z, vk_z are in public, and everybody can verify whether the π is legal or not.

$$Setup\left(1^{\lambda}\right) \rightarrow pp_z \tag{1}$$

$$KeyGen(C) \rightarrow (pk_z, vk_z) \tag{2}$$

$$GenProof\,(pk_z, C, K) \rightarrow \pi \tag{3}$$

$$VerProof\,(vk_z, C, \pi) \rightarrow result \tag{4}$$

$$Pr\left\{VerProof\,(vk_z, C, \pi) \rightarrow 1 \middle| \begin{array}{c} KeyGen(C) \rightarrow (pk_z, vk_z) \\ GenProof\,(pk_z, C, K) \rightarrow \pi \end{array}\right\} = 1 \tag{5}$$

3.2 CP-ABE

In CP-ABE scheme, our scheme is based on BSW07 scheme. Access policy is used to encrypt plaintext into ciphertext, and attribute set is used to generate key, so that data owners can make unique access structure according to their own data, so that members with attributes can access. Only the properties of the key that meet the ciphertext access structure can be decrypted.

In the CP-ABE scheme, we need to construct Bilinear pairing. Let $\mathbb{G}_0$ be a addition cyclic group of prime order q, let $\mathbb{G}_t$ be a factorial cyclic group of prime order q, let g be a generator of $\mathbb{G}_0$. In addition, let $\hat{e} : \mathbb{G}_0 \times \mathbb{G}_0 \rightarrow \mathbb{G}_1$ denote the bilinear map. In addition, $\mathbb{G}_0 \times \mathbb{G}_0 \neq 1_{\mathbb{G}_t}$, $\hat{e}(ag, bg) = \hat{e}(g, g)^{ab}$.

Access structure is the key to attribute base encryption. Currently, there are mainly the following three access structures. In the AND gate access structure, all threshold nodes are AND. The threshold node in the access tree is AND, OR, which is converted to polynomial in the design. The LSSS matrix transforms the threshold nodes in the access policy into matrix form. In this model we adopts the access tree as the access policy structure.

3.3 Blockchain Data Sharing Model

The blockchain data sharing scheme adopts the double chain structure in the alliance chain, one chain is a public chain, and the other is a private chain, as it is shown in Fig. 1. The public chain performs zero-knowledge verification and data storage address storage, and the public chain user nodes are first-level nodes. The private chain is a data chain that stores data shared by users. Only when data visitors pass the zero-knowledge verification of the public chain smart contract can they access the private chain. All data access records are recorded on the public chain. The verification of the proof of the user attribute set is carried out through the smart contract of the public chain. When constructing the block chain, set up the authoritative user node, and the user's property is set by the authoritative user node.

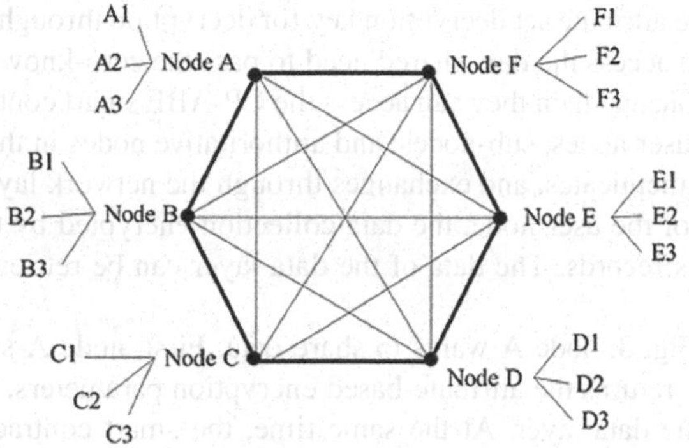

Fig. 1. Alliance chain

4 Privacy Protection Model for Blockchain Data Sharing

4.1 Privacy Protection Model

The model in this article is based on the blockchain model, the BSW07 scheme and the zk-SNARK scheme. The blockchain data sharing system adopts the structure of alliance chain, which is open to specific user nodes, and data can be shared among user nodes. User nodes can be divided into first-level nodes, second-level sub-nodes, and third-level sub-nodes. The child node can apply for the authorization to access the data of the upper node through the upper node, and the application and authorization are implemented based on the zero-knowledge proof technology, as it is shown in Fig. 2. The data of each node is encrypted by CP-ABE, and users who meet the attributes of the access policy can access and decrypt the data. All the user nodes on the block chain can verify and check the proof.

This model mainly has three layers, namely the smart contract layer, the network layer, and the data layer. The smart contract layer includes zero-knowledge proof smart contracts and ciphertext policy attribute-based encryption smart contracts. The zero-knowledge proof smart contract based on zk-SNARK technology is used to receive the

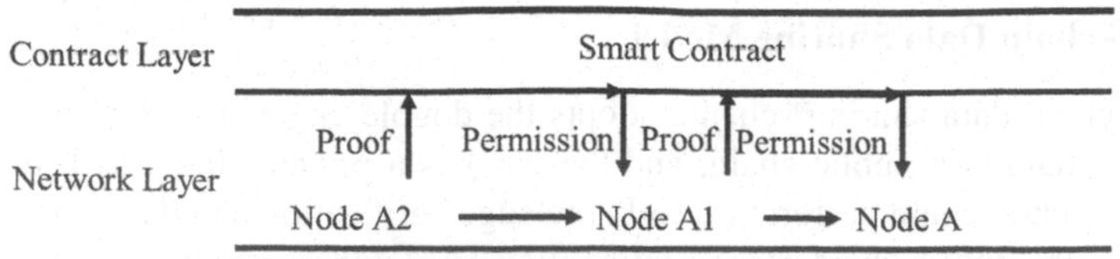

Fig. 2. User node structure

access strategy submitted by the user node to generate zero-knowledge proof parameters, and to verify the zero-knowledge proof file submitted by the node, and return the result to the node and broadcast it to the whole chain.

The attribute-based encryption smart contract plays the role of the attribute authority center. The user submits the attribute set to it, and the attribute-based encryption smart contract returns the attribute set decryption key for decryption through a secure channel. Users who want to access the data shared need to pass the zero-knowledge proof smart contract verification, and then they can access the CP-ABE smart contract. The network layer contains all user nodes, sub-nodes, and authoritative nodes in the blockchain, and communicates, authenticates, and exchanges through the network layer. The data layer contains the data of the user node, the data collection encrypted by the attribute base, and all data access records. The data of the data layer can be retrieved through smart contracts.

As shown in Fig. 3, node A wants to share data. First, node A submits the access structure and data, returns the attribute-based encryption parameters, encrypts the data, and places it in the data layer. At the same time, the smart contract generates zero-knowledge proof proof parameters based on the access structure. Node B who wants to obtain the data of node A obtains the zero-knowledge proof parameters from the smart contract layer and uses its own attribute set to generate the zero-knowledge proof. After the verification is passed, the smart contract layer distributes the private key to node B, and node B can use the private key Access the data shared by node A and broadcast the access record to the entire blockchain. In the whole process, the attribute set of node A and the access policies of node B have been protected, other nodes can get nothing about the attributes and the access policies. No nodes will know the sensitive attributes of node A since it only submit the proof of attributes to the blockchain and everybody can verify it. Access policy is used to encrypted the data, if the sensitive attributes are under protected, the access policy will be protected.

4.2 Privacy Protection Algorithms

In this model, we use zk-SNARK to design a general circuit and automatically generate the corresponding circuit when the user defines the access strategy, such as (Revenue > 5 million and Industry = Finance) or (Staff number > 1000), which is transformed into polynomial verification, generating certification parameter and verification parameter, and encrypting the data using an access strategy by CP-ABE.

Let's give an example, assume access policy (A or B) and (C or D). In this access policy, there are four kinds of user node can access, AC, AD, BC, BD. However, ABCD

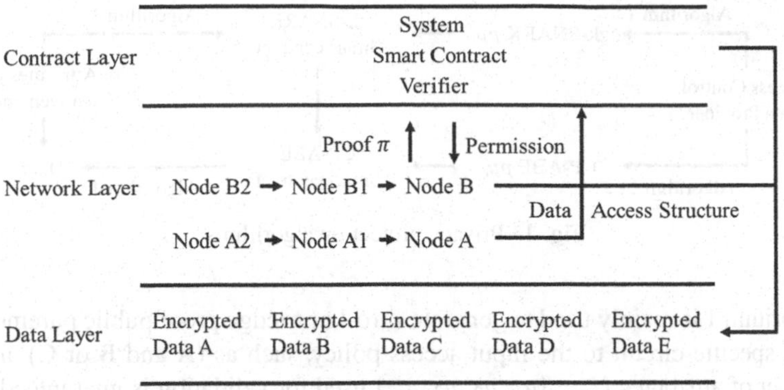

Fig. 3. Privacy protection model for blockchain data sharing

are sensitive attributes such as income expense, user node do not want to make this public. So we carry out zero-knowledge proof to protect this sensitive information.

First, we initialize the blockchain and establish a double-chain blockchain structure. We set up three level user node from first level to third level and set up the authority node. The authority nodes have the right to set up attributes for each level nodes, and these attributes are updatable. Second, we initialize zk-SNARK smart contract and ciphertext policy attribute-based encryption smart contract.

For CP-ABE smart contract, when the node user data provider submit the access policy, it first generates the master key (6), then uses master key generates public key (7), then generates the serect key (8) to the data provider.

$$MK = \left(\beta, g^{\alpha}\right) \tag{6}$$

$$PK = \mathbb{G}_0, g, h = g^{\beta}, f = g^{1/\beta}, e(g, g)^{\alpha} \tag{7}$$

$$SK = \left(D = g^{(a+r)/\beta}, \forall j \in S : D_j = g^r \cdot H(j)^{r_j}, D_j' = g^{r_j}\right) \tag{8}$$

For zk-SNARK smart contract. Given secuirty parameter, initializes the algorithm settings, generates PK_z, VK_z. The data demander uses proving key and attribute set to generate proof. The blockchain smart contract layer uses the verification key to verify the proof. The data provider uses the access policy to encrypt data. If the smart contract verify the proof is legal, the data demander will get the sercet key to decrypt the encrypted data.

In order to achieve the above requirements, we design the following four algorithms, and their relationship is shown in Fig. 4. The data provider inputs access policy to Algorithms 1 and 2, then Algorithms 1 and 2 generate parameters to the smart contract layer. The data demander obtains the zero-knowledge proof parameters through the smart contract layer and generates the proof with his own attribute set U. Then data demander submit the proof to the zk-SNARK smart contract layer for verification through algorithm 3. When the proof pass verification, the data demander can access and decrypt the data through algorithm 4.

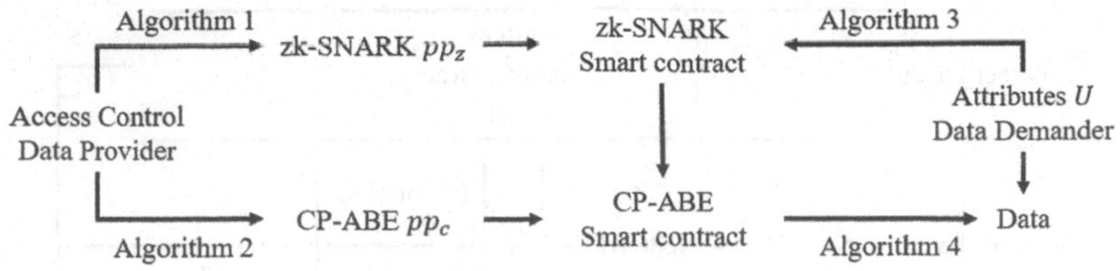

Fig. 4. Privacy protection algorithms

Algorithm 1 is mainly used to generate zero-knowledge proof public parameters. By setting a specific circuit to the input access policy, such as (A and B or C) and (D or F). The set of attributes $U = \{u_1, u_2, u_3, \ldots\}$ used for validation is guaranteed to meet the access policy requirements. The algorithm is part of the zk-SNARK smart contract, which is executed once when the data provider submits the access policy. The detailed process is as follows.

Algorithm 1 zk-SNARK Smart contract Setup

Inputs: Access Policy, A security parameter $\boldsymbol{\lambda}$
Outputs: zk-SNARK public parameters $\boldsymbol{pp_z}$
$\boldsymbol{pp_z}$ = **Function** Setup ($\mathbf{1}^{\boldsymbol{\lambda}}$)
For each policy $\boldsymbol{p}$ **in** Access Policy:
  Construct a circuit $\boldsymbol{C_p}$
  Compute $(\boldsymbol{pk_z}, \boldsymbol{vk_z}) = \boldsymbol{KeyGen}(\boldsymbol{C_p})$
Set $\boldsymbol{PK_z} = \cup \boldsymbol{pk_z}$ and $\boldsymbol{VK_z} = \cup \boldsymbol{vk_z}$
Output $\boldsymbol{pp_z} = (\boldsymbol{PK_z}, \boldsymbol{VK_z})$

Algorithm 2 is used to generate CP-ABE public parameter list, which is mainly used for attribute base encryption of data. This algorithm is part of the CP-ABE smart contract and is executed once when the data provider submits the access policy. The detailed process is as follows.

Algorithm 2 CP-ABE Smart contract Setup

Inputs: Access Control, A security parameter $\boldsymbol{\kappa}$
Outputs: CP-ABE parameters $\boldsymbol{pp_c}$
$\boldsymbol{pp_c}$ = Function Setup ($\mathbf{1}^{\boldsymbol{\kappa}}$)
For parameter $\mathbf{1}^{\boldsymbol{\kappa}}$:
  Compute $\boldsymbol{MK_c} = (\boldsymbol{\beta}, \boldsymbol{g^{\alpha}})$
  Compute $\boldsymbol{PK_c} = \mathbb{G}_0, \boldsymbol{g}, \boldsymbol{h} = \boldsymbol{g^{\beta}}, \boldsymbol{f} = \boldsymbol{g^{1/\beta}}, \boldsymbol{e(g,g)^{\alpha}}$
Output: $\boldsymbol{pp_c} = (\boldsymbol{PK_c})$

Algorithm 3 is used to generate and verify the proof in zk-SNARK. The data visitor obtains the proof parameters from the smart contract layer of zk-SNARK, and then input the attribute set and the zk-SNARK proof parameters to generate the proof, and submit the proof to the smart contract layer of zk-SNRAK for verification. After the smart contract layer verification is passed, the data visitor will get the right to access CP-ABE smart contract. The detailed process is as follows.

Algorithm 3 Proof Generation and Verification

Inputs: Attributes $\boldsymbol{U}, \boldsymbol{PK_z}, \boldsymbol{VK_z}$
Outputs: Proof, Permission or Denial
For $\boldsymbol{pk_z}$ **in** $\boldsymbol{PK_z}$:
  Compute $\boldsymbol{proof_z = Genproof(Attributes\ U, pk)}$
  $\boldsymbol{Proof_z = \cup\, proof_z}$
For $\boldsymbol{proof_z}$ **in** $\boldsymbol{Proof_z}$:
  Compute $\boldsymbol{Verproof(proof_z, vk_z)}$
  If $\boldsymbol{Verproof(proof_z, vk_z)}$ succeed:
    Give right to access CP-ABE Smart Contract
  If $\boldsymbol{Verproof(proof_z, vk_z)}$ fail:
    Deny access request
Output: Permission or Denial

Algorithm 4 is mainly used for data encryption and decryption. When the data provider access policy is submitted to the CP-ABE smart contract layer, it will obtain a PK that can be used for property base encryption of the data. After being verified by the zk-SNARK smart contract layer, the data visitor can submit the attribute set to the CP-ABE smart contract layer to obtain the SK, which is used to decryption the data stored in the private chain. The detailed process is as follows.

Algorithm 4 Data Encryption and Decryption

Inputs: Access policy, Permission, Attributes $\boldsymbol{U}$, Data
Outputs: Encrypted Data
For Data, $\boldsymbol{PK}$, Access policy:
  Compute $\boldsymbol{Encrypted\ data = Encrypt(PK, Access\ Policy, Data)}$
For Permission, $\boldsymbol{MK}$ and Attributes $\boldsymbol{U}$:
  Compute $\boldsymbol{SK_c = \left(D = g^{(a+r)/\beta}, \forall j \in S: D_j = g^r \cdot H(j)^{r_j}, D_j' = g^{r_j}\right)}$
For $\boldsymbol{SK}$ and Encrypted data:
  Compute $\boldsymbol{Data = Decrypt(SK,\ Encypted\ Data)}$
Output: Data

The above algorithm 1 to 4 realizes the protection of sensitive access policy and sensitive attributes in this process of blockchain data sharing. It also meets the requirements of confidentiality and zero-knowledge. For confidentiality, the ciphertext policy attribute encryption can effectively protect the data. Users who do not own the set of attributes specified by the access policy cannot obtain any information about the plaintext from the ciphertext. For zero-knowledge, nodes other than the relevant participating nodes cannot obtain any information about access policies and user attribute sets from the access records or certificates in the broadcast.

Table 1 shows the comparative analysis between our model and Wang Xiuli et al.'s Model. We can see that comparing to Wang Xiuli et al.'s Blockchain Data Sharing Model, our model protection the senitive attributes and access policy.

Table 1. Comparative analysis between our model and Wang Xiuli et al.'s model [20]

	Our model	Wang Xiuli et al.'s model
Blockchain structure	Alliance chain	Alliance chain
Access control	CP-ABE	CP-ABE
Sentitive attributes protection	Use zk-SNARK to protect	No protection
Access policy protection	Use zk-SNARK and CP-ABE to protect	No protection

5 Conclusion

In this paper, we proposed a privacy protection model to solve the problem that sensitive attributes and access policies are easily leaked in the process of sharing data on blockchain. Based on the blockchain data sharing model using CP-ABE encryption, this article uses zero-knowledge proof to propose a blockchain data sharing privacy protection model based on zk-SNARK technology. This model not only satisfies the fine-grained access control requirements of data sharing, but also effectively protects the privacy protection problems that occur when user nodes adopt ciphertext policy attribute-based encryption on blockchain data sharing, preventing the risk of sensitive access strategies and sensitive attribute leakage.

Acknowledgment. This work is supported by the National Key Research and Development Program of China (No. 2017YFB1400700).

References

1. Agrawal, S., Ganesh, C., Mohassel, P.: Non-interactive zero-knowledge proofs for composite statements. In: Shacham, H., Boldyreva, A. (eds.) CRYPTO 2018. LNCS, vol. 10993, pp. 643–673. Springer, Cham (2018). https://doi.org/10.1007/978-3-319-96878-0_22
2. Ben-Sasson, E., Bentov, I., Horesh, Y., Riabzev, M.: Scalable, transparent, and post-quantum secure computational integrity. Eprint. Iacr. Org. **693423**, 1–83 (2018)
3. Ben-Sasson, E., et al.: Zerocash: decentralized anonymous payments from bitcoin. In: Proceedings - IEEE Symposium on Security and Privacy, pp. 459–474 (2014)
4. Waters, B.: Ciphertext-policy attribute-based encryption: an expressive, efficient, and provably secure realization. In: International Workshop on Public Key Cryptography, pp. 53–70 (2011)

5. Bunz, B., Bootle, J., Boneh, D., Poelstra, A., Wuille, P., Maxwell, G.: Bulletproofs: short proofs for confidential transactions and more. In: Proceedings - IEEE Symposium on Security and Privacy, 2018-May, pp. 315–334 (2018)
6. Camenisch, J., Dubovitskaya, M., Enderlein, R.R., Neven, G.: Oblivious transfer with hidden access control from attribute-based encryption. In: Visconti, I., Prisco, R. (eds.) SCN 2012. LNCS, vol. 7485, pp. 559–579. Springer, Heidelberg (2012). https://doi.org/10.1007/978-3-642-32928-9_31
7. Chen, J., Gong, J., Wee, H.: Improved Inner-Product Encryption with Adaptive Security and Full Attribute-Hiding. In: Peyrin, T., Galbraith, S. (eds.) ASIACRYPT 2018. LNCS, vol. 11273, pp. 673–702. Springer, Cham (2018). https://doi.org/10.1007/978-3-030-03329-3_23
8. Cui, H., Deng, R.H., Lai, J., Yi, X., Nepal, S.: An efficient and expressive ciphertext-policy attribute-based encryption scheme with partially hidden access structures, revisited. Comput. Netw. **133**, 157–165 (2018)
9. Fu, Y.: Research on supply chain information sharing mechanism and management mode based on blockchain. Doctor level of thesis. Central University of Finance and Economics (2018)
10. Gabizon, A., Williamson, Z.J., Ciobotaru, O.: PLONK: permutations over lagrange-bases for oecumenical noninteractive arguments of knowledge. In: Stanford Blockchain Conference, pp. 1–33 (2020)
11. Goyal, V., Pandey, O., Sahai, A., Waters, B.: Attribute-based encryption for fine-grained access control of encrypted data. In: Proceedings of the ACM Conference on Computer and Communications Security, pp. 89–98 (2006)
12. Groth, J., Kohlweiss, M., Maller, M., Meiklejohn, S., Miers, I.: Updatable and Universal Common Reference Strings with Applications to zk-SNARKs. In: Shacham, H., Boldyreva, A. (eds.) CRYPTO 2018. LNCS, vol. 10993, pp. 698–728. Springer, Cham (2018). https://doi.org/10.1007/978-3-319-96878-0_24
13. Guan, Z.: Efficient Privacy-Preserving Account-Model Blockchain Based on Zero-Knowledge Proof. Master level of thesis. ShanDong University (2020)
14. Katz, J., Sahai, A., Waters, B.: Predicate encryption supporting disjunctions, polynomial equations, and inner products. Lecture Notes in Computer Science (Including Subseries Lecture Notes in Artificial Intelligence and Lecture Notes in Bioinformatics), 4965 LNCS (2006), 146–162 (2008)
15. Khalil, R., Zamyatin, A., Felley, G., Gervais, A., Moreno-sanchez, P.: Commit-Chains: Secure, Scalable Off-Chain Payments. IACR Cryptology EPrint Archive (2018)
16. Li, G., He, D., Guo, B., Lu, S.: Blockchain privacy protection algorithms based on zero knowledge proof. J. Huazhong Univ. Sci. Technol. (Natural Science Edition), **48**(7) (2020)
17. Petkus, M.: Why and How zk-SNARK Works. 1–65. http://arxiv.org/abs/1906.07221 (2019)
18. Sahai, A., Waters, B.R.: Fuzzy identity based encryption. In: Proceedings of the 24th Annual International Conference on Theory and Applications of Cryptographic Techniques, pp. 1–9 (2005)
19. Sandborn, W., et al.: Sonic. Am. J. Gastroenterol. **103**, S436 (2008)
20. Wang, X., Jiang, X., Li, Y.: Model for data access control and sharing based on blockchain. J. Softw. **30**(6), 1661–1669 (2019)
21. Wang, Y., Fan, K.: Effective CP-ABE with hidden access policy. J. Comput. Res. Develop. **56**(10), 2151–2159 (2019)
22. Wei, X.: Research and Implementation of Anonymous Electronic Survey System Based on Non-Interactive Zero Knowledge. Master level of thesis. Northeastern University (2015)
23. Zhang, S., Ling, J., Chen, J.: Traceable blockchain ledger privacy protection scheme. Comput. Eng. Appl. (2020)

Hierarchical Identity-Based Conditional Proxy Re-encryption Scheme Based RLWE and NTRU Variant

Chao Wang[1,2], Yiliang Han[1,2](✉), Xiaowei Duan[1,2], and KaiYang Guo[1,2]

[1] College of Cryptographic Engineering, Engineering University of PAP, Xi'an 710086, China
[2] Key Laboratory of PAP for Cryptology and Information Security, Xi'an 710086, China

Abstract. IB-PRE can perform fine-grained access control on the user's decryption rights based on the identity of the delegatee, while the proxy cannot obtain the identity information of the delegator and delegatee. The current identity-based proxy re-encryption scheme achieves the ciphertext conversion between users at the same level, while it does not further distinguish between different levels of user identity, which is not suitable for hierarchical user management system. This paper combines hierarchical identity encryption with proxy re-encryption, and uses RLWE encryption system and NTRU variant as the underlying encryption scheme. According to the difference of the key reversibility and the ciphertext share between the two systems, the control of the decryption authority of users' different levels was realized. The transformed cipher text still satisfies the rules of higher level to decrypt lower level cipher, and constructs a unidirectionality, collusion resistant, and non-interactive hierarchical identity-based conditional proxy re-encryption scheme(HIB-CPRE), which is IND-sid-CPA security under the RLWE difficult assumption.

Keywords: IB-PRE · RLWE · NTRU variant · Hierarchical identity encryption

1 Introduction

With the development of cloud storage and cloud access technology, the current data forwarding model is mainly based on the data owner upload data, cloud server updates and stores data, and users access data from the cloud server. The function of the model is constantly expanded. The data forwarding model based on identity encryption can effectively control the decryption ability of data visitors, but this control is rough and cannot meet the multi-level user management system. In reality, the decryption ability of users should match their level in the system. With the propose of hierarchical identity encryption [1], the fine-grained access control of the decryption ability of users is realized in hierarchical management system. The upper level users can decrypt the secret text of the lower level users, otherwise they cannot decrypt them.

In 1998, Blaze [2] et al. first proposed the concept of proxy re-encryption at the EUROCRYPT. After that, scholars have made in-depth research on its unidirectional, collusive, multi hop and security, and constantly improved the concept and function of

J. Zeng et al. (Eds.): ICPCSEE 2021, CCIS 1452, pp. 240–259, 2021.
https://doi.org/10.1007/978-981-16-5943-0_20

proxy re-encryption. Based on the traditional mathematical difficulties, a series of proxy re-encryption schemes [3, 4] are proposed. However, with the development of quantum computing, the security of the cryptosystem based on the traditional mathematical difficulties is challenged. It will be a frontier field for the research and development of proxy re-encryption system to combine the proxy re-encryption and lattice cryptosystem which can effectively resist quantum computing attacks. In 2010, xagawa [5] first constructed the proxy re-encryption scheme on the lattice. Then, according to the basic characteristics and function expansion of lattice proxy re-encryption system, three kinds of proxy re-encryption schemes are proposed, which are LWE type [6–8], RLWE type [9], NTRU type [10, 11].

In the cloud computing environment, the diversity and hierarchy of user's identity put forward higher application requirements for proxy re-encryption. A single proxy re-encryption can only realize proxy forwarding of data, cannot carry out fine-grained proxy forwarding of data according to the user's identity, and cannot effectively control the proxy's ability to transform data. In 2007, Green [12] et al. first Combine identity-based encryption with proxy re-encryption, the identity-based proxy re-encryption (IB-PRE) scheme is constructed. In 2013, Singh [13] et al. constructed the first identity-based proxy re-encryption scheme based on lattice. In 2019, Jiang [14] et al. proposed a single bit and bidirectional proxy re-encryption scheme based on LWE problem in standard model, and Hou [15] et al. proposed a multi bit and bidirectional proxy re-encryption scheme based on LWE problem. However, the structure of re-encryption key of the above two schemes does not satisfy the non-interaction and collusion resistant. Wu [16] et al. proposed an identity-based proxy re-encryption scheme with double private key mechanism based on LWE problem. The construction of re-encryption key is non-interactive and collusion resistant. In 2020, Dutta [17] et al. constructed a unidirectional identity-based proxy re-encryption scheme on lattice. In 2021, Wu [18] et al. proposed a verifiable identity-based proxy re-encryption scheme based on LWE problem by using lattice homomorphic signature technology. The delegatee can verify whether the proxy re-encrypts the ciphertext data correctly, which effectively controls the re-encryption ability of the proxy. However, the current identity-based proxy re encryption scheme cannot achieve fine-grained access control for multi-level identity and cannot be applied to hierarchical identity management system.

Aiming at the problem of the current identity-based proxy re-encryption scheme, we combine the hierarchical identity-based encryption and proxy re-encryption to achieve hierarchical management of identities. User identity information is divided into senior users and ordinary users. The ciphertext after proxy conversion still obeys the rules of decrypting the lower ciphertext by the higher level. We use two encryption schemes: the RLWE encryption system and the NTRU variant of CPA security. The proxy will use NTRU variant to re-encrypt ciphertext when authorizing to senior users. and use RLWE encryption to re-encrypt ciphertext when authorizing to ordinary users. According to the difference of the key reversibility and the ciphertext share between the two systems, realize the control of the decryption authority of different levels of users.

Aiming at the problem that the semi-trusted proxy can arbitrarily convert other ciphertext of delegator by using the re-encryption key to cause information leakage, we embed a ciphertext type into each pair of ciphertext and re-encryption key. the proxy will re-encrypt the ciphertext correctly only if the ciphertext type matches.

Our goal is to propose a hierarchical identity-based conditional proxy re-encryption scheme that satisfies unidirectionality, collusion resistant and non-interaction, and the scheme is under the assumption of RLWE difficulty achieve IND-sid-CPA security.

2 Preliminaries

2.1 RLWE

Definition 1 (Ring Learning with Errors Distribution). Defining polynomial rings with integer coefficients $R = \mathbb{Z}(x)/x^n + 1$, let $n \in \mathbb{Z}, n = 2^k$. a prime positive integer modulus q defining the quotient ring $R_q = R/qR = \mathbb{Z}_q(x)/x^n + 1$, choosing $a, s \in R_q$ uniformly at random. Choosing $e \leftarrow \chi$, χ is an error distribution over R. The RLWE distribution $A_{s,\chi}$ over $R_q \times R_q$ is $(a, b = as + e \bmod q)$.

Definition 2 (Ring Learning with Errors Decision Problem). Given m independent samples (a_i, b_i), whether it can be distinguished into RLWE distribution $A_{s,\chi}$ and uniformly random distribution $R_q \times R_q$ on the basis of non-negligible advantages.

2.2 Operation Definition

For a polynomial $X \in R_q$, sample polynomials $X_j \in R_2, j \in \{0, 1, \cdots, \lfloor \log_2 q \rfloor\}$ satisfies $X = \sum_{j=0}^{\lfloor \log_2 q \rfloor} X_j 2^j$, sample the $\lfloor \log_2 q \rfloor + 1$ dimensional polynomial vector $X' = \left[X_0, X_1, \cdots, X_{\lfloor \log_2 q \rfloor}\right] \in R_2^{\lfloor \log_2 q \rfloor}$, $Y = \left[Y_0, Y_1, \cdots, Y_{\lfloor \log_2 q \rfloor}\right] \in R_q^{\lfloor \log_2 q \rfloor}$, we define operation $\otimes$.

$$X' \otimes Y = \sum_{j=0}^{\lfloor \log_2 q \rfloor} X_j \cdot Y_j \tag{1}$$

2.3 HIBE-CPRE Model Definition

A hierarchical identity base conditional proxy re-encryption scheme should satisfy the following rules. The converted ciphertext of the proxy still obeys the rules that the higher level can decrypt the lower level user ciphertext, while the lower level cannot decrypt the upper level user ciphertext; the higher level user can authorize any level user, but the ordinary user can only authorize the same level user; when the ciphertext type is inconsistent, the proxy cannot re-encrypt the ciphertext correctly. When authorizing senior users, proxy re-encryption operations are performed using NTRU variants.

with reversible keys and single ciphertext share, combined with bit decomposition technology to effectively reduce noise. When authorizing ordinary users, a proxy re-encryption operation is performed using the RLWE encryption system with an irreversible key and a double ciphertext share.

A HIBE-CPRE scheme consists of the following 10 algorithms:

Setup(κ): On input security parameters κ, public parameters PP, master key MK.

KeyGen(PP, MK, id): On input public parameters PP, master key MK, user identity information id. When the user is an senior user, outputs the user key $(pk_{id}, sk_{N,id})$, When the user is an ordinary user, outputs the user key $(pk_{id}, sk_{R,id})$.

N.Enc(pk = b_A, m): On input the public key pk_A of the senior user A and a message $m \in M$, sample ciphertext type $t \leftarrow \mu_q$, outputs the ciphertext (C_A, t).

N.ReKeyGen(pk_B, sk_A, t): On input the public key pk_B of the senior user B, the private key sk_A and the ciphertext type t, outputs a re-encryption key $rk_{A \to B}$.

N.ReEnc($rk_{A \to B}$, C_A): On input a ciphertext C_A and a re-encryption key $rk_{A \to B}$, when the ciphertext type is consistent with the re-encryption key type, the proxy can output the correct re-encryption ciphertext C_B.

N.Dec(C_B, sk_B): On input a re-encryption ciphertext C_B and the secret key sk_B, outputs a message $m \in M$.

R.Enc(pk_H, m, t′, t″): On input the public key b_H of the senior user or the public key b_{H_i} of the ordinary user and a message $m \in M$, sample ciphertext type $t', t'' \leftarrow \mu_q$, when the delegator is the senior user, outputs a ciphertext $C_H = (C, C_0, t')$, when the delegator is the ordinary user, outputs a ciphertext $C_H = (C, C_0, t'')$.

R.ReKeyGen(pk_{B_i}, sk_H, t′, t″): When the delegator H is senior user, On input the public key pk_{B_i} of the ordinary user B_i, the private key sk_H of the delegator H and the ciphertext type t', outputs a re-encryption key $rk_{H \to B_i}$. When the delegator H is ordinary user, On input the public key pk_{B_i} of the ordinary user B_i, the private key sk_H of the delegator H and the ciphertext type t'', outputs a re-encryption key $rk_{H \to B_i}$.

R.ReEnc($rk_{H \to B_i}$, C_H): On input a ciphertext (C, C_0) and a re-encryption key $rk_{H \to B_i}$, when the ciphertext type is consistent with the re-encryption key type, the proxy can output the correct re-encryption ciphertext C_B.

R.Dec(C_{B_i}, sk_{B_i}): On input a re-encryption ciphertext C_{B_i} and the secret key sk_{B_i}, outputs a message $m \in M$.

2.4 HIBE-CPRE Security Model Definition

The IND-CPA security of the hierarchical identity-based conditional proxy re-encryption scheme is defined based on the indistinguishability of the three games under the RLWE difficult assumption.

Game0. This is IND-CPA real attack game under the standard model, all the queries of the attacker ϑ are answered by the real system. The game is divided into three phases.

(1) **Setup.**
The challenger ξ accesses the real system, inputs security parameters κ, and the real system outputs public parameters PP and sends to the attacker ϑ.

(2) **Query phase**
Private key queries. The attacker ϑ queries the private key of the user *id*. The challenger ξ enters the identity information of the user *id* into the real system. The real system runs ***KeyGen*** algorithm to generate the user's private key and sends it to the attacker ϑ.
Re-encryption key queries.
(3) The attacker ϑ querise the re-encryption key which can convert ciphertext of the senior user A to the senior user B. The challenger ξ enters the identity information of the two users into the real system, and the real system runs the ***N.ReKeyGen*** algorithm to generate the re-encryption key $rk_{A \to B}$ and sends it to the attacker ϑ.
(4) The attacker ϑ querise the re-encryption key which can convert ciphertext of the any user H(H can be senior user or ordinary user) to the ordinary user B_i. The challenger ξ enters the identity information of the two users into the real system, and the real system runs the ***R.ReKeyGen*** algorithm to generate the re-encryption key $rk_{H \to B_i}$ and sends it to the attacker ϑ.
Encryption queries.
(5) When the delegatee is a senior user, the attacker ϑ queries the ciphertext of the delegate A, the real system runs the ***N.Enc*** algorithm, outputs the ciphertext C_A and sends it to ϑ.
(6) When the delegatee is an ordinary user, the attacker ϑ queries the ciphertext of the delegate H, the real system runs the ***R.Enc*** algorithm, outputs the ciphertext C_H and sends it to ϑ.
(7) Challenge phase.

The attacker ϑ selects two plaintexts (m_0, m_1) and the target user A^* (the A^* user who is not inquired) to send to the challenger ξ, ξ randomly selects $b \leftarrow \{0, 1\}$, when the target user is an senior user, the real system runs ***N.Enc*** to generate the challenge ciphertext C_{A^*} and sends it to ϑ, when the target user is an ordinary user When the real system runs ***R.Enc*** to generate challenge ciphertext C_{A^*} and sends it to ϑ.

The attacker ϑ gives a guess value $b' \leftarrow \{0, 1\}$, if $b' = b$, declares the attacker ϑ to win, and defines the advantage of the attacker to break the HIB-CPRE algorithm as ε.

Game1. In Game1, the public parameters and private key are generated by the challenger ξ simulating the real system, and the other phase are the same as Game0.

Game2. In Game2, the challenge ciphertext are randomly selected by the challenger ξ simulating the real system, and the other phase are the same as Game1.

Based on the RLWE difficulty assumption, the three games are indistinguishable from the polynomial-time attacker ϑ. We can obtain that under the standard IND-CPA game, the advantage of the polynomial-time IND-CPA attacker is negligible. It is proved that the HIB-CPRE scheme is IND-sid-CPA security in the standard model.

3 HIB-CPRE Scheme

Setup(κ): On input security parameters κ, defining polynomial rings with integer coefficients $R = \mathbb{Z}(x)/f(x)$, let $f(x) = x^n + 1$, where $f(x)$ is a monic square free polynomial of degree n. selects a prime integer q and a small prime number p with respect q

($p \ll q$), defining the quotient ring $R_q = \mathbb{Z}_q(x)/f(x)$, the range of the effective coefficients of the polynomial on R_q is $\{-\lfloor 2/q \rfloor, \cdots, \lfloor 2/q \rfloor\}$. define the coefficient value space of the plaintext m as $M = \{0, \cdots, p-1\}^n$, define β_χ-bounded discrete gaussian distribution χ with distribution parameter α over R_q and β_μ-bounded discrete uniform random distribution μ_q over R_q, $\beta_\chi, \beta_\mu \gg 1$, uniformly random sampling of $2(m+1)$ polynomials $a_0, a_1, \cdots, a_m, b_0, b_1, \cdots, b_m \leftarrow \mu_q$, define user identity $id = \{id_1, id_2, \cdots, id_m\} \in \{0, 1\}^m$, outputs public parameters $PP = (n, p, q, \alpha)$, master key $MK = (a_0, a_1, \cdots, a_m, b_0, b_1, \cdots, b_m)$.

KeyGen(PP, MK, id): On input senior user A identity information $id(A) = \{id_1, id_2, \cdots, id_m\} \in \{0, 1\}^m$, public parameters PP, master key MK, compute $a_A = a_0 + \sum_{i=1}^{m} id_i a_i \in \mu_q$, $b_A = b_0 + \sum_{i=1}^{m} id_i b_i \in \mu_q$, sample polynomials $\overline{s}, e, e' \leftarrow \chi$, compute $s_{N,A} = p\overline{s} + 1$, require $s_{N,A}$ to meet $b_A = a_A s_{N,A}^{-1} + e$, $s_{N,A}^{-1} s_{N,A} \bmod q = 1$, outputs private key $s_{N,A}$ of senior user A.

On input the identity information of the ordinary user A_i managed by senior user A $id(A_i) = \{id_{A1}, id_{A2}, \cdots, id_{Am}\} \in \{0, 1\}^m$, compute $a_{A_i} = a_0 + \sum_{i=1}^{m} id_{Ai} a_i \in \mu_q$, $b_{A_i} = b_0 + \sum_{i=1}^{m} id_{Ai} b_i \in \mu_q$, compute $s_{R,A_i} = s_{N,A}^{-1} + e'$, require s_{R,A_i} to meet $b_{A_i} = a_{A_i} s_{R,A_i} + e \in R_q$, outputs private key s_{R,A_i} of ordinary user A_i.

N.HIB-CPRE: Proxy re-encrypt the original ciphertext to senior user

N.Enc(pk = b_A, m): On input public key pk and plaintext m, sample $r, e_0, t \leftarrow \chi$, if t is not invertible modulo q, resample. compute $\overline{C} = p(b_A r + e_0) + m$, where $\overline{C}_j = \{p(b_A r + e_0) + m\}_j$, $j = \{0, 1, \cdots, \lfloor \log_2 q \rfloor\}$, $\overline{C} = \sum_{j=0}^{\lfloor \log_2 q \rfloor} \overline{C}_j 2^j$, the coefficients of $\overline{C}_j$ is in the range $\{0, 1\}$, outputs the ciphertext (C_A, t).

$$C_A = \left(t^{-1}\overline{C}_0, \cdots, t^{-1}\overline{C}_{\lfloor \log_2 q \rfloor}\right) \tag{2}$$

N.ReKeyGen(pk_B, sk_A, t): The algorithm is non-interactive, and the whole process is only executed by the delegator A, on input public key $b_B = a_B s_{N,B}^{-1} + e$ of the senior user B, the private key $s_{N,A}$ of the user A, the ciphertext type t, outputs re-encryption key $rk_{A\to B} = \left((rk_{A\to B})_0, \cdots, (rk_{A\to B})_j, \cdots, (rk_{A\to B})_{\lfloor \log_2 q \rfloor}\right)$.

$$(rk_{A\to B})_j = t\left(p(b_B + e_2) + s_{N,A} \cdot 2^j\right) \tag{3}$$

N.ReEnc($rk_{A\to B}$, C_A): On input the re-encryption key $rk_{A\to B}$ and the ciphertext C_A, when the ciphertext type t is consistent, the proxy can re encrypt the ciphertext correctly, outputs re-encrypt ciphertext C_B.

$$C_B = rk_{A\to B} \otimes C_A = \sum_{j=0}^{\lfloor \log_2 q \rfloor} (rk_{A\to B})_j \cdot t^{-1}\overline{C}_j \tag{4}$$

N.Dec($C_B, s_{N,B}$): On input re-encrypt ciphertext C_B and the private key $s_{N,B}$, compute $m' = (s_{N,B}C_B) \bmod q$, $m = m' \bmod p$.

R.HIB-CPRE: Proxy re-encrypt the original ciphertext to ordinary user.

R.Enc(pk_H, m, t', t''): On input public key $pk = b_H$ or b_{H_i} and plaintext m, sample $t', t'', e_1 \leftarrow \chi$, outputs the ciphertext $C_H = (C, C_0)$ and Corresponding ciphertext type t', t''.

$$\text{When user } H \text{ is senior user, } C = p(b_H r + e_1) + m,\ C_0 = t'^{-1} p a_H r \tag{5}$$

$$\text{When user } H \text{ is ordinary user, } C = p(b_{H_i} r + e_1) + m,\ C_0 = t''^{-1} p a_{H_i} r \tag{6}$$

R.ReKeyGen(pk_{B_i}, sk, t', t''): On input the identity information of the ordinary user B_i:$id(B_i) = \{id_{B1}, id_{B2}, \cdots, id_{Bm}\} \in \{0,1\}^m$, compute $a_{B_i} = a_0 + \sum_{i=1}^{m} id_{Bi} a_i \in \mu_q$, $b_{B_i} = b_0 + \sum_{i=1}^{m} id_{Bi} b_i \in \mu_q$, according to the level of user H, select the private key:$sk = (s_{N,H} \text{ or } s_{R,H_i})$, select the appropriate ciphertext type t' or t'', sample $e_3 \leftarrow \chi$, outputs re-encryption key $rk_{H \to B_i} = (rk_1, rk_2)$.

$$\text{When user } H \text{ is senior user, } rk_{H \to B_i} = \left(t'\left(b_{B_i} - s_{N,H}^{-1} + pe_3\right), t' a_{B_i}\right) \tag{7}$$

$$\text{When user } H \text{ is ordinary user, } rk_{H_i \to B_i} = \left(t'\left(b_{B_i} - s_{R,H_i} + pe_3\right), t' a_{B_i}\right) \tag{8}$$

R.ReEnc($rk_{H \to B_i}, C_H$): On input the re-encryption key $rk_{H \to B_i}$ and the ciphertext C_H, When the ciphertext type is consistent, the proxy can re encrypt the ciphertext correctly, Outputs re-encrypt ciphertext $C_{B_i} = (C', C_0')$.

$$C' = C + rk_1 \cdot C_0 \tag{9}$$

$$C_0' = rk_2 \cdot C_0 \tag{10}$$

R.Decrypt(C_{B_i}, s_{N,B_i}): On input re-encrypt ciphertext $C_{B_i} = (C', C_0')$, the private key s_{R,B_i}, compute $m' = (C' - s_{R,B_i} C_0') \bmod q$, $m = m' \bmod p$. The workflow of our scheme is shown in Fig. 1.

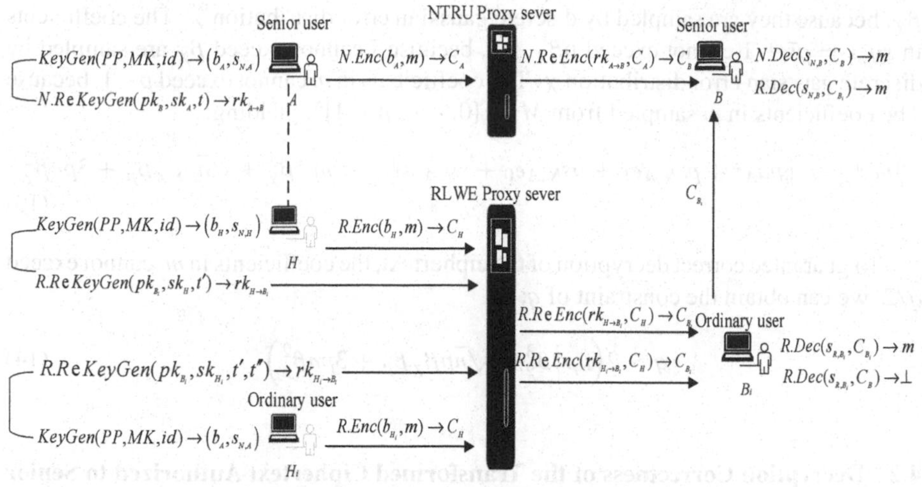

Fig. 1. The workflow of our proposed HIB-CPRE

4 Property Analysis

This chapter analyzes the correctness, unidirectionality and collusion resistance of our scheme.

4.1 Decryption Correctness of the Original Ciphertext Authorized to Senior Users

First compute $\overline{C}_j$: $tC_A = t\left(t^{-1}\overline{C}_0, \cdots, t^{-1}\overline{C}_{\lfloor \log_2 q \rfloor}\right) = \left(\overline{C}_0, \cdots, \overline{C}_{\lfloor \log_2 q \rfloor}\right)$, then compute

$$\begin{aligned} m' &= \left(s_{N,A}t \sum_{j=0}^{\lfloor \log_2 q \rfloor} \overline{C}_j 2^j\right) \bmod q \\ &= \left(s_{N,A}tt^{-1}(p(b_A r + e_0) + m)\right) \bmod q \\ &= \left(ps_{N,A}b_A r + ps_{N,A}e_0 + s_{N,A}m\right) \bmod q \\ &= \left(pa_A r + ps_{N,A}er + ps_{N,A}e_0 + s_{N,A}m\right) \bmod q \end{aligned} \tag{11}$$

Since $\|m'\|_\infty < 2/q$ then $m' = pa_A r + ps_{N,A}er + ps_{N,A}e_0 + s_{N,A}m$, compute

$$m' \bmod q = \left(pa_A r + ps_{N,A}er + ps_{N,A}e_0 + s_{N,A}m\right) \bmod p = \left(s_{N,A}m\right) \bmod p \tag{12}$$

Since $s_{N,A} = p\overline{s} + 1$ then $s_{N,A} \bmod p = 1$, compute $\left(s_{N,A}m\right) \bmod p = m \bmod p = m$.

The coefficients in a_A cannot exceed β_μ, because they are sampled by discrete uniform discrete uniform random distribution μ_q. The coefficients in r, e, e_0 cannot exceed

β_χ, because they are sampled by discrete gaussian error distribution χ. The coefficients in $s_{N,A} = p\overline{s} + 1$ cannot exceed $p\beta_\chi + 1$, because $\overline{s}$ cannot exceed β_χ are sampled by discrete gaussian error distribution χ. The coefficients in m cannot exceed $p-1$, because The coefficients in m sampled from $M = \{0, \cdots, p-1\}^n$, yielding:

$$\|m'\|_\infty = \|pa_Ar + ps_{N,A}er + ps_{N,A}e_0 + s_{N,A}m\|_\infty < np^2\beta_\chi^3 + \sqrt{n}p\beta_\chi\beta_\mu + 3pn\beta_\chi^2 \tag{13}$$

To guarantee correct decryption of the ciphertext, the coefficients in m' cannot exceed $q/2$, we can obtain the constraint of q:

$$q > 2\left(np^2\beta_\chi^3 + \sqrt{n}p\beta_\chi\beta_\mu + 3pn\beta_\chi^2\right) \tag{14}$$

4.2 Decryption Correctness of the Transformed Ciphertext Authorized to Senior Users

Compute:

$$\begin{aligned} m' &= \left(s_{N,B}C_B\right) \bmod q \\ &= \left(s_{N,B} \cdot \sum_{j=0}^{\lfloor \log_2 q \rfloor} (rk_{A\to B})_j \cdot t^{-1}\overline{C}_j\right) \bmod q \\ &= \left(s_{N,B} \cdot \sum_{j=0}^{\lfloor \log_2 q \rfloor} t\left(p(b_B + e_2) + s_{N,A} \cdot 2^j\right) \cdot t^{-1}\overline{C}_j\right) \bmod q \\ &= \left(\sum_{j=0}^{\lfloor \log_2 q \rfloor} ps_{N,B}(b_B + e_2)\overline{C}_j + \sum_{i=0}^{\lfloor \log_2 q \rfloor} s_{N,B}s_{N,A}\overline{C}_j \cdot 2^j\right) \bmod q \\ &= \left(\sum_{j=0}^{\lfloor \log_2 q \rfloor} p\left(a_B + s_{N,B}e + s_{N,B}e_2\right)\overline{C}_j + s_{N,B}s_{N,A}\overline{C}\right) \bmod q \end{aligned} \tag{15}$$

Since $\|m'\|_\infty < 2/q$ then $m' = \sum_{j=0}^{\lfloor \log_2 q \rfloor} p\left(a_B + s_{N,B}e + s_{N,B}e_2\right)\overline{C}_j + s_{N,B}s_{N,A}\overline{C}$, compute

$$m' \bmod p = \left(\sum_{j=0}^{\lfloor \log_2 q \rfloor} p\left(a_B + s_{N,B}e + s_{N,B}e_2\right)\overline{C}_j + s_{N,B}s_{N,A}\overline{C}\right) \bmod p = \left(s_{N,B}s_{N,A}m\right) \bmod p$$

Since $s_{N,A} \bmod p = 1$, $s_{N,B} \bmod p = 1$ then $\left(s_{N,A}s_{N,B}m\right) \bmod p = m \bmod p = m$.
The coefficients in a_A, a_B cannot exceed β_μ, because they are sampled by discrete uniform random distribution μ_q. The coefficients in r, e, e_0, e_2 cannot exceed β_χ, because they are sampled by discrete gaussian error distribution χ. The coefficients

in $s_{N,A}$, $s_{N,B}$ cannot exceed $p\beta_\chi + 1$. The coefficients in m cannot exceed $p - 1$. The coefficients in $\overline{C}_j$ cannot exceed 1, because the coefficients in m sampled from {0, 1}, yielding:

$$\left\| \sum_{j=0}^{\lfloor \log_2 q \rfloor} p\left(a_B + s_{N,B}e + s_{N,B}e_2\right)\overline{C}_j \right\|_\infty \leq pn\left(\lfloor \log_2 q \rfloor + 1\right)\left(\beta_\mu + 2p\beta_\chi^2 + 2\beta_\chi\right) \tag{16}$$

$$\left\| s_{N,B}s_{N,A}\overline{C} \right\|_\infty \leq n(p\beta_\chi + 1)\left(p\beta_\chi\beta_\mu + \sqrt{n}(p\beta_\chi + 1)\left(p\beta_\chi^2 + \beta_\chi + p - 1\right)\right) \tag{17}$$

We have

$$\begin{aligned} \|m'\|_\infty &\leq pn\left(\lfloor \log_2 q \rfloor + 1\right)\left(\beta_\mu + 2p\beta_\chi^2 + 2\beta_\chi\right) \\ &\quad + n(p\beta_\chi + 1)\left(p\beta_\chi\beta_\mu + \sqrt{n}(p\beta_\chi + 1)\left(p\beta_\chi^2 + \beta_\chi + p - 1\right)\right) \\ &< 2pn(p\beta_\chi + 1)\left(\lfloor \log_2 q \rfloor + 1\right)\left(\beta_\chi\beta_\mu + 3\sqrt{n}p\beta_\chi^3\right) \end{aligned} \tag{18}$$

To guarantee correct decryption of the ciphertext, the coefficients in m' cannot exceed $q/2$, we can obtain the constraint of q:

$$q > 4pn(p\beta_\chi + 1)\left(\lfloor \log_2 q \rfloor + 1\right)\left(\beta_\chi\beta_\mu + 3\sqrt{n}p\beta_\chi^3\right) \tag{19}$$

4.3 Decryption Correctness of the Original Ciphertext Authorized to Ordinary Users

1) Since user H is senior user then $b_H = a_H s_{N,H}^{-1} + e, C = p(b_H r + e_1) + m$, $C_0 = t' p a_H r$, compute

$$\begin{aligned} m' &= \left(C - s_{N,H}^{-1} t' C_0\right) \bmod q \\ &= \left(p(b_H r + e_1) + m - s_{N,H}^{-1} t' t'^{-1} p a_H r\right) \bmod q \\ &= \left(p a_H s_{N,H}^{-1} r + per + pe_1 + m - s_{N,H}^{-1} p a_H r\right) \bmod q \\ &= (per + pe_1 + m) \bmod q \end{aligned} \tag{20}$$

Since $\|m'\|_\infty < 2/q$ then $m' = per + pe_1 + m$, compute

$$m' \bmod p = (per + pe_1 + m) \bmod p = m \bmod p = m \tag{21}$$

To guarantee correct decryption of the ciphertext, the coefficients in m' cannot exceed $q/2$, we can obtain the constraint of q:

$$q \geq 2\left(p\sqrt{n}\beta_\chi^2 + p\beta_\chi + p - 1\right) \tag{22}$$

2) Since user H is ordinary user then $b_H = a_H s_{N,H}^{-1} + e$, $C = p(b_H r + e_1) + m$, $C_0 = t' p a_H r$, compute

$$\begin{aligned} m' &= \left(C - s_{R,H_i} t' C_0\right) \bmod q \\ &= \left(p(b_{H_i} r + e_1) + m - s_{R,H_i} t' t'^{-1} par\right) \bmod q \\ &= \left(pa_{H_i} s_{R,H_i} r + per + pe_1 + m - s_{R,H_i} pa_{H_i} r\right) \bmod q \\ &= (per + pe_1 + m) \bmod q = (per + pe_1 + m) \bmod q \end{aligned}$$

Since $\|m'\|_\infty < 2/q$ then $m' = per + pe_1 + m$, compute

$$m' \bmod p = (per + pe_1 + m) \bmod p = m \bmod p = m \tag{21}$$

To guarantee correct decryption of the ciphertext, the coefficients in m' cannot exceed $q/2$, we can obtain the constraint of q:

$$q \geq 2\left(p\sqrt{n}\beta_\chi^2 + p\beta_\chi + p - 1\right) \tag{22}$$

4.4 Decryption Correctness of the Transformed Ciphertext Authorized to Ordinary Users

1) Since user H is senior user then $rk_{H \to B_i} = \left(t'\left(b_{B_i} - s_{N,H}^{-1} + pe_3\right), t' a_{B_i}\right)$, compute

$$\begin{aligned} m' &= \left(C' - s_{R,B_i} C_0'\right) \bmod q \\ &= \left(C + rk_1 \cdot C_0 - s_{R,B_i} \cdot rk_2 \cdot C_0\right) \bmod q \\ &= \left(p(b_H r + e_1) + m + t'\left(b_{B_i} - s_{N,H}^{-1} + pe_3\right) \cdot t'^{-1} pa_H r - s_{R,B_i} \cdot t' a_{B_i} \cdot t'^{-1} pa_H r\right) \bmod q \\ &= (per + pe_1 + pa_H re + pa_H rpe_3 + m) \bmod q \end{aligned} \tag{23}$$

Since $\|m'\|_\infty < 2/q$ then $m' = per + pe_1 + pa_H re + pa_H rpe_3 + m$, compute

$$m' \bmod p = (per + pe_1 + pa_H re + pa_H rpe_3 + m) \bmod p = m \tag{24}$$

To guarantee correct decryption of the ciphertext, the coefficients in m' cannot exceed $q/2$, we can obtain the constraint of q:

$$q > 2pn\left((p+1)\beta_\mu \beta_\chi^2 + \beta_\chi^2 + \beta_\chi + 1\right) \tag{25}$$

2) When user H is senior user, analyze decryption correctness of the transformed ciphertext by the superior user B of the ordinary user B_i:

$$\begin{aligned} m' &= \left(C' - s_{N,B}^{-1} C_0'\right) \bmod q \\ &= \left(per + pe_1 + pa_H ra_{B_i} e' + pa_H re + pa_H rpe_3 + m\right) \bmod q \end{aligned} \tag{26}$$

Since $\|m'\|_{\infty} < 2/q$ then $m' = per + pe_1 + pa_H ra_{B_i} e' + pa_H re + pa_H rpe_3 + m$, compute

$$m' \bmod p = \left(per + pe_1 + pa_H ra_{B_i} e' + pa_H re + pa_H rpe_3 + m\right) \bmod p = m \tag{27}$$

To guarantee correct decryption of the ciphertext, the coefficients in m' cannot exceed $q/2$, we can obtain the constraint of q:

$$q > 2pn\left(n\beta_{\mu}^2\beta_{\chi}^2 + (p+1)\beta_{\mu}\beta_{\chi}^2 + \beta_{\chi}^2 + \beta_{\chi} + 1\right) \tag{28}$$

3) Since user H is ordinary user then $rk_{H_i \to B_i} = \left(t'\left(b_{B_i} - s_{R,H_i} + pe_3\right), t' a_{B_i}\right)$, compute

$$\begin{aligned} m' &= \left(C' - s_{R,B_i} C_0'\right) \bmod q \\ &= \left(C + rk_1 \cdot C_0 - s_{R,B_i} \cdot rk_2 \cdot C_0\right) \bmod q \\ &= \left(p(b_{H_i} r + e_1) + m + t''\left(b_{B_i} - s_{R,H_i}^{-1} + pe_3\right) \cdot t''^{-1} pa_{H_i} r - s_{R,B_i} \cdot t'' a_{B_i} \cdot t''^{-1} pa_{H_i} r\right) \bmod q \\ &= \left(per + pe_1 + pa_{H_i} re + pa_{H_i} rpe_3 + m\right) \bmod q \end{aligned} \tag{29}$$

Since $\|m'\|_{\infty} < 2/q$ then $m' = per + pe_1 + pa_{H_i} re + pa_{H_i} rpe_3 + m$, compute

$$m' \bmod p = \left(per + pe_1 + pa_{H_i} re + pa_{H_i} rpe_3 + m\right) \bmod p = m \tag{30}$$

To guarantee correct decryption of the ciphertext, the coefficients in m' cannot exceed $q/2$, we can obtain the constraint of q:

$$q > 2pn\left((p+1)\beta_{\mu}\beta_{\chi}^2 + \beta_{\chi}^2 + \beta_{\chi} + 1\right) \tag{31}$$

4) When user H is ordinary user, analyze decryption correctness of the transformed ciphertext by the superior user B of the ordinary user B_i:

$$\begin{aligned} m' &= \left(C' - s_{N,B}^{-1} C_0\right) \bmod q \\ &= \left(per + pe_1 + pa_{H_i} ra_{B_i} e' + pa_{H_i} re + pa_{H_i} rpe_3 + m\right) \bmod q \end{aligned} \tag{26}$$

Since $\|m'\|_{\infty} < 2/q$ then $m' = per + pe_1 + pa_{H_i} ra_{B_i} e' + pa_{H_i} re + pa_{H_i} rpe_3 + m$, compute

$$m' \bmod p = \left(per + pe_1 + pa_{H_i} ra_{B_i} e' + pa_{H_i} re + pa_{H_i} rpe_3 + m\right) \bmod p = m \tag{27}$$

To guarantee correct decryption of the ciphertext, the coefficients in m' cannot exceed $q/2$, we can obtain the constraint of q:

$$q > 2pn\left(n\beta_{\mu}^2\beta_{\chi}^2 + (p+1)\beta_{\mu}\beta_{\chi}^2 + \beta_{\chi}^2 + \beta_{\chi} + 1\right) \tag{28}$$

4.5 Unidirectionality

The Unidirectionality of the proxy re-encryption scheme means that the proxy cannot complete the bidirectional conversion through a re-encryption key, nor can it be compute $rk_{B\to A}$ from $rk_{A\to B}$. The three types of re-encryption keys generated by delegator in this scheme are composed of the delegatee's public key, the delegator's private key and the ciphertext type, and noise is added in the re-encryption key generation. Since the delegatee's private key is hidden from the proxy then the proxy cannot compute the re-encryption key converted from delegator to delegatee. Our scheme satisfies the unidirectionality.

4.6 Collusion Resistance

Collusion resistance means that when one of delegator or delegatee is a corrupt user, the proxy and the corrupt party cannot obtain the private key information of the other party. When the delegatee is a corrupt user, the public key of the delegatee is open to the proxy, but this is useless information for the proxy. Due to the existence of noise polynomials and ciphertext types, the proxy still cannot recover the private key information of the delegator from the re-encryption key. When the delegator is a corrupt user, because the re-encryption key only has the public key information of the delegatee, and it is difficult to recover the private key information from the public key, the proxy cannot collude with the delegator to obtain the private key information of the delegatee. The re-encryption keys generated in different situations in our scheme are all constructed based on the public key of the delegatee and the private key of the delegator, so they all meet collusion resistance.

5 Security Analysis

The IND-CPA security of the hierarchical identity-based conditional proxy re-encryption scheme is proved from the following two cases. One is the security analysis of the proxy re-encryption algorithm N.HIB-CPRE that authorizes ciphertext to senior users, and the other is authorization to ordinary users. Another one is the security analysis of proxy re-encryption algorithm R.HIB-CPRE that authorizes ciphertext to ordinary users. The IND-CPA security in the two cases is guaranteed based on the NTRU encryption variant and the RLWE encryption system under the RLWE difficult assumption.

5.1 N.HIB-CPRE Security Analysis

IND-CPA security analysis of the proxy re-encryption scheme N.HIB-CPRE (***KeyGen***(), ***N.Enc***(), ***N.ReKeyGen***(), ***N.ReEnc***(), ***N.Dec***()) that authorizes ciphertext to senior users.

Theorem 1. If RLWE decision assumption is difficult, then the N.HIB-CPRE scheme is IND-sid-CPA security under the standard model.

Proof. Define three games (Game0, Game1, Game2), proving the three games are indistinguishable from the polynomial time attacker that under the RLWE difficult assumption, and then obtain that the advantage of the IND-CPA attacker attacking the scheme in polynomial time can be negligible.

Game0. This is IND-CPA real attack game under the standard model.

Setup. The challenger ξ accesses the real system, inputs security parameters κ, and the real system outputs public parameters $PP = (n, p, q, \alpha, a_0, a_1, \cdots, a_m, b_0, b_1, \cdots, b_m)$ and sends to the attacker ϑ.

Query Phase

Private Key Queries. The attacker ϑ queries the private key of the senior user *id*. The challenger ξ enters the identity information of the senior user *id* into the real system. The real system compute $a_{id} = a_0 + \sum_{i=1}^{m} id_i a_i \in \mu_q$, $b_{id} = b_0 + \sum_{i=1}^{m} id_i b_i \in \mu_q$, sample polynomials $\overline{s}, e \leftarrow \chi$, compute $s_{N,id} = p\overline{s}+1$, require $s_{N,id}$ to meet $b_{id} = a_{id}s_{N,id}^{-1}+e$, $s_{N,id}^{-1}s_{N,id} \bmod q = 1$, outputs private key $s_{N,id}$ and sends to the attacker ϑ.

Re-encryption Key Queries. The attacker ϑ queries the re-encryption key of the senior user A to the senior user B. The challenger ξ enters the identity information of the two users into the real system, and the real system compute the re-encryption key $(rk_{A\to B})_j = t\left(p(b_B + e_2) + s_{N,A} \cdot 2^j\right)$ and sends $rk_{A\to B}= ((rk_{A\to B})_0, \cdots, (rk_{A\to B})_j, \cdots, (rk_{A\to B})_{\lfloor \log_2 q \rfloor})$ to the attacker ϑ.

Encryption Queries. The attacker ϑ queries the ciphertext of the delegate A, the real system runs the ***N.Enc*** algorithm, outputs the ciphertext $C_A = \left(t^{-1}\overline{C}_0, \cdots, t^{-1}\overline{C}_{\lfloor \log_2 q \rfloor}\right)$ and sends it to ϑ.

Challenge Phase. The attacker ϑ selects two plaintexts (m_0, m_1) and the target user A^* (the A^* user who is not inquired) to send to the challenger ξ, ξ randomly selects $b \leftarrow \{0, 1\}$, compute $\overline{C}^* = p(b_{A^*}r + e_0) + m_b$, let $\overline{C}_j^* = \{p(b_{A^*}r + e_0) + m_b\}_j$, $j = \{0, 1, \cdots, \lfloor \log_2 q \rfloor\}$. Outputs the challenge ciphertext $C_{A^*} = \left(t^{-1}\overline{C}_0^*, \cdots, t^{-1}\overline{C}_{\lfloor \log_2 q \rfloor}^*\right)$ and sends it to ϑ.

The attacker ϑ gives a guess value $b' \leftarrow \{0, 1\}$, if $b' = b$, declares the attacker ϑ to win, and defines the advantage of the attacker to break the N.HIB-CPRE algorithm as ε.

Game1. In Game1, the public parameters and private key are generated by the challenger ξ simulating the real system, and the other phase are the same as Game0.

Setup. The challenger ξ uniformly random sampling polynomial $\overline{s}, e \leftarrow \chi$ from discrete Gaussian distribution χ, uniformly random sampling of $m + 1$ polynomials $a_0, a_1, \cdots, a_m \leftarrow \mu_q$, compute $s_{N,id} = p\overline{s}+1$, $b_i = a_i s_{N,id}^{-1} + e$, output $m+1$ polynomials $b_0, b_1, \cdots, b_m$. Finally send $PP = (n, p, q, \alpha, a_0, a_1, \cdots, a_m, b_0, b_1, \cdots, b_m)$ to ϑ.

Query Phase

Private Key Queries. The attacker ϑ queries the private key of the senior user *id*. The challenger ξ Compute $s_{N,id} = p\overline{s} + 1$, outputs the private key $s_{N,id}$ and sends to the attacker ϑ.

Re-encryption Key Queries. The attacker ϑ queries the re-encryption key of the senior user A to the senior user B. The challenger ξ computes the keys of the two users according to the above two algorithms, and computes the re-encryption key $(rk_{A\rightarrow B})_j = t\left(p(b_B + e_2) + s_{N,A} \cdot 2^j\right)$ and sends $rk_{A\rightarrow B} = ((rk_{A\rightarrow B})_0, \cdots, (rk_{A\rightarrow B})_j, \cdots, (rk_{A\rightarrow B})_{\lfloor \log_2 q \rfloor})$ to the attacker ϑ.

The encryption queries and challenge phase are the same as Game0, we will not be described again.

Game1 and Game0 are indistinguishable from polynomial time attackers ϑ.

In Game1, $a_0, a_1, \cdots, a_m$ is selected uniformly and randomly from R_q, and the distribution of $b_i = a_i s_{N,id}^{-1} + e$ is calculated to be close to a uniform distribution. The public parameters $b_0, b_1, \cdots, b_m$ and which is uniformly random selected from R_q in Game0 are indistinguishable for the attacker ϑ. $s_{N,id}$ is computed in Game1, since $b_i = a_i s_{N,id}^{-1} + e$, $b_{id} = b_0 + \sum_{i=1}^{m} id_i b_i$, $a_{id} = a_0 + \sum_{i=1}^{m} id_i a_i \in \mu_q$ then $b_{id} = a_{id} s_{N,id}^{-1} + e$, therefore, the private key queries in Game1 and the private key queries in Game0 is indistinguishable for the attacker ϑ. The calculation method of the challenger in the re-encryption queries, encryption queries, and challenge phase is the same as that of Game0. In summary, Game1 and Game0 are indistinguishable for polynomial time attackers.

Game2. In Game2, the challenge ciphertext generation process is different from Game1, and the rest of the process is the same.

In Game1, the challenger computes $\overline{C}^* = p(b_{A^*} r + e_0) + m_b$ and divides $\overline{C}^*$ to get the challenge ciphertext C_{A^*}, but in Game2, $\overline{C}^*$ is randomly sampled from R_q.

Game1 and Game0 are indistinguishable from polynomial time attackers ϑ.

Lemma 1. Assume the attacker ϑ can distinguish Game2 and Game1 with a non-negligible advantage ε, then there must be a polynomial time algorithm η that can solve the RLWE decision problem with the advantage $\frac{\varepsilon}{2}$.

For the instance $\left(b_{A^*}, \overline{C}^{*\prime}\right)$, when ϑ guesses b correctly, the polynomial time algorithm η outputs 1 and judges it as an RLWE instance, otherwise judges it as a instance which is uniformly random sampled from $R_q \times R_q$. According to the different generation algorithms of the instance $\left(b_{A^*}, \overline{C}^{*\prime}\right)$, the discussion can be divided into the following two situations:

1) $\left(b_{A^*}, \overline{C}^{*\prime}\right)$ is uniformly random sampled from $R_q \times R_q$, then $\overline{C}^* = p\overline{C}^{*\prime} + m_b$ obeys uniform distribution, the challenge ciphertext C_{A^*} also obeys uniform distribution. At this point, ϑ guesses the right b by the probability of $\frac{1}{2}$, therefore, the probability that η outputs 1 is also $\frac{1}{2}$.
2) $\left(b_{A^*}, \overline{C}^{*\prime}\right)$ is RLWE instance, let $\overline{C}^{*\prime} = b_{A^*} r + e_0$, $\overline{C}^* = p\overline{C}^{*\prime} + m_b$, then the challenge ciphertext C_{A^*} obeys uniform distribution. At this point, ϑ guesses the right b by the probability of $\frac{1}{2} + \varepsilon$, therefore, the probability that η outputs 1 is also $\frac{1}{2} + \varepsilon$.

Therefore η solves the RLWE decision problem with at least the probability of $\frac{1}{2}(\frac{1}{2}+\varepsilon+\frac{1}{2})=\frac{1}{2}+\frac{\varepsilon}{2}$, the advantage is $\frac{1}{2}+\frac{\varepsilon}{2}-\frac{1}{2}=\frac{\varepsilon}{2}$. Since the RLWE decision problem is difficult, the advantage of solving in polynomial time is negligible. Therefore, assume the attacker ϑ can distinguish the advantage of Game2 and Game1 is negligible, Game2 and Game1 are indistinguishable for η.

In summary, based on the RLWE decision assumption, the advantage of the polynomial time IND-CPA attacker in Game0 is negligible, and the N.HIB-CPRE scheme is IND-CPA security under the standard model.

5.2 R.HIB-CPRE Security Analysis

IND-CPA security analysis of the proxy re-encryption scheme R.HIB-CPRE (***KeyGen***(), ***R.Enc***(), ***R.ReKeyGen***(), ***R.ReEnc***(), ***R.Dec***()) that authorizes ciphertext to ordinary users.

Theorem 2. If RLWE decision assumption is difficult, then the R.HIB-CPRE scheme is IND-sid-CPA security under the standard model.

Proof. Define three games (Game0, Game1, Game2), proving the three games are indistinguishable from the polynomial time attacker that under the RLWE difficult assumption, and then obtain that the advantage of the IND-CPA attacker attacking the scheme in polynomial time can be negligible.

Game0. This is IND-CPA real attack game under the standard model.

Setup. Same as the setup phase in Game 0 of Sect. 5.1.

Query phase.
Private key queries.

1) The attacker ϑ queries the private key of the senior user *id*, same as the private key queries in Game 0 of Sect. 5.1.
2) The attacker ϑ queries the private key of the ordinary user *id'*. The challenger ξ enters the identity information of the senior user *id* into the real system. The real system compute $a_{id'}=a_0+\sum_{i=1}^{m} id'_i a_i \in \mu_q$, $b_{id'}=b_0+\sum_{i=1}^{m} id'_i b_i \in \mu_q$, sample polynomials $e' \leftarrow \chi$, compute $s_{R,id'}=s_{N,id}^{-1}+e'$, require $s_{R,id'}$ to meet $b_{id'}=a_{id'}s_{R,id'}+e$, outputs private key $s_{R,id'}$ and sends to the attacker ϑ.

Re-encryption Key Queries. The attacker ϑ queries the re-encryption key of the user H to the senior user B_i. The challenger ξ enters the identity information of the two users into the real system, since H is the senior user then the real system compute the re-encryption key $rk_{H\to B_i}=\left(t'\left(b_{B_i}-s_{N,H}^{-1}+pe_3\right),t'a_{B_i}\right)$, since H is the senior user then the real system compute the re-encryption key $rk_{H_i\to B_i}=\left(t'\left(b_{B_i}-s_{R,H_i}+pe_3\right),t'a_{B_i}\right)$ and sends to the attacker ϑ.

Encryption Queries. The attacker ϑ queries the ciphertext of the delegate H, the real system runs the ***R.Enc*** algorithm, outputs the ciphertext $C_H = (C, C_0)$ and sends it to ϑ.

Challenge Phase. The attacker ϑ selects two plaintexts (m_0, m_1) and the target user H^* (the H^* user who is not inquired) to send to the challenger ξ, ξ randomly selects $b \leftarrow \{0, 1\}$, according to the identity of user H, sample polynomials $t', t'' \leftarrow \chi$,compute $C^* = p(b_{H^*}r + e_0) + m_b$, $C_0^* = t'pa_{H^*}r$ or $C_0^* = t''pa_{H^*}r$, outputs the challenge ciphertext $C_{H^*} = \left(C^*, C_0^*\right)$ and sends it to ϑ.

The attacker ϑ gives a guess value $b' \leftarrow \{0, 1\}$, if $b' = b$, declares the attacker ϑ to win, and defines the advantage of the attacker to break the R.HIB-CPRE algorithm as ε.

Game1. In Game1, the public parameters and private key are generated by the challenger ξ simulating the real system, and the other phase are the same as Game0.

Setup. The challenger ξ uniformly random sampling polynomial $\bar{s}, e \leftarrow \chi$ from discrete Gaussian distribution χ, uniformly random sampling of $m + 1$ polynomials $a_0, a_1, \cdots, a_m \leftarrow \mu_q$, compute $s_{N,id} = p\bar{s} + 1$, $b_i = a_i s_{N,id}^{-1} + e$, output $m + 1$ polynomials $b_0, b_1, \cdots, b_m$. Finally send $PP = (n, p, q, \alpha, a_0, a_1, \cdots, a_m, b_0, b_1, \cdots, b_m)$ to ϑ.

Query phase

Private Key Queries. The attacker ϑ queries the private key of the senior user *id* and the ordinary user *id'*. The challenger ξ Compute $s_{N,id} = p\bar{s} + 1$, $s_{R,id'} = s_{N,id}^{-1} + e'$ outputs the private key $s_{N,id}$ of the senior user *id* and the private key $s_{R,id'}$ of the ordinary user *id'* then sends them to the attacker ϑ.

The re-encryption key queries, encryption queries and challenge phase are the same as Game0, we will not be described again.

Game1 and Game0 are indistinguishable from polynomial time attackers ϑ.

The setup phase of Game1 and the private key query of the senior users are the same as the analysis in the previous section. This section only analyzes the private key query of ordinary users.

For private key query phase of the ordinary users, according to $b_i = a_i s_{N,id}^{-1} + e$, $b_{id'} = b_0 + \sum_{i=1}^{m} id'_i b_i$, $a_{id'} = a_0 + \sum_{i=1}^{m} id'_i a_i$ we can obtain $b_{id'} = a_{id'} s_{N,id}^{-1} + e$, according to $a_i s_{R,id'} + e = a_i s_{N,id}^{-1} + e + a_i e'$, we can obtain $a_{id'} s_{R,id'} + e = a_{id'} s_{N,id}^{-1} + a_{id'} e' + e = b_{id'} + a_{id'} e'$, therefore $s_{R,id'}$ approximately satisfies $b_{id'} = a_{id'} s_{R,id'} + e$, this approximation is indistinguishable to the attacker ϑ, so the ordinary user's private key $s_{R,id'}$ is indistinguishable to the attacker ϑ compared to Game0.

Game2. In Game2, the challenge ciphertext generation process is different from Game1, and the rest of the process is the same.

In Game2, $C_{H^*} = \left(C^*, C_0^*\right)$ is randomly sampled from $R_q \times R_q$.

Game1 and Game0 are indistinguishable from polynomial time attackers ϑ.

Lemma 2. Assume the attacker ϑ can distinguish Game2 and Game1 with a non-negligible advantage ε, then there must be a polynomial time algorithm η that can solve the RLWE decision problem with the advantage $\frac{\varepsilon}{2}$.

For the instance $(b_{H^*}, C^{*\prime})$, when the attacker guesses b correctly, the polynomial time algorithm η outputs 1 and judges it as an RLWE instance, otherwise judges it as a instance which is uniformly random sampled from $R_q \times R_q$. According to the different generation algorithms of the instance $(b_{H^*}, C^{*\prime})$, the discussion can be divided into the following two situations:

1) $(b_{H^*}, C^{*\prime})$ is uniformly random sampled from $R_q \times R_q$, then $C^* = pC^{*\prime} + m_b$ obeys uniform distribution, the challenge ciphertext $C_{H^*} = (C^*, C_0^*)$ also obeys uniform distribution. At this point, ϑ guesses the right b by the probability of $\frac{1}{2}$, therefore, the probability that η outputs 1 is also $\frac{1}{2}$.
2) $(b_{H^*}, C^{*\prime})$ is RLWE instance, let $C^{*\prime} = b_{H^*}r + e_0$, $C^* = p(b_{H^*}r + e_0) + m_b$, then the challenge ciphertext $C_{H^*} = (C^*, C_0^*)$ obeys uniform distribution. At this point, ϑ guesses the right b by the probability of $\frac{1}{2} + \varepsilon$, therefore, the probability that η outputs 1 is also $\frac{1}{2} + \varepsilon$.

Therefore η solves the RLWE decision problem with at least the probability of $\frac{1}{2}(\frac{1}{2} + \varepsilon + \frac{1}{2}) = \frac{1}{2} + \frac{\varepsilon}{2}$, the advantage is $\frac{1}{2} + \frac{\varepsilon}{2} - \frac{1}{2} = \frac{\varepsilon}{2}$. Since the RLWE decision problem is difficult, the advantage of solving in polynomial time is negligible. Therefore, assume the attacker ϑ can distinguish the advantage of Game2 and Game1 is negligible, Game2 and Game1 are indistinguishable for η.

In summary, based on the RLWE decision assumption, the advantage of the polynomial time IND-CPA attacker in Game0 is negligible, and the R.HIB-CPRE scheme is IND-CPA security under the standard model.

6 Comparison

Compared with the existing identity-based proxy re encryption [13–17] Based on lattice difficulty problem, this scheme realizes non interactivity, unidirectionality and collusion resistance. At the same time, it refines the user's identity into different levels and adds the function of hierarchical encryption, which is more suitable for the current application scenario of user hierarchical management. The specific comparison is shown in the following Table 1.

The computation unit of our scheme is polynomial, and the main computational cost is polynomial multiplication. Calling Number Theoretic Transforms (NTT) for polynomial multiplication can effectively speed up the operation efficiency. The computational cost of the identity-based proxy re-encryption scheme based on the LWE [13–17] mainly comes from the matrix multiplication operation, and the computational complexity of NTT is usually lower than that of the matrix operation, therefore our scheme is compared with the existing LWE identity-based proxy re-encryption scheme will be more efficient. Moreover, the user private key, the size of the re-encryption key and the size of the ciphertext of our scheme are smaller than the existing LWE-based identity-based proxy re-encryption scheme. The specific comparison is shown in the following Table 2.

Table 1. Scheme functional comparison

Cryptosystem	Difficult assumption	Non-interactivity	Directionality	collusion resistance	Message length (bit)	Hierarchical encryption
[13]	LWE	×	×	×	Singe-bit	×
[14]	LWE	×	×	×	Multi-bit	×
[15]	LWE	√	√	√	Multi-bit	×
[16]	LWE	√	√	√	Multi-bit	×
[17]	LWE	√	√	√	Multi-bit	×
Our scheme	RLWE	√	√	√	Multi-bit	√

Table 2. Computational complexity and size comparison

Cryptosystem	User private key size (bit)	Re-encryption key size (bit)	Ciphertext size (bit)	Computational complexity
[13]	$m \log q$	$m \log q$	$(m+1)\log q$	nm/m
[14]	$mn \log q$	$mn \log q$	$2m \log q$	nm/m^2
[15]	$2m(2m+l)\log q$	$4mm \log q$	$(2m+l)\log q$	2 nm/4 m^2
[16]	$2mnl \log q$	$(m+2nl)^2 \log q$	$m + 2nl \log 2q$	$mn + 2n^2 l/(m+2nl)^2$
[17]	$2ml \log q$	$(2mt+l)(2m+l)\log q$	$(2m+l)\log q$	2 nm/4 nm
N.HIB-CPRE	$n \log q$	$n \log q$	$2n \log q$	1 NTT/t NTT
R.HIB-CPRE	$n \log q$	$2n \log q$	$3n \log q$	2 NTT/2 NTT

Note: $m > 6n \log q$, $t = \lceil \log q \rceil$, $l = O(\log q)$, Computation complexity: the multiplication computation contained in each encryption/re-encryption, the computation complexity of the scheme [13–17] is the number of multiplication computation in matrix operations, the computation complexity of our scheme is the number of times NTT is called for polynomial multiplication

7 Conclusions

This paper combines the idea of hierarchical identity encryption with proxy re-encryption, uses NTRU encryption variants and RLWE public key encryption as the underlying encryption scheme, divide users into advanced users and ordinary users. According to the difference of the key reversibility and the ciphertext share between the two systems, realize the control of the decryption authority of different levels of users. so that the transformed ciphertext still satisfies the rules of hierarchical identity encryption, our scheme is more applicable than the existing identity-based proxy re-encryption scheme, and is more suitable for the current hierarchical user management system, and its computational complexity and size are better than the LWE-type identity-based proxy re-encryption scheme. In the next step, we will try to implement the scheme by software and extend its functionality and construct a verifiable, CCA security HIB-CPRE scheme.

References

1. Horwitz, J., Lynn, B.: Towards hierarchical identity-based encryption. In: Advances in Cryptology-EUROCRYPT 2002, pp. 466–481. Springer, Heidelberg (2002). Doi: https://doi.org/10.1007/3-540-46035-7_31
2. Blaze, M., Bleumer, G., Strauss, M.: Divertible protocols and atomic proxy cryptography. In: Nyberg, K. (ed.) EUROCRYPT 1998. LNCS, vol. 1403, pp. 127–144. Springer, Heidelberg (1998). https://doi.org/10.1007/BFb0054122
3. Ateniese, G., Fu, K., Green, M., et al.: Improved proxy re-encryption schemes with applications to secure distributed storage. ACM Trans. Inf. Syst. Secur. (TISSEC) **9**(1), 1–30 (2006)
4. Weng, J., Chen, M., Yang, Y., et al.: CCA-secure unidirectional proxy re-encryption in the adaptive corruption model without random oracles. SCIENCE CHINA Inf. Sci. **53**(3), 593–606 (2010)
5. Xagawa, K., Tanaka, K.: Proxy re-encryption based on learning with errors. In: Proceedings of the 2010 Symposium on Cryptography and Information Security—SCIS 2010, pp. 29–35 (2010)
6. Wang, X., Hu, A., Fang, H.: Improved collusion-resistant unidirectional proxy re-encryption scheme from lattice. IET Inf. Secur. **14**(3), 342–351 (2006)
7. Wang, X., Hu, A., Fang, H.: Multi-user file-sharing systems based on LWE. China Commun. **17**(7), 166–182 (2020)
8. Li, J., Ma, C., Gu, Z.: Multi-use deterministic public key proxy re-encryption from Lattices in the auxiliary-input setting. Int. J. Found. Comput. Sci. **31**(05), 551–567 (2020)
9. Polyakov, Y., Kurt, R., et al.: Fast proxy re-encryption for publish subscribe systems. ACM Trans. Privacy Secur. **20**(4), 1–31 (2017)
10. Nuñez, D., Agudo, I., Lopez, J.: NTRUReEncrypt: an efficient proxy re-encryption scheme based on NTRU. In: Proceedings of the 10th ACM Symposium on Information, Computer and Communications Security- ASIA CCS 2015, pp. 179–189. ACM (2015)
11. Zhang, M., Du, L.: A collusion-resistant and uni-directional proxy re-encryption scheme based on NTRU. J. Cryptologic Res. **7**(2), 187–196 (2020)
12. Green, M., Ateniese, G.: Identity-based proxy re-encryption. In: International Conference on Applied Cryptography and Network Security, pp. 288–306. Springer, Berlin, Heidelberg (2007). Doi: https://doi.org/10.1007/978-3-540-72738-5_19
13. Singh K, Rangan C P, Banerjee A K: Lattice based identity based proxy re-encryption scheme. J. Internet Serv. Inf. Secur. (JISIS) **3**(3/4), 38–51 (2013)
14. Jiang, M., Guo, Y., Yu, L., et al.: Efficient identity-based proxy re-encryption on lattice in the standard model. J. Electron. Inf. Technol. **41**(1), 61–66 (2019)
15. Hou, J., Jiang, M., Guo, Y., et al.: Efficient identity-based multi-bit proxy re-encryption over lattice in the standard model. J. Inf. Secur. Appl. **47**, 329–334 (2019)
16. Wu, L., Yang, X., Zhang, M., et al.: New identity based proxy re-encryption scheme from lattices. China Commun. **16**(10), 174–190 (2019)
17. Dutta, P., Susilo, W., Duong, D.H., et al.: Identity-based unidirectional proxy re-encryption in standard model: a lattice-based construction. In: You, I. (eds.) WISA 2020. LNCS, vol. 12583, pp. 245–257. Springer, Cham (2020). Doi:https://doi.org/10.1007/978-3-030-65299-9_19
18. Wu, L., Yang, X., Zhang, M., et al.: IB-VPRE: adaptively secure identity-based proxy re-encryption scheme from LWE with re-encryption verifiability. J. Ambient Intell. Humanized Comput., 1–14 (2021)

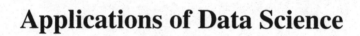

Applications of Data Science

Intelligent Service Robot for High-Speed Railway Passengers

Ruyu Sheng, Yanqing Wang(✉), and Longfei Huang

College of Information Engineering, Nanjing Xiaozhuang University, Nanjing 211171, China

Abstract. With the rapid development of road traffic, the number of high-speed rail passengers is huge, and the flow of people is dense. In epidemic situation, it is prone to intensive infection in high-speed rail carriages, which is not conducive to national prevention and control work. Based on face recognition technology, the intelligent service robot for high-speed rail passengers walks in accordance with the set route and detects the face mask of high-speed rail passengers. The face database of high-speed rail passengers is compared in real time. The passengers who do not wear masks are reminded in time to reduce the risk of infection. Moreover, the robot can accurately remind the passengers of leaving the station in time, and has the functions of automatic selling and student ticket checking. The experimental result is shown to promote the further development of high-speed rail services.

Keywords: High-speed rail service robot · Face mask recognition detection · Neural network · Action recognition

1 Introduction

In this paper, based on face recognition technology and action recognition technology, high-speed rail passengers can only serve robots. The system will promote the further development of face recognition technology and action recognition technology in China's epidemic prevention, transportation, security, e-commerce and other fields, make up for the blank of the domestic market and break the monopoly of foreign technology. It can effectively alleviate the existing problems of high-speed rail, such as passengers missing the departure time, the existence of escape tickets, ticket purchase information does not meet the current identity, sitting wrong position. At the critical moment when the rapid and large-scale expansion of high-speed rail network drives economic development, the development of epidemic prevention and the improvement of service quality in high-speed rail are also urgent. The intelligent service robot for high-speed rail passengers will make the high-speed rail service abroad more humanized and promote the intelligent development of high-speed rail.

J. Zeng et al. (Eds.): ICPCSEE 2021, CCIS 1452, pp. 263–271, 2021.
https://doi.org/10.1007/978-981-16-5943-0_21

2 The Working Process of Robot

2.1 Set the Robot Route and Pass the Passenger Seat Number and Face Information

The position coordinates of the robot are set according to the size of the car. The robot establishes a map for positioning, moves back and forth within the set range, and collects face information through the camera. The passenger seat number and face information of the train are stored in advance in the high-speed railway passenger intelligent service robot database, which provides data basis for the next face comparison.

2.2 Collect Current Position Face Information and Compare with Passenger Information

After the robot opens the camera, the gray processing of each frame obtained by the camera is carried out, and the collected face information is compared with the passenger information of the robot database. If the current passenger does not conform to the current position, the voice broadcast warning is carried out to solve the problem of passenger sitting wrong position and timely prevent the emergence of ticket evasion.

2.3 Detect Whether the Passenger Wears a Mask and Accurate Voice Broadcast Reminders

The python deep learning framework TensorFlow and Keras are mainly used for deep learning training [1], and the processed images are judged. If the passenger does not wear a mask [2], the voice broadcast warning is carried out.

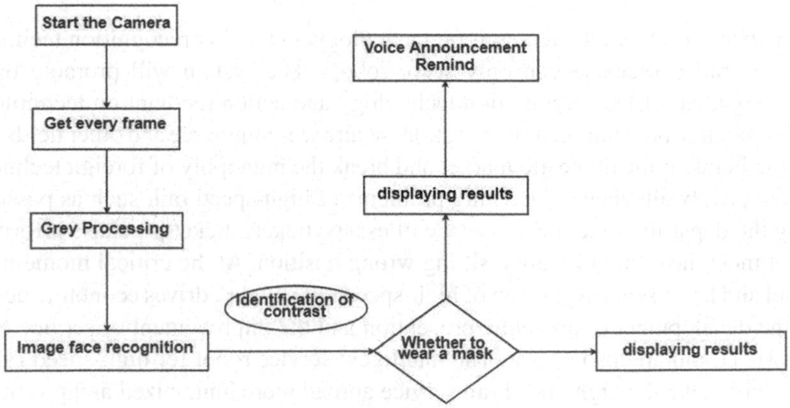

Fig. 1. Face mask recognition and detection process

3 The Realization Function of Robot

(1) Face mask recognition and detection: the intelligent service robot for high-speed rail passengers proposed in this paper can accurately remind passengers who does not wear a mask, and track the face of the diagnosed patients who have travelled through high-speed rail. When tracking the patient, face recognition is also carried out on the contact of the diagnosed patients on the high-speed rail, and the contact information is output in form.
(2) Intelligent reminding high-speed railway passengers to get to the station: the passenger's state is judged by face positioning technology and action recognition. If the passenger who should leave the bus does not have the relevant action to leave the station, the voice broadcast is used to remind the passenger to leave the station.
(3) To solve the problem of passenger position mismatch and flight tickets: the intelligent service-oriented robot of high-speed railway passengers detects and collects the passenger information in the positioning direction through its pre-set displacement route and position information. If the passenger does not meet the position, it will be reminded. If there is no change in passenger information, the flight ticket warning voice will be issued.
(4) Automatic ticket checking: the passengers who buy the student tickets check the tickets themselves. The students put the student certificate in the identification place at the general control panel for the detection of the student certificate stamp checking module. The robot extracts and detects the seals of different colors or types and compares the face information to check the tickets, without the need for staff to check the information one by one. Passengers who purchase adult tickets are controlled by the background for random checking. Passengers only need to put the identity card in the identification place out of the general control panel for identity card detection and face comparison.
(5) Mobile self-sale food: the automatic sale function based on face recognition can meet the different payment methods of passengers. When entering the face payment mode, the robot will automatically shut down the data transmission of video stream and switch to the camera interface. After the successful comparison, the serial video transmission will be restored. Finally, the payment of passengers is completed. Under the action of electromagnetic relay, the goods are successfully sold (Fig. 2).

4 Introduction and Analysis of Key Technologies

4.1 Target Detection Based on Deep Learning R-CNN Model

The R-CNN model [3] based on deep learning is used for target detection, and TensorFlow is used as the deep learning framework [4] to train through the COCO dataset. The R-CNN algorithm process is the input image, and each image generates 1000 to 2000 candidate regions. The convolutional neural network (CNN) is used to extract features for each candidate region, and the features are sent to the SVM classifier for category judgment. Finally, the regression is used to finely correct the position of the candidate box.

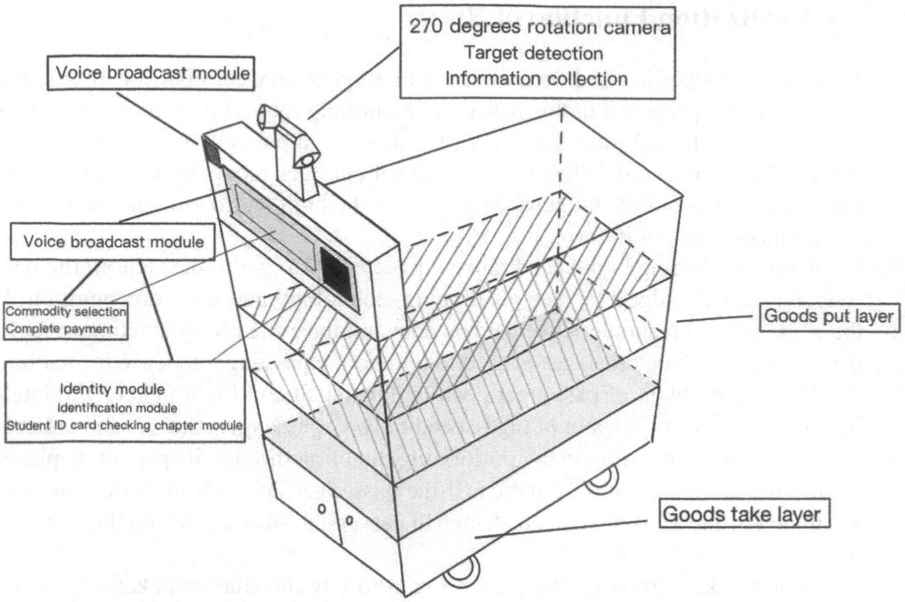

Fig. 2. Intelligent service robot for high-speed railway passengers

When generating candidate regions, the selected search element method is adopted. Firstly, the image is segmented into a large number of small regions. Then, according to the color (color histogram) similar, texture (gradient histogram) similar, the total area after merging is small, and the priority regions with a large proportion of the total area in its BBOX are merged. Finally, all regions that have existed, namely candidate regions, are output. The candidate regions were normalized to the same size before feature extraction, and then trained using CNN model. Category judgment is based on linear SVM classifier. Position refinement is performed by training a linear regression model.

4.2 Face Mask Recognition and Detection Based on Deep Learning

The detection technology mainly uses python deep learning framework TensorFlow and Keras to conduct deep learning training through network images as face mask data. The detection model is relatively small, the backbone network has only eight convolution layers, positioning and classification layer has 16 layers. Although for long-distance face detection effect is general, but for ordinary face can be detected successfully and at the same time can be multiple face recognition judgment, the model specific operation process as Fig. 1. The continuous movement of high-speed railway passenger intelligent service robot along a fixed route can solve the problem of instability of long-distance face mask detection [5]. The model connects the positioning classification layer on five convolution layers, and its size and anchor setting information are shown in Table 1. For example, the size of the first convolution layer feature map is 33 * 33, and the input image point corresponding to the center point of the sliding window of 33 * 33 is selected as anchor point. The size of anchor is (0.04, 0.056), and the aspect ratio under each area size is (1, 0.62, 0.42). The calculation formula of anchor size is:

$$S_k = S_{min} * 100 + \frac{S_{max} * 100 - S_{min} * 100}{m - 1} * (k - 1) \tag{1}$$

It is the ratio of the prior box size of each feature layer to the original image size, and the maximum and minimum ratios, respectively. m is the number of feature layers, k is the number of layers of the current feature layer. The model is set to 0.002 and 0.0004 respectively. Because the model connects the location classification layer on five convolution layers, m is 5.

The face mask detection code running environment is *anaconda3* + *python3.6* + *tensorflow2.0.0* + *opencv4.5.1* + *keras2.3.1*, and its running effect Fig. 3.

Table 1. Convolutional layer size and anchor information settings

Convolutional layer	Size of feature map	Size of anchor	Anchor aspect ration
The first layer	33 * 33	0.04, 0.056	1, 0.62, 0.42
The second layer	17 * 17	0.08, 0.11	1, 0.62, 0.42
The third layer	9 * 9	0.16, 0.22	1, 0.62, 0.42
The fourth layer	5* 5	0.32, 0.45	1, 0.62, 0.42
The fifth layer	3 * 3	0.64, 0.72	1, 0.62, 0.42

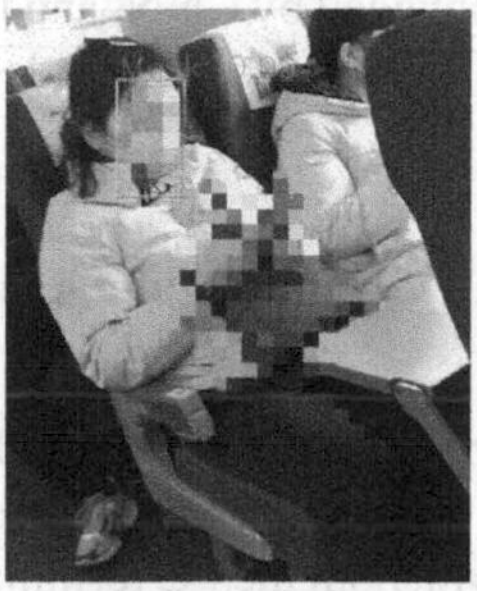

Fig. 3. Detection effect of face mask

4.3 Face Recognition Technology Based on Opencv

(1) Face acquisition data: the face data collected by the face gate before entering the station are used here.
(2) Identify human face and eyes in the camera: use opencv built-in classifier for scanning.

(3) Training data: the data of each frame of the camera are saved, and the camera recording is withdrawn after 1200 samples. Then, the collected face data are trained to extract features [6].
(4) Face test: the face in the camera is compared with the features in the model to determine whether it is the person and show the relevant information. After the recognition is successful, the classified tickets are divided into adult tickets and student tickets. Passengers provide different certificates, such as identity cards and student certificates, and put them in the recognition place of the general control panel of the robot. The robot compares the extracted face information with the certificate information provided by the passengers to complete the ticket inspection (Fig. 4).

4.4 Behavioral Action Recognition Technology Using Convolutional Neural Network or Cyclic Neural Network and its Extended Model

Convolutional neural network: the video is divided into static frame data stream and inter-frame dynamic data stream. The static frame data stream can use single frame data, and the inter-frame dynamic data stream uses optical flow data. In each data, deep convolutional neural network is used for feature extraction [7, 8], and SVM is used for action recognition. Only the relevant data of the joint part of the human posture is used for feature extraction by deep convolution network. Finally, the whole video is converted into a feature vector by statistical method, and SVM is used to train and identify the final classification model. Loop neural network and its extended model: this network combines CNN and LSTM [9, 10] to extract features from video data. The image information of a single frame is obtained by CNN, and then the output of CNN is passed through LSTM in chronological order, so that the video data is finally characterized in spatial and temporal dimensions.

4.5 Robot Self-service Vending Machine Management and Control System Based on JAVA SMM Framework

The automatic receiving system needs to meet the mobile payment function, and it needs to be managed and controlled in the background. Therefore, when selecting the framework, it needs to be compatible with both the PC terminal and the mobile phone terminal [11]. Therefore, the cross-platform framework based on JAVA SMM framework is proposed. The SMM framework set includes Spring framework, SpringMVC framework and MyBatis framework. The SMM framework is divided into Controller layer, Service layer and Dao layer. Each layer is responsible for the control of the specific business module process, the logic application design of the business module, the interaction design with the database, and the display of the front jsp page. Service layer is built on the Dao layer, under the Controller layer. So the Service layer calls both the Dao layer interface and the class interface to the Controller layer (Fig. 5).

Compared with Struts2, SpringMVC module has more advantages in terms of project management and security [12], and the number of referred files is relatively small. When dealing with Ajax requests, it can directly obtain the returned data of Ajax.

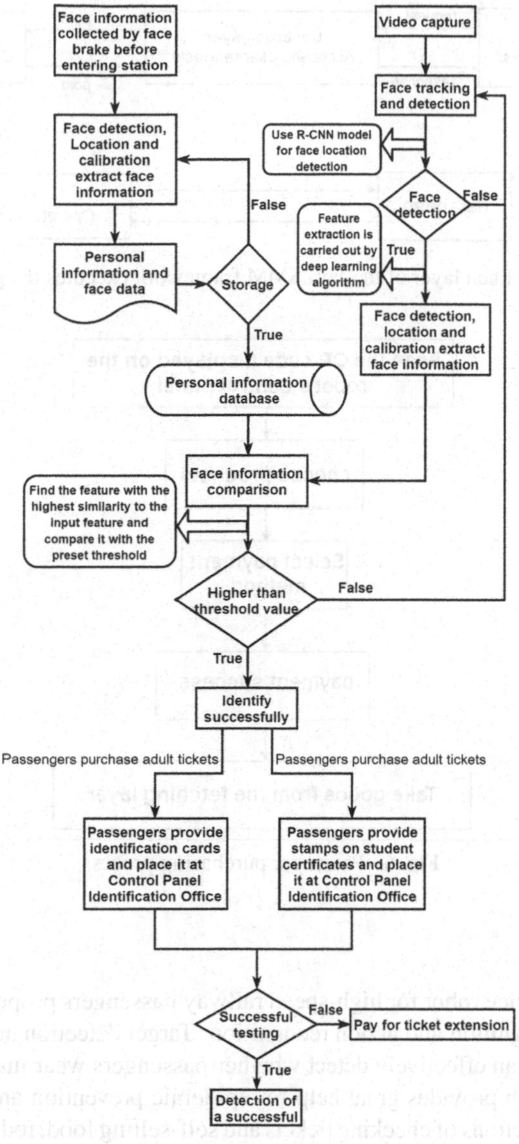

Fig. 4. Face recognition technology process

MyBatis [13] supports flexible SQL statements, supports stored procedures, and has flexible and convenient operation and maintenance.

The system security architecture based on Apache Shiro [14] is powerful and easy to use. This Java security framework can perform authentication, authorization, encryption [15] and session management (Fig. 6).

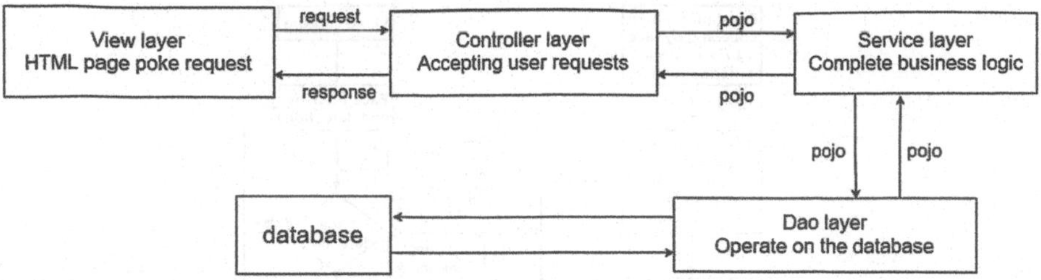

Fig. 5. Each layer of the Java SMM framework executes the process

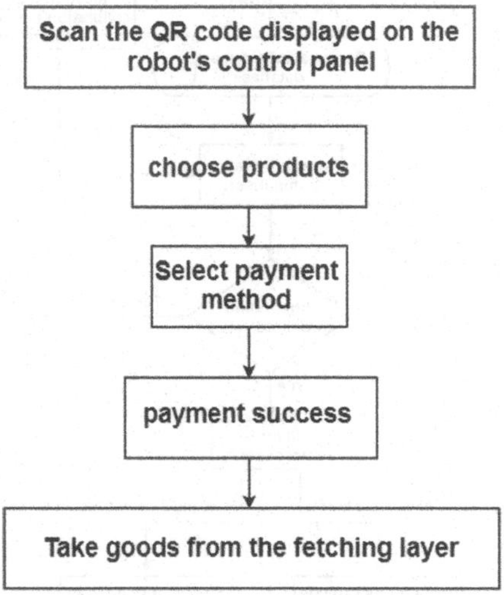

Fig. 6. Passenger purchasing process

5 Conclusion

The intelligent service robot for high-speed railway passengers proposed in this paper is based on face recognition and action recognition. Target detection and passenger information collection can effectively detect whether passengers wear masks and accurately remind them, which provides great help for epidemic prevention and control. Furthermore, it has the functions of checking tickets and self-selling food, reducing the workload of crews, avoiding ticket evasion, improving passengers' sense of experience and so on.

Acknowledgement. This article is supported by the 2020 Innovation and Entrepreneurship Training Program for College Students in Jiangsu Province (Project name: high-speed railway passenger behavior management system, No. 202011460091T).

This article is supported by the National Natural Science Foundation of China Youth Science Foundation project (Project name: research on Deep Discriminant Spares Representation Learning Method for Feature Extraction, No. 61806098).

This article is supported by Scientific Research Project of Nanjing Xiaozhuang University (Project name: multi-robot collaborative system, No. 2017NXY16).

References

1. Weijun, X.: Application of deep learning in computer vision field. Electron. Technol. **50**(05), 20–21 (2021)
2. Peng, Z., Zhikun, Y., Kaisong, L.: An occlusion face recognition algorithm based on feature restructuring and optimization training. J. Dalian Jiaotong Univ. **42**(03), 116–120 (2021)
3. Jialiang, M., Bin, C., Xiaofei, S.: A general target detection framework based on improved Faster R-CNN. Comput. Appl. 1–9 (2021)
4. Changjian, W., Yong, D., Pancheng, L.: Faster R-CNN target detection by fusion and improved correlation network. Faster FPN and Faster CNN target detection. Comput. Eng. 1–13 (2021)
5. Zhuqing, X., Xingyuan, W., Nan, X.: Research on face selection optimization algorithm in surveillance video. Electron. Technol. Softw. Eng. (09), 104–105 (2021)
6. Shu, L.: Application research of face recognition system based on OpenCV. Inf. Syst. Eng. (12), 85–87 (2020)
7. Yuanpan, Z., Guangyang, L., Ye, L.: Overview of deep learning application in image recognition. Comput. Eng. Appl. **55**(12), 20–36 (2019)
8. Hongtao, L., Qinchuan, Z.: Research overview of deep convolutional neural network in computer vision. Data Acquisition Processing **31**(01), 1–17 (2016)
9. Tongtong, S., Huazhi, S., ChuMei, M.: Human behavior recognition based on cyclic neural network. J. Tianjin Normal Univ. Nat. Sci. Ed. **38**(06), 58–62 (2018)
10. Yang Li, W., Yuqian, W.J.: Review of research on cyclic neural networks. Comput. Appl. **38**(S2), 1–6 (2018)
11. Changping, Z., Ou, J.: Vending machine based on mobile payment. Mod. Electron. Technol. (17) 38–40 (2004)
12. Yuxin, Z.: Design and implementation of patriotic education website based on SpringMVC framework. Mech. Electr. Inf. (20), 131–133 (2020)
13. Huijun, D.: Design and implementation of reading website based on Spring and MyBatis framework. Inf. Comput. (Theor. Ed.) **33**(04), 97–99 (2021)
14. Qinghua, L., Anming, H.: Research on the security application of Apache Shiro framework in web system. Comput. Knowl. Technol. **17**(06), 52–53 (2021)
15. Wei, W.: Security intelligence application based on face recognition technology. Digit. Technol. Appl. **39**(04), 146–148 (2021)

An M-ALO-SVR Model to Predict Output of Industry-University-Research Collaboration Network in China

Ruiqiong Zhong[1](✉), Ben Wang[2], and Gege Feng[3]

[1] Jinan University, Guangzhou 510632, Guangdong, People's Republic of China
200911843@oamail.gdufs.edu.cn
[2] Guangdong University of Foreign Studies, Guangzhou 510006, Guangdong, People's Republic of China
[3] The Hong Kong Polytechnic University, Hong Kong, Hong Kong

Abstract. The output prediction of industry-university-research cooperation network is a prerequisite for optimization of network resource allocation and improvement of network innovation performance. Accurate prediction of network output can provide data for feedback systems, offer methods and reference to government macro-level control, and avoid resource wastes caused by improper input of capital and humans. In this paper, a prediction model based on Ant Lion Optimizer and Support Vector Regression is proposed. First, the M-ALO-SVR model was built. Then, Pareto function and regulatory factors were applied to accelerate the convergence of ALO-SVR optimization, improving the global search ability of the algorithm. Finally, the empirical research of the industry-university-research cooperation network was implemented, and simulation experiments were conducted with samples of China Statistical Yearbooks. The results show that the M-ALO-SVR model performs well in the innovation network output prediction. The predictive goodness of fit of the model reaches 99.7%, improved by 0.02% and 3.3% respectively compared with that of ALO-SVR model and SVM. The running time of the model is two seconds fewer than that of the ALO-SVR model. In addition, the optimizing function of the model converges at higher speed and its MSE is optimal.

Keywords: Support vector regression (SVR) · Ant Lion Optimizer (ALO) · Innovation network · Prediction

1 Introduction

In 2017, China's R&D expenditure accounted for 2.15% of GDP, equal to that of moderately developed countries. But the transfer rate of innovation achievements in China, which was less than 10%, was much lower than that in developed countries with 70%. Meanwhile, the transfer benefit of technology in Chinese universities, scientific research institutions and enterprises was generally low. According to the 2018 China Patent survey report, the total transfer rate of valid patents in China was 36.3%, among which the

J. Zeng et al. (Eds.): ICPCSEE 2021, CCIS 1452, pp. 272–282, 2021.
https://doi.org/10.1007/978-981-16-5943-0_22

enterprises took up the highest percent, with 46.0%, and the universities accounted for the lowest, with only 2.7% [1].

With a variety of cooperation, such as patent application, joint publication of papers, co-construction of R&D centers or laboratories, and movement of R&D personnel, the universities, enterprises and scientific research institutes built two-way connections among each other, forming the industry-university-research cooperation network. But despite the increasing R&D investment year by year, the patents measuring the quality of scientific and technological innovation of the universities, enterprises and scientific research institutes in China were still much less than that of the United States and other developed countries, the transfer rate of scientific and technological achievements was low, and the effect of industry-university-research cooperation was far from the expected level. On these grounds, it is necessary to predict the output of the industry-university-research cooperation network, optimize the allocation of resources, and improve the innovation performance of the network. Accurate prediction of network output can provide data basis for the feedback system, offer methods and reference to government macro-level control, and avoid resource wastes caused by improper input of capital and humans.

In recent years, the prediction of innovation output mainly includes patent prediction based on Support Vector Regression [2] as well as science and technology output prediction based on BP neural network [3]. The prediction of industry-university-research cooperation network mainly focuses on potential partner prediction based on link prediction method [4]. Nevertheless, there is no much research about output prediction of industry-university-research cooperation network. Meanwhile, the prediction accuracy of the Support Vector Regression is only 77.27%, which is not ideal; prone to overfitting, the BP neural network converges at low speed and easily gets into local extremum, and to adjust its parameters is a complex issue [5, 6].

In this paper, the Modified Ant Lion Optimizer-Support Vector Regression (M-ALO-SVR, Abbreviation as MAS) prediction model is proposed based on the dynamic and nonlinear industry-university-research cooperation network. To start with, build the M-ALO-SVR model. And then, accelerate the convergence for global optimization of the ALO-SVR (AS) with Pareto function and regulatory factors. Finally, conduct positive research on the innovation network of industry-university-research cooperation, and compare the performance of MAS with that of AS and SVM.

The remaining sections are arranged as follows: Sect. 2 introduces work related to building the M-ALO-SVR model, Sect. 3 shows the complete M-ALO-SVR model, Sect. 4 demonstrates the empirical research, including the three experiments and discussion, and the last section presents the conclusion and the prospect for future research.

2 Related Work

2.1 SVR Algorithm

Support Vector Regression (SVR) is an important application of Support Vector Machine (SVM), whose modeling capability with small sample data meets the needs of innovation network output prediction.

The decision function is built as follow [7]:

$$f(X) = \omega^T \phi(X) + b \tag{1}$$

where ω is weight vector; $\phi(X)$ is a set of nonlinear mapping functions; b is deviation, which determines the distance between the hyperplane and the origin. On the premise that the total deviation is ε, the slack variable $\xi_i^* \geq 0$ $\xi_i \geq 0$ and the penalty factor C are added for adjustment. The optimization problem can be described as follow [7]:

$$R(\omega, \xi, \xi^*) = \min[\frac{1}{2}\|\omega\|^2 + C\sum_{i=1}^{l}(\xi_i + \xi_i^*)] \tag{2}$$

where the penalty factor C weighs the fitting ability and the prediction ability of test samples.

2.2 Ant Lion Optimizer (ALO)

Ant Lion Optimizer [8] is a continuously iterative process to search for the optimal solution. It is a bionic optimization algorithm proposed in 2015. The main steps are as follows:

(1) Ants walk randomly [9]:

$$X_t = [0, \cdots, cumsum(2r(t_k) - 1)] \tag{3}$$

where *cumsum* is cumulative sum, t step size of the random walk, t_k the number of iterations, and $r(t_k)$ the random function defined below [9]:

$$r(t_k) = \begin{cases} 1 \text{ } if \text{ } rand > 0.5 \\ 0 \text{ } if \text{ } rand \leq 0.5 \end{cases} \tag{4}$$

(2) Ants fall into the trap: to ensure the randomness of ants' walk in the search space, position of the ants is normalized.
(3) Ants fall into the center of the trap.
(4) Rebuild the trap: if the fitness of the updated ant after each of the iteration is greater than that of the selected ant lion, the ant will be captured.
(5) Elite strategy: the ant's random walk is affected by the global optimal ant lion. The position is updated according to the average of the result of roulette and the position of the elite ant lion.

3 Prediction Model for Innovation Output

3.1 Prediction Model

Based on SVR algorithm, the parameters of kernel function σ and the penalty factor C are needed for prediction. In this paper, Modified Ant Lion Optimizer (M-ALO) is used to optimize the parameters of SVR. The MAS model is shown as follow.

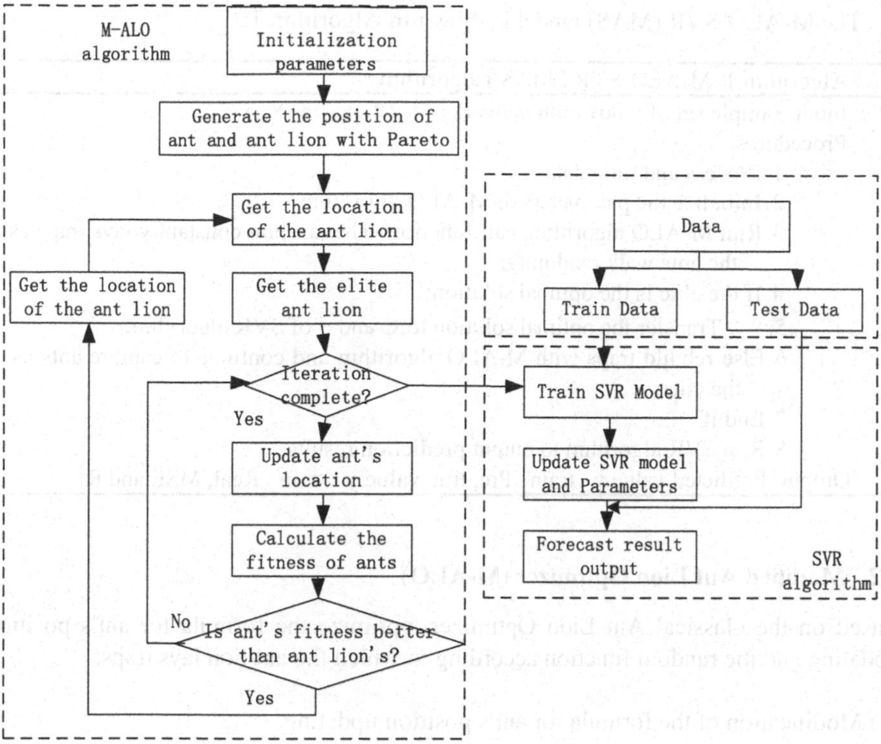

Fig. 1. M-ALO-SVR (MAS) Model

The major steps are as follows:

(1) Process data: preprocess and normalize the innovation network structure data and input data. Make the data produced in the kth loop as input, and that in the k + 1th loop as output.
(2) Set parameters for the model: the number of ant lions is 500, the maximum number of iterations 10000, the biggest σ 0.9 and the smallest 0.000001; the upper limit for optimization of the penalty factor C is 300 and the lower limit is 2.
(3) Apply M-ALO algorithm to optimize the parameters for the model: M-ALO algorithm is used to optimize SVR parameters.
(4) Predict innovation network output: validate the existing model with test data.

The M-ALO-SVR (MAS) model is shown in Algorithm 1:

Algorithm 1 M-ALO-SVR (MAS) algorithm

Input: sample set of innovation network $D = \{x_1, x_2, \cdots, x_m\}$

Procedures:

1. Process and train data;
2. Initialize the parameters of M-ALO algorithm;
3. Run M-ALO algorithm, call function. The ant lion constantly lays traps as the ants walk randomly;
4. If the elite is the optimal solution;
5. Transfer the optimal solution to C and σ of SVR algorithm;
6. Else rebuild traps with M-ALO algorithm and continue to capture ants as the elite;
7. End if
8. Run SVR algorithm to output prediction results;

Output: Predicted value y_ train_ Pre, true value y_ train_ Real, MSE and R^2

3.2 Modified Ant Lion Optimizer (M-ALO)

Based on the classical Ant Lion Optimizer, optimize the formula for ant's position updating and the random function according to which the ant lion lays traps.

(1) Modification of the formula for ant's position updating.

Since the ant lion is used to find a better solution near the suboptimal solution, the step size of the algorithm in the early stage should be larger than that in the later stage. In this paper, the adjusting factor $\frac{mc}{100e^{\lg(t)}}$ is introduced into the step size of the ant lion walk [10]:

$$R_E^t = R_E^t + (2rand - 1)\frac{mc}{100e^{\lg(t)}} \tag{5}$$

where mc is the maximum number of loops. As the number of iterations increases, the adjusting factor decreases from $(mc)/100$ to 0.01 so as to achieve automatic adjustment of the step size.

(2) Modification of the random function according to which the ant lion lays traps.

Function (4) generates random numbers between 0 and 1 with uniform distribution function. The normalized Pareto distribution probability density function with formal parameter $k \neq 0$, scale parameter σ and threshold parameter θ is shown as follow [11]:

$$y = f(x|k,\sigma,\theta) = \frac{1}{\sigma}(1 + k\frac{(x-\theta)}{\sigma})^{-1-\frac{1}{k}} \tag{6}$$

where $\theta < x$ when $k > 0$ and $\theta < x < \left(\theta - \sigma/_k\right)$ when $k < 0$.

This paper adopts the probability density function of Function (6) to replace the random function of Function (2) and (3).

The algorithm of M-ALO is shown in Algorithm 2.

Algorithm 2 pseudo code of Modified Ant Lion Optimizer (M-ALO)

Input: Ant Lion Optimizer parameters (colony_size, min_values, max_values, iterations, function)

Procedures:

Generate ants' position $\mathbf{M_{Ant}}$ and ant lion's position $\mathbf{M_{Antlion}}$ with Pareto distribution function, **count** = 0;

Do

If (**count** $\neq$ **0**)

Update ant lion's position $\mathbf{Antlion_j^t}$ **as Ant position** $\mathbf{Ant_i^t}$;

End If

Obtain the position of the ant lion $\mathbf{M_{Antlion}}$, and evaluate its position with fitness function $\mathbf{M_{OAL}}$;

Obtain the elite and keep it;

While (k> Ni iteration completed) do

Apply adjusting factor $\frac{\mathbf{mc}}{\mathbf{100e^{lg(t)}}}$, to update ants' position $\mathbf{X_i^t}$;

Calculate and keep the fitness of ants in matrix$\mathbf{M_{OA}}$;

End While

Mark **count as 1**;

While ($\mathbf{f(Ant_i^t) > f(Antlion_j^t)}$)

Output: the elite and the optimal position; output parameter C and σ to VAR algorithm

4 Empirical Research

4.1 Sample Selection and Index System

Based on results of Research on the Structure and Co evolution of Regional Innovation Network [12], and combined with the influencing factors of innovation performance in literature [13–15], the index system of innovation output prediction of the innovation network is shown in Table 1.

Table 1. Index system

Variable	Index type	Primary index	Secondary index
Dependent variable	Output index of innovation network	Technical performance	y1: Quantity of patent authorization
		Science and technology performance	y2: Published scientific and technical papers (piece)
		Industry performance	y3: Sales revenue of new high tech products (10000-yuan)

(*continued*)

Table 1. (*continued*)

Variable	Index type	Primary index	Secondary index
Independent variable	Investment index of innovation network	Personnel input	x1: Full time equivalent of R&D personnel (man-year)
		Capital investment	x2:Internal expenditure of Research and Development (R&D) (10000-yuan)
	Structure index of innovation network	Network openness	x3: Foreign investment (10000-dollar)
			x4: Contract amount of foreign technology import (10000-dollar
		Network centrality	x5: Technology market turnover (10000-yuan)
		Network link	x6: Enterprise funds in scientific and technical activities of colleges and universities (10000-yuan)
			x7: Government funds in scientific and technical activities (10000-yuan)
	Knowledge flow index	Knowledge transfer	x8: Revenue from transfer and license of patent ownership of scientific and technological activities in colleges and universities (10000-yuan)
		Knowledge overflow	x9: Science Citation Index (piece)
			x10: Engineering Index (piece)

The 310 samples from 2009 to 2018 include data about innovation network input, network structure and knowledge flow of 30 provinces, autonomous regions and the whole country. The purpose of this paper is to predict the performance of the innovation network with sample data of innovation input, network structure and knowledge flow in industry-university-research cooperation. Finally, the number of patent authorization is selected as the prediction content. The data and panel indices are from China Statistical Yearbook.

4.2 Performance Evaluation of the Model

MAS Algorithm and AS Algorithm. In this paper, R^2 and the Mean Square Error (MSE) function are applied to compare the performance of AS and MAS algorithm. The number of feature dimensions represents the number of secondary indices of independent variables. For example, feature dimension of 2 means cross validation based on variable x1 and x2, and feature dimension of 5 means cross validation based on variable x1, x2, x3, x4 and x5. All dimensions were tested when running the two algorithms. The average value of R^2 and MSE of the model were calculated by 10-fold cross validation, as shown in Table 2 and Table 3.

Table 2. R^2 and MSE of AS

Feature dimensions	R^2	MSE
2	0.992791768	387836030.0
7	0.996987931	162063170.4
10	0.996576628	184193129.7

Table 3. R^2 and MSE of MAS

Feature dimensions	R^2	MSE
2	0.992801562349606	387309023.188192
7	0.996991639	161863655.3
10	0.996577337	184155009.7

As shown in Table 2 and Table 3, the R2 and MSE of both AS and MAS reached the optimization with feature dimension of 7. In other words, the R2 values of AS with 0.996987931 and MAS with 0.996991639 indicated that these two algorithms achieved the best prediction fitness and accuracy when the seven factors were considered.

Comparison of Optimization Convergence Curves. The fitting curves of predicted values and true values of AS and MAS with 7 feature dimensions and the optimization convergence curves of different loops are shown in Fig. 2 and 3.

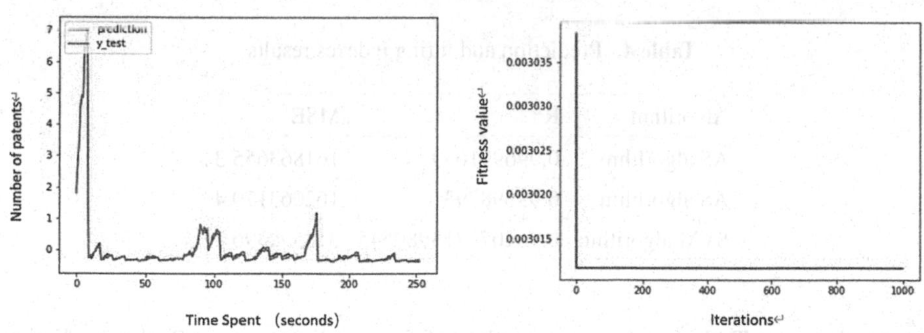

Fig. 2. Curves of AS optimization and function convergence with dimensions of 7

As shown on the left of Fig. 1, when the number of iterations is 150 or 200 around, there is deviation between the two curves. When the number is 50, the fitness on the right of the figure is 0.0032 and the curve converges at this value.

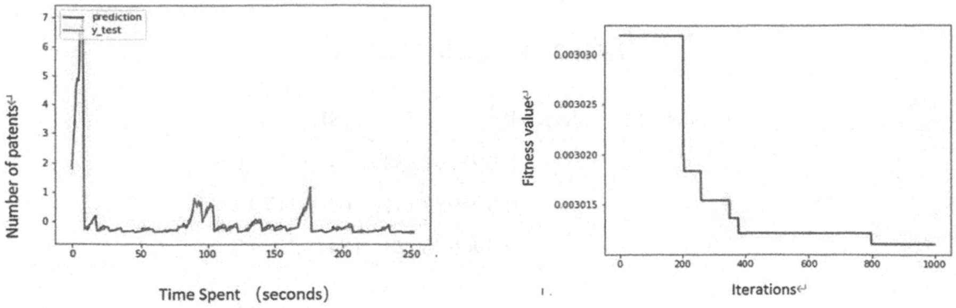

Fig. 3. Curves of MAS optimization and function convergence with dimensions of 7

As shown on the left of Fig. 2, the predicted values fit the real values by and large. When the number of iterations on the right is 200, the fitness value is 0.003014. As the number gradually decreases to 650 around, the fitness value remains unchanged. When the number of iterations is close to 1000, the fitness converges.

Figure 2 and 3 show the comparison of prediction results between AS and MAS. When the number of dimensions is 7, the fitting curves of the predicted value and the real value of the two methods show good fitting effect. As the number of iteration increases, the fitness value will decrease in varying degrees. The MAS makes a full global search available in the early stage and reduces the probability of falling into local optimum. Its curve enjoys faster and more accurate convergence.

4.3 Comparison of Different Models

Take 7 feature dimensions as input, run MAS algorithm, AS algorithm and SVM algorithm. Their R-square values are shown as Table 4.

Table 4. Prediction and fitting indexes results

Algorithm	R^2	MSE
AS algorithm	0.996991639	161863655.3
AS algorithm	0.996987931	162063170.4
SVM algorithm	0.964078713980545	318592396.8

As shown in Table 3, when the number of feature dimension is 7, the prediction fitness of MAS algorithm based on generating countermeasure network is 0.9972, and the MSE value of the model is 3.41282328993821e−23, while the fitness of RBF and

SVM model is 0.974, and the MSE value of mean square deviation is 318592396.8. The prediction fitness of MAS algorithm based on generated countermeasure network is 3.3% higher than that of SVM algorithm.

The running time of MAS algorithm, AS algorithm and SVM algorithm is 12.6 s, 14.5 s and 0.48 s respectively.

4.4 Discussion

The prediction results in Table 5 and Table 6 show that when the number of feature dimension is 7, the values of main evaluation indices of MAS model are different from that of other models as follows: (1) the R2 of MAS model is improved by 0.02% and 3.3% respectively compared with that of AS model, SVM model; (2) the MSE of MAS model is the lowest while that of SVM model is the highest; (3) the running time of MAS model is 2 s fewer than that of AS model. According to the comparison and analysis of the main indices, the prediction performance of MAS model is better than that of other models in the research.

The prediction results also show that the SVM model running RBF has the worst prediction performance in terms of accuracy and Mean Square Error because parameters have a great impact on the learning performance and generalization ability of SVM algorithm; the prediction performance of AS model is between that of SVM model and MAS model because Ant Lion Optimizer can find the optimal solution for SVR model, and hence the parameters of SVR model are optimized; the prediction accuracy, Mean Square Error, running time and generalization ability of MAS algorithm are better than AS algorithm, because the Pareto function and adjusting factors help the function converge and find the optimal solution faster.

The data of this paper is from China Statistical Yearbook on Science and Technology. The data volume is 310 and the volume of each province and city is different. The data of some provinces and cities in Tibet and Qinghai are even missing. Therefore, the MAS model of this paper is more suitable for small sample and nonlinear prediction based on small sample data.

5 Conclusion

In this paper, a deep learning model is built by combining Support Vector Regression and Ant Lion Optimizer, and advantages of the proposed model are analyzed and proved in small sample and unbalanced data prediction by predicting the innovation output of China industry-university-research cooperation innovation network. The proposed model can effectively solve the problems in the prediction of industry-university-research cooperation network. With high accuracy of prediction and generalization ability that meets the requirements of deep learning models, it can be widely used in practice. In the future, the author will further explore different effects of different features on the innovation output, and adopt the prediction model to design a feedback mechanism for innovation network so as to improve the interpretability of deep learning models and provide strategies and methods for improving innovation performance.

References

1. Intellectual property development research center of State Intellectual Property Office. China patent investigation report in 2018 (2020). http://www.iprdaily.cn/article_20784.html.
2. Xinyuan, W.: Research on core patent prediction from the perspective of technology domain segmentation. Shanxi University, Taiyuan (2018)
3. Zewen, H., Wuyishan.: Analysis and prediction of influencing factors of science and technology output-based on multiple regression and BP neural network. Sci. sci. sci. **30** (07), 992–1004 (2012)
4. Feifei, L.W., Wanzhao, L., Chenran, J., Yawen, H.: Research on potential cooperation opportunities of industry university research based on paper Patent Organization Cooperation Network. Inf. Sci. **37**(09), 9–16 (2019)
5. Chuang, M., Daiqi, Z., Ye, Z.: Prediction method of water resources demand based on BP neural network with improved whale algorithm. Comput. Sci. **47**(11), 32–36 (2020)
6. Juan, T.J.: Research and application of pso optimization neural network algorithm. Jiangsu University, Zhenjiang (2013)
7. Sun, H.: fTGARCH model and GARCH-SVR model to predict the volatility of financial time series. Dalian University of technology, Dalian (2020)
8. Leaveager: Ant Lion Algorithm.22 oct 2018. https://blog.csdn.net/qq_30142403/article/details/83268157.
9. Wang, R., Zhou, Y.W., Han, B., Li, J.F., Liu, Q.: Levy flying ant lion optimization guided by adaptive boundary and optimization. Microelectron. Comput. **35**(09), 20–25 (2018)
10. Huang, C., Zhao, K.: Three dimensional path planning of UAV with improved ant lion optimizer. Acta Electron. Sin. **40**(07), 1532–1538 (2018)
11. Subhashini, K.R., Satapathy, J.K.: Development of an enhanced ant lion optimization algorithm and its application in antenna array synthesis [J]. Appl. Soft Comput. **59**, 153–173 (2017)
12. Liangbing, W.: Research on the structure and co evolution of regional innovation network [D]. University of Science and Technology of China (2014)
13. Huiyan, W., Xinyun, L., Yinliang, X.: Research on performance evaluation and influencing factors of high quality economic development driven by scientific and technological innovation. Economist, 64–74 (2019)
14. Fengchao, L., Na, Z., Liangshi, Z.: Research on innovation efficiency evaluation of high tech manufacturing industry in Northeast China based on two-stage network DEA model. Manage. Rev. **32**(04), 90–103 (2020)
15. Xiyun, N.: Research on innovation efficiency of science and technology in innovative cities based on Improved DEA model. Hefei University of Technology (2019)

Prediction of Enzyme Species by Graph Neural Network

Tingyang Zhao(✉), Lina Jin, and Yinshan Jia

Liaoning Petrochemical University, Fushun, Liaoning, China

Abstract. Choosing an effective classification and recognition method in a large protein database plays a crucial role in the classification of enzymes. In previous studies on enzyme classification, only node characteristic information of amino acid were generally considered in the process of model training. The characteristics of amino acid nodes and topological structure in enzyme protein structure are proposed in this paper. The model was trained by graph neural network. By comparing with K nearest neighbor, support vector machine, random forest and multi-layer perceptron, it is shown that the graph neural network method has great advantages. The accuracy obtained by graph neural network is obviously higher than others.

Keywords: Classification of enzymes · Graph neural network · Receiver operating characteristic curve

1 Introduction

As a catalyst, enzymes play an important role in biomedical field, and the classification of enzymes is an important content in the field of enzymology. Therefore, how to achieve rapid and accurate classification of enzymes is very important [1]. With the development of science and technology, more and more enzymes with new functions have been discovered. In 2018, the International Federation of Biochemistry and Molecular Biology added a new enzyme to the original six enzymes, which is also called translocation enzyme. Therefore, the types of enzymes can be divided into seven categories—oxidation-reductase, translocation enzyme, hydrolase, lyase, isomerase, ligase and translocation enzyme [2]. In this paper, human protein data in UniProt database is taken as data set, which contains 18639 protein information. In the past, Support Vector Machines (SVM), K Nearest Neighbor (KNN), random forest classification model (RF) have been mostly used to predict enzyme classification [3]. Although relatively accurate results can be achieved, these methods do not use the topological structure information among molecules in the data. Graph neural network is selected to classify and predict the problem, which make full use of node features and spatial structure features to train the model more effectively [4].

Supported by the Scientific Research Fund of Liaoning Provincial Education Department (L2019048), and Talent Scientific Research Fund of LSHU (2016XJJ-033) of China.

J. Zeng et al. (Eds.): ICPCSEE 2021, CCIS 1452, pp. 283–292, 2021.
https://doi.org/10.1007/978-981-16-5943-0_23

2 Data Source and Preprocessing

2.1 Data Source

Raw protein data are downloaded from Protein Data Bank (http://www.rcsb.org/) and their corresponding labels are obtained from Uniprot (https://www.uniprot.org/). The protein data are labelled with 7 main classes, respectively oxidoreductases (marked with EC1), transferases (marked with EC2), hydrolases (marked with EC3), lyases (marked with EC4), isomerases (marked with EC5), ligases (marked with EC6) and Translocases (marked with EC7). Proteins in humans are selected and numbers of data samples from each class are collected as follows: EC1: 2238 samples, EC2: 10050 sample, EC3: 7231 samples, EC4: 1313 samples, EC5: 760 samples, EC6: 276 samples, EC7: 83 samples.

2.2 Data Preprocessing

Raw data are pdb format that contains basic information of protein, details of protein sequence and atoms marked with 3D coordinate. Instead of using atoms as unit, we choose amino acids as unit to extract features for training models [5,6]. Mean value of 3D coordinates of atoms belonging to the same amino acid are calculated as 3D points of amino acid. Next, two matrices, respectively feature matrix and adjacent matrix, are calculated. As for feature matrix, only class of amino acid is applied as unit feature. Amino acid class is represented with 21 class one-hot encoding (there are 21 kinds of amino acids in human body). As for adjacent matrix, we calculated Euclidean Distance between two units [7–9]. Distance threshold of 6 is set. For pairs' Euclidean Distance that are below the threshold, corresponding area in adjacent matrix is set to 1, and vice versa.

3 Experimental Method and Specific Content

3.1 Graph Neural Network

Traditional deep learning methods have been successfully applied in feature extraction of Euclidean structural data, but data are formed with non-Euclidean structure in many real-world scenes. For non-Euclidean structural data, traditional deep learning methods are hard to be directly applied and achieve satisfactory predictions. Graph Neural Networks (GNNs) are designed to extract effective features from non-Euclidean structural data [10]. Graph Convolution Network (GCN), as one kind of GNN, has been used in molecular graph structural data and achieve good performance. GCN extends traditional image data to graph data. Its core idea is to learn a function, with which nodes in one graph can be aggregated with their neighbors to a new feature representation [11]. Combined with traditional machine learning methods, better performances are usually obtained with the new feature extracted [12]. In the process of deep learning, feature acquisition is a very important task, and the topological structure features in enzyme structure are considered in the training process of graph neural network. The primary structure of enzyme protein is the sequence of amino acid residues in polypeptide chain, so we think that the topological structure of amino acids in enzyme protein is definitely related to the function and type of enzyme to some extent. The Graph Neural Networks method used in this paper is as Fig. 1, among them, amino acids are the original information of enzyme protein molecules, and all the node features and topological structure features of enzyme protein molecules are extracted, coded separately and input into the network to be represented by Graph representation layer. Then, the graph embedding layer is introduced to embed the graphical representation of enzyme protein molecules, and the graphical representation will be sent to multiple gcl layers for feature extraction and learning. Finally, the full connected layer predicts the enzyme protein species.

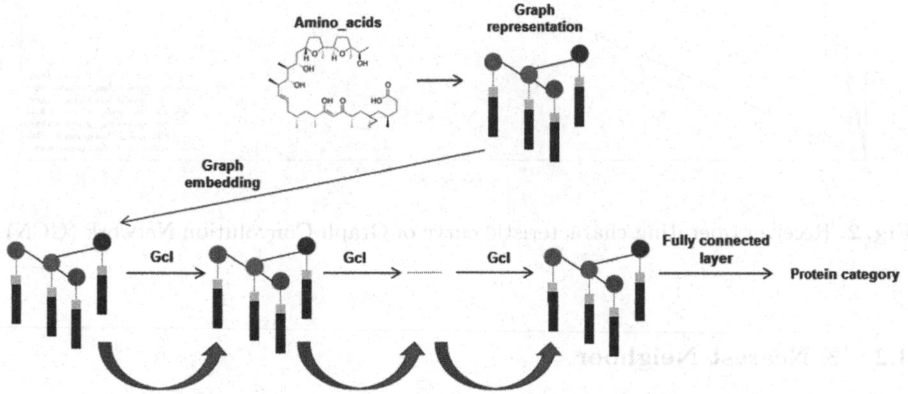

Fig. 1. Proposed graph neural network.

Based on GCN, the proposed network architecture is shown in Fig. 1. Two inputs of the network are respectively feature matrix F, $F \in R^{m \times n}$, where m is the dimension of node feature and n is the number of nodes, and adjacent matrix A, $A \in R^{n \times n}$, X is the matrix after dimension reduction. in the graph, After obtaining the input of the model, feature matrix F goes through dimension reduction process by:

$$X = F \cdot W + b \tag{1}$$

To represent the space connection of each input, X' is calculated as:

$$X' = \begin{cases} X_j, A_{i,j} = 1 \\ 0, A_{i,j} = 0 \end{cases} \tag{2}$$

X' represent features taking both node features and edge features in account, i and j in the formula represent the i and j amino acids in the enzyme protein molecule. Next, flatten layer is used followed by two fully connect layers. Finally, softmax activation function is added after fully-connected layers. Cross entropy loss is applied as loss function, and the number of convolution layers of graph neural network is six, and the learning rate is 5e−5. Figure 2 is the receiver operating characteristic curve obtained by predicting enzyme classification by Graph Convolution Network (GCN):

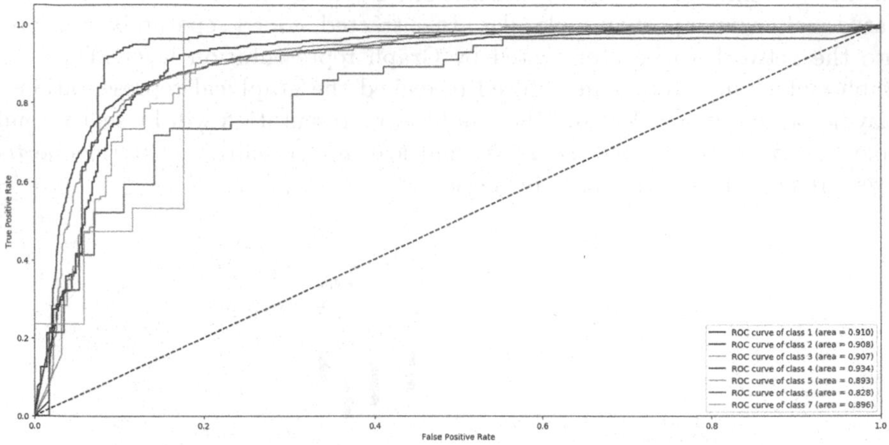

Fig. 2. Receiver operating characteristic curve of Graph Convolution Network (GCN).

3.2 K Nearest Neighbor

KNN, known as the k-nearest neighbor algorithm, essentially selects the nearest K labeled samples around the test samples, and then calculates the category with the most categories in these samples, which is regarded as the category of the test samples. Usually, the Euclidean distance is used in KNN algorithm. Take two-dimensional space as an example. The Euclidean distance between two points in two-dimensional space is calculated as follows:

$$\rho = \sqrt{(x_2 - x_1)^2 + (y_2 - y_1)^2} \tag{3}$$

Expand to multidimensional space, the formula becomes like this:

$$d(x, y) := \sqrt{(x_1 - y_1)^2 + (x_2 - y_2)^2 + \cdots + (x_n - y_n)^2} = \sqrt{\sum_{i=1}^{n} (x_i - y_i)^2} \tag{4}$$

In KNN, the choice of K value is very important. K value is one of the few super parameters in KNN algorithm, and the choice of K value also directly affects the performance of the model [13]. If we set the value of k small, that means a more complex and accurate model, and it is easier to overfit; On the contrary, if the value of K is larger, the model chance is simpler. A very extreme example is that if the value of K is set equal to the number of training samples, that is, K = N, then the final test result of test samples of whatever category will be the class with the largest number of test samples. In this paper, the parameter of KNN is set the value of K to 100.

3.3 Support Vector Machines

Support Vector Machines (SVM), also known as support vector networks, are supervised learning models and related learning algorithms that use classification and regression analysis to analyze data in machine learning. Given a set of training samples, each training sample is labeled as belonging to one or the other of two categories. Support Vector Machine (SVM) training algorithms create a model that assigns new samples to one of the two categories, making it an improbabilistic binary linear classifier [14]. SVM models represent samples as mapped points in space, so that samples with a single category can be separated as clearly as possible. All such new samples are mapped into the same space and can be predicted based on which side of the interval they fall into. Use SVC in sklearn to classify enzyme protein molecules, and set penalty parameter c to 1.

3.4 Random Forest

Random Forest is a classification algorithm proposed by Leo Breiman. It uses the self-service resampling technique to repeatedly and randomly select N samples from the original training sample set N to generate a new training sample set to train decision trees. Then, according to the above steps, M decision trees are generated to form a random forest. The classification results of new data depend on the scores formed by the voting number of classification trees. Use random forest in sklearn to classify enzyme protein molecules, and set the max depth to 2. Other parameters are default values.

The classification ability of a single tree may be very small, but after a large number of decision trees are randomly generated, a test sample can select the most possible classification by counting the classification results of each tree. The general process of random forest is as follows:

(1) Selecte n samples from random sample with return in that sample set;
(2) Randomly select k features from all the features, and use these features to build a decision tree for the selected samples;
(3) Repeat the above two steps for m times to generate m decision trees and form a random forest;
(4) For the new data, after each tree makes a decision, it finally votes to confirm which category it belongs to.

3.5 Multi-layer Perception

Neural network is widely used in the field of machine learning at present, for example, image recognition and speech recognition can be carried out by using neural network, which can be extended and applied to autonomous driving vehicles [15]. It is a highly parallel information processing system, has a strong adaptive ability to learn, does not depend on the object of study the mathematical model of the controlled object of the system parameter variations and external disturbance has good robustness, can deal with complex multi-input and multi-output nonlinear systems, neural networks to solve the basic problem is a classification problem. Multilayer perceptron usually has many levels and can be regarded as the basic form of neural network [16]. Use MLPclassifier in sklearn to classify enzyme protein molecules, and set the hidden layer sizes to 30. Select relu as the activation function and set the max iter to 100. Other parameters are default values.

4 Experimental Results and Discussion

This paper divide the experimental data into 7 groups according to the type of enzyme. For each group, we used 80% of the data as training data and 20% of the data as test data. Train the training set data through the network. After optimizing, adjusting and confirming the parameters in the network, the test set data is input into the model. The final prediction classification result is obtained and compared with the real value. Figure 3, 4, 5 and Fig. 6 is the receiver operating characteristic curve obtained by Support Vector Machines, K Nearest Neighbor, Multi-layer Perception and Random Forest classification model.

The accuracy, Sensitive and Specificity of seven enzyme classifications obtained by Graph Neural Network, K Nearest Neighbor, Support Vector Machin, Random Forest and Multi-layer Perceptron are displayed in Table 1. Most of the highest values of these evaluation criteria appear in Graph Neural Network. Moreover, by comparing the receiver operating characteristic curve of these five methods, it is obvious that Graph Neural Networks is obviously superior to other methods. The AUC value of some enzymes is not the best in graph neural network, because the data sample size is too small and the training and test samples in the prediction process are randomly selected, so there are some unstable factors that lead to this result. So it can conclude that Graph Neural Network can play a more effective role in enzyme classification and recognition.

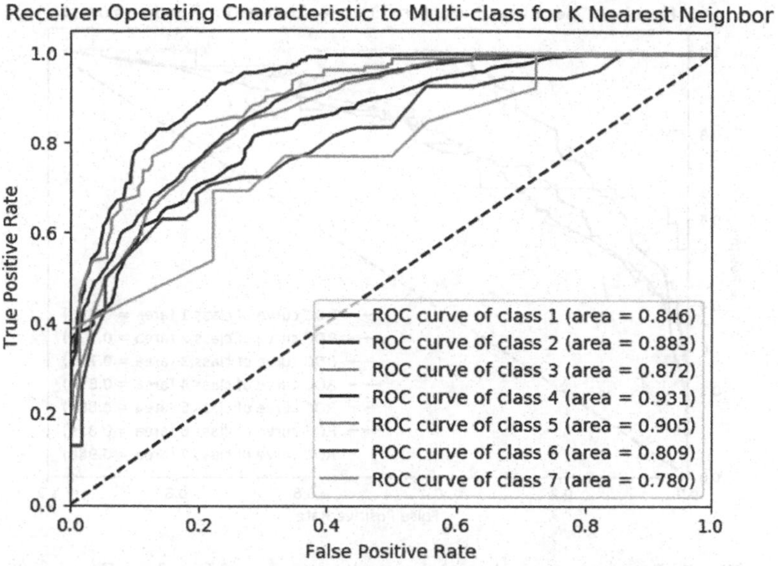

Fig. 3. Receiver operating characteristic curve of K Nearest Neighbor.

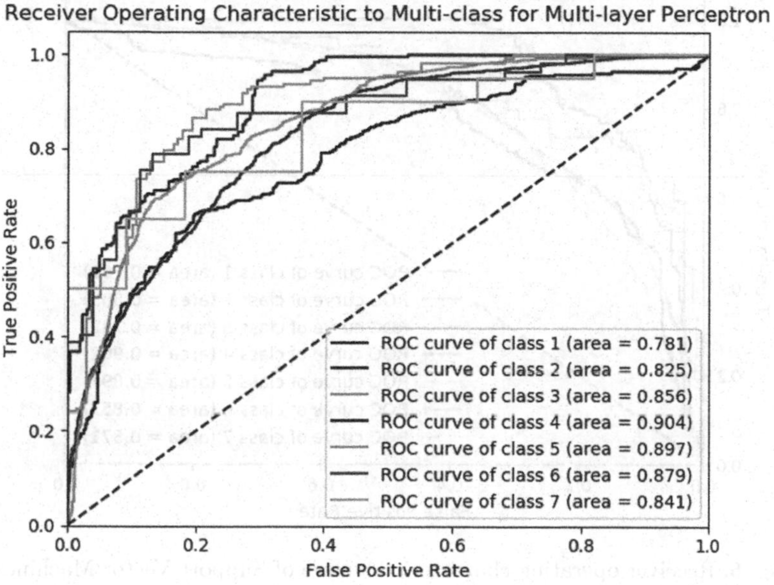

Fig. 4. Receiver operating characteristic curve of Multi-layer Perception.

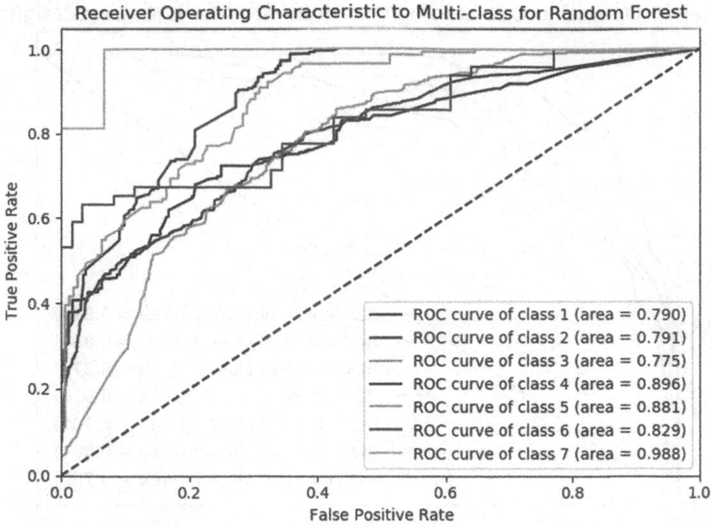

Fig. 5. Receiver operating characteristic curve of Random Forest.

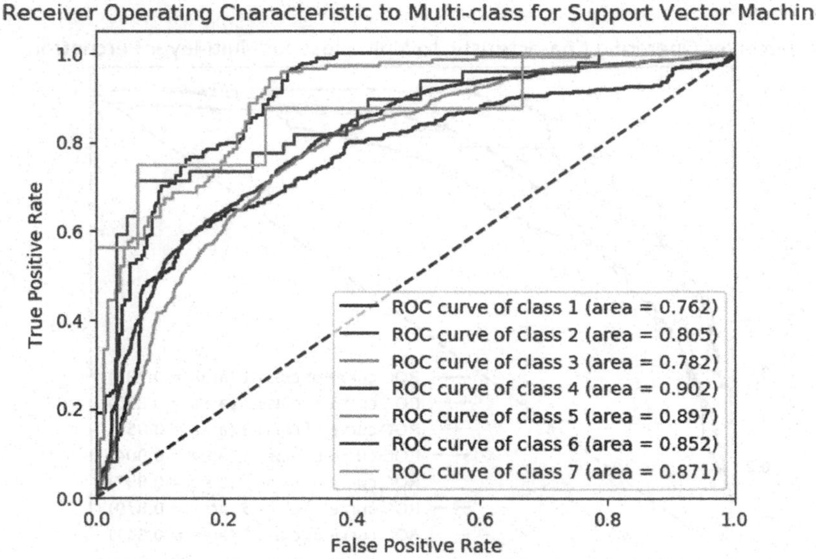

Fig. 6. Receiver operating characteristic curve of Support Vector Machines.

Table 1. The accuracy, sensitive and specificity obtained by various methods.

	GCN			KNN			RF			MLP			SVM		
class	ACC	Sen	Spe	ACC	Sen	Spe	ACC	Sen	Spe	ACC	Sen	Spe	ACC	Sen	Spe
EC1	87. 28%	84. 15%	90. 40%	74. 04%	89. 72%	54. 49%	73. 60%	74. 89%	72. 22%	72. 92%	80. 23%	65. 93%	70. 11%	75. 63%	64. 84%
EC2	85. 80%	88. 41%	83. 18%	78. 14%	81. 97%	74. 33%	69. 43%	80. 74%	58. 22%	67. 68%	92. 60%	41. 84%	72. 97%	77. 50%	68. 28%
EC3	85. 07%	86. 10%	84. 04%	78. 28%	74. 74%	81. 56%	71. 30%	57. 80%	83. 75%	73. 65%	90. 81%	57. 56%	72. 57%	70. 58%	74. 44%
EC4	91. 63%	88. 97%	94. 30%	83. 78%	83. 73%	84. 96%	81. 30%	64. 26%	98. 47%	81. 87%	68. 28%	96. 09%	80. 34%	73. 13%	87. 89%
EC5	85. 20%	81. 58%	88. 82%	80. 46%	74. 65%	85. 63%	76. 50%	77. 85%	75%	73. 18%	45%	97. 53%	81. 79%	82. 86%	80. 86%
EC6	78. 57%	85. 71%	71. 43%	73. 64%	89. 29%	57. 41%	75. 00%	67. 21%	69. 39%	63. 36%	100%	35. 09%	80. 91%	92. 45%	70. 18%
EC7	91. 18%	82. 35%	100%	54. 84%	22. 22%	100%	90. 32%	100%	81. 25%	77. 42%	81. 81%	75%	83. 87%	72. 73%	90%

5 Future Work

In this paper, the advanced graph neural network structure is used to represent protein data, which not only uses the characteristic information of amino acids, but also adopts the spatial structure information of amino acids to train the model more effectively. Experimental results show that this method is superior to other methods, and can predict the classification of enzymes more effectively and accurately. In the future work, this method can be considered to predict the secondary structure classification of enzymes.

References

1. Chang, I., Baldi, P.: A unifying kinetic framework for modeling oxidoreductase-catalyzed reactions. Bioinformatics **29**(10), 1299–1307 (2013)
2. Chou, K.-C., Elrod, D.W.: Prediction of enzyme family classes. J. Proteome Res. **2**(2), 183–190 (2003)
3. Tan, J.-X., Lv, H., Wang, F., Dao, F.-Y., Chen, W., Ding, H.: A survey for predicting enzyme family classes using machine learning methods. Curr. Drug Targets **20**(5), 540–550 (2019)
4. Chou, K.-C.: Prediction of protein cellular attributes using pseudo-amino acid composition. Proteins Struct. Funct. Bioinf. **43**(3), 246–255 (2001)
5. Soudy, M., et al.: UniprotR: retrieving and visualizing protein sequence and functional information from Universal Protein Resource (UniProt knowledgebase). J. Proteomics **213**, 103613 (2020)
6. Zhao, D., Duan, S., Yan, Z., Alippi, C.: Advances in deep neural information processing. Neurocomputing **408**, 80–81 (2020)
7. Wang, X., Li, Z., Jiang, M., Wang, S., Zhang, S., Wei, Z.: Molecule property prediction based on spatial graph embedding. J. Chem. Inf. Model. **59**(9), 3817–3828 (2019)
8. Park, C., Park, J., Park, S.: AGCN: attention-based graph convolutional networks for drug-drug interaction extraction. Expert Syst. Appl. **159**, 113538 (2020)
9. Zhang, S., Tong, H., Xu, J., Maciejewski, R.: Graph convolutional networks: a comprehensive review. Comput. Soc. Netw. **6**(1), 1–23 (2019)
10. Spinelli, I., Scardapane, S., Uncini, A.: Adaptive propagation graph convolutional network. IEEE Trans. Neural Netw. Learn. Syst. (2020)
11. Yang, L., Guo, Y., Gu, J., Jin, D., Yang, B., Cao, X.: Probabilistic graph convolutional network via topology-constrained latent space model. IEEE Trans. Cybern., 1–14 (2020)

12. Bao, D., Zheng, W., Hu, W.: Hybrid graph convolutional networks for semi-supervised classification. In: Proceedings of 2019 the 9th International Workshop on Computer Science and Engineering, WCSE 2019, Hong Kong, China. SCIence and Engineering Institute (SCIEI) (2019)
13. Huang, W.-L., Chen, H.-M., Hwang, S.-F., Ho, S.-Y.: Accurate prediction of enzyme subfamily class using an adaptive fuzzy k-nearest neighbor method. Bio Syst. **90**(2), 405–413 (2007)
14. Zhou, X.-B., Chen, C., Li, Z.-C., Zou, X.-Y.: Using Chou's amphiphilic pseudo-amino acid composition and support vector machine for prediction of enzyme subfamily classes. J. Theor. Biol. **248**(3), 46–551 (2007)
15. Torrisi, M., Pollastri, G., Le, Q.: Deep learning methods in protein structure prediction. Comput. Struct. Biotechnol. J. **18**, 1301–1310 (2020)
16. Gao, R., et al.: Prediction of enzyme function based on three parallel deep CNN and amino acid mutation. Int. J. Mol. Sci. **20**(11), 2845 (2019)

Construction of the Integration System of Cultural Resources and Products in Li Miao Region Tourism

Liping Chen(✉), Xu Wang, and Xiaomei Yang

Sanya Aviation and Tourism College, Sanya, Hainan, China

Abstract. The conventional tourism resource product fusion system applied to the tourism culture in Li Miao region has the deficiency of low integration rate and poor integrity of construction. To solve the problem, the fusion system of cultural resources and products in Li Miao region tourism is proposed. Based on their determination of the mathematical relationship, the cultural resources and product information in Li Miao region tourism was introduced, and their fusion can be realized. Based on the confirmation of the fusion mechanism of cultural resources and products and the procedure of the fusion system, the proposed fusion system was constructed. The experimental data show that the proposed tourism resource product fusion system is 38.9% higher than the conventional system, and the construction integrity is improved by 31.7%. It is proved the effectiveness of the construction of the fusion system of cultural resources and products in Li Miao region tourism.

Keywords: Li Miao region tourism · Cultural resources · Product integration · System construction

1 Introduction

The conventional tourism resource product integration system is based on the multivariate analysis of tourism resources to realize the construction of tourism resource product integration system [1]. When it is applied to the tourism culture in Li Miao region, due to the limitations of conventional system and mechanism, there is a deficiency of low integration rate and poor integrity of construction, which is not suitable for the integration of cultural resources and products in Li Miao region [2, 3]. Therefore, the fusion system of cultural resources and products in Li Miao region tourism is proposed. By using the MARE equation, the mathematical relationship between cultural resources and product integration was determined, and the cultural resources and product information in Li Miao region tourism was introduced. The fusion mechanism of cultural resources and products is determined by means of depth-priority fusion, breadth-priority fusion and non-repeated grabbing fusion [4]. Based on the confirmation of the fusion system procedure, the fusion system construction of cultural resources and products in Li Miao region tourism is completed. In order to ensure the effectiveness of the construction of

J. Zeng et al. (Eds.): ICPCSEE 2021, CCIS 1452, pp. 293–302, 2021.
https://doi.org/10.1007/978-981-16-5943-0_24

tourism resource product integration system, the experimental environment of tourism resource product integration system was simulated. Two different tourism resource product integration systems are used to conduct the tourism resource product integration rate and build the integrity simulation test. The experimental results show that the proposed tourism resource product fusion system is highly effective.

2 Characterization of Integration of Cultural Resources and Products in Li Miao Region Tourism

2.1 Construction of Mathematical Relationship Between Cultural Resources and Products

The mathematical relationship between cultural resources and product integration is constructed. Firstly, cultural resources and products are quantified. Cultural resources are real matrix A of order n. There is a non-negative matrix E of order n, make the $\mathrm{A} = \lambda I_n - E, \lambda \geq \rho(E)$. A is the m-matrix of cultural resources. I_n stands for the product, n stands for the identity matrix, and $\rho(.)$ stands for the spectral radius of a matrix. And $\lambda > \rho(E), A$ is non-singular product m-matrix. If $\lambda = \rho(E), A$ is singular product m-matrix [5]. A nonsingular m-matrix A, having non-positive diagonal elements and non-negative diagonal elements. For a non-singular or irreducible m-matrix A, its diagonal elements must be positive. So, we get the Sylvester equation.

$$AX + XB = C \tag{1}$$

A, B has positive diagonal and non-positive non-diagonal elements, C is a non-negative matrix. The sylvite equation of the above type is called m-matrix sylvite equation, also known as MSE equation [6].

MSE equations often appear in an iterative method of m-matrix riccati equations. The riccarti equation of m-matrix algebra (also known as the MARE equation) is defined as follows:

$$XDX - AX - XB + C = 0 \tag{2}$$

Where A, B, C, D are matrices. Their order is determined by the following matrix division.

$$M = \begin{array}{c c} & \begin{array}{cc} m & n \end{array} \\ \begin{array}{c} m \\ n \end{array} & \begin{pmatrix} B & -D \\ -C & A \end{pmatrix} \end{array} \tag{3}$$

The smallest non-negative solution to the MARE equation is obvious. A very important application of the MARE equation is the representation of the mathematical relationship between cultural resources and product integration [7]. Every component in a mathematical relation has a physical meaning, and it is very important to calculate every tiny component of the solution precisely. In order to ensure this, the MSE equation is solved by using the iterative method, and the mathematical relationship between cultural resources and product integration is constructed.

2.2 Introduction of Cultural Resources and Product Information in Li Miao Region Tourism

The introduction of cultural resources and product information in Li Miao region tourism mainly includes four parts: the introduction of Li Miao resources, Li Miao culture, natural ecological resources and cultural ecological resources.

In Li Miao Nationality region, whether natural or human resources are unique, so this area has strong attraction to the market [8].

First, there are some original ecological natural resources in ethnic areas. It is a tropical Marine monsoon climate with bright sunshine, abundant rainfall and fresh air, and the annual average temperature is 23–25 °C [9]. This region has a wide variety of tropical plants with a forest coverage rate of more than 70%. There are abundant mountain hot spring and virgin forest resources, as well as unique mountain and pastoral scenery.

Second, Li Miao's traditional cultural resources are characteristic [10]. Firstly, Li Miao culture covers all aspects of material culture, spiritual culture, behavioral culture and method culture [11]. Secondly, it is of a strong original nature. Since Li Miao region has been in a development state of close isolation for a long time, its culture is less impacted by the outside world. Therefore, most cultures still maintain their original characteristics. The Li Miao culture, especially the Li Nationality culture, is irreplaceable and monopolistic. Li Nationality has a history of more than 3,000 years. From the perspective of the whole Chinese nation, the central areas of Hainan have the largest population of Li Nationality with the most distinctive culture and the least change, hence its monopoly status of Li Nationality culture is unshakable [12]. The integration of natural resources and cultural ecology has laid a resource foundation for the development of Li Miao cultural ecological tourism.

Natural ecological resources include mountains, Wuzhishan nature reserve, Qixianling scenic spot, more than 2,000 species of plants, above 500 species of animals, and over 70% of the forest cover [13].

Cultural ecological resources include clothing, food, utensils, musical instruments and so on. The ship-shaped house is the representative residence, the dragon quilt is the tributary of the past dynasties, and the gourd boat is the living fossil for the study of prehistoric traffic. Three-color rice is the most characteristic food of Miao Nationality. In terms of spiritual culture, the core of Li Nationality is honesty and trustworthiness, while religious belief mainly covers worship to nature, ancestors and totems. Miao religious belief is mainly worship to nature, ancestors and spirits [14].

In terms of behavioral culture, Li Nationality is the only nationality that still uses plant dyeing. And Miao Nationality's characteristic is the use of batik and embroidery [15].

Based on the determination of the mathematical relationship between cultural resources and product integration and the introduction of cultural resources and product information in Li Miao region tourism, the fusion of tourism cultural resources and products in Li Miao region is realized.

3 Construction of Integration System of Cultural Resources and Products in Li Miao Region Tourism

3.1 The Determination of the Integration Mechanism of Cultural Resources and Products

The fusion mechanism of cultural resources and products is realized by means of depth-priority fusion, breadth-priority fusion and non-repeated grabbing fusion. The fusion strategy of depth-priority fusion is similar to the strategy of Chinese family succession, like the typical case of the inheritance of feudal emperors. Usually for the eldest son, if the eldest son dies, the second eldest son inherits first. If the grandchild dies and the grandchild has no child, then the second grandchild inherits. This inherited priority relationship is called "depth-priority fusion", ABEH fusion path as shown in Fig. 1.

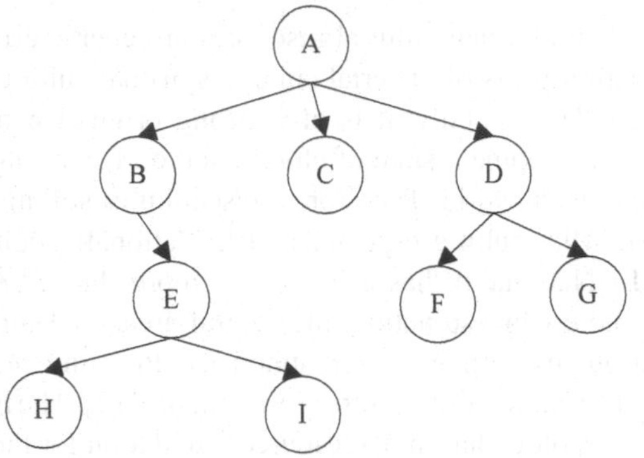

Fig. 1. Schematic diagram of fusion mechanism

The "depth-priority fusion" strategy is to fuse as far as possible to the furthest path until it is impossible to fuse. The disadvantage of this approach is that many nodes are repeatedly merged, therefore, in the actual process, it is necessary to judge whether to continue depth-priority fusion. The total link library needs to be checked before fusing the data.

Breadth-priority fusion strategy refers to the process of fetching, the next level of fusion after completing the current level of fusion. The design and implementation of the algorithm is relatively simple. In order to cover as many tourism and cultural product data as possible, breadth-priority fusion strategy is generally used. The breadth-priority fusion strategy is also known as "hierarchy priority", where the first is highest, and so on. In each layer, the internal priorities are arranged from left to right. As shown in Fig. 1, ABCDEFG….fusion paths.

There are three advantages using breadth-priority fusion: firstly, the algorithm of breadth-priority fusion is relatively simple, and the data of Li Miao tourism and cultural products closer to the seed site would be more important. Secondly, depth-priority fusion is not as deep as people's thought, and many channels can be reached in breadth-priority fusion. Thirdly, the strategy of again breadth-priority fusion is very suitable for multi-cooperative fusion. However, the strategy of breadth-priority fusion may encounter dead chain, which will lead to repetitive grabbing of Li Miao tourism and cultural product data and waste of resources.

The key to not repeat fetching fusion is to remember history. The way to record history is to hash table. If the data of a certain Li Miao tourism cultural product has been captured at some time in the past, the value of its corresponding slot is set to 1; let's put a 0 here.

Each hash table is a byte (8 bits), large amounts of data require at least 10 GB of capacity. However, the 32-bit operating system has a maximum addressing space of 4 GB (including 1 GB of space occupied by the Linux kernel), so the memory cannot allocate such a large hash table. To improve the large amount of data stored in the hash table, the MD5 signature function is constructed. In order to load the whole hash table into memory, the data structure of Bitmap is usually used to compress the memory consumption. Bitmap needs to touch the four basic operations of pressing in (&) and pressing or (|), and moving left (<<) and right (>>), as shown in Fig. 2.

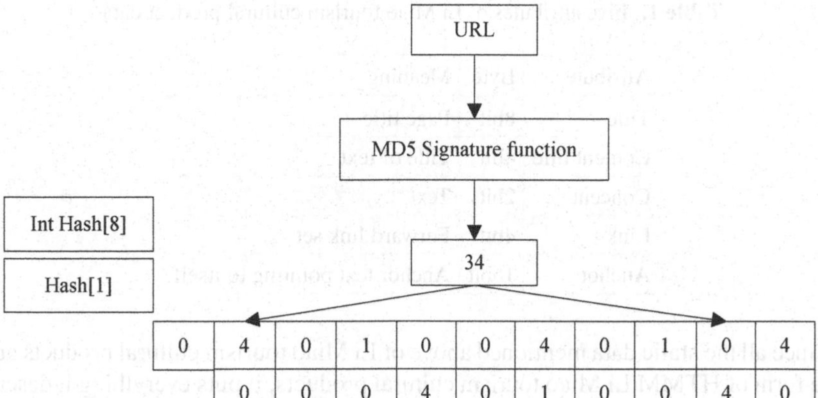

Fig. 2. A hash table of a Bitmap structure

Define a Hash table, or Int Hash. Perform MD5 signature on the URL, assuming the resulting value is 34. If you divide 34 into 32, you get the quotient of 1, and then you take the module of 8 and you get 1, which means that the slot is in the Hash. By dividing 34 by 32, the remainder is 2, which means that 34 maps to the third bit of the whole 32-bit integer of Hash. If you take this bit position 1, then the hash is 4. By the above way, determine the integration mechanism of cultural resources and products.

3.2 Determination of Fusion System Procedures

Firstly, some valuable attributes, such as anchor text, title and text, can be extracted from the semi-structured Li Miao tourism cultural product data. These attributes are combined to form a data object of Li Miao tourism cultural products. The structured data of Li Miao tourism cultural products aims to extract the basic attributes of Li Miao's tourism cultural products. It can describe the tourism cultural product data of Li Miao and package it into a tourism cultural product data of Li Miao. Anchor text is the data link text of Li Miao's tourism and cultural products. The fusion engine can judge the content attributes of the Li Miao tourism cultural product data according to the anchor text description that points to a certain Li Miao tourism cultural product data.

Data title of Li Miao's tourism and cultural products specifically refers to the text in the HTMM identification language <title></title>. These words are compiled by the data producer of Li Miao's tourism and cultural products and they express the basic meaning of Li Miao's tourism and cultural products data. The text title of Li Miao's tourism and cultural products refers to the extraction of appropriate text in the text.

The text of Li Miao's tourism and cultural product data refers to the subject content of Li Miao's tourism and cultural product data. It fully expresses the basic information of Li Miao's tourism cultural product data, which generally appears in HTMM labels such as <div>. The positive link of Li Miao tourism cultural product data is written by Li Miao tourism cultural product data writer, and is also the anchor text of other Li Miao tourism cultural product data. Table 1 is the 5 attributes that are generally included in the data of Li Miao's tourism and cultural products:

Table 1. Five attributes of Li Miao tourism cultural product data

Attribute	Byte	Meaning
Title	8bit	Page title
Content title	4bit	Title of text
Concent	2bit	Text
Link	4bit	Forward link set
Anchor	16bit	Anchor text pointing to itself

Since all the static data mentioned above of Li Miao tourism cultural products are all in the form of HTMM Li Miao tourism cultural products, it puts everything it describes into the tag in HTMM syntax. In order to describe the organization structure of Li Miao's tourism cultural product data more clearly, the labels in Li Miao's tourism cultural product data are sorted out in the order of occurrence and recorded by appropriate structure. Because of the nesting of tags, the collation results in a tree structure are as shown in Fig. 3.

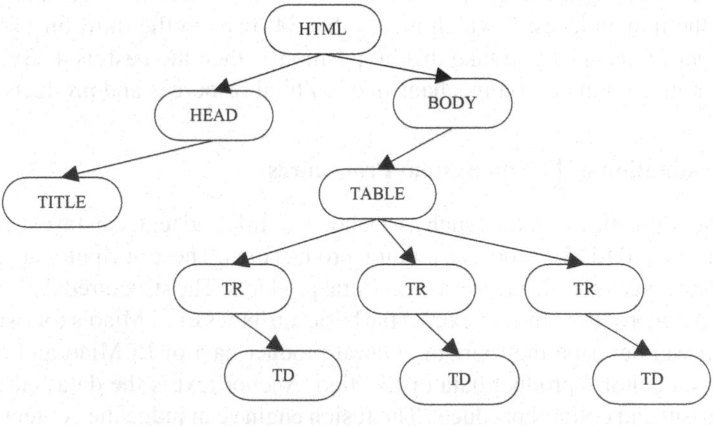

Fig. 3. HTMM tag tree

Then, the text is obtained through a specific representation. This introduces the concept of text blocks, which are considered to be text blocks between tags. Define a set of rules that allow them to rate text blocks. The higher the score, the more important the text block is. Figure 4 is the general algorithm of simple specific characterization algorithm.

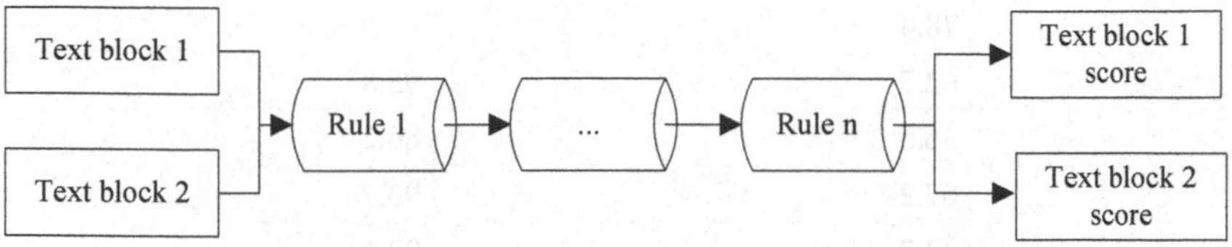

Fig. 4. Simple specific characterization algorithm

Finally, recheck the data of Li Miao's tourism and cultural products. The repeated or repeated data of Li Miao's tourism and cultural products will greatly load the fusion engine, resulting in the decline of the efficiency of the fusion engine. The first step of rechecking the data of Li Miao tourism cultural products is to judge whether the data of Li Miao tourism cultural products are duplicated or similar. The key of weight loss is that non-repeated tourism and cultural product data is left. That is, the data of Li Miao's tourism and cultural products are collected and discarded. This method is simple to realize and largely guarantees the principle of keeping the original first. Therefore, the fusion system procedure is determined to complete the fusion system construction of cultural resources and products in Li Miao region tourism.

4 Experimental Results and Analysis

In order to ensure the effectiveness of the fusion system construction of cultural resources and products in Li Miao region tourism proposed in this paper, a simulation experiment was conducted. During the test, different tourism resource product fusion systems were used as test objects to conduct the fusion rate and build the integrity simulation test, simulating tourism resource product fusion system. In order to ensure the effectiveness of the experiment, the conventional tourism resource product fusion system is used as the object of comparison. Compare the results of two simulation tests and present the test data in the same data chart.

4.1 Comparison of Fusion Rate

During the experiment, two different tourism resource product fusion systems were used to carry out work in the simulated environment. The change of fusion rate data of ten groups of tourism resource product fusion system was analyzed. The comparison of test results is shown in Table 2.

Table 2. Comparison of fusion rates

Case type number	Conventional system construction method/%	Proposed system construction method/%
1#	48.6	89.6
2#	58.6	95.2
3#	78.4	87.6
4#	62.7	95.8
5#	36.5	86.2
6#	67.2	95.7
7#	43.2	92.4
8#	56.8	95.6
9#	72.3	89.6
10#	62.4	88.4

The fusion rates of the proposed tourism resource product fusion system and the conventional tourism resource product fusion system are processed by arithmetic mean value. It is concluded that the fusion rate of the conventional tourism resource product fusion system is 52.7%. The proposed tourism resource product fusion system has a fusion rate of 91.6%. It is suggested that the fusion rate of tourism resource products is increased by 38.9% compared with that of conventional tourism resource products.

4.2 Comparison of Build Integrity

During the experiment, two different tourism resource product fusion systems were also used to carry out work in the simulated environment, and the changes in the integrity of

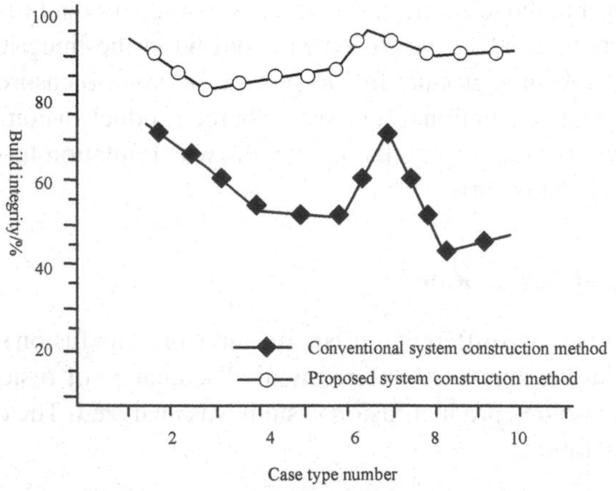

Fig. 5. Comparison of build integrity

the tourism resource product integration system were analyzed. The comparison curve of the test results is shown in Fig. 5.

The construction integrity of the proposed tourism resource product fusion system and the conventional tourism resource product fusion system is processed by arithmetic mean value. It is concluded that the average construction integrity of the conventional tourism resource product fusion system is 58.6%. The proposed tourism resource product integration system has an average integrity of 90.3%. It is concluded that the proposed tourism resource product integration system is 31.7% more than the conventional tourism resource product integration system. It is suitable for the construction of the fusion system of cultural resources and products in Li Miao region tourism.

5 Conclusions

This paper proposes the construction of the integration system of cultural resources and products in Li Miao region tourism. Based on the characterization of the fusion of tourism cultural resources and products in Li Miao region and the determination of the fusion mechanism and procedures, the research was implemented. The experimental data show that the method presented in this paper is highly effective, and theoretical basis is provided for the construction of the fusion system of cultural resources and products in Li Miao region tourism.

Acknowledgment. 2018 Plan Project of Philosophy and Social Sciences of Hainan Province (No. HNSK (QN)18–101).

References

1. Liao, Q.H., Zheng, X., Sun, Y., Chen, T.: The evolutionary logic and future trend of urban cultural tourism integration policy in China. Urban Stud. **28**(5), 7–11 (2021)
2. Liu, H.M.: Mechanism and path of high-quality development of ethnic village tourism to guide rural revitalization. Soc. Sci. **4**(288), 58–63 (2021)
3. Ying, L.: Modeling and simulation of attraction of domestic tourism to overseas tourists. Comput. Simul. **33**(10), 429–432 (2016)
4. Tian, L., Zhang, P.Y.: The literature review and frame construction about integration of tourism industry. J. Tech. Econ. Manag. **9**, 119–123 (2016)
5. Cheng, X., Shi, D.: The evaluation of cultural resources abundance in international tourism culture demonstration zone of the southern Anhui province. Sci. Geogr. Sinica **14**(2), 231–233 (2017)
6. Lin, L., Hao, Y.L., et al.: Research on the fusion mechanism of local music cultural resources and tourism industry—taking Tai'an region for example. J. Dezhou Univ. **14**(7), 221–222 (2016)
7. Prieto, L.C.H., Vega, M.G.: Cultural resources as a factor in cultural tourism attraction: technical efficiency estimation of regional destinations in Spain. Tour. Econ. Bus. Finance Tour. Recreat. **23**(2), 260–280 (2017)
8. Yao, Y.Y.: Protective tourism development of ethnic minority intangible cultural heritage under the background of cultural and tourism integration. Soc. Sci. **4**(288), 64–69 (2021)

9. Cheng, X., Tang, H., Zhou, Z., et al.: Study on the construction of tourism resources evaluation system in sports intangible cultural heritage—take Puningyingge as an example. Bull. Sport Sci. Technol. **10**(3), 111–112 (2018)
10. Lu, H.Q.: A Study on the Development of Yangshang Salt Merchants Cultural Tourism Products from the Perspective of Tourism Integration. Northern Econ. Trade **23**(6), 145–146 (2017)
11. Wang, K.: Analysis of cultural tourism resources in the middle region of Hainan. J. Popular Lit. **4**, 250 (2019)
12. Tian, C.Y.: On countermeasures of tourism leading the protection and development of three mountains and five imperial gardens culture. J. Beijing Union Univ. **45**(9), 156–157 (2016)
13. Zhong, M.M.: On integration of cultural tourism resources of Ming Dynasty in Chuzhou and Nanjing: a perspective of Nanjing urban agglomerations. J. Wuhan Bus. Univ. **27**(1), 111–113 (2016)
14. Weng, G.M., Ling-Yan, L.I.: The coupling coordination degree and spatial correlation analysis on integrational development of tourism industry and cultural industry in China. Econ. Geogr. **45**(4), 233–234 (2016)
15. Luo, W., Zhang, J., Wu, Q., et al.: Spatial integration analysis of provincial historical and cultural heritage resources based on geographic information system (GIS)—a case study of spatial integration analysis of historical and cultural heritage resources in Zhejiang province. Int. Arch. Photogramm. Remote Sens. Spat. Inf. Sci. **XLI-2** (W5), 483–488 (2017)

Research on Non Intrusive Intelligent Monitoring System for Elevator State

Yao Ma, Yu Yang, Qianshan Wang, Xuepeng Li, Zeng Xu, and Haifang Li(✉)

College of Information and Computer,
Taiyuan University of Technology, Taiyuan 030600, China
lihaifang@tyut.edu.cn

Abstract. The current elevator status monitoring systems basically realize the monitoring of elevator status by collecting signals from the main board of the elevator. However it is costly and lacks universality, which also requires invasive installation. To address the above problems, a non-invasive intelligent monitoring method is proposed in this paper for elevator operation status. The method decomposes the acceleration signal into vertical and horizontal components, estimates the dynamics of the elevator using Kalman filter, performs vibration analysis on the horizontal components, establishes a baseline for normal operation, and automatically calibrates the sensors by combining the operating characteristics of the elevator. The traceless Kalman filter based on fused SLAM was performed to couple the sensor information and track the real-time position of the elevator. The effectiveness and robustness of the method are verified in actual operation, and the problem of elevator position error accumulation is basically solved without installing fiducials. The designed non-intrusive elevator status intelligent monitoring method is low-cost and universal, which is of practical significance for promoting on-demand elevator maintenance.

Keywords: Non-invasive condition monitor · Signal decomposition · Information fusion · Kalman filter · Sensor auto correction

1 Introduction

In order to strengthen elevator quality and safety work, the elevator industry is vigorously promoting on-demand maintenance, and to promote on-demand maintenance, the first problem to be solved is the monitoring of elevator operation status. Only on the basis of condition monitoring can we further estimate and predict the health status of elevators. At present, only a few leading elevator manufacturers in the industry, such as Hitachi and Otis, have corresponding elevator condition monitoring systems for their own elevators. The way they use is to collect the signal from the control board to monitor the elevator status, which requires invasive installation.

Among the current studies on elevator condition monitoring and fault diagnosis, invasive methods are used: literature [1] and literature [2] collected door control signals to monitor and predict the faults of elevator door systems; literature [3] designed a

J. Zeng et al. (Eds.): ICPCSEE 2021, CCIS 1452, pp. 303–318, 2021.
https://doi.org/10.1007/978-981-16-5943-0_25

remote and highly concurrent monitoring system that collected signals from elevator control chips; literature [4] mined the relationship between motor voltage and elevator operation faults; literature [5] carried out the monitoring and prediction of elevator faults by measuring traction machine current for early prediction of elevator faults; literature [6] analyzed the fault information and related factors of multiple faulty elevators by collecting main board signals using data mining; literature [7] collected motor code signals as well as regulator signals to estimate the key performance indicators of elevators described by ISO 18738–1:2012 standard.

The current study has used more non-invasive methods. The literature [8–11] collected the vibration signals of the main components of the elevator such as bearings and traction machines outside the car. Literature [12] collected car vibration signals and noise signals for fault classification; literature [13] proposed an easy-to-install non-contact elevator intelligent sensing node to track the position of the elevator car and monitor abnormal stops through the analysis of acceleration; literature [14] combined unsupervised learning and supervised learning methods to extract the vibration data from elevator inspection big data for fault diagnosis and prediction. The literature [15] monitors the operation status of elevators by measuring acceleration signals.

Due to the differences in objective factors such as many elevator brands in the market, operating hours, and operating environment, each elevator has its own normal operation mode. In order to interoperate with all elevators without considering their own sensors and control systems, this paper adopts a non-invasive approach. And most of the current related researches cannot meet the actual demand in real time, and need to install fiducials in order to solve the long-time error accumulation problem. For this reason, this paper designs a real-time condition monitoring system without installing beacons, easy to deploy and move, universal and does not affect the working status of elevators through decomposition of acceleration signals, estimation of elevator acceleration and velocity, automatic calibration of sensors, and information fusion based on traceless Kalman filtering.

2 Hardware Platform and Technology Route

This paper builds a hardware platform of Raspberry Pi 3B+ equipped with GMP180 (barometric pressure sensor) and MPU6050 (acceleration sensor). The two sensors are bound to one core of Raspberry Pi CPU, and the parallel execution of data acquisition is ensured by multi-process running mode. The calculation results are transmitted to the cloud via socket.

The technical route is shown in Fig. 1, where the data is collected by the acceleration sensor and the barometric pressure sensor. Firstly, the two sensors are initialized and the initial elevation of the gravitational acceleration and barometric pressure sensors are updated. As the collected acceleration is triaxial data, the update of gravitational acceleration g is performed by Weiszfeld algorithm to find the median center. Based on this, the collected elevator car acceleration is decomposed into gravity direction and horizontal direction by the updated g. The acceleration in the gravity direction is the acceleration in the car motion direction, and the estimation of the elevator acceleration and velocity is obtained by combining the Kalman filter with the motion characteristics of the elevator, while the offset from the baseline is calculated by vibration analysis based on the acceleration signal in the horizontal direction. And the running position of the elevator is tracked using the traceless Kalman filter algorithm coupled with acceleration and air pressure sensor information by fusing SLAM. Finally, the real-time monitoring information is sent to the cloud.

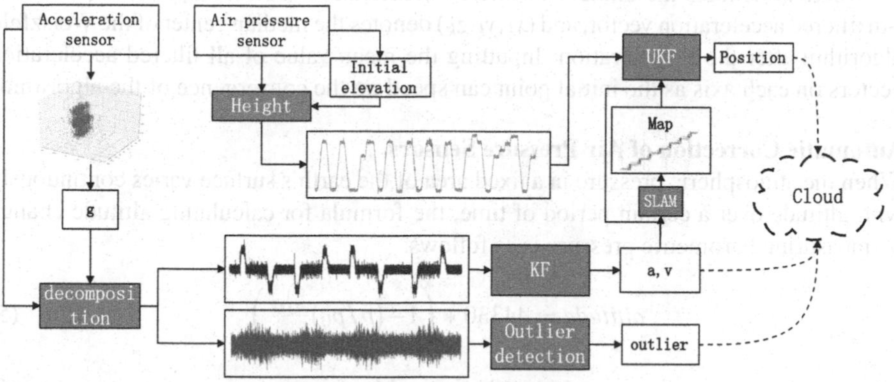

Fig. 1. Technology roadmap

3 Feature Extraction and Pre-processing

3.1 Automatic Sensor Correction

Automatic correction of the sensors includes initializing the gravitational acceleration vector measured by the accelerometer and the initial elevation of the barometric sensor, as well as performing automatic updates to both.

Automatic Correction of Acceleration Sensors
When the system first starts operation (which requires the elevator to be at standstill) or when the elevator is detected to be in the standstill phase during operation, a sliding window (with a width of 2 s and a step length of 0.5 s) is used to process the acquired triaxial acceleration data sequence and filter the outliers based on the mode of the acquired acceleration vectors. The Weiszfeld algorithm is then used to find the median center of the collected triaxial acceleration data, i.e., the point with the smallest sum of Euclidean distances to all the filtered acceleration vectors, which is used as the gravitational acceleration vector used by the system after initialization or update. The Weiszfeld algorithm [16] used in this paper is as follows.

$$x_{k+1} = \sum_{j=1}^{m} a_i/d_i(x_k, y_k, z_k) \Big/ \sum_{j=1}^{m} 1/d_i(x_k, y_k, z_k) \tag{1}$$

$$y_{k+1} = \sum_{j=1}^{m} b_i/d_i(x_k, y_k, z_k) \Big/ \sum_{j=1}^{m} 1/d_i(x_k, y_k, z_k) \tag{2}$$

$$z_{k+1} = \sum_{j=1}^{m} c_i/d_i(x_k, y_k, z_k) \Big/ \sum_{j=1}^{m} 1/d_i(x_k, y_k, z_k) \quad (3)$$

$$d_i(x, y, z) = \|(a_i, b_i, c_i) - (x, y, z)\| \quad (4)$$

where m denotes the number of filtered acceleration vectors, (a_i, b_i, c_i) denotes the i-th filtered acceleration vector, and (x_k, y_k, z_k) denotes the median center of the Weiszfeld algorithm after the k-th iteration. Inputting the mean value of all filtered acceleration vectors on each axis as the initial point can speed up the convergence of the algorithm.

Automatic Correction of Air Pressure Sensors

When the atmospheric pressure in a fixed area of the earth's surface varies continuously with altitude over a certain period of time, the formula for calculating altitude change by measuring barometric pressure is as follows.

$$altitude = 44330 * \left(1 - (p/p_0)^{\frac{1}{5.255}}\right) \quad (5)$$

$$h = altitude - altitude_0 \quad (6)$$

Where altitude, p and p_0 are the current altitude and current sensor measured altitude respectively. When the system just starts to run (the elevator should be stationary and in the reference position), or when the elevator is detected to be in the stationary stage during operation, the current elevation sequence of the elevator is recorded by means of sliding window through the barometric pressure sensor until the elevator is detected to be in motion, using a sliding window with a width of 2 s and a step length of 1 s. The 3σ principle is used to filter the abnormal values and then the average value is used as the initial or updated initial elevation of the barometric pressure sensor.

3.2 Acceleration Signal Decomposition

The collected acceleration signal is subtracted from the calibrated gravitational acceleration vector, and the residual quantity is decomposed into the direction of gravity and the direction perpendicular to gravity by the inner product operation. The acceleration signal decomposed to the gravity direction is the sum of the acceleration of the elevator moving along the vertical direction (the actual direction of elevator movement) and the noise (the vibration of the elevator in the vertical direction); the acceleration signal decomposed to the direction perpendicular to gravity is the vibration signal of the elevator in the horizontal direction. The calculation formula for the decomposition process of the acceleration signal is as follows.

$$I_{residual} = I - g \quad (7)$$

$$a_{vertical} = (I_{residual} \bullet g)/|g| \quad (8)$$

$$I_{vertical} = a_{vertical} * g \quad (9)$$

$$I_{horizontal} = I_{residual} - I_{vertical} \tag{10}$$

Where g denotes the gravitational acceleration vector obtained after calibration, I denotes the collected triaxial acceleration, $I_{residual}$ is the residual amount after removing the gravitational acceleration, $a_{vertical}$ is the mode of the acceleration component in the gravitational direction, $I_{vetical}$ is the acceleration component in the gravitational direction, and $I_{horizontal}$ denotes the acceleration component in the horizontal direction. Figure 2 shows an example of signal decomposition, the left figure is the source data; the right figure is the decomposed data, where the vertical line is the acceleration data decomposed to the gravity direction. By decomposing the acceleration signal and combining it with the automatic calibration of the sensor, the robustness of the system can be enhanced by using the installation without caring about the direction of the z-axis of the acceleration sensor.

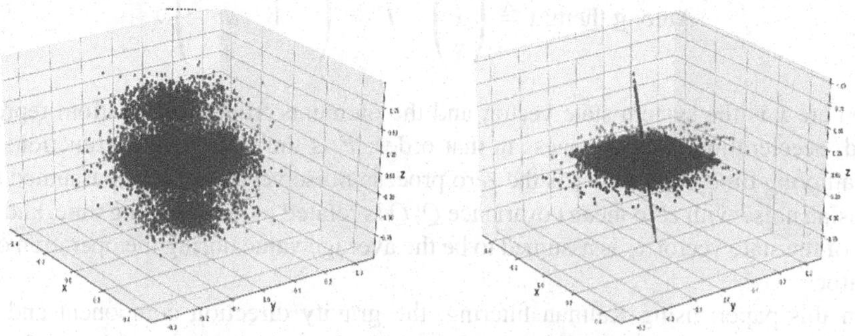

Fig. 2. Signal decomposition diagram

4 Estimation of Elevator Dynamics Characteristics

The motion process of the elevator can be divided into: stationary phase, acceleration phase, uniform speed, and deceleration phase. The acceleration and velocity characteristics of the elevator operation process are shown in Fig. 3.

In the ideal state, the change of acceleration in the acceleration and deceleration phases can be approximately divided into three sub-stages: increasing uniformly, maintaining and decreasing uniformly [6]. It can be seen that in the ideal state, the derivatives of acceleration are maintained at constant values except for the transition points between stages. Accordingly, the hypothesis of ideal motion of the elevator is proposed.

$$\dot{a}_k = \dot{a}_{k-1} \tag{11}$$

The elevator motion model is simplified by assuming that the derivative of the acceleration at the current time step is equal to the derivative of the acceleration at the previous time step by Eq. (11). Then the dynamics of the elevator can be characterized by the following state-space model.

$$x_k = Fx_{k-1} + w_k \tag{12}$$

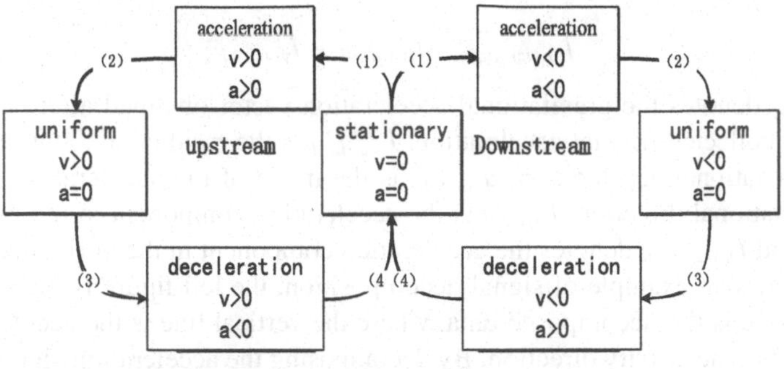

Fig. 3. Ideal motion characteristics of elevator

$$\text{Among them: } x \triangleq \begin{pmatrix} x \\ \dot{x} \\ \ddot{x} \end{pmatrix} \quad F \triangleq \begin{pmatrix} 1 & dt & dt^2/2 \\ & 1 & dt \\ & & 1 \end{pmatrix}.$$

where x is the system state vector, and the elements from top to bottom represent speed, acceleration, and sharpness in that order; F is the state transfer function; dt is the sampling time interval; w_k is the zero process noise vector, which is assumed to be Gaussian noise with zero mean covariance Q; Q is related to the sampling time, and each state of the state vector x_k is assumed to be the average value during the operation of the elevator.

In this paper, using Kalman filtering, the gravity direction component and time stamp obtained by signal decomposition from real-time measurements of the acceleration sensor are used as input, and the system state vector x is used as output, and the connection between the input and output is established through the prediction step and update step for optimal estimation, with the following main steps.

First, the sampling time interval dt is calculated, the state transfer function F and the process noise covariance Q_k are adjusted according to dt, the state of the next time step is predicted according to the system state model (12), and the system covariance is estimated.

$$\overline{P}_k = F_k P_{k-1} F_k^{\mathrm{T}} + Q_k \tag{13}$$

The Kalman gain K is calculated from the estimated system covariance, H is the measurement function $H = (0,0,1,0)^{\mathrm{T}}$, and R is the measurement noise covariance.

$$K_k = \overline{P}_k H^T \left(H \overline{P}_k H^{\mathrm{T}} + R \right)^{-1} \tag{14}$$

The innovation, i.e. the residual between the predicted and measured values, is calculated from the inputs to the system.

$$x_k = \overline{x}_k + K_k y_k \tag{15}$$

$$P_k = (I - K_k H) \overline{P}_k (I - K_k H)^{\mathrm{T}} + K_k R K_k^{\mathrm{T}} \tag{16}$$

As seen in Fig. 3, ideally, when the elevator is at standstill or in the uniform speed phase, the acceleration is zero, but it is basically impossible to get a zero value due to the noise of the sensor itself and the presence of process noise such as vibration in the elevator operation, so set the observation window the observation window w = [avertical^{k-n+1}, avertical^{k-n+2}......avertical^{k}] and the threshold value for the gravity direction component of the system input.avertical^{k} indicates the current acceleration, n is the observation window width, when all the values in the window are less than the threshold value, the acceleration will be corrected to 0. Conversely, when the acceleration is zero, the elevator is in the stationary or uniform velocity phase. Therefore, when the acceleration is zero and the magnitude of the speed is less than the set threshold, the speed can be corrected to zero, and at the same time, it can be determined that the elevator is at standstill for automatic calibration of the acceleration sensor.

The acceleration and velocity characteristics monitored during one test run using the above method are shown in Fig. 4 below, from which the four phases of elevator operation can be clearly seen. Nodes (1), (2), (3) and (4) correspond to the four phases of the elevator described in Fig. 3.

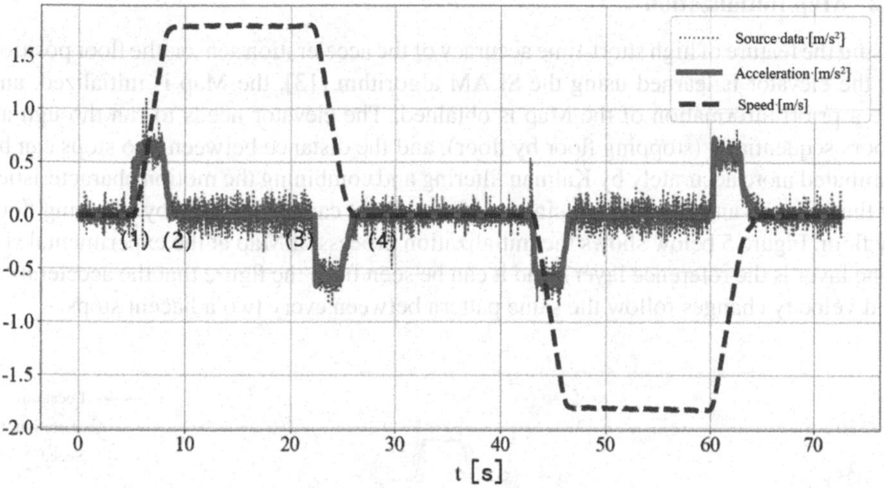

Fig. 4. Example of acceleration velocity estimation

5 Elevator Operating Position Estimation

5.1 State Transition Function

The acceleration sensor has high short-term accuracy, but the estimation error of displacement will accumulate over time due to the cumulative error generated by two integrations, and the short-term accuracy is high, but the long-term error will increase to unacceptable. While the estimation error of displacement by the air pressure sensor accumulates relatively slowly, the two can be integrated to improve the accuracy of the estimation of elevator running position. According to Newton's formula, at moment k the operating position h_k of the elevator relative to the reference floor can be expressed as:

$$h_k = h_{k-1} + B_k u_k + w_k \tag{17}$$

Where u_k is the control vector $u_k = (x, \dot{x})^T$, which is the velocity and acceleration estimated using the acceleration sensor in the previous section. $B_k = (dt, dt^2/2)$. Due to the special operation mechanism of the elevator, the height of the car relative to the reference point is a fixed set of discrete values when the elevator is in the stationary phase. It is described by the list Map $Map \triangleq (fh_1..fh_i.fh_n)$, where fh_i refers to the height of the ith floor relative to the reference point and n is the number of floors that can be reached by the elevator. The state transition function of the system can be expressed as:

$$h_k = f(h_{k-1}) = \begin{cases} h_{k-1} + B_k u_k + w_k \text{ Elevator movement} \\ fh_i \; h_{k-1} \in U(fh_i, \delta_{fh_i}) \text{ Elevator static} \end{cases} \tag{18}$$

δ_{fh_i} denotes the a priori measurement error.

5.2 Map Initialization

Using the feature of high short-time accuracy of the acceleration sensor, the floor position of the elevator is learned using the SLAM algorithm [13], the Map is initialized, and the a priori information of the Map is obtained. The elevator needs to run through all floors sequentially (stopping floor by floor), and the distance between two stops can be estimated more accurately by Kalman filtering and combining the motion characteristics of the elevator, and the floor spacing and floor height can be obtained by scanning floor by floor. Figure 5 below shows the initialization process of Map at the experimental site (one layer is the reference layer), and it can be seen from the figure that the acceleration and velocity changes follow the same pattern between every two adjacent stops.

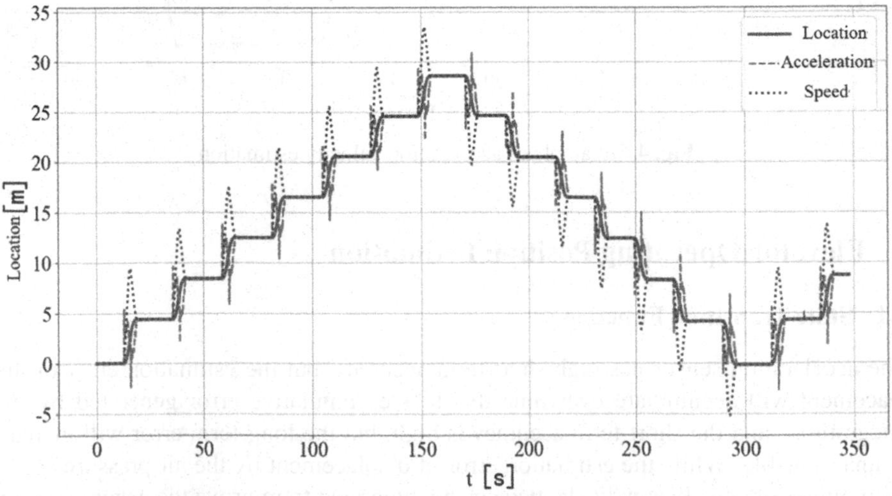

Fig. 5. Map initialization

5.3 Unscented Kalman Filter

The computational complexity of Unscented Kalman filter is the same as that of extended Kalman filter, which can achieve at least the second order accuracy [17]. Compared with the extended Kalman filter, the unscented Kalman filter does not need to know the specific expressions of the state transition function and the measurement function, and does not require the differentiability of the function. Therefore, this paper uses unscented Kalman filter to estimate the elevator position.

The core of Unscented Kalman filter is unscented transform, which approximates the nonlinear distribution by nonlinear variation through deterministic sampling. After unscented transform, Kalman filter computing framework is used to update the state. The specific calculation framework of UKF used in this paper is shown in Fig. 6.

Since the input and output of the system have the same meaning and are one-dimensional vectors, a simplified UKF formula is used in this paper.

First, a symmetric sampling strategy [18] is used to obtain 2n + 1 sampling points and the corresponding weights; all the sampling points are substituted into the state transfer function, and the results are weighted according to the UT transform to obtain the state prediction values and the prediction covariance.

The input and output of the system are the same, so there is no need to perform the UT transform of the measurement step, and the innovation and Kalman gain are calculated based on the measurement value and the measurement noise covariance after obtaining the measurement information, and then the update of the state estimate and the corresponding covariance is performed.

In this paper, an adaptive method based on the new interest and process noise is used. Let σ be the square root of P. When the innovation is greater than 2σ, i.e., the predicted value is too far away from the observed value, it means that the prediction function and the statistical properties of the noise are not adapted to the actual motion at this time step, so a fixed value sf is increased for Q to increase the weight of the measured value in the Kalman filter. No decreases Q by sf until it reverts to the initial value.

If the elevator is stationary, the nearest floor fhi from the current position will be detected according to Map, and if $h_k \in U(fh_i, \delta_{fh_i})$, the Map will be updated and the automatic sensor calibration procedure will be executed; otherwise, it indicates that the elevator may stop at an abnormal position.

6 Vibration Analysis

The vibration of the car during elevator operation is extracted as an important feature for non-intrusive monitoring of elevator operation status. Since the vibration signal in the vertical direction contains the acceleration of the elevator movement during the acceleration and deceleration phases, the horizontal vibration signal is used to detect abnormal vibration. A baseline representing the elevator's health condition is determined from the energy of the horizontal direction vibration; the data collected in real time is then observed for deviations from the baseline. If it deviates, it indicates that the elevator may be abnormal. The method of baseline generation [19] used in this paper is as follows.

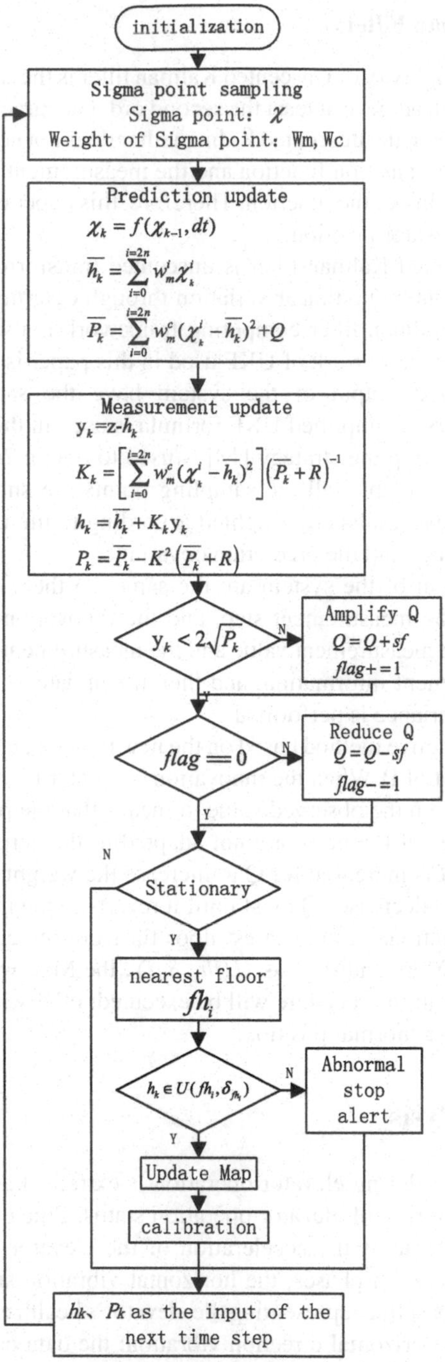

Fig. 6. UKF calculation process

The moving average (MA) is chosen as the indicator of convergence, and n is the number of collected signals.

$$MA_n = \left(\sum_{i=1}^{n} \left| I_{horizontal_i} \right|^2 \right) \Big/ n \tag{19}$$

The moving average error EMA is calculated, and the slope k of the points m-1 apart is used as the medium- and long-term convergence indicator. The stopping criterion is $k \approx 0$. Convergence is completed when the stopping criterion is satisfied, and if it does not converge, the data collection continues to calculate the k value.

$$EMA_j = \left| MA_j - MA_{j-1} \right| \big/ MA_{j-1} * 100 \tag{20}$$

$$k = \operatorname{atan}\left((EMA_n - EMA_{n-m}) / (x_n - x_{n-m}) \right) \tag{21}$$

After convergence is completed, the mean value μ of the recorded horizontal vibration energy is used as the baseline and the standard deviation σ is calculated. Based on the statistical principle, $\mu + 3\sigma$ is classified as the alarm line and a discriminant model is established for online monitoring. The new data collected is input into the discriminant model, and if it falls in the area between the baseline and the alarm line, the status is normal; if it exceeds the alarm line, it indicates that the operation may be abnormal and alarm processing is performed.

7 Experimental Results and Analysis

The elevator in the school of information and computer science is selected as the experimental environment to verify the performance of the method. The hardware platform is placed in the southwest corner of the car for trial operation, as shown in Fig. 7. The

Fig. 7. Experimental site

sensor bmp180 adopts ultra-high linearity mode and the sampling frequency is 39.2 Hz. The range of mpu6050 is ±2G and the sampling frequency is 200 Hz. The cloud server is based on the IPv6 cloud service platform of cel network.

Figure 8 shows the estimation results of the acceleration characteristics and velocity characteristics of the test elevator during the trial run, and the detection of the direction of operation and the phase of movement (described in Fig. 3) of the elevator in combination with the acceleration characteristics and velocity characteristics can be correct up to 100%. Among them, Figs. (1) and (2) show the quality parameters calculated according to the elevator technical conditions "GB/T10058–2009": maximum acceleration, maximum deceleration, and a95 (acceleration and deceleration of 95% of the sampled data) for each operation of the test elevator. Among them, the operating interval of speed is [−1.91, 1.80], the maximum acceleration and deceleration velocity are all less than 1.5 m/s^2, and the acceleration and deceleration velocity a95 are all greater than 0.5 m/s^2, which is in line with the national standard. The quality parameters calculated according to GB/T10058-2009 can reflect the health status of the elevator system and can also be used as data characteristics for predictive maintenance of elevators.

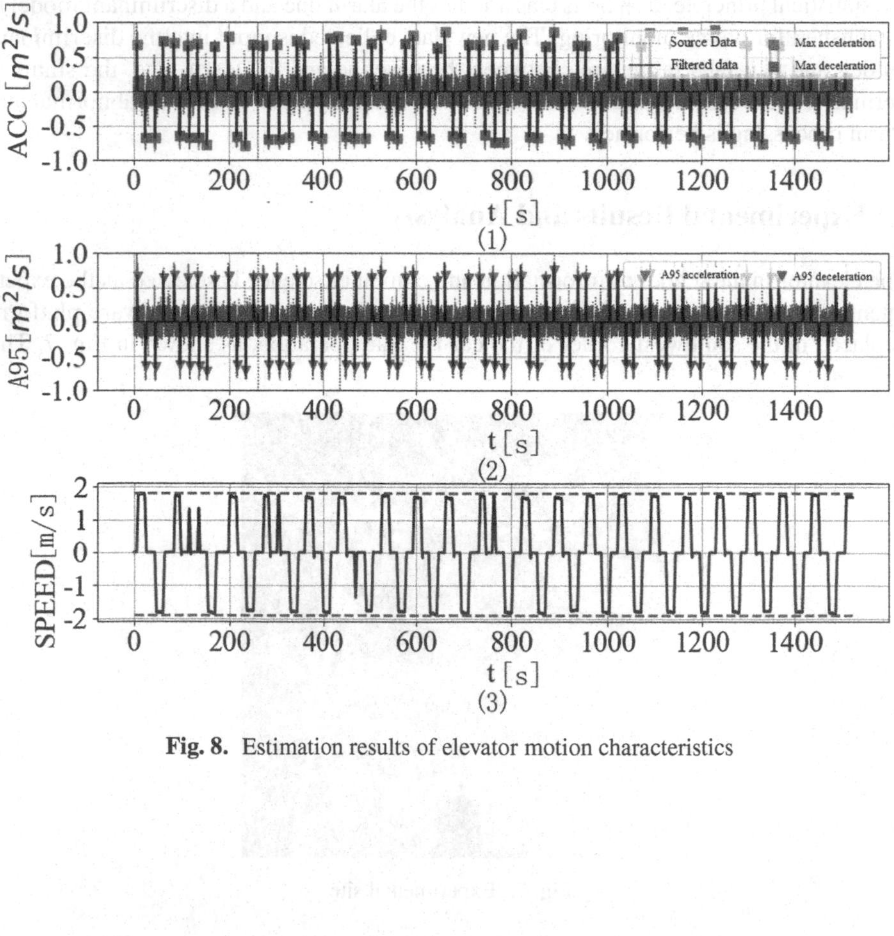

Fig. 8. Estimation results of elevator motion characteristics

The tracking results of the elevator running position by fusing the information of the two sensors through the traceless Kalman filter are shown in Fig. 9. As seen from the figure, since the UKF algorithm cannot correct the errors accumulated by the measurement in time, it leads to an increasing accumulation of errors, and the state estimation of the UKF algorithm significantly deviates from the true state as the running time grows. In the UKF algorithm, after the introduction of automatic sensor calibration, the accumulated errors are well controlled and the long-time state estimation performance is improved significantly. The optimization of the state transfer function combined with the SLAM algorithm further corrects the operating position of the elevator, which further reduces the error of the UKF state estimation.

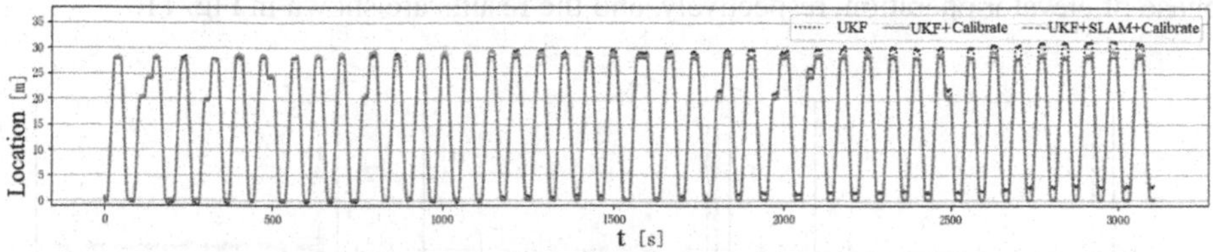

Fig. 9. Elevator position tracking results

The specific error analysis can be seen in Fig. 10. From the error accumulation probability distribution graph, it can be seen that the UKF algorithm integrating automatic calibration and SLAM can effectively control the accumulation of errors, and the average error of position is 0.098 m and the root mean square error is 0.143 m which converges faster and has higher filtering accuracy. The method designed in this paper basically solves the accumulation problem of position error and can better estimate the operating position of the elevator.

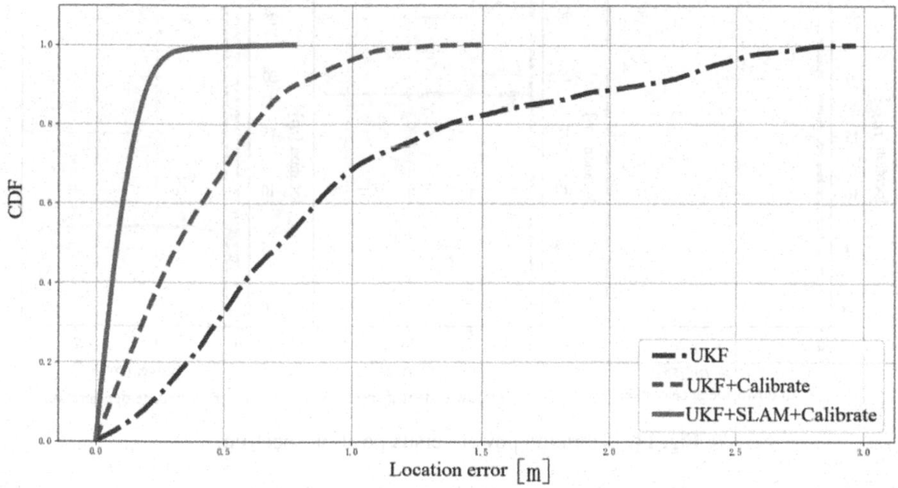

Fig. 10. Cumulative probability distribution of position error

When problems occur in the traction system of the elevator and the misalignment and deformation of the guideway will lead to an abnormal increase of vibration energy. On the basis of obtaining the real-time acceleration and position of the elevator, an elevator vibration energy position spectrum can be established to reflect the working condition of the elevator. In order to exclude the interference of the car's own acceleration during the acceleration and deceleration stages of the elevator, the acceleration signal in the horizontal direction is used to establish the vibration energy position spectrum. The problem of elevator traction system or guideway system is simulated by making a large jump during the operation of the elevator [15]. The large jumps were performed in the upward acceleration phase, upward uniformity phase, upward deceleration phase, downward acceleration phase, downward uniformity phase, and downward deceleration phase of elevator operation, respectively, and the results are shown in Fig. 11.

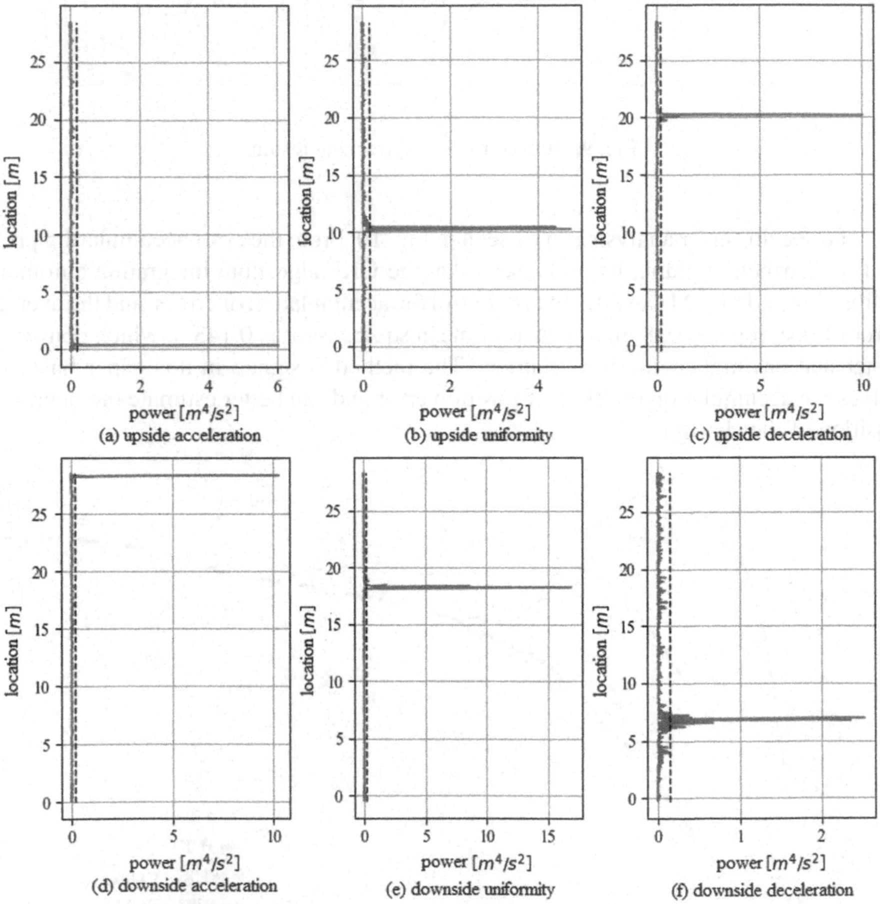

Fig. 11. Vibration power versus position spectrums

The vertical axis in the figure is the elevator running position, the horizontal axis is the vibration energy at the corresponding position, and the dotted line is the alert line generated by training with the baseline generation algorithm. The spikes in the spectrum are the abnormal vibrations generated by the simulation. It can be seen that the occurrence of simulated faults can be clearly detected at all stages of elevator operation, and the location of the car at the time of the fault can be located, providing important information for the maintenance of the elevator system and the rescue of trapped people.

8 Conclusion

This paper realizes real-time monitoring of elevator operation status and position. The dynamics characteristics of the elevator are estimated by decomposing the acceleration signal, auto-calibrating the sensor and using Kalman in combination with the motion characteristics of the elevator, and calculating the relevant quality parameters of GB/T10058-2009. The position tracking was performed using a traceless Kalman filter incorporating the SLAM algorithm, and the problem of position error accumulation was solved. Tests were conducted in the actual elevator operating environment, and the average positioning error was 0.098 m after one hour of continuous operation without installing beacons. It implements vibration analysis of the horizontal component of the elevator's acceleration, establishes the baseline and vibration energy position spectrum of the elevator operation, and verifies it by simulating faults. The non-intrusive elevator condition monitoring system designed in this paper is inexpensive, simple and convenient to install and move, and realizes plug-and-play, universally and completely decoupled from the elevator's own control system, which will not affect the control and use of the elevator. In future work, this paper plans to study the detection function of the elevator door system, conduct experiments in elevators with known faults, and performs further fault analysis.

Acknowledgments. This work was supported by the Next Generation Internet Technology Innovation Project Of Celtic Network (No. NGII20181206) , the National Natural Science Foundation of China (No.61976150), the Key R & D Projects of Shanxi Province (No. 201803D31038).

References

1. Wen, P., Zhi, M., Zhang, G., et al.: Fault prediction of elevator door system based on PSO-BP neural network. Engineering **8**(11), 761–766 (2016)
2. Feng, Y., Xing, H., Chen, Y.: Elevator door fault warning system based on microcomputer. Machinery **56**(10), 99–101 (2018)
3. Wang, X., Ge, H., Wang, R., et al.: Design of high concurrent communication server of elevator remote monitoring system. Comput. Sci. **44**(4), 157–160 (2017)
4. Zhao, H., Wu, Y., He, P., et al.: Research on fault detection method of elevator traction machine brake. Mach. Tool Hydraul. **046**(001), 185–188 (2018)
5. Zheng, X., Fu, X., Tang, X.: Research on elevator fault monitoring method based on power quality analyzer. Electron. Test **15**, 95–96 (2018)

6. Zhang, X., Zhao, Y.: Research on elevator fault prediction technology. China Elevator **030**(005), 20–23 (2019)
7. Esteban, E., Salgado, O.: Model-based approach for elevator performance estimation. Mech. Syst. Sig. Process. **68**(69), 125–137 (2016)
8. Taplak, H., Erkaya, S., Yildirim, Ş., et al.: The use of neural network predictors for analyzing the elevator vibrations. Arab. J. Sci. Eng. **39**(2), 1157–1170 (2014)
9. Liu, C., Zhou, S., Liu, X., et al.: The elevator fault diagnosis method based on sequential probability ratio test (SPRT). Autom. Control Intell. Syst. **5**(4), 5583–5591 (2017)
10. Zhou, Q., Ding, S., Feng, Y., et al.: The elevator brake intelligent monitoring and fault early warning system based on SVM. China Spec. Equip. Saf. **34**(5), 22–27 (2018)
11. Xu, J., Xu, L., Wang, H., et al.: Condition monitoring of elevators based on vibration analysis. J. Mech. Electric. Eng. **36**(3), 279–283 (2019)
12. Yi, S., Wan, Z., Tao, R., et al.: Diagnosis of elevator faults with SVM based on optimization by GA. Comput. Digital Eng. **44**(1), 166–170 (2016)
13. Skog, I., Karagiannis, I., Bergsten, A.B., et al.: A smart sensor node for the internet-of-elevators-non-invasive condition and fault monitoring. IEEE Sens. J. **17**(16), 5198–5208 (2017)
14. Chen, Z., Zn, W., Zhang, G., et al.: Research of big-data-based elevator fault diagnosis and predication. J. Mech. Electric. Eng. **1**, 90–94 (2019)
15. Hao, Z., Ge, W., Hao, J., et al.: Research on embedded elevator running condition monitoring system. J. Electron. Meas. Instrum. **33**(8), 187–193 (2019)
16. Weiszfeld, E.: Sur le point pour lequel la somme des distances de n points donnes estminimum Faseb. J. Official Publ. Feder. Am. Soc. Exp. Biol. **20**(3), 559–566 (2006)
17. Yang, F., Zheng, L., Ji, W., et al.: Double layer unscented Kalman filter. Acta Autom. Sin. **45**(7), 1386–1391 (2019)
18. Merwe, R., Wan, E.: Sigma-point Kalman filters for probabilistic inference in dynamic state-space models. Doctoral Dissertation, OGI School of Science and Engineering (2004)
19. Romero, A., Lage, Y., Soua, S., et al.: Vestas V90-3MW wind turbine gearbox health assessment using a vibration-based condition monitoring. Syst. Shock Vibr. (6), 1–18 (2016)

Research on Application of Frontier Technologies at Smart Airport

Kang Liu(✉), Lei Chen, and Xia Liu

Sanya Aviation and Tourism College, Sanya 572000, Hainan, China

Abstract. Throughout the development of the global airport industry, increasing new technologies are applied to airport industry. In this paper, the concept of smart airport is proposed. Firstly, the development history of smart airport was introduced, and the construction needs of China's smart airports were studied. Then it analyzed the application of RFID technology, biometric technology, interactive wayfinding technology, big data and artificial intelligence technology at smart airports. The analytical results show that the application of the frontier technologies make the airport operation more efficient, decision-making more collaborative, control more accurate, and travel more convenient, thus helping the airport to accelerate the construction goal of intelligent airport.

Keywords: Smart airport · Frontier technologies · Application

1 Definition of Smart Airport

With the emergence of new technologies such as the Internet of Things, artificial intelligence and blockchain, the definition of a smart airport is constantly changing. The essence is to use a new generation of information technology to sense, analyze and integrate the real-time status of the airport, so that different types of personnel, such as operators, managers, air passengers and cargo owners, can make timely and intelligent decisions, and enjoy convenient services in various activities such as operation, safety service, administration and business [1].

Under the conditions of knowledge and information society, the smart airport makes it possible and realistic to achieve optimal operational efficiency and highest service quality. The smart airport is based on the construction of smart infrastructure and its overall goal is improving smart operation, smart service and smart management. It integrates the concept and technology of "wisdom" into people, things and objects to make life "smart" and sustainable [2].

J. Zeng et al. (Eds.): ICPCSEE 2021, CCIS 1452, pp. 319–330, 2021.
https://doi.org/10.1007/978-981-16-5943-0_26

2 The Current Development State of Smart Airports

Since entering the 21st century, the global aviation industry has shown explosive growth. Under this situation, the operation mode and business model of airports have also entered a rapid development stage. The United States, Europe and other developed countries adjusted and changed the aviation field first. The developed countries discarded old ideas that did not adapt to the development of the times and created a new era of aviation. They improved the efficiency of airport operations, increased airport traffic, and enriched the diversity of airport service businesses, while ensuring safe airport operations. Throughout the development of the global airport industry, the evolution of airports to date has gone through three stages: basic airport operations, agile basic and intelligent airport operations, and has begun to develop into a fully artificially intelligent airport (Table 1).

Table 1. The development of smart airports

Type	Main features	Mainland airport representatives
Airport 1.0 Basic Airport Operation	□Operating as a "landlord" □Focus on efficient and safe landings, departures and other aircraft-related operations	√Sanya Phoenix International Airport √Guilin Liangjiang International Airport
Airport 2.0 Agile Airport Operation	□Centralized and shared services strategy □Technology-driven collaboration □High operational efficiency □Achieved faster turnaround times and improved passenger experience	√Shenzhen Baoan International Airport √Shanghai Pudong International Airport √Tianjin Binhai International Airport √...
Airport 3.0 Smart Airport Operations	□Become a "virtual service provider" □Advanced sense-analyze-feedback capabilities □Real-time data analysis and mining □Comprehensive visualization of proactive services □Intelligent security detection and analysis □Business process integration and synergy for all parties involved	√Beijing Daxing International Airport √Chengdu Tianfu International Airport √Qingdao Jiaodong International Airport √...

(*continued*)

Table 1. *(continued)*

Type	Main features	Mainland airport representatives
Airport 4.0 Full Artificial Intelligence Airport	□To autonomously sense and identify business status relying on AI technology, and autonomously complete complex event processing □To predict airport situation autonomously and complete the assessment and analysis of special situation □To analyze business big data comprehensively, and learning to -make decision autonomous…	√Airports of the Future

2.1 Basic Airport Operation

In the 1980s, the global airport industry was in the basic airport phase, which was safety-oriented and focused on the necessary safety facilities, aircraft takeoffs and landings, and other flight operations. Airports provided the most basic passenger services, including ticketing, boarding, security screening, baggage transportation, and simple things to drink vending.

2.2 Agile Airport Operation

In the 1990s, the global airport industry entered the agile airport phase, which focused more on the flexibility and agility of airport operations. Airports were able to adapt to changes in the environment and the rapid pace of operations. New technologies were continuously integrated into airport business applications, and interdepartmental cooperation became closer.

2.3 Smart Airport Operations

In the 21st century, with the introduction of "smart cities", some well-known airports have started to achieve real-time information interaction, collaborative decision-making and process integration, by leveraging emerging and mature technologies [3]. Through the process integration with other partners such as airlines, tenants, field units, fuel companies and catering companies, a complete passenger service value chain is formed, creating new profit growth for all parties.

2.4 Full Artificial Intelligence Airport

With the further development of artificial intelligence and big data, the future airports will perceive and identify business status independently, complete complex event processing autonomously, be able to make decision by self-learning, and predict about the airport situation automatically. This will greatly reduce the operating costs and error rates of airports, airlines and other business units, while enhancing the intelligence of airports, improving work efficiency, and improving passenger service levels.

3 Construction Needs of Domestic Smart Airport

3.1 The Own Development Needs of Airports

As the second largest aviation market in the world, China faces significant development opportunities and serious challenges for airport transport, due to its high growth demand. Based on public data from the China Air Transport Association (CATA), this paper shows statistics on the passenger throughput and landings of the top 15 domestic airports in 2019 compared to those in 2018 (Table 2).

Table 2. Comparison of passenger throughput and landings at the top 15 domestic airports

Airport	Passenger throughput (persons)		Landings and takeoffs (sorties)	
	2019 Year	2018 Year	2019 Year	2018 Year
Beijing/Capital	100,013,642	100,983,290	594,329	614,022
Shanghai/Pudong	76,153,455	74,006,331	511,846	504,794
Guangzhou/Baiyun	73,378,475	69,720,403	491,249	477,364
Chengdu/Shuangliu	55,858,552	52,950,529	366,887	352,124
Shenzhen/Baoan	52,931,925	49,348,950	370,180	355,907
Kunming/Changshui	48,075,978	47,088,140	357,080	360,785
Xi'an/Xianyang	47,220,547	44,653,311	345,748	330,477
Shanghai/Hongqiao	45,637,882	43,628,004	272,928	266,790
Chongqing/Jiangbei	44,786,722	41,595,887	318,398	300,745
Hangzhou/Xiaoshan	40,108,405	38,241,630	290,919	284,893
Nanjing/Lukou	30,581,685	28,581,546	234,869	220,849
Zhengzhou/Xinzheng	29,129,328	27,334,730	216,399	209,646
Xiamen/Gaoqi	27,413,363	26,553,438	192,929	193,385
Wuhan/Tianhe	27,150,246	24,500,356	203,131	187,699
Changsha/Huang Hua	26,911,393	25,266,251	196,213	186,772

As can be seen from the table, the visitor throughput and flight movements of the top 15 domestic airports are increasing to varying degrees, except for Capital Airport. The growth rates of the two (see Fig. 1) are different, with the growth rate of passenger throughput at the top 15 airports in 2019 being 2 to 3% points higher than the growth rate of takeoffs and landings, indicating that the growth rate of transportation demand is higher than the growth rate of capacity supply. As passenger throughput continues to increase, the flight areas, terminals and ground transportation resources will be further strained in many large domestic airports. Meanwhile, during the airport operation, there are information in a scattered state, decentralized data structure, high cost and low efficiency of cross-sectoral communication, etc., which results difficulty to make effective decisions through comprehensive assessment [4]. Therefore, while airports are constantly expanding their hardware, the need for intelligent solutions is even more urgent.

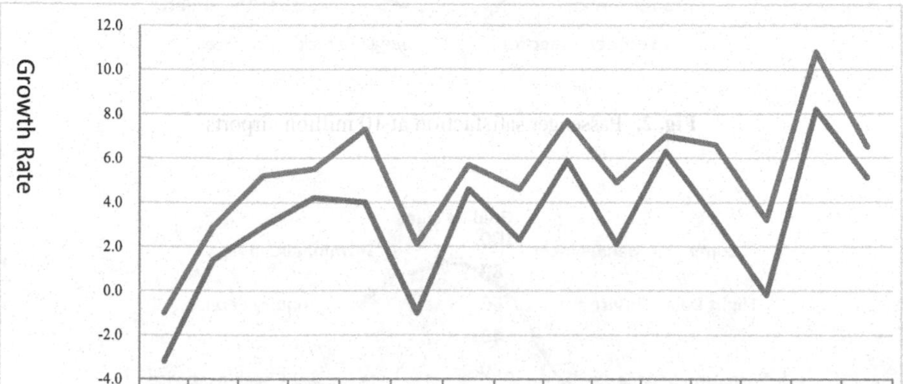

	Beijing / Capital	Shanghai / Pudong	Guangzhou / Baiyun	Chengdu / Shuangliu	Shenzhen / Baoan	Kunming/Changshui	Xi'an / Xianyang	Shanghai/Hongqiao	Chongqing/ Jiangbei	Hangzhou / Xiaoshan	Nanjing / Lukou	Zhengzhou/Xinzheng	Xiamen / Gaoqi	Wuhan / Tianhe	Changsha/Huang Hua
Series1	-1.0	2.9	5.2	5.5	7.3	2.1	5.7	4.6	7.7	4.9	7.0	6.6	3.2	10.8	6.5
Series2	-3.2	1.4	2.9	4.2	4.0	-1.0	4.6	2.3	5.9	2.1	6.3	3.2	-0.2	8.2	5.1

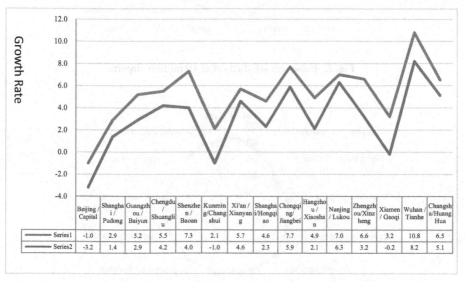

Fig. 1. Growth rate of passenger throughput and flight movements in 2019 at the top 15 domestic airports compared

3.2 Passenger Travel Demand

Key factors affecting passenger experience at the airport include on time flight, baggage issues, and service details. According to the data published by China Civil Airport Association (CCAA), there are 14 indicators of passenger satisfaction with the airport experience in terms of signage, terminal facilities, inquiry service, check-in service, security check service, boarding service, baggage claim service, transportation in and out of the airport, transit service, airport commercial catering, information service, flight delay service, and complaint management. The results of the 14 indicators of passenger satisfaction between airports with 10 million and airports with millions of passengers are shown in Figs. 2 and 3.

From the results, the passenger satisfaction of airport signage, check-in procedures, terminal facilities are higher than those of other indicators, both for 10 million and million airports. On the other hand, information service, airport catering service and flight delay

Fig. 2. Passenger satisfaction at 10 million airports

Fig. 3. Passenger satisfaction at 1 million airports

service are rated the lowest. This shows that passengers are quite sensitive to information, especially when it comes to flight delays, which requires airports to strengthen the transparency of information and provide passengers with timely information on the flight delay reasons, flight delay times, and follow-up arrangements [5]. It has always been a pressing need for domestic travelers to find a suitable restaurant at the airport according to their needs, and until today, complaints about airport catering are still at the top of the list of airport complaints. Service pain points still exist. The airports needs to be customer-centric, combining big data, artificial intelligence and other technologies to explore the personalized needs of customers. At the same time, the airport should enrich the business model of catering, and accelerate the realization of airport catering "same quality and price in the same city", so as to improve the satisfaction of airport passengers.

3.3 Airline Demand

Since airlines are important customers of airports, the facilities, business collaboration, system interoperability and data sharing should meet the operational needs of airlines. Although the requirements vary between different airlines, the basic requirements for airports include real-time information access and delivery, transparent operational status, collaboration-based AOC (Airport Operation Management Center), and on-time performance, turnaround speed and connectivity. One of the high turnaround speeds is the airline's desire to keep turnaround times (time between takeoff and landing) low to improve aircraft utilization (in the air). Therefore the shorter the turnaround time, the faster the next flight will depart. The more stable and predictable the turnaround state is, the shorter the wheel block time is and the less buffer time the airline needs [3]. Connectivity is the flexibility to connect transit passengers quickly, which allows for the shortest possible MCT (Minimum Connecting Time), and therefore shorter turnaround times and better on-time performance metrics.

3.4 Needs of Other Stakeholder Units

Other stakeholders include regulators, merchants, hotels, etc. Their needs for the airport are as follows: First, sustainability, i.e., stable and sustained business growth (including throughput of passengers and flight movements, turnover, profit) [7]; Second, stability. This includes organization, operation and customer. Any changes in these metrics could cause investors to lose confidence who don't like any surprises; and third, efficiency/productivity. The continuous investment in operating costs and the "evaporation" of assets require increased productivity and an increasingly efficient use of airport equipment, personnel and fixed assets (infrastructure). Thus, the higher the utilization rate, the later the expansion investment will be.

4 The Application of Frontier Technologies in Domestic Smart Airports

The construction of smart airports is inseparable from the application of new technologies, and their development cannot be achieved without the support of technology. With the rapid development and update of current science and technology, the application of new technology is one of the effective means to improve the competitiveness of airports.

4.1 Application of Permanent Luggage Tags

Radio frequency identification (RFID) is a wireless communication technology that provides accurate baggage tracking and positioning in airport applications, facilitating the automation of check-in and boarding processes. It can enhance the passenger experience, provide more targeted passenger services, and reduce manual input [8]. But the technology needs a unified standard, and airports should be equipped with relevant systems to read baggage tags. In view of this, frequent flyer program could be considered during the construction of smart airports, using RFID technology to create frequent flyer cards, which stores personal biometric information, etc. Travelers holding the frequent flyer card can enjoy priorities such as fast self-service check in, discounts at stores and discounted airport parking fees. The card should support common use in both airports. In addition to the physical card, with the rapid popularity of smartphone terminals, an electronic frequent flyer card can be introduced and installed on passengers' mobile smart terminals, which can save the production and issuance cost of cards, and is easier for passengers to carry and use. And smartphone terminals can be used to bring more functions, and introduce comprehensive information services such as electronic passenger guidance and boarding.

4.2 Application of Biometric Identification Technology

Biometric identification technology mainly refers to a technology that carries out identity authentication through human biometric features. It completes biometric identification through a series of processes such as information collection, template creation, feature identification and feature verification. The biometric identification technology is characterized by universality, uniqueness, stability and irreproducibility, etc. It is more secure, confidential and convenient than traditional identification methods. And it has the advantages of not being easily forgotten, good anti-counterfeit performance, not easily stolen, and more available at anytime, anywhere, etc. [9]. But sample collection and safe storage are the key points. In the airport industry, biometrics are commonly used in the following applications.

1) Border control. Use international standards for passenger entry and exit security management. It collects and stores human biological information such as fingerprints of outbound and inbound passengers, and compares them when conducting outbound and inbound border control, which can effectively screen the identity of outbound and inbound passengers.
2) Check-in and boarding. Biometric travel document reconciliation systems ensure that the passenger checking in and boarding is the same person to ensure flight safety. Biometrics collected at self-service check-in terminals can also be compared with "blacklist" to identify suspect passengers.
3) Passenger Security Screening. The biometric information of passengers is collected and compared with that of international and domestic wanted criminal suspects or terrorists at each control point of the airport to improve airport security.
4) Access control for airport employees. For the office area of the airport, security control is particularly important. The biometric information of employees is collected in advance and stored in the system, which can effectively and easily manage the access control of the office area.
5) Other systems. For example, employee background checks, logical access control (including the use of biotechnology login, password management and encryption control, etc.).

4.3 Application of Interactive Wayfinding Technology

Based on the presentation of ordinary static maps, Interactive wayfinding adds interaction with the user. The user can look for location and its route, and control the way it is presented. The traditional method of wayfinding is to provide static maps in the airport terminal, such as small atlases and large screens, and then passengers read the maps and find out the route by themselves. Interactive wayfinding technology provides passengers with a convenient and interesting wayfinding experience. Users can query the destination they want to reach, and the system calculates the route and presents it to them in a realistic and visual way. However, the application process requires multi-system coordination, such as GIS system, high-bandwidth communication system and other positioning system [10]. The application of interactive wayfinding technology in airports mainly includes the following three aspects:

1) Information kiosks with manual assistance. The kiosk interacts with passengers via video (connected to customer service), where passengers can upload scanned documents, ask questions, get answers to their questions and print them out. In the future, this manual help process can also be replaced by computerized customer service with artificial intelligence.
2) Self-service kiosks. Passengers are guided through interactive means to refine and model the questions they want to submit (e.g., wayfinding needs), then query for answers and print them. The query results can be stored in a barcode or QR code (Quick Response code), which the traveler scans at the next kiosk to obtain updated information (e.g., updated route directions). Passenger identification and storage can help optimize this service process and enhance the passenger experience, so the kiosk can also add the function of passenger identification to prepare for future airport collection of passenger data and thus optimize operations.
3) Mobile phone information access. Cell phones can be used to scan QR codes or barcodes to obtain information already stored in the code, or to use cell phone applications to provide passenger information and wayfinding services, similar to the above-mentioned "self-service kiosks". The difference is that the cell phone can obtain data in real time and follow the passenger's movement, so the passenger can get the latest guidance in real time. In the future, mobile applications could also be extended to virtual reality (e.g., retail service delivery), as many cities organically combine airports, railroads, subways, buses and cabs at airports, to form a large-scale transportation hub. So when travelers have a need to transfer, they are likely to get lost in the mass of signage. This is where interactive wayfinding comes in handy. The most basic functions include multi-language selection, in-building facilities or transfer directions, flight/high-speed rail/subway inquiries and route printing, etc. Auxiliary functions such as augmented reality, weather forecasts and municipal information can also be added.

4.4 Self-service Baggage Handling Technology

Self-service baggage handling is for passengers who have completed self-check-in, providing them with faster and centralized baggage check-in, eliminating the need to wait in line at the check-in counter. The self-service baggage handling system provides users with a unified user interface and is connected to multiple airline departure hosts. This allows the same counter/check-in terminal to handle multiple airline baggage checks [11]. There are three types of self-service baggage handling: public baggage check-in counters, self-printing baggage labels and self-service check-in. For passengers, it can simplify the check-in and check-in process and save the time of searching for a particular airline; for airlines, the shared service can save staff and equipment overhead and reduce costs. For airports, it saves space and provides the opportunity to redesign check-in halls, to provide better service to passengers while having the potential opportunity to increase revenue. However, it is not easy to implement and requires planning at the early stages of terminal construction, and the functional layout of the check-in hall design baggage handling system design to be considered.

4.5 Big Data and Artificial Intelligence Technologies

For smart airports, big data and artificial intelligence mainly to establish the thematic event model knowledge base, by combining the airport business rules, event nature and expert knowledge base. When similar events occur, scientific decision basis can be provided through model comparison.

1) Establishing an industry knowledge base. Industry knowledge base makes information and knowledge orderly, which is the primary contribution of knowledge base to the organization. To build a knowledge base, it is necessary to collect and sort out the original information and knowledge on a large scale, classify and save it according to certain methods, and provide corresponding retrieval means. After such a process, a large amount of implicit knowledge is encoded and digitized, and then, the information and knowledge become ordered from the original chaotic state. This facilitates the retrieval of information and knowledge, and lays the foundation for effective use.
   The construction of airport knowledge base mainly includes: First, the construction of model knowledge base, which can help information systems of enterprises to connect and collaborate effectively. When knowledge and information are organized, the time for finding and using them is greatly reduced, and its flow is naturally accelerated. Second, the construction of the safety and security knowledge base, it can provide scientific decision-making basis for airport security. The expert group collects and organizes cases of typical security incidents at airports, collects, cleans and stores data according to the whole link of the origin, process and result of the incident, and establishes a knowledge base of incident models in combination with the nature of business [12]. Eventually, through the model knowledge base of different scenarios, different types and different natures, the early warning of security topics beforehand, the processing decision during the event and the scientific tracing afterwards are realized. Third is the construction of passenger service knowledge base, which can help enterprises realize the effective management of frequent flyers. Information management of airport frequent flyers has been a relatively complex task, which involves factors such as the type of service, service awareness and service quality of the airport, but with different services of different staff, it will directly affect the satisfaction level of passengers, which may lead to the loss of frequent flyers as a result. Therefore, an important element of passenger service knowledge base is to save all the information of customers in order to facilitate the staff to provide targeted and personalized services at any time according to the travel

habits and consumption patterns of passengers, and at the same time to supervise the whole process of staff's services [13].

2) Intelligent analysis model. The knowledge base established through big data information technology should have three capabilities: the ability to convert data into knowledge, the ability to organize knowledge by classification and self-classification, and the ability to mine knowledge, which all require a powerful platform based on cloud computing. The platform should have self-learning capability and be able to process text data in depth, while human-machine collaboration is needed to correct the deviation of machine learning through human intervention. Finally, the airport needs to build a platform for knowledge mining and utilization, using technologies such as industry intelligence, data warehouses, and artificial intelligence, to be able to acquire and discover new industry artificial intelligence through braking relationships and automatic reasoning [14].

5 Conclusion

Since IBM first proposed the concept of "Smart Earth" in November 2008, more and more new technologies have been widely used in many industries and industries, and from the perspective of the main new technologies applied in domestic smart airports, we regret to see that these new technologies are actually more mature in other industries and even in foreign airports in developed countries. Mature applications, the reasons for this, and the domestic airport industry closed, data monopoly and other factors related. Of course, as China becomes more and more open, and especially as China proposes the strategic plan to build a "four characteristics airport" (safe, green, intelligent and humanistic), we believe that these new technologies will be increasingly applied to the construction of domestic airports [15]. It is believed that these new technologies will be increasingly used in the construction of domestic airports.

Acknowledgment. This research was financially supported by Foundation for Teaching and Research Team of Ideological and Political Theory Course in Hainan Province, and Philosophy and Social Science Planning Subjects of Sanya (Project Number: SYSK2021-05).

References

1. Jinjun, K., Hao, S.: Exploring the construction of smart airports. China Sci. Technol. Inf. (20), 32–33 (2019)
2. Xuan, Y.: Innovation is the new driving force for smart airports. China Civ. Aviat. J. (005), 02 (2019)
3. Jiezhuo, R., Hui, H., Xuange, Z.: Digital information technology in smart airports. Modern manuf. (36), 80–81(2020)
4. Bin, H.A.O.: Research on large airports operation control system based on total airport management. J. Civ. Aviat. (05), 38–41(2021)
5. Xinliang, S.: Comparison of irregular flight services between China and EU and suggestions for improvement. Civ. Aviat. Manage. (11), 73–76(2020)

6. Lanqing, L.: China's smart airport construction status and development prospects. Air Trans. Bus. (05), 11–16(2018)
7. Haiyang, W.: Study on efficiency management improvement model of civil aviation airports. Contemp. Econ. (02), 109–111(2020)
8. Xue, Y.: Application of RFID in baggage transportation in civil aviation. J. Civ. Aviat. (04), 85–88 (2020)
9. Zhou, Z.: Technologies speed up and enhance airport security check. Bull. Sci. Technol. (36), 112–115 (2020)
10. Xinhui, R.,Huiting, H.: Influencing factors of airport terminal wayfinding based on an integrated model. J. Trans. Inf. Saf. (39), 153–160(2021)
11. Meng, X.: Research on the development of smart airport in the new era. Civ. Aviat. Manage. (06), 39–42 (2020)
12. Diange, S.: Research on the construction of civil airport unsafe event analysis expert system. Saf. Secur. (41), 52–57 (2020)
13. Xiangdong, J.: Design and research of smart airport management platform based on big data technology. China CIO News (07), 40–41(2020)
14. Yiye, Y., Chongjun, F., Aizhi, A.: The application research of artificial intelligence in the construction of intelligent civil aviation. Intell. Comp. Appl. (10), 214–215(2020)
15. Xin, W.: Study on the synergistic integration of four characteristics airport. Intell. Build. (03), 47–50 (2019)

Design and Evaluation of an Exergame System of Knee with the Azure Kinect

Guangjun Wang[1,2,3], Ming Cheng[3], Xueshu Wang[3], Yi Fan[3], Xin Chen[3], Liangliang Yao[3], Hanyuan Zhang[1,2,4], and Zuchang Ma[1,2](✉)

[1] Hefei Institutes of Physical Science, Chinese Academy of Sciences, Hefei 230031, China
zcma@iim.ac.cn
[2] University of Science and Technology of China, Hefei 230026, China
[3] The University Key Laboratory of Intelligent Perception and Computing of Anhui Province, Anqing Normal University, Anqing 246013, China
[4] Department of Sports Medicine and Arthroscopic Surgery, The First Affiliated Hospital of Anhui Medical University, Hefei 230022, China

Abstract. Timely and effective knee function evaluation and knee exercises promote the prevention and self-management of knee diseases. In this paper, a Kinect-based exergame system is proposed to assess and train the knee function. Azure Kinect was used to capture and generate 3D models of the user and immerse them in an interactive virtual environment. The software included three functional modules: knee function evaluation, Knee exercises game, and Comprehensive evaluation. The stand, step, leg lift, and squat were selected for knee function evaluation and exercises. Twenty volunteers participated in the experiment. Intra-class correlation coefficients (ICCs) were calculated to assess the reliability of kinematic measurements of knee angles during the movements. The ICC of these movements were stand (ICC = 0.987), step (ICC = 0.997), left leg lift (ICC = 0.981), right leg lift (ICC = 0.990), stand (ICC= 0.998). The results show that the test-retest reliability is high. It means that the motion capture data is effective and the data obtained by Kinect is stable.

Keywords: Azure Kinect · Exergame system · Knee function evaluation

1 Introduction

With the growth of age, the knee gradually degenerates and produces knee diseases, such as knee osteoarthritis, etc. [1]. In daily life, inappropriate exercise, unhealthy living habits, and falls are easy to cause different degrees of knee injury [2, 3]. Knee diseases are the most common joint diseases, which significantly affect people's quality of life. Timely and effective knee function evaluation and knee exercises could promote the prevention and self-management of knee diseases.

At present, there are many scales for knee evaluation, such as WOMAC, Lysholm, AKS, etc. [4, 5], which have been widely used in clinical. However, these scales are mainly based on subjective judgment, so it is difficult to obtain the activity information of the knee quantitatively. In recent years, motion capture has been gradually applied

J. Zeng et al. (Eds.): ICPCSEE 2021, CCIS 1452, pp. 331–342, 2021.
https://doi.org/10.1007/978-981-16-5943-0_27

to evaluating human joint function and disease. The motion capture system obtains the position or angle information of joint motion through sensors accurately. The Kinect depth camera launched by Microsoft can obtain the motion information of the main joint position through the bone tracking system. Also, the Kinect is portable, cheap and non-wearable, and has been gradually applied to the research of knee and lower limb. Wochatz M et al. verified the reliability and validity of the Kinect V2 for the evaluation of lower extremity rehabilitation exercises [6]; Tanaka R et al. verified the reliability of the Kinect V2 to measure the trunk, hip, and knee angle during walking on flatland and a treadmill [7]; Gh A et al. analyzed the concurrent validity of evaluating knee kinematics using the Kinect system during rehabilitation exercise [8]. These studies show that Kinect can obtain knee motion information accurately. Kinect obtains data through visuals. If there is an occlusion in joint parts, it is easy to cause the loss or inaccurate local joint information. Therefore, some simple movements can be obtained accurately, while movements with occlusion were difficult to obtain information accurately.

Based on the research and experiment on the possibility of Kinect obtaining joint information related to the knee, this study analyzed knee function evaluation movements. Finally, four movements, namely **stand, step, leg lift, and squat**, were selected for knee function evaluation. A knee exercise game based on these four movements was designed and implemented. This paper introduces the design of knee function evaluation, knee exercises game and also analyzes the effectiveness of the system for obtaining motion data and the reliability of retesting.

2 System Design

2.1 Overall Design

The purpose of the system was to evaluate the function of the knee through simple devices and standardized movements. The exergame opinions according to the evaluation results. The overall functional design of the system as shown in Fig. 1.

The system was composed of a computer and an Azure Kinect sensor, respectively used in data processing and motion interaction [9]. The computer requirements were as follows: 7th generation Intel® Core™ i5 processor (quad-core 2.4 GHz or faster), 4 GB of RAM, NVIDIA GeForce GTX 1050 or equivalent, dedicated USB3 port. The system was developed in the Unity 2019.3 version and C# programming language, and the database was designed by MySQL. The bone data was obtained using the Azure Kinect Examples for Unity 1.9.1 development kit available online. The system can be operated at room temperature.

2.2 The Azure Kinect and Human Bone Tracking

The system used the bone tracking function of Azure Kinect to track the movement of the user in real-time. Azure Kinect was also used as an interactive device to perform cognition and movement training [10]. The Azure Kinect sensor and human bone tracking structure as shown in Fig. 2.

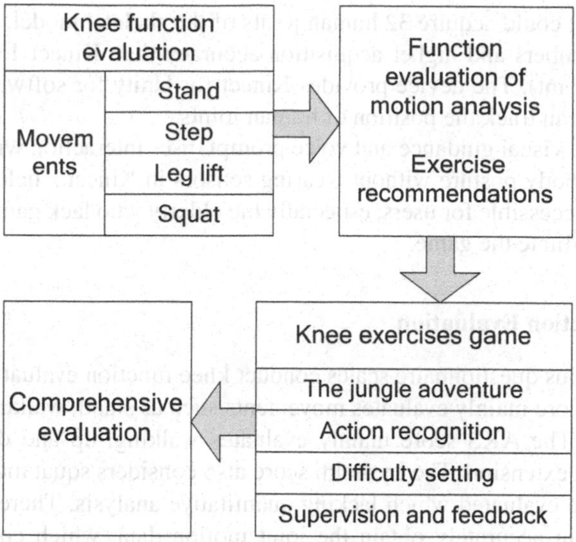

Fig. 1. The overall functional design of the system

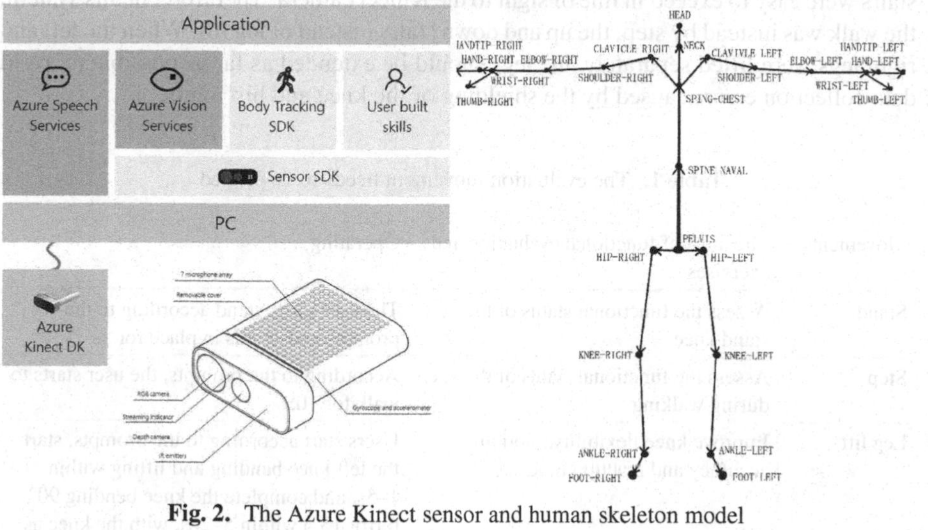

Fig. 2. The Azure Kinect sensor and human skeleton model

Azure Kinect is a new generation of depth cameras based on TOF released by Microsoft in 2019. It includes a 1-megapixel depth sensor, wide and narrow field Angle (FOV) options, a 12-megapixel RGB camera, which provides color image data and depth image. Microphone array for far-field speech and sound capture; Includes both accelerometer and gyroscope (IMU) for sensor direction and space tracking; External sync pins allow easy synchronization of sensor data streams from multiple Kinect devices.

Azure Kinect could acquire 32 human joints of the skeletal model, which has more joint capture numbers and higher acquisition accuracy than Kinect 1.0 (20-joint) and Kinect 2.0 (25-joint). The device provides Kinect for Unity for software development in Unity, which can track the position of human joints.

According to visual guidance and voice prompt, user interaction with the computer system through body posture without wearing sensors in Kinect's field of vision. This pattern is more accessible for users, especially the elderly who lack game experience, to evaluate and facilitate the game.

2.3 Knee Function Evaluation

At present, Various questionnaire scales conduct knee function evaluation subjectively. The WOMAC score mainly evaluates movements such as stand, walking, up and down stairs, bending. The AKS score mainly evaluates walking, up and down stairs, and knee flexion and extension. The Lysholm score also considers squat movements. These movements were evaluated which lacking quantitative analysis. Therefore, the Kinect depth camera can accurately obtain the joint motion data, which could be used for quantitative analysis and evaluation.

Through the experiment, we found that the movements such as walk up and down stairs were easy to exceed in line of sight to the Kinect camera. Therefore, in this system, the walk was instead by step, the up and down stairs instead of leg lift. When the left and right legs were lifted separately, the legs should be extended as far as possible to avoid data collection errors caused by the shielding of the knee and hip joints.

Table 1. The evaluation movement needs to be probed.

Movements	The aims of functional evaluation and exercises	Operating
Stand	Assess the functional status of the stand knee	The user starts stand according to the prompts and stands in place for 5s
Step	Assess the functional status of the knee during walking	According to the prompts, the user starts to walk for 10S
Leg lift	Improve knee flexibility, motion accuracy and reaction time	Users start according to the prompts, start the left knee bending and lifting within 1–5s, and complete the knee bending 90° (error 15°) within 5–20s, with the knee as far as possible and the right leg upright. The right leg bends the knee and lifts the leg
Squat	Training knee muscle group strength, body balance and postural control	Users start according to the prompts, 1–5S start squat knee flexion movement, and within 10–30s will be adjusted to the knee Angle of 90°–120°, knee as far as possible, and then stand up and restore the stand position

After analyzing the extracted characteristic parameters of each movement and combining them with the opinions of 3 knee surgeons, the four movements, namely stand, step, leg lift, and squat, were finally selected as our systematic evaluation and training movements. These movements were design as shown in Table 1.

Through the system development and design, the evaluation effect of knee motor function as shown in Fig. 3.

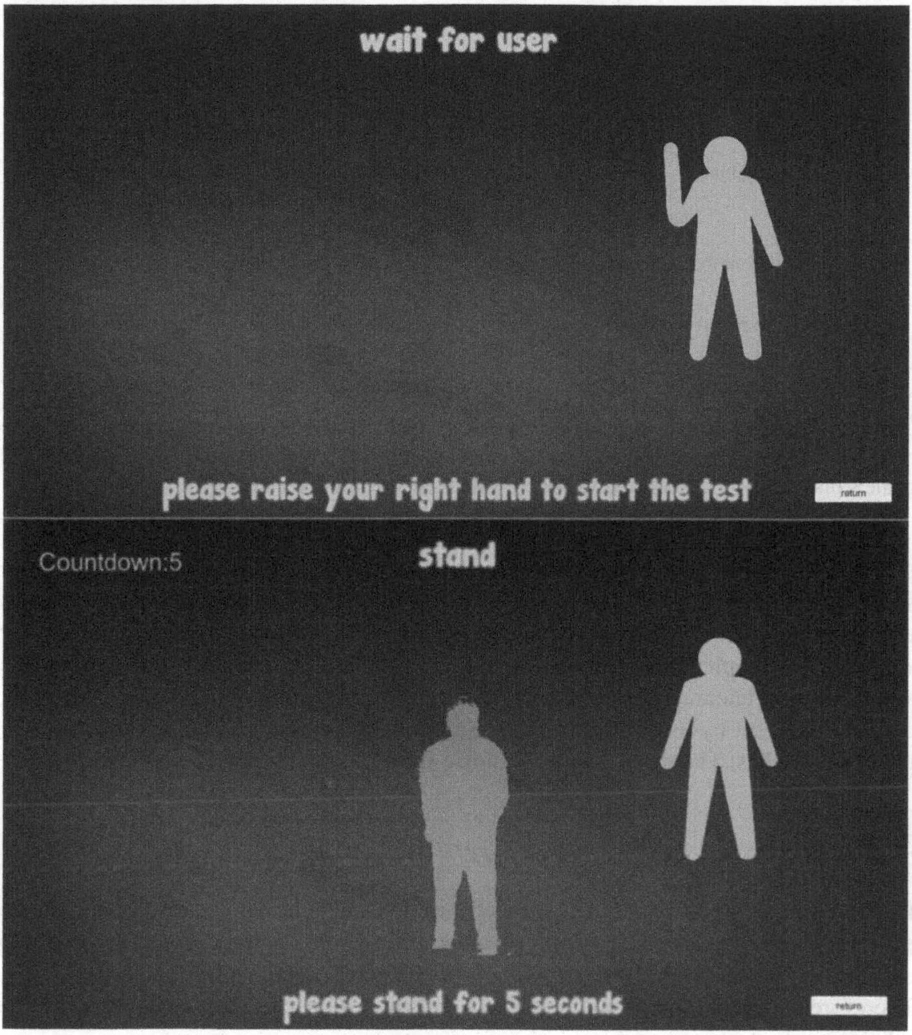

Fig. 3. The effects of knee function evaluation system

2.4 Knee Exergame Design

We developed a game called "Jungle Adventure". The game is simulated by the user explorer adventure in the jungle. The user incarnated the role in the specified time through the specified path through the jungle. The path is divided into multiple walkways for switching; the walkway arranged obstacles and precious stones, etc.; the role touches the obstacle life value was reduced and touches the precious stones to get score. The main operation modes as shown in Table 2.

Table 2. The movement interaction design of the exergame.

Movements	User interaction	Design principles
Stand	Walking	The game character sets the initial walking speed. Stand in place indicates that the character is walking forward*.
Step	Running	In the game, the character moves into running motion and moves faster
Leg lift	Turn left and right to change lanes	Characters can turn left and right to gain more gold, or avoid obstacles
Squat	Squat to avoid obstacles	Characters squat to avoid obstacles in the game

* In consideration of the applicability of elderly users, set the stand movement as the base movement to avoid continuous fatigue of other movements. At the same time, to improve the game's playability, the character was set to walk forward automatically.

The difficulty parameters of the game can be set, including playtime, health, gem score, type of obstacle set, frequency, etc. The setting of the training movement was also graded as none (the user was unable to complete the movement), slow (the user was slow to complete the movement), and fast model. The difficulty of the game was automatically generated based on the suggested movement after the function evaluation, or it could be set by itself based on the recommended parameter settings according to the user's wishes. The game effects as shown in Fig. 4.

3 Methods

In this paper, the knee motion data were acquired through the somatosensory interactive equipment to conduct system function evaluation and game training. It was necessary to evaluate the feasibility and reliability of obtaining human knee motion data with Kinect. Therefore, this paper mainly analyzes the feasibility of knee motion acquisition and the reliability of retest.

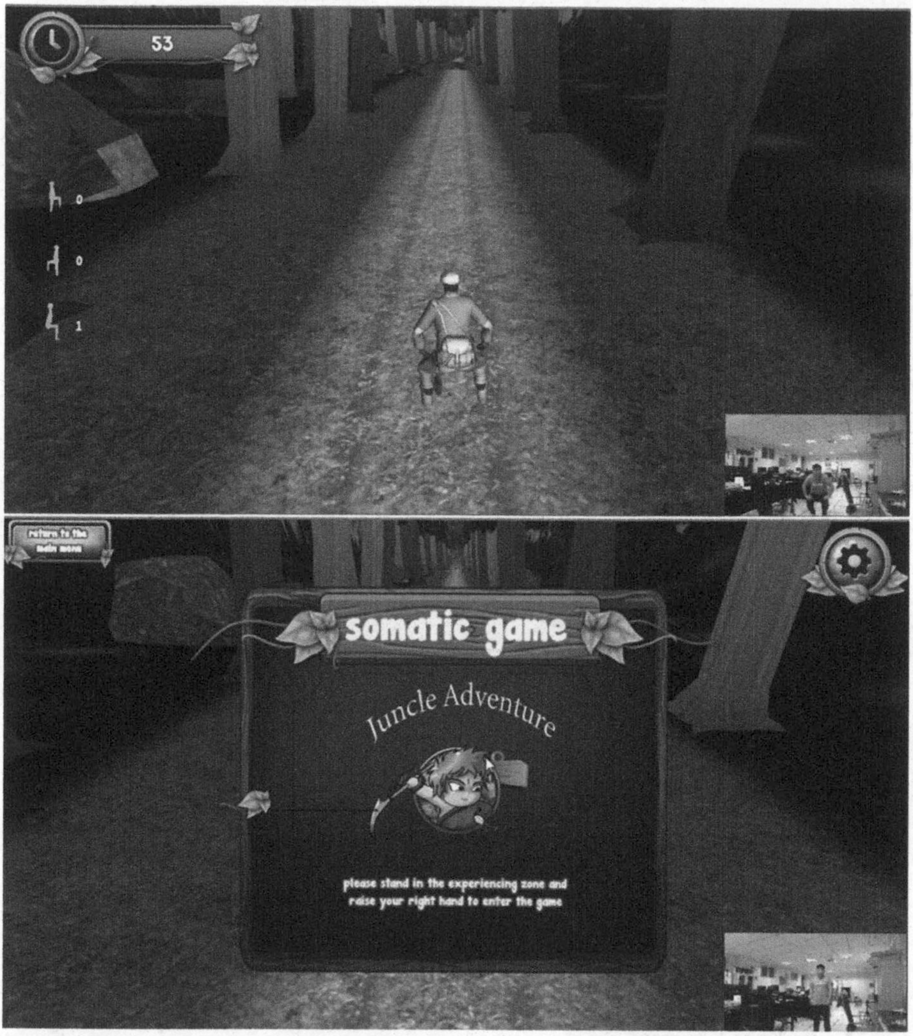

Fig. 4. The effects of knee exergame

3.1 Participants

Twenty participants (9 females, 11 males; age: 60 ± 5 years) without any acute knee complaints or acute infections were included in the study. Each participant received an oral and written explanation of the purpose and the study design. Before enrolment, all participants signed a written consent form. Study approval was received by the ethics committee where the study was conducted.

3.2 Instrumentation and Procedure

To assess the validity of motion data acquired and the test-retest reliability by Kinect, we designed experiments and defined characteristic parameters for evaluation. Previous studies [11] used the extracted hip, knee, and ankle coordinate data to estimate the knee Angle using the Microsoft Kinect official software development kit (SDK). The Euler angles calculated knee angles. Knee angles were defined by vector conventions (knee joint: hip-knee and knee-ankle).

The bending and extension of the knee were the angles projected on the sagittal plane, respectively by the angle formed by the line connecting the hip and knee and the line connecting the knee and ankle. This experiment can be simplified as the projection Angle of the hip - knee - ankle joint on the sagittal plane. On the other hand, the ankle force line's angle was an important index to measure the varus of the knee. This experiment could be simplified as the angle of the projection of the hip joint, knee joint, and ankle joint on the coronal plane. Referring to the Wochatz M study define the beginning of the movement and the maximum knee angle during the movement in the sagittal plane or the coronal plane. It was shown in Table 3.

Table 3. Definition of calculated joint angles and joint positions during evaluation movement.

	Parameters	Design principles
	Knee-L-S-bas	Knee-Left angle in sagittal plane at the beginning of the movement
	Knee-L-S -max	Maximum-Left knee angle in sagittal plane during the movement
	Knee-R-S -bas	Knee-right angle in sagittal plane at the beginning of the movement
	Knee-R-S -max	Maximum-right knee angle in sagittal plane during the movement
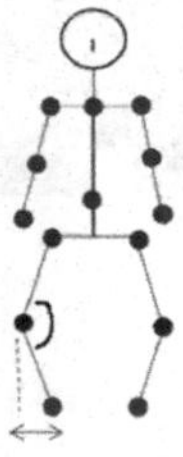	Knee-L-C-bas	Knee-Left angle in coronal plane at the beginning of the movement
	Knee-L-C-max	Maximum-Left knee angle in coronal plane during the movement
	Knee-R-C-bas	Knee-right angle in coronal plane at the beginning of the movement
	Knee-R-C-max	Maximum-right knee angle in coronal plane during the movement

The recorded data were manually synchronized by the start and end position of the movement (distinct movement cues) without any interpolation of the data.

Parameters were averaged over these movement repetitions. Each participant performed two experiments on each of the four movements. The study was conducted in a test-retest design with measurement sessions being separated by a mean of 7 days. The Kinect camera was placed 2.5 m in front of the participant and was elevated by 1.2 m for an optimal field of view [12, 13]. Movements were collected with a sampling frequency of 30 Hz for the Kinect system.

3.3 Data Analysis and Statistical Testing

In this study, the test-retest reliability of the system represents the similarity of Kinect system data when the same person completes a movement. ICC was calculated to evaluate the test-retest reliability. The intra-group correlation coefficient (ICC) was an index of reliability coefficient for measuring and evaluating test-retest reliability.

Firstly, the ICC value of each movement completed in each test was calculated, and then the ICC value of all test subjects was averaged to calculate the reliability of the system for acting. Similarly, the system reliability for each movement can be calculated.

MATLAB software was used for data preprocessing and joint Angle calculation, and the feasibility of each movement was analyzed by drawing and displaying. Statistical analysis was then completed using IBM SPSS Statistics 22.

4 Result

Accurate data can be obtained by calculating the knee Angle of each test subject, which can be observed and analyzed through the view. Figure 5 shows the curve of the angle change of the knee in some movements.

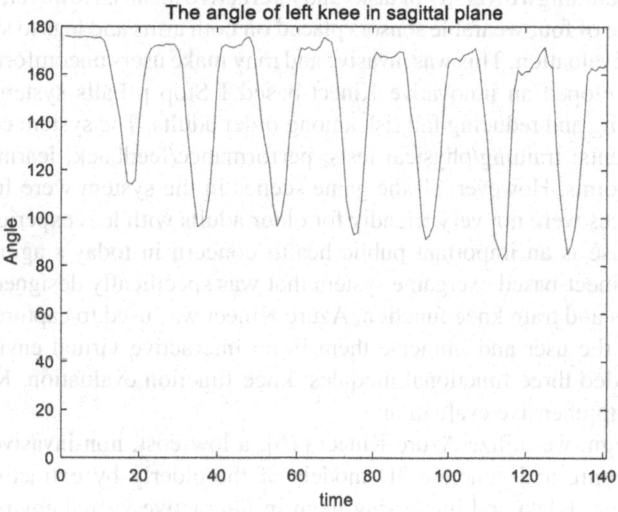

Fig. 5. Knee movements angle change curve (step)

The average value of the system ICC straight completed by each movement was calculated, as shown in Table 4.

Table 4. Intra-class correlation coefficient of knee angle measured of the movements.

	ICC	95% confidence interval		Use F test of truth value 0			
		Lower	Upper	Value	df1	df2	Sig
Stand	0.987	.943	.998	547.4	7	7	.000
Step	0.997	.987	.999	3048.3	7	7	.000
Left leg lift	0.981	.894	.996	1038.547	7	7	.000
Right leg lift	0.990	.955	.998	1750.036	7	7	.000
Squat	0.998	.979	1.000	3326.103	7	7	.000

ICC: Intra-class correlation coefficient, Diff: Difference between time 1 and 2, Lower: lower 95% confidence interval, Upper: upper 95% confidence interval.

5 Discussion

Previous studies mainly adopted commercial Kinect exergame interventions that were mostly designed for entertainment purposes rather than for the knee exercises of the elderly. Only a few exergames were designed specifically for the elderly. Guimarães et al. [14] described an exergame for cognitive-motor training and fall prevention in older adults, including a diversity of tasks and interactive means. However, this exergame required the use of four wearable sensors placed on both arms and legs to support motion capturing and evaluation. This was invasive and may make users uncomfortable. Marston et al. [15] developed an innovative Kinect-based I Stop p-Falls system for continuously monitoring and reducing fall risk among older adults. The system comprised four main components: training/physical tests, performance/feedback, learning/education, and meeting points. However, all the game scenes in the system were fully synthetic. Virtual interfaces were not very friendly for older adults with less experience in games.

Knee disease is an important public health concern in today's aging society. We developed a Kinect-based exergame system that was specifically designed for the older adults to assess and train knee function. Azure Kinect was used to capture and generate 3D models of the user and immerse them in an interactive virtual environment. The software included three functional modules: knee function evaluation, Knee exercises game and Comprehensive evaluation.

In our design, we utilize Azure Kinect [16], a low-cost, non-invasive, marker-less camera, to capture and generate 3D models of the elderly by extracting them from the entire captured data and immersing them in interactive virtual environments. This approach was non-invasive because no external devices or sensors need to be worn by the elderly. Furthermore, using such live human models/avatars helps in understand the emotional experiences of participants involved in the exergame. Besides, the system highlights the synchronous interventions of cognition and motor. Traditional exercise therapy was mostly a single-factor intervention, while augmented reality therapy can carry out multi-factor interventions such as exercise and cognition. Some explicit cognitive training tasks for attention, memory and executive functions were added to each game in the system.

In this paper, the knee motion data were acquired through the somatosensory interactive equipment to conduct system function evaluation and game training. It was necessary to evaluate the feasibility and reliability of obtaining human knee motion data with Kinect. Therefore, this paper mainly analyzes the feasibility of knee motion acquisition and the reliability of retest. The stand, step, leg lift, and squat were selected for knee function evaluation. This study was made to confirm the test-retest reliability of the system using Azure Kinect during the evaluation actions. Twenty injury-free individuals volunteered to participate. Intra-class correlation coefficients were calculated to assess the reliability of kinematic measurements of knee angles during the movement. The ICC of these movement were stand (ICC = 0.987), step (ICC = 0.997), left leg lift (ICC = 0.981), right leg lift (ICC = 0.990), stand (ICC = 0.998).The results show that the test-retest reliability was high. It means that the motion capture data was effective, and the data obtained by Kinect was stable.

6 Conclusion

We developed a Kinect-based exergame system to assess and train knee function among the elderly from the user experience. Based on the test-retest reliability results, the motion capture data was effective, and the data obtained by Kinect was stable. The study's major limitation was that the data obtained by Kinect did not carry out comparative experiments. Later, comparative analysis can be conducted in combination with other high-precision motion capture technologies to verify the accuracy of the data.

Funding. This research was funded by the Key Project on Anhui Provincial Natural Science Study by Colleges and Universities under Grant "Key technical research of knee function evaluation and rehabilitation training" (No. KJ2019A0555), Key project of Science and Technology Service Network Program of Chinese Academy of Sciences "Construction of chronic disease risk prevention and control service system based on big data" (No. KFJ-STS-ZDTP-079), Major Science and Technology Projects of Anhui Province" Research and demonstration of key technologies of non-medical sexual health promotion services" (No. 18030801133).

References

1. Wolf, B.R., Gulbrandsen, T.R.: Degenerative meniscus injury in older athletes. Clin. Sports Med. **39**(1), 197–209 (2020)
2. Blake, M.H., Johnson, D.L.: Knee meniscus injuries: common problems and solutions. Clin. Sports Med. **37**(2), 293–306 (2018)
3. Xie, Y., Yu, Y., Wang, J.X., et al.: Health-related quality of life and its influencing factors in Chinese with knee osteoarthritis. Qual. Life Res. **29**, 2395–2402 (2020)
4. Buul, G., Stanclik, J., Stok, J., et al.: Focal articular surface replacement of knee lesions after failed cartilage repair using focal metallic implants: a series of 132 cases with 4-year follow-up. Knee **29**, 134–141 (2021)

5. Tagliero, A.J., Kennedy, N.I., Leland, D.P., et al.: Meniscus repairs in the adolescent population-safe and reliable outcomes: a systematic review. Knee Surg. Sports Traumatol. Arthroscopy **28**(11), 3587–3596 (2020)
6. Wochatz, M., Tilgner, N., Mueller, S., et al.: Reliability and validity of the Kinect V2 for the assessment of lower extremity rehabilitation exercises. Gait Posture **70**, 330–335 (2019)
7. Tanaka, R., Tamura, H., Kawanishi, H.: Reliability of a markerless motion capture system to measure the trunk, hip and knee angle during walking on a flatland and a treadmill. J. Biomech. **109**, 109929 (2020)
8. Guojiong, H., Wenli, W., Bin, C., et al.: Concurrent validity of evaluating knee kinematics using Kinect system during rehabilitation exercise. Med. Novel Technol. Devices **9**, 100068 (2021)
9. Zhu, Y., Lu, W., Gan, W., et al.: A contactless method to measure real-time finger motion using depth-based pose estimation. Comp. Biol. Med. **12**, 104282 (2021)
10. Cerqueira, T., Moura, J., Lira, J., et al.: Cognitive and motor effects of kinect-based games training in people with and without Parkinson disease: a preliminary study. Physiother. Res. Int. **25**(1), 1807 (2020)
11. Eltoukhy, M., Oh, J., Kuenze, C., et al.: Improved kinect-based spatiotemporal and kinematic treadmill gait assessment. Gait Posture **51**, 77–83 (2017)
12. Pagliari, D., Pinto, L.: Calibration of kinect for Xbox one and comparison between the two generations of microsoft sensors. Sensors (Switzerland) **15**, 14248220 (2015)
13. Khoshelham, K., Elberink, S.O.: Accuracy and resolution of kinect depth data for indoor mapping applications. Sensors **12**, 1437–1454 (2012)
14. Guimares, V., Pereira, A., Oliveira, E., et al.: Design and evaluation of an exergame for motor-cognitive training and fall prevention in older adults. In: The 4th EAI International Conference, pp. 202–207 (2018)
15. Marston, H.R., Woodbury, A., Gschwind, Y.J., et al.: The design of a purpose-built exergame for fall prediction and prevention for older people. Eur. Rev. Aging Phys. Act. **12**(13), 1574 (2015)
16. Tlgyessy, M., Dekan, M., Chovanec, U.: Skeleton tracking accuracy and precision evaluation of kinect V1, kinect V2, and the azure kinect. Appl. Sci. **11**(12), 5756 (2021)

Study on the Influencing Factors and Countermeasures of Tourism Development in Sanya City Based on Grey Relational Analysis

Yuanhui Li(✉), Dan Luo, Lei Chen, Dan Ren, and Xia Liu

Sanya Aviation and Tourism College, Sanya 572000, Hainan, China

Abstract. The development of tourism is closely related to Sanya's economic growth. In this paper, the correlation between tourism development and economic growth in Sanya city is studied. It was first measured with grey correlation analysis. Then, tourism factors in Sanya was indicated, including tourism foreign exchange income, total value of accommodation and catering, the number of A-level and above scenic spots, tourism turnover, overnight domestic tourists, total passenger flow of airport and railway station and other factors. It is found that these factors have a significant impact on the output value of tourism in Sanya City. The analytical results show that tourism factors and economic growth in Sanya city is highly correlated. Improvement measures are put forward accordingly.

Keywords: Tourism development in Sanya city · Grey correlation analysis · Countermeasure research

1 Introduction

Sanya, as a crucial city in Hainan and a pivot city of the national strategy "Belt and Road Initiative", performs a special role and shoulders significant responsibility in the whole pattern of Hainan development. Under the background of developing Hainan Free Trade Zone, Sanya gets an unprecedented opportunity for development. It will provide better service and integrate into both domestic and international circulation.

By the end of 2020, the number of overnight tourists received in the city had added up to 17.1440 million, and the annual total tourism revenue was 42.474 billion yuan. There were 262 hotels listed in the statistics of the city, including 14 five-star hotels, 16 four-star hotels and 6 three-star hotels. There are altogether 14 A-level or above scenic spots in the city, among which there are three 5A scenic spots and six 4A scenic spots. The development of leisure agricultural tourism has been accelerated. Two tourist routes, such as Rose Valley and Family Trip, have been rated as China's Beautiful Rural High-quality Tourism Routes. Bohou Village has been rated as China's Beautiful Leisure Village, and seven industrial towns with provincial characteristic have been newly recognized. In 2012, the tourism income of Sanya City was 19.2 billion yuan, while in 2020, the tourism income was 42.474 billion yuan, which has increased by 2.21 times in 9 years, realizing a leap-forward growth.

J. Zeng et al. (Eds.): ICPCSEE 2021, CCIS 1452, pp. 343–356, 2021.
https://doi.org/10.1007/978-981-16-5943-0_28

In 2020, the gross domestic product (GDP) of Sanya is 69.541 billion yuan, and the total revenue of tourism accounts for 61.08% of GDP. Tourism industry is playing an increasingly important role in the development of Sanya's national economy and its development trend is in good shape. However, leisure tourism still dominates the tourism industry while high-end tourism projects are insufficient. In addition to Sanya International Duty-free Mall, it lacks attractive tourism consumption places, which means tourism consumption service, as well as crucial tourism service support, such as catering, accommodation, traveling, etc. need to be promoted. The development of tourism industry is affected by multiple factors such as the layout of tourism industry, the types and quality of tourism products, the number of tourists, the construction of tourism infrastructure and the quality of tourism services. Which are the key factors affecting the development and upgrading of Sanya's tourism industry is a special concern of Sanya Municipal Government and the related tourism development departments. It also affects the orientation of Sanya's construction of an international boutique tourism city. In this paper, the grey correlation analysis method is used to analyze the factors that affect the development of Sanya tourism industry, so as to explore the key factors that affect the development of Sanya tourism industry, and provide scientific basis for improving the tourism income and tourism competitiveness of Sanya city.

2 Introduction of Grey Correlation Analysis Method

Grey correlation analysis is to compare or describe the relative changes over time between systems or among various factors in the system with specified quantities in the development process, that is, to analyze the set shape of time series curves, and to measure the correlation between them with the approaching degree of their change level, direction, speed, etc. In the case of collecting less data and insufficient information to study a certain kind of problem, the grey correlation analysis has a special advantage, so it has been adopted and used in many fields. It obtains the evolution law through the given information in the system, and judges the influence degree of the factors on the system according to the value of the correlation coefficient of the two factors in the analysis system. The law is that the greater the correlation coefficient is, the greater the influence on the system will be, while the smaller the correlation coefficient is, the slighter the influence on the system will be. By using the grey correlation analysis method, we can know which factors are the main factors that promote the development of the system, and which factors are the factors that hinder the development of the system. Grasping the development direction of the system can make development strategies and methods more targeted, purposeful and accurate [1].

In addition, because the grey correlation analysis method has fewer requirements and restrictions on experimental observation data, and less need of computing, it can better eliminate the contradiction between qualitative and quantitative results and overcome the defects that may occur when using other mathematical statistics methods.

The analytical procedure of the grey correlation analysis method include: (1) Determining the reference number series which can reflect system's overall behavior, and the comparative number series which can influence system behavior; (2) Dimensionless treatment is conducted on reference number series and comparative number series.

The dimensionless treatment usually uses the initial value method, the extreme value method, the mean value method and the standard deviation method. (3) Calculate the absolute value of the corresponding component difference between the initial value of the comparative number series and the reference number series, and get the maximum and minimum difference between the two poles.(4) Calculate the grey correlation coefficient between reference number series and comparative number series. (5) Calculate the grey correlation degree between the comparative number series and reference number series. The closer the value is to 1, the higher the correlation is. (6) Rank the grey correlation degree. Sequence the correlation degree between comparative number series and the same reference number series according to the value, and form a relational order. If $x_i > x_j$, it means that comparative number series X_i has greater influence on the same reference number series than that of comparative number series X_j [2].

3 Grey Correlation Analysis of the Influence Factors of Tourism Development in Sanya City

3.1 Data Selection

According to the grey system theory, this paper takes the tourism industry in Sanya City as a whole system, and various influence factors of tourism development are the components of this system. These factors influence each other and interact with each other under the action of tourism environment at home and abroad, and they form a unified whole. In this system, tourism industry has obvious dynamic characteristics and uncertainty. The data selected in this paper is limited data, which is fuzzy to a certain extent and belongs to grey data. Therefore, grey correlation analysis method can be used here. This paper selects the relevant data of tourism industry in Sanya City from 2012 to 2020 to do the grey correlation analysis. Income is an important criterion to measure the development of an industry. In this paper, the general tourism income from 2012 to 2020 in Sanya City is selected as reference number series, while the foreign exchange earnings from tourism, total value of the accommodation and catering, the number of star hotels, the number of grade A and above scenic spots, tourist volume of circular flow, the overnight tourists from home and abroad, overnight entry visitors, average occupancy of hotels, the total passenger flow at airport and railway station, the number of tourism colleges, the number of students in colleges and universities as comparative number series.

3.2 Grey correlation data calculation

1) Data Presentation

The reference number series and comparative number series that influence the development of tourism in Sanya are listed. Take X_0 as the reference number series, X_1 to X_{11} as the comparative number series, and the specific data is shown in Table 1 [3].

Table 1. Grey correlation reference number series and comparative number series of tourism in Sanya City

Index	2012	2013	2014	2015	2016
x_0	19.200	23.272	26.974	30.231	32.240
x_1	266.670	260.194	191.372	169.675	254.761
x_2	3.409	3.714	4.731	4.668	4.845
x_3	235	241	243	252	250
x_4	7	8	14	17	16
x_5	75.1478	57.9639	72.1792	99.5733	111.576
x_6	10.5408	11.7772	13.1390	14.5991	16.0669
x_7	48.14	48.00	38.86	35.82	44.89
x_8	58.74	56.70	61.40	64.50	66.02
x_9	17.5834	19.9269	22.7546	24.8128	27.9175
x_{10}	5	5	5	5	5
x_{11}	44228	41768	45757	47925	48504
Index	2017	2018	2019	2020	
x_0	40.634	49.243	63.319	42.474	
x_1	536.846	547.345	810.750	101.580	
x_2	5.239	5.596	6.627	5.605	
x_3	252	255	252	262	
x_4	16	14	14	14	
x_5	129.610	140.491	140.368	91.802	
x_6	17.6169	20.7118	23.0570	16.9900	
x_7	69.28	79.34	90.63	15.41	
x_8	69.57	71.46	71.81	50.89	
x_9	30.3593	31.6610	33.9709	24.7760	
x_{10}	6	6	6	6	
x_{11}	49311	50859	55601	60954	

2) Dimensionless treatment

The data of reference number series and comparative number series are dimensionless processed by the initial value method, and the initial value data of X_1 is shown in Table 2 [4].

Table 2. Data table of grey correlation initial value of tourism in Sanya City

x_0	1	1.2121	1.4049	1.5745	1.6792
x_1	1	0.9757	0.7176	0.6363	0.9553
x_2	1	1.0895	1.3878	1.3693	1.4212

(continued)

Table 2. (*continued*)

x_3	1	1.0255	1.0340	1.0723	1.0638
x_4	1	1.1429	2.0000	2.4286	2.2857
x_5	1	0.7713	0.9605	1.3250	1.4848
x_6	1	1.1173	1.2465	1.3850	1.5243
x_7	1	0.9971	0.8072	0.7440	0.9323
x_8	1	0.9653	1.0453	1.0981	1.1239
x_9	1	1.1333	1.2941	1.4112	1.5877
x_{10}	1	1.0000	1.0000	1.0000	1.0000
x_{11}	1	0.9444	1.0346	1.0836	1.0967
x_0	2.1164	2.5647	3.2979	2.2122	
x_1	2.0131	2.0525	3.0403	0.3809	
x_2	1.5368	1.6415	1.9440	1.6442	
x_3	1.0723	1.0851	1.0723	1.1149	
x_4	2.2857	2.0000	2.0000	2.0000	
x_5	1.7247	1.8695	1.8679	1.2216	
x_6	1.6713	1.9649	2.1874	1.6118	
x_7	1.4390	1.6480	1.8825	0.3201	
x_8	1.1844	1.2165	1.2225	0.8664	
x_9	1.7266	1.8006	1.9320	1.4091	
x_{10}	1.2000	1.2000	1.2000	1.2000	
x_{11}	1.1149	1.1499	1.2571	1.3782	

3) Absolute value sequence

From the formula$\Delta_i(k) = |x_0(k) - x_i(k)|$, the absolute value sequence matrix of the initial value of $x_1, x_2 \ldots, X_{11}$ and X_0 and their corresponding component difference is deduced, as shown in Table 3 [5].

Table 3. Grey correlation absolute value sequence matrix of tourism in Sanya City

$\Delta_1(k)$	0	0.2364	0.6873	0.9383	0.7238
$\Delta_2(k)$	0	0.1226	0.0171	0.2052	0.2579
$\Delta_3(k)$	0	0.1866	0.3709	0.5022	0.6153
$\Delta_4(k)$	0	0.0692	0.5951	0.8540	0.6065

(*continued*)

Table 3. (*continued*)

$\Delta_5(k)$	0	0.4408	0.4444	0.2495	0.1944
$\Delta_6(k)$	0	0.0948	0.1584	0.1895	0.1549
$\Delta_7(k)$	0	0.2150	0.5977	0.8305	0.7468
$\Delta_8(k)$	0	0.2468	0.3596	0.4765	0.5552
$\Delta_9(k)$	0	0.0788	0.1108	0.1634	0.0914
$\Delta_{10}(k)$	0	0.2121	0.4049	0.5745	0.6792
$\Delta_{11}(k)$	0	0.2677	0.3703	0.4909	0.5825
$\Delta_1(k)$	0.1032	0.5122	0.2576	1.8313	
$\Delta_2(k)$	0.5795	0.9232	1.3539	0.5680	
$\Delta_3(k)$	1.0440	1.4796	2.2255	1.0973	
$\Delta_4(k)$	0.1694	0.5647	1.2979	0.2122	
$\Delta_5(k)$	0.3916	0.6952	1.4300	0.9906	
$\Delta_6(k)$	0.4450	0.5998	1.1105	0.6004	
$\Delta_7(k)$	0.6773	0.9168	1.4154	1.8921	
$\Delta_8(k)$	0.9320	1.3482	2.0754	1.3458	
$\Delta_9(k)$	0.3898	0.7641	1.3659	0.8031	
$\Delta_{10}(k)$	0.9164	1.3647	2.0979	1.0122	
$\Delta_{11}(k)$	1.0014	1.4148	2.0407	0.8340	

4) Maximum difference and minimum difference

According to formula $M = \max_i \max_k \Delta_i(k)$, $m = \min_i \min_k \Delta_i(k)$, the maximum and minimum difference between the poles can be deduced as follows: $M = 2.225524$, $m = 0$.

5) Grey correlation coefficient matrix

According to the formula $\gamma(x_0(k), x_i(k)) = \frac{m+\xi\times M}{|x_0(k)-x_i(k)|+\xi\times M}$, take $\xi 0.5$, the correlation coefficient matrix between Sanya tourism industry and the foreign exchange earnings from tourism, total value of the accommodation and catering, the number of star hotels, the number of grade A and above scenic spots, tourist volume of circular flow, the overnight tourists from home and abroad, overnight entry visitors, average occupancy of hotels, the total passenger flow at airport and railway station, the number of tourism colleges, the number of students in colleges and universities can be deduced, as shown in Table 4.

Table 4. Grey correlation coefficient matrix of tourism in Sanya City

γ_{1k}	1	0.8248	0.6182	0.5425	0.6059
γ_{2k}	1	0.9007	0.9849	0.8443	0.8118
γ_{3k}	1	0.8564	0.7500	0.6890	0.6439
γ_{4k}	1	0.9414	0.6516	0.5658	0.6472
γ_{5k}	1	0.7163	0.7146	0.8168	0.8513
γ_{6k}	1	0.9215	0.8754	0.8545	0.8778
γ_{7k}	1	0.8381	0.6506	0.5726	0.5984
γ_{8k}	1	0.8185	0.7558	0.7002	0.6671
γ_{9k}	1	0.9339	0.9094	0.8720	0.9241
γ_{10k}	1	0.8399	0.7332	0.6595	0.6210
γ_{11k}	1	0.8061	0.7503	0.6939	0.6564
γ_{1k}	0.9151	0.6848	0.8120	0.3780	
γ_{2k}	0.6575	0.5466	0.4511	0.6621	
γ_{3k}	0.5159	0.4292	0.3333	0.5035	
γ_{4k}	0.8679	0.6633	0.4616	0.8399	
γ_{5k}	0.7397	0.6155	0.4376	0.5290	
γ_{6k}	0.7143	0.6498	0.5005	0.6496	
γ_{7k}	0.6216	0.5483	0.4402	0.3703	
γ_{8k}	0.5442	0.4522	0.3490	0.4526	
γ_{9k}	0.7406	0.5929	0.4489	0.5808	
γ_{10k}	0.5484	0.4491	0.3466	0.5237	
γ_{11k}	0.5263	0.4402	0.3529	0.5716	

6) Grey correlation degree and correlation order

According to the formula $\gamma(x_0, x_i) = \frac{1}{n}\sum \gamma(x_0(k), x_i(k))$, grey correlation degree and correlation order are calculated between Sanya tourism industry and the foreign exchange earnings from tourism, total value of the accommodation and catering, the number of star hotels, the number of grade A and above scenic spots, tourist volume of circular flow, the overnight tourists from home and abroad, overnight entry visitors, average occupancy of hotels, the total passenger flow at airport and railway station, the number of tourism colleges, the number of students in colleges and universities, as shown in Table 5 [6].

3.3 Grey Correlation Analysis of Marine Tourism Industry

It can be seen from Table 5 that the grey correlation degree of the indicators in the comparative series is all greater than 0.5, indicating that the indicators selected are reasonable and scientific to a certain extent, and it has a significant impact on the output value of

Table 5. Grey correlation order of Sanya tourism and its influence factors

Influence factors	B_1	B_2	B_3	B_4	B_5	B_6
Grey relational degree	0.7090	0.7621	0.6357	0.7376	0.7134	0.7826
Correlation order	6	3	9	4	5	1
Influence factors	B_7	B_8	B_9	B_{10}	B_{11}	
Grey reltional degree	0.6267	0.6377	0.7781	0.6357	0.6442	
Correlation order	11s	8	2	10	7	

B_1: Foreign exchange earnings from tourism
B_2: Total value of catering and accom-modation
B_3: Number of star hotels
B_4: Number of grade A and above scenic spots
B_5: Tourism turnover
B_6: Overnight domestic tourists
B_7: Overnight entry tourists
B_8: Average occupancy rate of hotels
B_9: Passenger flow at the airport and railway station
B_{10}: Number of Tourism Colleges
B_{11}: Number of university students

the tourism industry in Sanya City. Among the influence factors, the grey correlation degree between the total tourism income of Sanya City and foreign exchange earnings from tourism, total value of the accommodation and catering, the number of grade A and above scenic spots, tourist volume of circular flow, the overnight tourists from home and abroad, average occupancy of hotels, the total passenger flow at airport and railway station are greater than 0.7, indicating that improving the quality of Sanya tourism services, completing the training work of high-rank and medium-rank tourism talents, enhancing the management ability and service awareness, expanding the international market on the basis of rooting in domestic market, are the key to further promote the development of tourism in Sanya. The grey correlation coefficient of number of star hotels, overnight entry visitors, average occupancy of hotels, the number of tourism colleges, as well as the number of students in colleges and universities are all greater than 0.6, indicating that the development of tourist facilities, exploring international market on the basis of fully developing domestic tourism market, increasing the number of entry tourists, completing the training work of high-rank and medium-rank tourism service talents are the important basis of benign development of Sanya tourism.

4 Evaluation of Comprehensive Strength of Tourism in Sanya City

4.1 Data Selection

The comprehensive strength of regional tourism is affected by multiple factors. According to the principles of comprehensiveness, systematicness, simplicity, comparability and operability, this paper constructs the evaluation index system framework to evaluate the comprehensive strength of tourism in Sanya City. On the basis of maintaining the consistency with the above analysis, all the relevant data except the total income of tourism in Sanya City are retained. In order to facilitate the calculation of future data, the data of these 11 indexes in Sanya City over the years are processed by min-max standardization, and the formula is $x'_{ij} = \frac{x_{ij} - x_{min}}{x_{max} - x_{min}} (i = 1, 2, \ldots, n; j = 1, 2 \ldots, m)$.

Table 6. Standardized value of comprehensive strength evaluation index of tourism in Sanya City

Index	2012	2013	2014	2015	2016
Foreign exchange earning from tourism	23.28	22.37	12.66	9.60	21.60
Total value of catering and accommodation	0.00	9.48	41.08	39.12	44.62
Number of hotels	0.00	22.22	29.63	62.96	55.56
Number of grade A and above scenic spots	0.00	10.00	70.00	100.00	90.00
Tourism turnover	20.82	0.00	17.22	50.42	64.96
Overnight domestic tourists	0.00	9.88	20.76	32.42	44.15
Overnight entry tourists	43.52	43.33	31.18	27.13	39.19
Average occupancy of hotels	37.52	27.77	50.24	65.06	72.32
Total passenger flow at airport and railway station	0.00	14.30	31.56	44.12	63.06
Number of colleges	0.00	0.00	0.00	0.00	0.00
Number of college students	12.82	0.00	20.79	32.09	35.11
Index	2017	2018	2019	2020	
Foreign exchange earning from tourism	61.38	62.86	100.00	0.00	
Total value of catering and accommodation	56.87	67.96	100.00	68.24	
Number of hotels	62.96	74.07	62.96	100.00	
Number of grade A and above scenic spots	90.00	70.00	70.00	70.00	
Tourism turnover	86.81	100.00	99.85	41.00	
Overnight domestic tourists	56.54	81.26	100.00	51.53	
Overnight entry tourists	71.62	84.99	100.00	0.00	
Average occupancy of hotels	89.29	98.33	100.00	0.00	
Total passenger flow at airport and railway station	77.96	85.90	100.00	43.89	
Number of colleges	100.00	100.00	100.00	100.00	
Number of college students	39.32	47.38	72.10	100.00	

Among them: x'_{ij} is the standardized data; x_{ij} is the original data of the index, x_{min} is the minimum value of sample data, and x_{max} is the maximum value of sample data [7].

In order to avoid the comprehensive evaluation score is too low, the standardized value is expanded by 100 times, as shown in Table 6.

4.2 Determination of Index Weight

Entropy value method is used to calculate the weight of evaluation index. Entropy value method is a kind of objective weighting method, it judges the utility value of index through the inherent information of evaluation index. It can avoid deviation caused by subjective human factors. Entropy method is used to calculate the weight of all indexes, to convert the standardized data to proportion value, and then calculate the entropy value e_j and weight W_j according to the proportion value.

The formula is $e_j = -k \cdot \sum_{i=1}^{n} P_{ij}\ln(P_{ij})$, usually $k = 1/\ln n$; $W_j = \frac{1-e_j}{\sum_{j=1}^{m}(1-e_j)}$, $j = 1, 2 \ldots, m$.

Table 7. Evaluation index system of comprehensive strength of tourism industry in Sanya City

Variable	B_1	B_2	B_3	B_4	B_5	B_6
Proportion	11.69%	7.18%	5.96%	6.24%	7.91%	8.60%
Variable	B_7	B_8	B_9	B_{10}	B_{11}	
Proportion	6.53%	5.85%	7.19%	24.33%	8.51%	

4.3 Comprehensive Strength Evaluation and Analysis

The total score of the comprehensive strength evaluation of the tourism industry in Sanya City is calculated. Table 7 shows the index weight and the index score value after standardized treatment. In the formula $A = \sum_{j=1}^{m} B_j \cdot W_j$, A is the total score of the comprehensive strength evaluation of Sanya City in a certain year, B_j is the evaluation score of the jth index in Sanya City in a certain year, W_j is the weight of the jth evaluation index, and m is the number of indexes. The total scores of the comprehensive strength evaluation of the tourism industry in Sanya City are shown in Table 8 and Fig. 1.

It can be seen from the evaluation results of the comprehensive strength of tourism industry in Sanya City that the overall strength of the tourism industry in Sanya City has been showing a trend of continuous growth except in 2020, and it has made a great leap during 2016 and 2017. The comprehensive evaluation score of the tourism industry in Sanya City has increased from 10.50 in 2012 to 93.53 in 2019, and the comprehensive strength has been significantly enhanced. Due to the impact of epidemic in 2020, multiple indexes were significantly affected, so the comprehensive evaluation score in 2020 dropped to 58.90.

Table 8. Evaluation results of comprehensive strength of tourism in Sanya City

Year	The total score of comprehensive strength assessment
2012	10.50
2013	11.58
2014	22.73
2015	32.18
2016	37.91
2017	75.54
2018	81.76
2019	93.53
2020	58.90

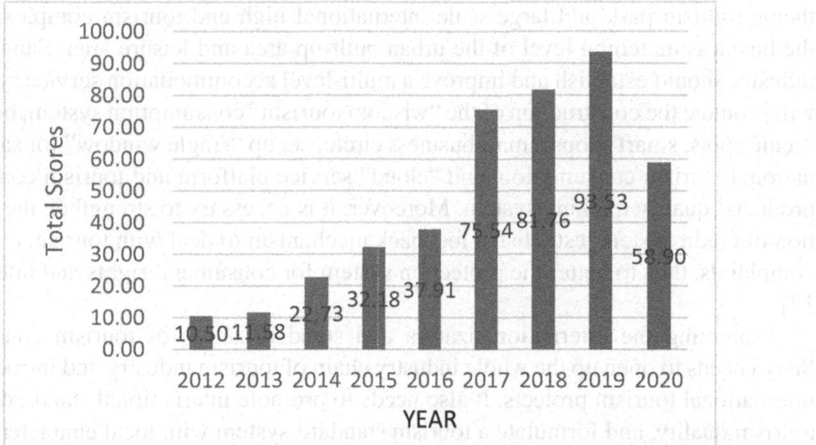

Fig. 1. Schematic diagram of the evaluation results of comprehensive strength of tourism in Sanya

5 Suggestions on Tourism Development in Sanya City

Sanya tourism development should take the construction of world-class tourist destination of consumption as core mission, comprehensively upgrade tourism product state and consumption services, promote the leaping upgrade of tourism industry, realize high-quality tourism development, thus to build a top world-class tourism and consumption destination and become the mainstay of constructing international tourism and consumption center in Hainan Free Trade Zone [8].

Sanya's tourism development should adapt to tourism industry change in the new development pattern, promote the transformation and upgrading of tourism consumption, improve the quality of tourism consumption service. It should also further open the field of tourism consumption, build full-domain, full-season tourism consumption system, and build it into “a romantic place with good mountains and waters”, as well as international tourism consumption resort with rich brands, comfortable environment, distinctive characteristic and pleasant ecological environment [9].

Fostering new forms of business and new hot spot in tourism consumption.It is necessary to speed up the construction of Sanya’s cruise home-port, vigorously expand cruise routes, carry out pilot cruise routes on international waters, and build a pilot cruise tourism zone. We will vigorously develop low-altitude tourism and promote the development of aviation industry chain. We will develop new projects, such as “tourism + culture,” “tourism + creativity,” and “tourism + sports”. We will build a number of world-class sports training and competition bases, developing beach sports, on-water sports and other tourism projects. Moreover, we will build national sports tourism demonstration zones and the core area of sports training bases in southern China [10].

Creating an international first-class tourism consumption environment. In the future, Sanya will aim to enhance the serving level of catering, accommodation, traveling, shopping and entertainment, and construct large high-standard comprehensive consumer business circle with international influence. It will also introduce some world famous large theme tourism park and large-scale international high-end tourism complex, enhance the business matching level of the urban built-up area and leisure area. Sanya tourism industry should establish and improve a multi-level accommodation service system. We will promote the construction of the “wisdom tourism” consumption system, build smart scenic spots, smart shops, smart business circle, set up “single window” for sanya international tourism consumption, and “cloud” service platform and tourism consumption products’ quality tracking system. Moreover, it is necessary to strengthen the construction of credit system, establish a feedback mechanism to deal with tourism consumers’ complaints, thus to better the protection system for consumers’ rights and interests [11, 12].

Promoting the internationalization and standardization of tourism consumption. Sanya needs to open up the whole industry chain of tourism industry and introduce large international tourism projects. It also needs to promote international standardization of tourism quality, and formulate a tourism standard system with local characteristics that conforms to international prevailing paradigms. It will strengthen the cooperation with international organization for standardization and other standardization organizations at home and abroad, and encourage more enterprises to carry out international certification of standardized quality and environmental management systems. Sanya also needs to attach importance to improving its tourism services, training foreign language talents and improving their foreign language service. It needs to promote the construction of multilingual public signs in tourist scenic spots and urban streets. It also needs to offer convenient service of mobile payment, foreign currency exchange and international traveling. We will actively carry out trials of seventh traffic rights, carry out refined airspace reform, and increase the number of international routes and non-stop flights to major passenger sources. We will innovate the international marketing of tourism, strengthen cooperation with overseas mainstream media, and strive to build Sanya’s tourism into an international brand [13, 14].

5.1 Developing a Seasonal Tourism Development Plan

According to the seasonal characteristics of Sanya's tourist flow over the years, the seasonal tourist reception plan for 2021 should be reasonably formulated as soon as possible to avoid other impacts caused by the dense tourist flow in the peak season. The tourism service resources should be reasonably distributed, and different tourist groups should be dispersed in an orderly way, and the reception pressure of the seasonal tourist population on Sanya should be reduced. Besides, tourism industry will be guided to adjust policies and plans. Sanya should combine the relevant state policy for poverty alleviation with completing traditional tourist scenic spot, and alleviate poverty through tourism, such as: scientific and reasonable development of tropical seaside tourism, development of rural tourism. In addition, in the current situation, sanya's hotel industry shoulders relatively heavy burden, the relevant government departments should introduce policies to guide the industry to take sanya's seasonal characters into consideration and adjust their way of promotion, transform the tourists in off-season to residents, lower the room rates, and stimulate domestic demand and attract local people to check in, which not only can help the hotel run smoothly but also increase local people's happiness index [15].

References

1. Wang, X.: Calculation Program of Grey System Analysis and use, 1st edn, HuaZhong University of Technology Press, Wuhan (2001)
2. Xie, S.: Discussion of the ways to increase farmers' income in Dalian based on grey correlation analysis. J. Agric. Econ. (05), 64–66 (2017)
3. Zhang, J.-F.: Study on influential factors of Shaanxi tourism revenue in recent 10 years based on grey relational analysis. J. Baoji Univ. Arts Sci. (Nat. Sci.) **40**(04), 67–72 (2020)
4. Chen, Y.: Analytical study of urban traffic conditions based on K-Means clustering and grey correlation analysis. J. Shandong Jiaotong Univ. **28**(4), 38–45 (2020)
5. Liu, L.-Y.: Research on regional science and technology innovation capability of Hunan Province based on DEA and grey correlation analysis. J. Hunan Univ. Finance Econ. **37**(01), 39–46 (2021)
6. Tian, M.: Study on tourism development and economic growth based on grey relational analysis: take Shandong Province as an example. J. Chongqing Univ. Technol. (Nat. Sci) **33**(2), 208–215 (2019)
7. Xi, C.: Evaluation of urban innovation capability based on panel data grey association analysis: taking the Yangtze River Delta urban agglomeration as an example. J. China Sci. Technol. Res. Rev. **53**(01), 69–76 (2021)
8. Fang, M.: A comparative analysis on the projected brand personalities of cultural tourism: Sanya versus Xiamen. J. Tourism Res. **11**(03), 57–69 (2019)
9. Liu, X.: Tourist flow forecast of Sanya city based on XGBoost and GM model. J. Sci. Technol. Econ. **32**(06), 46–50 (2019)
10. Luo, J.: Practical path of smart tourism in Sanya and the promotion strategy. J. Bus. Econ. (10), 12–23 (2020)
11. Chen, H.-Y.: Study on structure and evolution characteristics of inbound tourism market in Sanya City. J. Res. Dev. Mark. **35**(02), 274–279 (2019)
12. Bao, F.: Research on spatial structure evolution of inbound tourist sources of Sanya. J. Luoyang Inst. Sci. Technol. (Soc. Sci. Ed) **34**(05), 26–30 (2019)

13. Fei, S.: Analysis of the mechanism and process of the demarginalization of marginal regions by promoting tourism——a case study of Sanya in Hainan. J. Sichuan Tourism Univ. (01), 72–76 (2020)
14. Qin, M.: Baidu index, mixed-frequency model and Sanya tourism demand. J. Tourism Tribune **34**(10), 116–126 (2019)
15. Lai, W.: Difference in thermal comfort at the Province scale: a case study of the Nanshan cultural tourism Zone. J. Trop. Geogr. **40**(06), 1127–1135 (2020)

Blockchain-Based Power-Out Insurance Design and Practical Exploration

Chao Deng[1], Yong Yan[2,3], Keyuan Wang[4,5](✉), Hongtao Xia[3], Yang Yang[6], and Yi Sun[4,5]

[1] State Grid Ningbo Electric Power Co. Ltd., Ningbo 310009, China
[2] College of Energy and Electrical Engineering, Hohai University, Nanjing 211100, China
[3] Research Institute, State Grid Zhejiang Electric Power Co., Ltd., Hangzhou 310009, China
[4] Institute of Computing Technology, Chinese Academy of Sciences, Beijing 100190, China
wangkeyuan@ict.ac.cn
[5] School of Computer Science and Technology, University of Chinese Academy of Sciences, Beijing 100049, China
[6] State Grid Zhejiang Electric Power Co. Ltd., Hangzhou 310017, China

Abstract. The rapid development of blockchain technology has spawned more innovative applications in the insurance industry, and its multi-entity participation system and subversive trust mechanism will accelerate the construction of new ecosystems in the insurance industry. In this paper, blockchain is utilized to build an efficient mutual trust application platform with its technical characteristics such as decentralization, immutability, trustlessness, etc. And a new business model in the power industry of precise calculation and automatic claim settlement according to customers' electrical properties and terminal power outage signals was designed. An effective hedge of power failure happening to the customers was realized, and the power-out insurance was created. Moreover, the value of power data was realized, a new type of industry-finance business model was set up, and a profit growth for both the power companies and the insurance companies was created. It shows the validity of the 'blockchain + insurance' mode in constructing a mutual trust, mutual benefit and win-win power-finance ecosphere.

Keywords: Blockchain · Power-out insurance · Smart contract · Combined industry with finance

1 Introduction

Higher requirements have been put forward from the whole society for the power companies to improve their service standard of power supply. With the acceleration of economic construction, the government and power customers highly valued the standard of power supply services and looked forward to a optimized business environment with it. The government has deployed a series of targeted policies and measures to enhance the

This work was supported by Science and Technology Project of State Grid Corporation of China (No.5100-201957475A-0-0-00).

J. Zeng et al. (Eds.): ICPCSEE 2021, CCIS 1452, pp. 357–372, 2021.
https://doi.org/10.1007/978-981-16-5943-0_29

development confidence and competitiveness of SMEs. Power companies are required to adhere to the principle of focusing on customer service, comprehensively improve service capability and standard, sustainably create value for customers, let people get a sense of gain and happiness, serve the high-quality development of the economy and society in a more substantial, secure and better way. China has also put forward new requirements for the practice of blockchain technology. In the 18th collective study of the Political Bureau of the CPC Central Committee, General Secretary Xi emphasized to accelerate the development of blockchain technology and industrial innovation, and actively promote the integration development of blockchain and economic society. The open, shared, and collaborative technology form of blockchain is highly consistent with the strategic goal of "World's top Energy Internet" of State Grid Corporation of China. State Grid Corporation of China has attached great importance to the practice of blockchain technology and actively explores the typical applications of the combination of blockchain technology and the power industry to facilitate the blockchain technology in power fields from virtual to reality.

In 2019, the State Grid Xiamen Power Supply Company and Yingda Taihe Property and Insurance Xiamen Branch jointly launched household electrical appliances security insurance, electricity personal accident insurance and comprehensive insurance for the property of power users' equipment to reduce the losses of electricity customers' electrical appliances, property, and personal security etc. caused by accidents such as natural disasters [1]. However, these three types of insurance are mainly for personal injury and equipment damage, but didn't cover the loss of working time or production due to power outages. Some other industries have delay insurance as frontier. For instance, the transportation industry has introduced flight delay and train delay insurance to reduce the losses caused by the flight or train delays. Therefore, according to the characteristics of the power industry, in this paper, new type of insurance is deeply researched to reduce the loss of working time or production caused by power outages.

At present, insurance companies are facing problems such as inefficient information exchange, complex reinsurance responsibility assessment, scattered data sources, necessity of middlemen, manual claim review and processing environment [2]. If the policy, billing, information of customer and claim are added onto the blockchain, then the distributed ledger of blockchain can be used to realize the secure storage of the electronic insurance policy. Compared with the traditional storage system, the characteristics like authenticity, immutability and traceability of the electronic insurance policy can be guaranteed. Once the insurance meets the corresponding conditions, the system could trace the information on-chain to review the transaction process, and realize automatic claim settlement according to the previously set smart contract, which reduces the participation of third-party intermediaries such as insurance agents, reduces the data loss and other risks that occur in review or claim process, helps insurance companies save costs of claims, eliminates insurance fraud, and makes insurance business processes more standardized. In January 2017, ZA Tech released the "Annchain Cloud" platform based on blockchain technology [3]. Alibaba combines blockchain and medical insurance, On October 16 in 2018, launched a major illness mutual aid service "Xiang Hu Bao" on the Alipay App. Members who encounter a major illness can enjoy a guarantee fund ranging from 100,000 or 300,000 Yuan as compensation, and the cost is shared by all members. "Xiang Hu Bao" utilizes blockchain technology features such as open, transparent, and non-tamperable to ensure that there will be no insurance or debt fraud. The claim submission process is changed from just registration to a complete process of evaluation and payment. Having a simplified (and secure) environment to automate these tasks will fundamentally reduce fraud and provide a better customer experience.

2 Power-Out Insurance

2.1 Requirements Analyses of Power-Out Insurance

Insurance means that the policyholder pays insurance fee to the insurer in accordance with the contract, and the insurer shall be liable for compensation for the property losses caused by the occurrence of the possible accidents stipulated in the contract or the death, disability, illness of the policyholder or a commercial insurance act that assumes the responsibility of paying insurance when the age, time limit or other conditions agreed in the contract are met. From an economic perspective, insurance is a financial arrangement for apportioning accidental losses; from a legal point of view, insurance is a contractual act, a contractual arrangement in which one party agrees to compensate the other party's losses; from a social aspect, insurance is an important part of the social economic security system, a "exquisite stabilizer" of social production and social life; from the perspective of risk management, insurance is a method of handling risk. The emergence of power-out insurance is beneficial to power users and grid companies. Therefore, we analyzed the requirements of power users and grid companies for power-out insurance business.

(1) Enterprise users: When a power outage occurs, Enterprise power customers may have various kind of risks such as potential equipment damage, product scrapping at the moment of power outage, labor costs loss, or failure to deliver products on time. Therefore, enterprise users have a wide range of demands for purchasing commercial insurance to avoid enterprise losses.
(2) Resident users: The power outage will cause the risk of unavailability of air conditioning and electric heating in high or low temperature weather, and the risk of deterioration of food in refrigerator. At the same time, similar to the flight delay insurance, the power-out insurance will make up customers' emotional compensation to a certain extent. Since the group of resident users is huge, the flow benefits could be great.
(3) Grid requirements: Solving the limited power outage compensation problem. Since the main entity of the policyholder is the electricity customers, power-out insurance can be an extension of the existing power supply liability insurance and the enterprise's "three outage" loss insurance. Cultivating new business mode that combine industry and finance, which can drive the development of power finance and insurance field, develop a practical business model and realize a win-win situation for power customers, power supply companies, and financial institutions with the unique power data and power customer resource advantages of power supply companies.

2.2 Power-Out Insurance Business in Traditional Way

According to the traditional insurance process, the business logic of power-out insurance is shown in Fig. 1. The power users purchase the power-out insurance products from the insurance companies. Once a power outage occurred, the insurance company would check the information from the power supply company. After the insurance company receives the power-out information from the power supply company, the insurance company then settles the claim to the user. The whole process is supervised by the Regulatory authorities. The process may also be a process that requires the power user to actively contact the insurance company after a power outage, and then urge the insurance company to move on checking the power-out information from the power supply company. During the whole process, both power-out information verification and power-out information feedback require personnel communication and endorsement of the institutes, which is inefficient compared with the ideal automatic execution of the claim settlement.

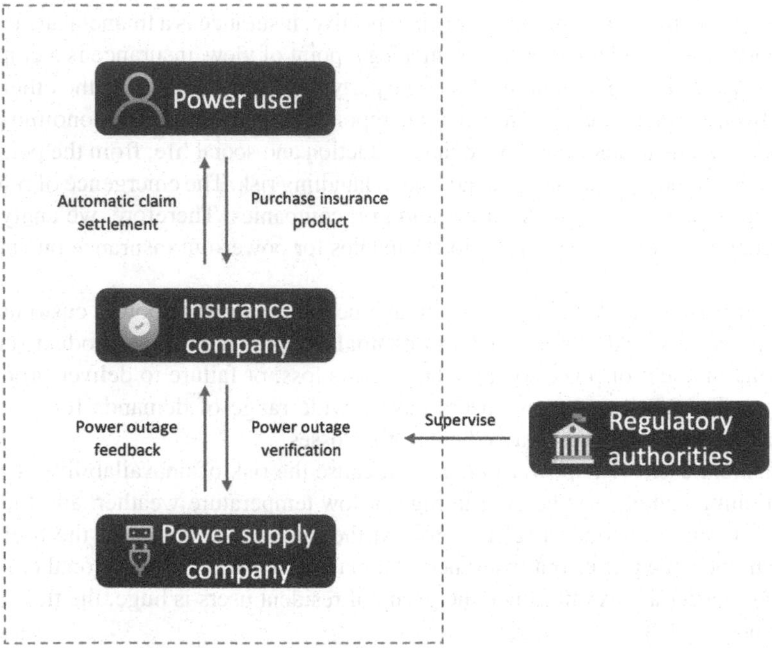

Fig. 1. Logic scheme of power-out insurance in traditional method

Power-out insurance business in traditional way, specifically, has the following problems.

(1) Data trust/security issues

In the traditional insurance method, the power outage data is maintained on the side of the power supply companies, and the power supply companies will only provide part of the power-out information when the insurance companies send people to verify the power-out information. During the time lag from the occurrence of a power outage to the time when the insurance company sends someone to check the information about it, the power supply company has long enough time to decide whether to give back the true information or not based on their own benefits. Besides, since power-out information is always maintained on the side of the power supply companies, no matter what kind of feedback information the power supply company endorses, the insurance company can only choose to believe it. Similarly, electricity users can only trust the verification information endorsed by the insurance company, because the verification process is only maintained by the insurance company. With blockchain technology, the power supply companies should submit power-out information as soon as possible without manual verification. After that, the power-out information is maintained by multiple parties and cannot be tampered with, which can solve data trust/security issues.

(2) Process efficiency issues

In the traditional business process, from the occurrence of a power outage to the compensation of the insurance company, there are three steps that may take a long time: from the occurrence of a power outage to the power user's feedback to the insurance company, the insurance company sends people to the power supply company to check information, and the power supply company gives back the power-out information to the insurance company. Since these business processes are executed serially, it is important to improve the efficiency of each step. Introducing blockchain technology, the whole business process is going to be moved onto the blockchain to achieve automatic execution. It can eliminate the necessity for power users to actively report the situation to the insurance company, and complete automatic insurance claims through the automatic execution of smart contracts on the blockchain, which can greatly improve the process efficiency.

(3) High claim costs

High claim costs issue is reflected in time costs and labor costs. Time costs refer to the long period of time from the occurrence of a power outage to the end of the claim settlement, which brings a poor experience for power users; labor costs refer to the necessity for the people to participate in communication in each step of the traditional business process, and the need for mutual recognition of entities Endorsement. For insurance companies, sufficient employees are required to deal with the verification of every power outage case. With blockchain technology, the automatic execution of business processes through smart contracts can significantly save time costs and labor costs.

3 Blockchain Technology Overview

Blockchain is a distributed ledger technology [4] that utilizes cryptographic technology to append consensus-confirmed blocks in order. It technically solves the security problems brought by the centralized trust model for the first time. It guarantees secure value transfer based on cryptographic algorithms, data traceablility and non-tamperable characteristics based on the hash chain structure and time stamp mechanism, and the consistency of block records between nodes based on the consensus algorithm. From a macro view, blockchain is not only a simple multi-machine/multi-location distributed system, but also requires operating the same system and sharing the same data in a distributed network. The participating nodes of the entire network collaborate to complete transaction verification and storage, which endows blockchain with the characteristics of transparency, sharing, openness, and co-construction. From a micro view, the blockchain orders the blocks in time sequence after the consensus verification of the whole network, and uses the hash value of the previous block to connect them in sequence to form a chain-like structure, thereby ensuring the timing and non-tamperable characteristics; A collection of transaction records is stored in each block, and each record contains the digital signature of the initiator to ensure the unforgeability and non-repudiation of every operation. Based on this designed structure, its distributed, transparent, and secure features, blockchain enables participants to exchange value conveniently, quickly, and at low cost on the Internet, which is the cornerstone of the Internet of Value. Blockchain has been rapidly applied since Bitcoin [5] was proposed in 2008. Initially, blockchain was mainly used in the field of digital currency, and various Bitcoin-like digital currencies have been continuously proposed, such as Litecoin and Dogecoin. Since 2015, blockchain has been gradually applied to finance and other related fields, setting off an upsurge in blockchain application and research around the world. For example, a blockchain clique jointly established by London Stock Exchange, London Clearing House, Société Générale, Union Bank of Switzerland, and European Clearing Center applies blockchain technology to the securities field and explores how to change the settlement of securities transaction through blockchain technology [6]; Factom Inc. applies blockchain technology to the field of digital storage, trying to innovate the data management and recording methods of business, society and government through blockchain technology [7]; Chronicled Inc. applies blockchain technology to the field of supply chain traceability, to ensure the authenticity of products and protect the rights and interests of consumers [8]; Guardtime Inc. uses blockchain to establish residents' medical records across Estonia to ensure the non-tamperable and anti-counterfeiting characteristics of medical records [9]. With the development of blockchain technology, more and more fields have taken a trial to combine with blockchain, giving birth to broader development prospects. Figure 2 is a series of fields that have been associated with blockchain in recent years [10].

At present, in the insurance industry, blockchain technology has been used to promote data sharing, optimize business processes, reduce operating costs, improve collaboration efficiency, and explore to build a credible system [11–13]. But for the specific energy industry insurance business, no matter the corresponding products launched in the insurance industry or the combination of blockchain and energy insurance, there still exists a lack of academic research.

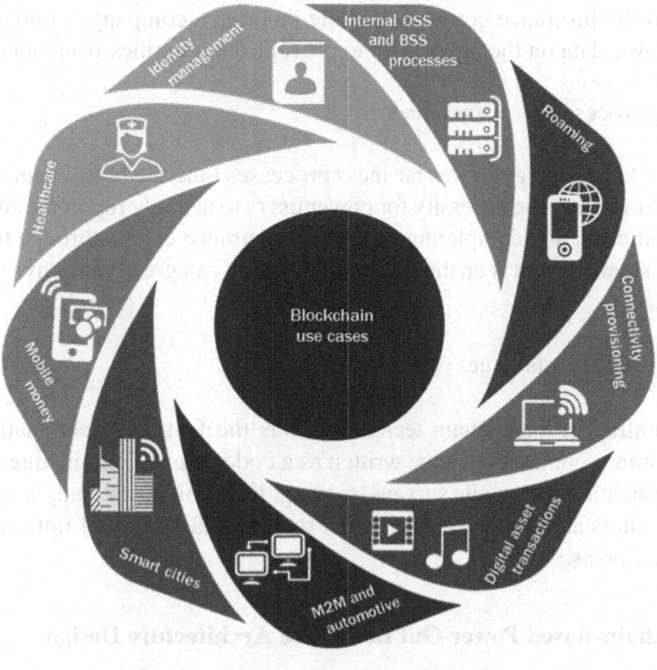

Fig. 2. Application fields of blockchain

4 Blockchain-Based Power-Out Insurance

4.1 Value of Blockchain in Power-Out Insurance

Data trust/security issues, process efficiency issues, and claim costs issues can be solved, with introducing blockchain technology to move traditional insurance business processes onto the blockchain for automatic execution.

(1) Solving data trust/security issues

Blockchain technology has features such as distributed storage and non-tamperable. The server nodes of the platform are deployed in insurance companies, power supply companies, and regulatory authorities. This solves the problem that insurance companies and power customers have low-level mutual trust in the power data involved in the business. Decentralized management does not require a central server, which avoids expensive operation and maintenance costs, and at the same time evades the security problems of centralized servers being attacked or data hard to recover.

Data security for power supply companies: Blockchain technology effectively integrates multiple data security technologies. Typically, the CA authority mechanism and multi-party data storage ensure the data security of power supply companies.

Data security for power users: Users' personal data, including power consumption data and personal financial data are encrypted then stored on the blockchain. The user authorizes to the insurance company and the insurance company is obliged to protect the user's power data on the blockchain to prevent other entities from obtaining it.

(2) Solving process efficiency issues

Blockchain technology moves business processes onto the blockchain for automatic execution, eliminating the necessity for power users to actively report the situation to the insurance company, and completing automatic insurance claims through the automatic execution of smart contracts on the blockchain, which can greatly improve the efficiency of the process.

(3) Solving claim costs issues

Smart contract in blockchain technology has the feature of automatic triggering. Specific insurance contract could be written as a code snippet, eliminating the necessity for claim applications or on-site surveys, solving the problems of complicated insurance claims procedures and high operating costs, realizing fast and real-time claims, saving time and labor costs.

4.2 Blockchain-Based Power-Out Insurance Architecture Design

Figure 3 shows the overall architecture of power-out insurance blockchain system.

This architecture involves four business entities, namely power users, power supply companies, insurance companies and the China Banking and Insurance Regulatory Commission. Among of which, power supply companies, insurance companies, and the China Banking and Insurance Regulatory Commission form a consortium blockchain to participate in the blockchain consensus, and the power users participate in business operations through the user service module of the insurance companies. The design of power users operating through insurance companies rather than directly participating in the blockchain consensus is based on the following considerations:

(1) Power users actually only have business interactions with insurance companies, including purchasing insurance products, querying their own information, and obtaining claims;
(2) To simplify the users' operation and separate the users from the details of the operation on the blockchain;
(3) To avoid the users from directly participating in the affairs on the blockchain, which can protect the data security of other users, insurance products and power-out information.

Without blockchain technology, insurance companies should physically submit the products to the China Banking and Insurance Regulatory Commission for review after the products being designed. This process is not automatic and requires manual recording, which usually requires a long period of time; Once a claim is applied by user, it will take a long time for the insurance company to obtain evidence from the power supply company outside the blockchain. In addition to the long verification period, it's also

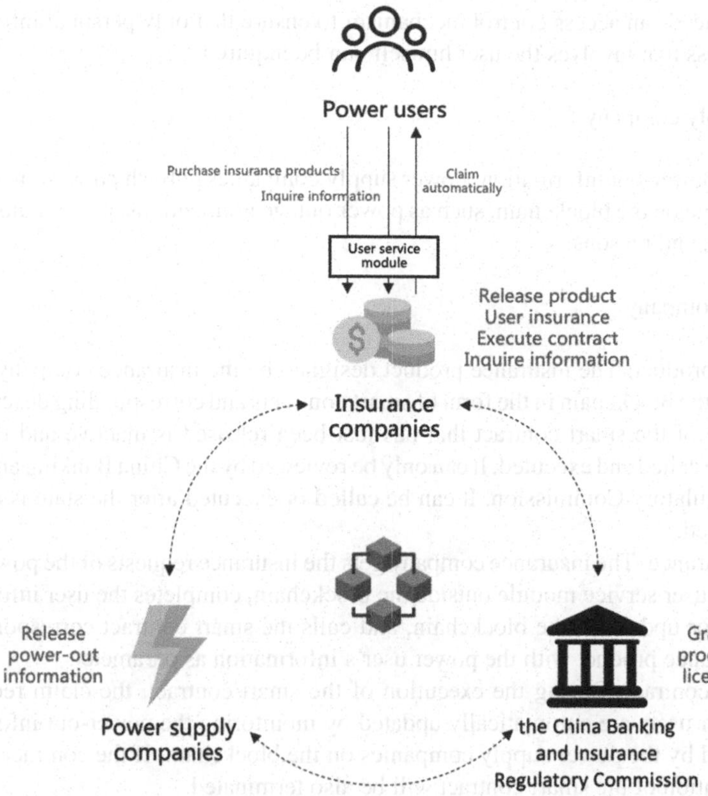

Fig. 3. Overall architecture of blockchain-based power-out insurance

about the trust among the power users, the insurance companies and the power supply companies during the insurance companies' verification process. When the power supply companies, insurance companies, and the China Banking and Insurance Regulatory Commission join into a consortium blockchain network, the data on the blockchain is shared among the participants and cannot be tampered with. The China Banking Regulatory Commission could review the insurance products of the insurance company in time. The insurance companies can automatically obtain the power-out information issued by the power supply companies, which greatly shortens the evidence collection period. The proposed blockchain-based power-out insurance will not only save the labor and time costs of the business process, but also significantly improve the user experience.

Functions of each entity:

Power user

- Purchase insurance products: Power users interact with the role user through the user service module provided by the insurance companies outside the blockchain, where power users submit their information and select insurance products;
- Inquire information: Power users interact with the role user through the user service module provided by the insurance companies outside the blockchain. The user service

module needs an access control mechanism to ensure that only personal information or business that involves the user himself can be inquired.

Power supply company

- Release power-out information: Power supply companies publish power-out information in time on the blockchain, such as power outage notifications, power outage time, coverage, and reasons.

Insurance company

- Release product: The insurance product designed by the insurance company is published to the blockchain in the form of smart contracts and corresponding descriptions. The state of the smart contract that has just been released is inactive and therefore cannot be called and executed. It can only be reviewed by the China Banking and Insurance Regulatory Commission. It can be called or executed after the state is changed to activated.
- User insurance: The insurance company gets the insurance requests of the power users from the user service module outside the blockchain, completes the user information creation or update on the blockchain, and calls the smart contract corresponding to the insurance product with the power user's information as parameter.
- Execute contract: During the execution of the smart contract, the claim records of the power users are automatically updated by monitoring the power-out information published by the power supply companies on the blockchain. If the contract expires, the execution of the smart contract will be also terminated.
- Inquire information: The insurance company gets the power users' information query requests from the user service module outside the blockchain, and returns corresponding user information, contract information, and claim records, etc.
- Claim automatically: The insurance company monitors the update of the users' claim records on the blockchain, and completes the claim automatically in time off-chain through the user service module.

China Banking and Insurance Regulatory Commission

- Grant product license: Review the insurance products released by insurance companies on the blockchain, that is, the content of the smart contract. The smart contract will be activated if it's passed. Only the China Banking Regulatory Commission can modify the state of the contract, and only the activated contract can be called or executed.

Whole business process of the designed architecture:

- The insurance company designs a power-out insurance and publishes the corresponding contract onto the blockchain in the form of smart contract and instructions;
- The China Banking and Insurance Regulatory Commission reviews the description of the deployed smart contract and modifies the state into activated if it's ok;

- Power user registers user information through the user service module of the insurance company and selects insurance product;
- The insurance company creates the information of the power user on the blockchain, and calls the corresponding smart contract according to the product selected by the power user;
- In the event of a power outage, the power supply company will upload the outage time and coverage information to the blockchain as soon as possible;
- The smart contract in execution monitors whether it meets a claim condition for the policyholder. When it meets, it will automatically update the pending claim record in the user information;
- The off-chain user service module of the insurance company monitors the update of pending claim records of users on the blockchain, and completes the claims automatically in time.

Specifically, according to the design of power-out insurance products, power supply companies, insurance companies, and the China Banking and Insurance Regulatory Commission join into a consortium blockchain and upload the data they can provide. For instance, power supply companies upload the start time, end time, and coverage of each power outage in real time, the number of the impacted users, the times of power outages, and the reasons for them; insurance companies need to publish insurance contracts to the blockchain in the form of smart contract when releasing products, and maintain data such as power users' personal information and history claim records.

Blockchain-based power-out insurance has greatly increased the automation of the claims process and improved business efficiency, which is reflected in the following three aspects.

(1) Automatic calculation of claim amount

Insurance calculation: When customers apply for insurance, electricity capacity, industry classification, load level, and power supply quantity etc. of the policyholders are read and save. When power outage occurs, compensation will be calculated automatically according to the start and end time, the power supply area, the insurance plan and so on.

(2) Automatic judgment of claim conditions.

In the preliminary design, claims can be settled after excluding internal faults on the customer side and considering yellow and above warning weather conditions.

Judge by power type: As for high-voltage users, distribution and transformation account numbers, types of power outages, time of power outages, and property rights of faulty equipment should be considered. As for low-voltage users, their own public transformer terminals, concentrator shutdown/power-on time, and the most recent power-out record of the meter should be considered.

(3) Automatically trigger claim settlement

The above-mentioned automatic calculation of claim amount and automatic judgment of claim conditions are based on smart contract technology, and the entire process does not require claim application and manual loss verification, which improves the automation and execution to be more efficient.

5 Social Benefits Analyses

At present, "Access to electricity" has become one of the important evaluation indicators for the World Bank's business environment report, and governments at all levels and State Grid Corporation have attached great importance to it [14]. "Access to electricity" is mainly evaluated by 4 secondary indicators, each with a weight of 25%. These 4 secondary indicators are: the number of procedures for obtaining electricity, the connection time, the proportion of expenses in national income per capita, and Power supply reliability and electricity transparency. Power supply reliability and electricity transparency involve indicators of financial containment measures to limit power outages. Figure 4 is an indicator map of power supply reliability and electricity transparency.

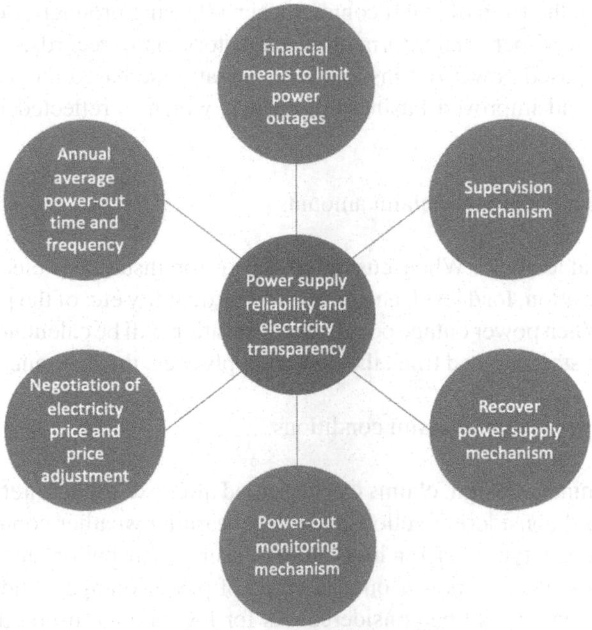

Fig. 4. Indicator map of power supply reliability and electricity transparency

Through the blockchain power-out insurance project, Ningbo Company took advance in exploring a penalty mechanism for power outages exceeding a certain frequency, and made full use of power outage data to study and formulate methods to provide certain financial compensation for users who experience power outages. These cater to the management requirements of "Financial containment measures to restrict power outages" of the government, filled the domestic gap in this field and realized the social benefits of optimizing the business environment.

6 Blockchain Application Challenges in Power-Out Insurance

6.1 Power Law Compensation Issues

According to the laws, regulations, and normative documents such as "Electricity Law" [15], "Electricity Supply and Use Regulations" [16], "Power Supply Business Rules" [17], and "Contract Law" [18] etc., power supply companies and users should sign power supply/use contracts equally, voluntarily, and under consensus. In the power supply/use contract, the rights and obligations of both entities are clarified. The power supply company shall be liable for compensation in the following situations: (1) The power supply quality of the power supply company does not meet the national standards; (2) The power supply company fails to notify the user in advance in accordance with the relevant national regulations when it is necessary to stop the power supply due to the maintenance of the power supply facilities, the legal power cut and the user's misuse of electricity etc.; (3) The power supply company also has the risk of liability for compensation due to a power failure caused by a third party, but the power supply company can claim compensation from the infringing party after taking the responsibility.

In addition, the following risks may exist during the operation of the power-out insurance blockchain platform: (1) The policyholder and beneficiary of the power-out insurance should be aware of, strictly speaking, that the policyholder and beneficiary of power-out insurance should be users, and there are also situations where the policyholder and the beneficiary are separated under certain circumstances. The specific provisions in insurance contract for power-out insurance should be complete enough to effectively solve the problem in the situation where the policyholder and the beneficiary are not the same. (2) The compensation problem due to the leakage of user data and other commercial secrets. If the leakage of user data and other commercial secrets causes losses to users due to technical problems or other reasons, users may demand compensation. (3) Users may claim compensation when errors occur in the process of data operation due to blockchain platform problems.

Although the above-mentioned laws, regulations and normative documents make specific provisions on matters concerning the responsibility of power supply enterprises, the causes of specific power outages are complex and diverse, and it is difficult to prove and claim for compensation. Moreover, domestic customers have not yet formed a broad awareness of their rights concerning to power outage.

According to the "Electricity Supply and Use Regulations", power supply companies have the obligation to notify customers in advance, including: (1) when power supply facilities need to be disconnected for regular maintenance, power supply companies shall notify customers or make announcements 7 days in advance; (2) when power supply facilities need to be disconnected for temporary maintenance, power supply enterprises shall notify customers 24 h in advance.

At present, blockchain-based power-out insurance is mainly offered to high-reliability power consuming companies, and the customer group is relatively fixed. In principle, simultaneous large-scale user claims can rarely happen, but attention should still be paid to the liability of power supply companies that may be arose by the usage of blockchain technology. It is suggested that in the next step more insurance market research should be dug into and in-depth study of the impact of the launch of blockchain-based power-out insurance on users' awareness of their rights should be carried out.

6.2 Blockchain Technical Problems to Be Solved

(1) Blockchain performance bottleneck

Although blockchain has been applied to many fields, the existing blockchain can only meet low-frequency transaction requirement. And in the face of high-frequency transaction scenarios, the throughput of blockchain is seriously insufficient to meet the practical demand. Thus, improving blockchain throughput has become an urgent issue for today's blockchain applications.

In this paper, we adopt the consortium blockchain technology architecture. The power supply companies will upload and store the confirmed business data on the blockchain, such as outage duration, outage users and outage types. User data will be encrypted and then uploaded to the blockchain through the insurance companies, which ensures efficient data transmission and storage on the blockchain and also enhances the privacy of data and effectively protects the security of user data transmission and storage process. Whenever a power outage event or a power outage claim event occurs, there will be corresponding data uploading, querying and other related operations, so the requirements for performance indicators such as throughput are not high, which current consortium blockchain's performance can meet. Consortium blockchain can provide customizable smart contract service, which can help easily formulate smart contracts based on power-out insurance business, realize automatic claim triggering, verification and execution, and improve the efficiency of insurance claim business.

Subsequently, as the marketization of power-out insurance deeply develops, there will be higher requirements for the performance of blockchain, if larger scale applications are required, such as the case that power supply companies and insurance companies or even all power users are connected on the blockchain. And we should seek out the optimization of the scaling of the blockchain architecture itself to solve the current performance bottleneck.

(2) Security still needs to be improved

Asymmetric encryption algorithms ensure the data security of blockchain to a certain extent, but most of the currently operating asymmetric encryption algorithms also face the challenge of deciphering by future quantum computing technology. In addition, the blockchain technology is applied in the way of network-wide dissemination, which also improves the possibility of information leakage to some extent.

(3) The lag of regulation

Blockchain integrates various and usually incompatible technologies to meet different needs of products, which requires communication and coordination among participants in financial industry, but the lag of regulation increases the risk of uncertainty during the application of blockchain technology.

7 Conclusion

This paper designed a blockchain-based power-out insurance, which is a new emerging business of State Grid Corporation. It complies with relevant laws and regulations, and solves the problem of limited power outage compensation for users. The overall architecture of the power-out insurance platform was designed based on blockchain technology, which can fix the issues of data security etc., save time and labor costs in traditional insurance industry, and take advantages of power-out insurance. But as for different industries, different power supply reliability of the certain region, corresponding countermeasures are needed. Designers need to take "early exploration, rapid iteration" as concept and fully utilize the unique competitiveness of power data resources. It is suggested combining industry and finance, and using technology to realize innovation, evaluation and promotion of business model according to the market demand, so as to promote the implementation of the project and achieve valuable results.

References

1. Yingfang, Z.: The first ubiquitous power Internet of things industry finance insurance launched, 29 June 2019. https://www.deelplumbing.com/cesa/news/show-745874.html. (in Chinese)
2. Xiaochun, L.: Fit analysis between blockchain technology and the insurance industry. Mod. Bus. **26**, 154–156 (2020). (in Chinese)
3. ZA Tech: ZA Tech releases "Annchain Cloud" to build blockchain economic model, 5 May 2017. https://www.zhongan.com/channel/about/1010985.html. (in Chinese)
4. The International Organization for Standardization. Blockchain and distributed ledger technologies — vocabulary: ISO 22739:2020. Blockchain and distributed ledger technologies (2020)
5. Nakamoto, S.: Bitcoin: a peer-to-peer electronic cash system, 31 Oct 2008. https://bitcoin.org/bitcoin.pdf.
6. Peter, R.: Financial giants jointly set up blockchain group, 17 Nov 2017. https://www.8btc.com/article/73326. (in Chinese)
7. Factom: Factom Blockchain, 1 June 2021. https://www.factom.com
8. Ajit, K.: Blockchain and ERP systems: the integrated future of supply chain management, 20 Apr 2018. https://blog.chronicled.com/blockchain-versus-erp-systems-why-one-is-superior-for-supply-chain-management-4486c12d56b2
9. Guardtime: Estonian and Hungarian governments partner with guardtime to develop an AI training range for health-care assured AI, 2 June 2020. https://guardtime.com/blog/estonian-and-hungarian-governments-partner-with-guardtime-to-develop-an-ai-training-range-for-health-care-assured-ai.

10. Swan, M.: Blockchain: Blueprint for a New Economy . Sebastopol, O'Reilly Media (2015)
11. Niao, Y.: Analysis of the applicability of blockchain technology in the insurance industry. Insur. Theory Pract. **7**, 44–53 (2020). (in Chinese)
12. Peng, W.: The technical application in the China Great Bay Area insurance cross-border service platform in the digital age - practical exploration of the integrated development of the cross-border insurance industry based on blockchain. Fintech Era **7**, 19–22 (2020). (in Chinese)
13. Weihua, X.: Exploration of insurance innovation development model empowered by blockchain technology - take mutual insurance as an example. Financ. Acc. Mon. **21**, 128–133 (2020). (in Chinese)
14. State Grid introduced 9 measures to improve the service level of "access to electricity". Rural Electric. **28**(11), 3. (2020). (in Chinese)
15. The National People's Congress of the People's Republic of China. Electricity Law, 1 June 2019. http://www.npc.gov.cn/npc/c30834/201901/f7508bc71261404680b8255a2bbcf839.shtml. (in Chinese)
16. The Central People's Government of the People's Republic of China. Electricity Supply and Use Regulations, 3 Feb 2019. http://www.gov.cn/gongbao/content/2019/content_5468920.htm. (in Chinese)
17. National Energy Administration. Power Supply Business Rules, 1 Mar 2012. http://www.nea.gov.cn/2012-01/04/c_131262676.htm. (in Chinese)
18. The National People's Congress of the People's Republic of China. Contract Law, 15 Mar 1999. http://www.npc.gov.cn/wxzl/wxzl/2000-12/06/content_4732.htm. (in Chinese)

Education Research, Methods and Materials for Data Science and Engineering

"Golden Course" Construction of Software Engineering Basic Course Based on Blended Learning

Zufeng Wu, Tian Lan, Ruijin Wang(✉), Xiaohua Wu, and Qiao Liu

University of Electronic Science and Technology of China, Chengdu, China
ruijinwang@uestc.edu.cn

Abstract. "Golden Courses" is the current driving force for deepening the reform of higher education and accelerating the construction of first-class undergraduate education. This paper is based on the education and teaching philosophy of higher education of software, and for the purpose of the national first-class curriculum construction of blended teaching. Taking the course of software engineering as an example, it explored to establish a framework of "Four-Phase Iteration" and "Dual Elements and Combination" curriculum, along with the combination of online and offline teaching methods and practices. Exploration and practice of the curriculum construction was implemented from the aspects of the curriculum design, teaching content, course resources, teaching methods, diversified evaluation and feedback, etc. It is proved to realize the development of "golden course" construction of blended teaching and deepen the reform in education.

Keywords: Software Engineering Foundation · Online and offline · Blended teaching · Golden course

1 Introduction

With the development of information technology and the change of students' sources of information acquisition, teaching models and methods also need adaptive changes. On June 21, 2018, Minister Chen Baosheng of the Ministry of Education proposed for the first time at the Undergraduate Education Work Conference of China's Higher Education Institutions to effectively "increase the burden" on college students and truly transform the "nonsensical class" into a deep, difficult, and effective "golden lesson" of challenge [1, 2]. In November of the same year, Wu Yan from the Department of Higher Education of the Ministry of Education proposed to create five types of "golden courses" in a speech at the "11th University Education Forum" held in Guangzhou. That includes "golden courses" of offline, online, hybrid of online and offline, virtual simulation and social practice [3], and he put forward the standard of Golden Lessons: high-level, innovative, and challenging [4]. How to fulfill the exemplary role of golden courses, to promote the education environment of undergraduate education, to eliminate "nonsensical courses", to stimulate teachers' teaching motivation, to improve students'

J. Zeng et al. (Eds.): ICPCSEE 2021, CCIS 1452, pp. 375–384, 2021.
https://doi.org/10.1007/978-981-16-5943-0_30

enthusiasm for learning, and to improve students' creativity are essential question faced by every teacher and needed to be ponder seriously [5, 6].

The hybrid of online and offline teaching is one of the important directions of the golden class construction. "Online + offline" hybrid teaching refers to a teaching mode that integrates the use of network information technology and traditional teaching methods, also a new mode that combines the advantages of online teaching and offline teaching. Through the organic combination of the two forms of teaching, it can not only stimulate the autonomy of students, but also stimulate the guiding role of teachers [7]. And some teachers and researchers try to propose a model that combines online and offline teaching. Like Yuan Ye [8] mentioned a SPOC-based online and offline teaching mode to convey the course content of professional theory and practice to students. Wang and Shi et al. also mentioned a way or an application to mix online and offline teaching [9, 10].

The so-called online and offline hybrid teaching is the process of "one-in-two" and "two-in-one" of courses. "One-in-two" refers to dividing the course content into online and offline teaching, and complementing the two. We can use it to improve the teaching effect of the course and create a golden course [11]. Therefore, how to combine the advantages of online and offline teaching, to maximize the strengths, to avoid weaknesses, and to improve the maximum teaching effectiveness are the questions that teachers should think about. Also, the basic courses like software engineering, as the establishment of knowledge systems of a subject, should be paid more attention.

2 Problems and Research Status

The difficulty and credit intensity of basic engineering courses in higher education are decreasing year by year, which is a worldwide trend of development. Instead, practical activities and general education curriculum are taking the position, through which the popularity and effect are obvious to all [12–14]. But for higher education, the cultivation of basic abilities is the driving force for the sustainable development of students, so the basic courses of disciplines should have an important position in the university training program. The course of "Software Engineering Fundamentals", as a basic course of software engineering discipline, has the characteristics of wide knowledge content, large subject span, and large differences in teaching requirements for different knowledge points.

In the teaching process of this kind of courses, problems such as too many conceptual contents and easy to be overlooked, insufficient in depth of the technical application explanation, and short of repetitive appeal of practical content, are easy to get into. Students have weak willingness to learn actively and get unsatisfactory learning result. At the same time, the traditional classroom teaching almost all led by teachers and students passively involved according to the teacher's plan. Students' different individual needs are often ignored.

According to the teaching experience of this course over the past years, the disadvantages of the basic courses of the discipline are summarized as follows:

1. The asymmetry of teacher-student resources leads to unclear learning goals and inability to meet differentiated learning. The teaching goals and requirements of

the traditional classroom teaching are usually left behind by the students after the first class of explanation, leading to goalless learning. Besides, due to the limitation of university teaching methods, the communication channel between teachers and students is single, usually in fixed Q&A time, so the solutions are not timely sufficient. In addition, individualized problems and common problems are not treated separately, it causes the needs of students at different levels cannot be individually met.

2. Both online and offline teaching have shortcomings, which will certainly affect the teaching effects if a single method is applied. Traditional classroom teaching is not targeted, and most students' active communication is insufficient, and individual tutoring content cannot form general knowledge radiation. On the other hand, although during the COVID-19 pandemic online courses proved to have a relatively good effect to a certain extent, they still have many shortcomings. Online teaching usually uses network text to interact, lacking the support of intuitive cases interpretation and other means. In-depth explanation and common problems cannot be solved after class. And how to effectively supervise students in online teaching environment is a tricky problem [15]. Therefore, the single teaching mode has its own shortcomings, which affects the teaching effect.
3. Consolidation after class is insufficient and continuous feedback is poor. The homework of traditional classroom teaching is not very targeted, and after-school questions and answers are mostly concentrated in a fixed period of time, and real-time feedback is insufficient. Online teaching provides timely real-time feedback, but in-depth explanations after feedback can only be conducted through the Internet, and the effect will be insufficient.
4. The content of practice is all a piece, and the learning effect of team members is uneven. The previous topics were mostly based on preset experience, which have not changed for many years, and it is difficult to control students' speculation and laziness. At the same time, it is often happens that individual students in the team complete all tasks while other students have uneven results, and the learning effects within the team vary greatly.
5. The assessment model is single and does not form the same direction as teaching. Traditional classroom assessments are mechanical, mostly using one-time final assessments such as final exam and experimental results. So usually the scores arc relatively low, also, assessment and teaching are separated. However, the evaluation of online teaching is based on data, which unable to achieve process supervision.

In order to improve the quality of course teaching and try to make up for the shortcomings of current teaching, and based on the characteristics of the basic courses of the discipline and the ability of students, this course group has designed an online and offline hybrid teaching model, based on years of online and offline teaching experience. The assessment model will give full play to their respective advantages to create a golden blended learning class.

3 Design of "Four-Phase Iteration" Mixed Teaching Model and "Dual Elements and Combination" Assessment Model

3.1 "Four-Phase Iteration" Mixed Teaching Model

Aiming at the teaching difficulties of the basic courses of a discipline, relying on the summary of many years of teaching experience of "Software Engineering Foundation", and combined with the advantages of online and offline teaching, we carried out a "Four-Phase Iteration" mixed teaching model The updated model is promoted step by step through the whole teaching and learning process in software engineering course includes four phases. The first phase of online teaching objectives includes pre-class explanation and preview understanding, the second phase of offline involves classification content knowledge internalization, the third phase of online includes after-class externalization and continuous improvement, and the fourth phase of offline contains practice of ability improvement. Online, offline, online, and offline corresponding implementations form a four-phase mix of pre-learning understanding, knowledge internalization, externalization consolidation, and ability improvement process. It can be easily implemented, giving full play to its fragmented learning advantages through preview understanding, online learning, timely feedback and consolidation after class four-phase form. Nevertheless, through offline learning, continuous feedback, targeted in-depth lectures with classified content, and practical ability training to efficiently solve key and difficult problems. It not only retains the effective advantages of offline teaching accumulated for years of teaching, but also gives full play to the advantages of online teaching to improve the teaching effect. The specific teaching plan design is shown in the figure below (Fig. 1).

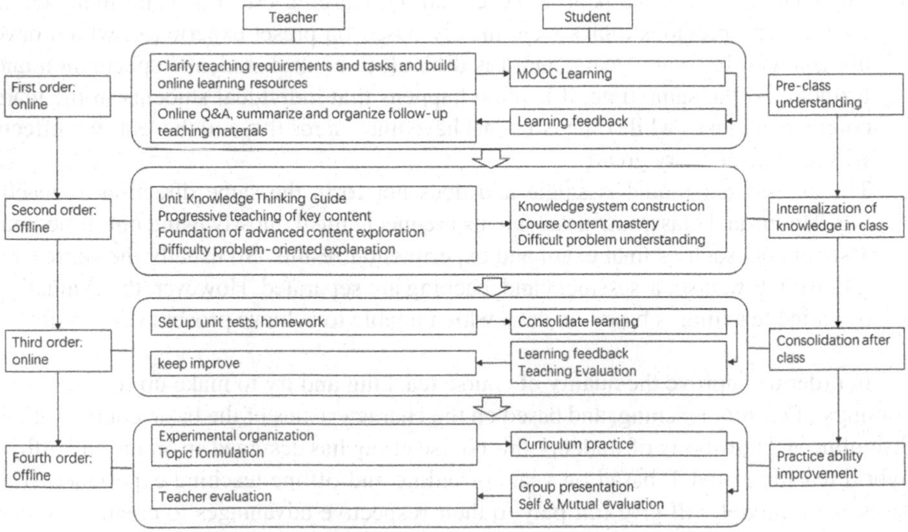

Fig. 1. "Four-Phase Iteration" mixed teaching mode

Phase 1: Teachers clarify the teaching goals and requirements online, guide the pre-class preview to understand the goals. The teaching requirements at the top of each page can be consulted at any time, and build a wealthy supplementary teaching resources for pre-learning and understanding before class. At the same time, students can provide feedbacks, doubts, difficulties they encounter through the online platform at any time, as well as opinions and suggestions of teaching to solve the problem of information asymmetry. In response to students' online feedback, the instructor summarizes and categorizes the questions and provides personalized online answers. Instructors can easily summarize and analyze the general problems, collect the teaching materials in the classroom, and classify the questions to meet the individual needs and differentiated needs of students. Such as:

Phase 2: Internalization of knowledge during offline courses, and the carrying of offline courses or open courses can solve the problems of traditional offline objectives and insufficient online courses. According to the teaching objectives and feedback, we have classified the problems encountered and prepare the courses accordingly to solve four types of problems. The first is to use heuristic teaching methods to guide students to build a knowledge system structure, draw a mind map to sort out the system, and inspire learning tasks and goals through cases and help students sort out the curriculum system to avoid biases. The second is to teach key methods step by step, and guide the understanding of technology applications through case deduction, application analysis, and intuitive technical method roadshows through important content such as program flow charts, case diagrams, and logical coverage tests to strengthen the training of technical application ability. The third is to organize and summarize the difficult points, problems, and common problems of previous students in online learning, in order to give targeted lectures to deepen understanding by using discussion-based teaching, case teaching, etc. The fourth is to carry out inquiry-based teaching, after-school homework expansion, etc., for the follow-up course content, and to establish a foundation for the advanced content.

Phase 3: After-class online focuses on externalization and consolidation, using online platforms to give out tests and homework, and supervising the consolidation with preset priorities, difficulties, and ability-building tasks. Task-driven teaching requirements urge students to consolidate the content of pre-class preview and classroom lectures. Through the feedback of assessments and assignments, and the evaluations, suggestions and opinions of teaching, continuous improvement will be achieved.

Phase 4: Practice of offline helps in capability improvement. Concerning the fact that many technical methods of this course are not the only solution, we set up a course practice group of 3–5 people. After thorough discussion, we designed a group of questions within familiar field, and double-checked with the previous two rounds of questions to avoid practical content to be all a piece, inspire students' initiative. It is required to clarify the small groups of labor and set up a stage explanation link. Teachers randomly select group members to explain, to promote all members to participate, and to avoid uneven learning effects within the group.

Giving full play to the advantages of online and offline teaching, needs of different stages of teaching will be met, so that to maximize strengths and avoid weaknesses, and to solve the problems of information asymmetry, insufficient teacher-student interaction, insufficient continuous improvement, insufficient initiative mobilization in the practice process, etc.

3.2 "Dual Elements and Combination" Assessment Model Design

Based on the progressively four-phase online and offline hybrid teaching model, a reasonable assessment design can effectively guide and strongly promote the students to study purposefully, actively and efficiently. This course group draws on the engineering certification concept of continuous improvement, follows the characteristics of online and offline mixed teaching, and takes into account the process and object factors to make assessment. The assessment is based on the core elements of teaching, which are teaching objects and teaching process, namely dual-element assessment. Among them, the teaching object is aimed at the subject of teaching, the teachers, and the subject of learning, the students. From the perspective of the objects of teaching and learning, the evaluation of teaching and learning is combined to avoid subject-guest inversion. The evaluation of the teaching process adds the process assessment based on the original final evaluation, to avoid the once-for-all deal on the score, to strengthen the objectivity of assessment, and at the same time to be used for teaching feedback and improvement. That is, the combination of process assessment and result assessment, namely dual-combination. The design of the assessment mode is shown in Fig. 2.

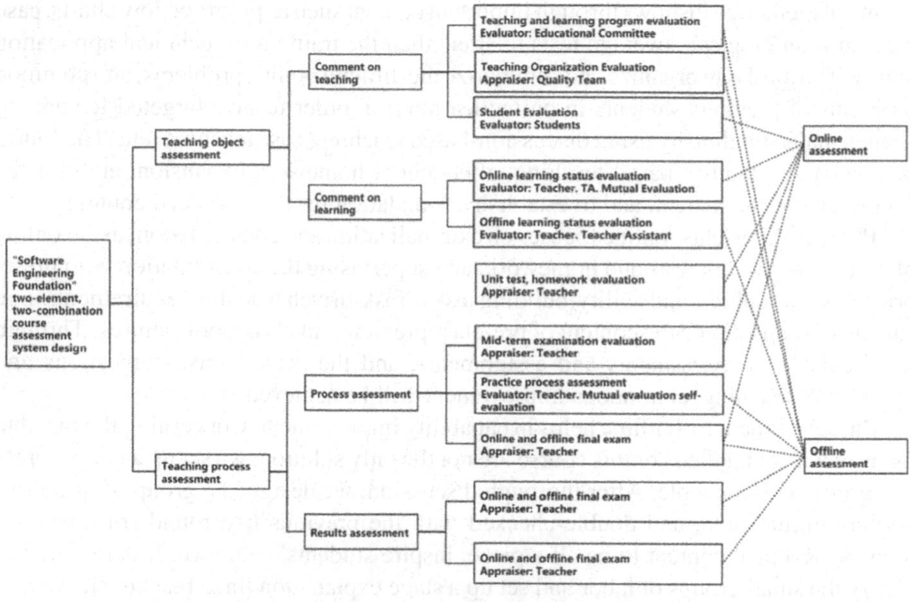

Fig. 2. Curriculum assessment model with dual-element and dual-combination

Through the assessment scheme of dual elements and combinations, comprehensive consideration of the teaching object and teaching process, avoiding point-to-area, and strengthening process assessment, to not only solve the shortcomings of once-for-all deal, but also form the homonymous incentive with the teaching process.

4 Analysis of Teaching Practice Results

4.1 Comparison of OBE Achievement Scale

The major of this course has passed Professional Engineering Certification twice in a row, the OBE achievement scale has been recognized by experts, and the graduation goals and course goals are affirmed as reasonable and effective, as shown in Table 1.

Table 1. "Software Engineering Foundation" OBE course goal support

Course targets	Weight coefficient	Learning results
CO1	0.2	CM1: The concept and characteristics of software, the software crisis and current status, and the definition and development process of software engineering
		CM2: Software life cycle, software process concept, and several common software process models
CO2	0.3	CM3: Introduction of the traditional and object-oriented requirements analysis methods
		CM4: Introduction of the traditional and object-oriented system design methods
		CM5: Programming specification, version management, CMMI model, etc
CO3	0.3	CM6: The concept of quality and quality assurance, software review, software reliability, ISO9000 quality standards, software testing strategies and techniques, etc.
		CM7:The concept and classification of software maintenance, problems that should be paid attention to, software maintenance process model and software maintenance technology, etc.
		CM8: Management of system development, engineering management principles and economic decision-making methods
CO4	0.2	CM9: Select a software project and complete the needs analysis, system design, quality assurance and project management experiments of the selected project in groups

For this course goal, the comparison data of this round of iteration and the previous round of goal achievement is shown in Table 2.

Table 2. Comparison and analysis table of the recent two rounds of achievement

	Course targets	Previous round achievement	Current round achievement	Year-on-year increase or decrease
Course achievement scale analysis	CO1	0.786	0.9	14.5%
	CO2	0.674	0.747	9.77%
	CO3	0.653	0.737	12.86%
	CO4	0.877	0.868	−1.03%
	Overall CO	0.731	0.7988	9.27%

The data shows that the overall curriculum goal has increased by 9.27%. Except for CO4, the sub-items have increased by at least close to 10%. However, CO4 is slightly decreased by 1.03% due to the increase in process assessment, which increases the difficulty of assessment. From the analysis of OBE achievement scale, the hybrid has achieved the expected results.

4.2 Comparative Analysis of Test Scores

According to the OBE standard, the course objectives are relatively stable, and the assessment content framework is basically steady. At the same time, the applications of methods such as separation of examination and teaching, team proposition, multi-stage examination, and streamline marking are applied to ensure the stability of the examination difficulty. Under the premise that the difficulty of the test is relatively stable, the score distribution of the test is of comparative significance. The two rounds of test score distribution is shown in Fig. 3.

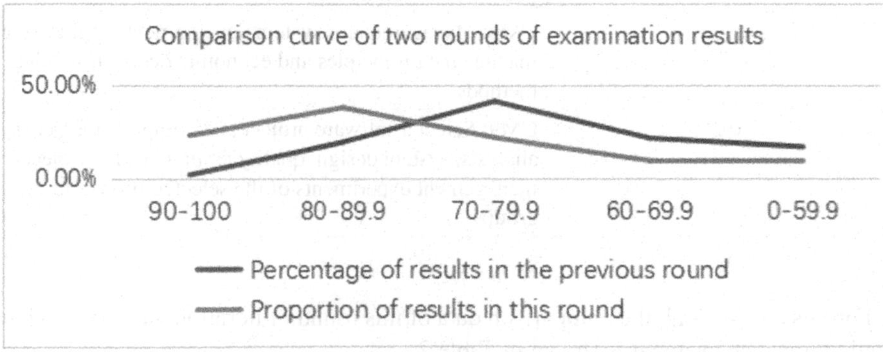

Fig. 3. Comparison curve of two rounds of examination results

The data shows that compared with the previous round, the scores of the current round of examinations have greatly increased the proportion of high grades, where 39.4% of the students got the scores higher than 80. Yet, low grades (below than 80) decreased by the same percentage. The overall grades of the class show a significant increase. From the perspective of score distribution, the hybrid teaching strategy has achieved the expected effect.

5 Conclusion

In this paper, we first mentioned the problems and challenges in the courses of Software Engineering Fundamentals. Then, to solve these problems, we carry out a four-phase mixed iteration teaching model and dual elements combination assessment model. The four phase model helps our teachers clarify the teaching goals and requirements, solves the problems of traditional offline objectives and insufficient online courses and helps students focus on externalization and consolidation and capability improvement. The assessment model helps students achieve personalization in their learning and solve the problem of uneven learning. From the two standards of OBE course goal achievement and score ratio distribution, the teaching method we proposed has obvious effects compared with the previous round. And the course has won the approval of the China University Computer Education MOOC Alliance Construction Curriculum Project, the CMOOC Alliance Online and Offline Hybrid Teaching Reform Project, the Ministry of Education's Industry-University Education Project, the Sichuan Province Excellent Online Open Course, and the University of Electronic Science and Technology of China's recognitions for teaching materials, large-scale open course projects, etc.

Acknowledgement. This article is the phased research result of the online and offline hybrid teaching reform project (project number B190202) of the Chinese University Computer Education MOOC Alliance in 2019. Cooperation Education Program (201802095001, 201702010025).

References

1. Yan, W.: Construction of china's gold course. Chin. Univ. Teach. (12), 4–9 (2018)
2. Jinli, X., Xiao Fangliang, X., Jianhua, Y.W.: Research on the construction of "golden course" for ship navigation courses under the new engineering background. Marit. Educ. Res. **36**(02), 59–62 (2019). (in Chinese)
3. Haihua, Z.: The exploration and practice of creating "Water Quality Engineering I" offline gold course. Educ. Teach. Forum (6), 211–212 (2020)
4. Zhang, D., Chen, R., Yuan, Y.: Construction and application of comprehensive evaluation model of "Golden Classroom". In: Proceedings of the 5th International Conference on Machine Vision and Information Technology (CMVIT 2021), pp. 272–277. IOP Publishing (2021)
5. Xiaomei, L., Xiaowei, L., Kui, H.: Research and practice on the teaching reform of business administration courses under the background of "Internet + education". Heilongjiang Educ. Higher Educ. Res. Eval. (8), 1–2 (2018)
6. Houren, Z.: Research on the teaching quality evaluation system of theoretical courses of business administration. Heilongjiang Educ. Higher Educ. Res. Eval. (9), 21–23 (2019)
7. Zhongmei, X, Qiong, W., Shumin, X.: The application research of "online + offline" hybrid teaching mode in the course of "software testing". Educ. World (8), 140–141 (2020)

8. Ye, Y.: Application of SPOC-based online and offline teaching mode in higher vocational computer courses. In: Proceedings of 2019 Asia-Pacific Conference on Advance in Education, Learning and Teaching (ACAELT 2019), UK, pp. 206–209. Francis Academic Press (2019)
9. Juehui, W., Xiangjun, Z.: Online and offline mixed teaching. Henan Educ. (Vocational Adult Education) (10), 39–40 (2019)
10. Changzheng, S, Kai, S.: Discussion on the application of online and offline mixed teaching mode in software technology curriculums. Educ. Teach. Forum (46), 214–215 (2019)
11. Li, Q., Junxian, L., Rong, Z.: Research on the teaching strategy of online and offline hybrid golden courses for college students' innovation and entrepreneurship education. Fin. Theory Teach. (4), 112–114 (2020)
12. Zhenhong, L.: Implementation plan design of online and offline "hybrid" teaching mode. Joint J. Tianjin Vocat. Coll. (9), 45–48+57 (2018)
13. Ji, W., Haibo, H., Longtao, X., Yangyang, Z.: Basic curriculum issues in the construction of engineering disciplines. In: 2018 National Solid Mechanics Conference, Harbin, China, vol. 11 (2018)
14. Ruijin, W., Shuhua, W., Shijie, Z., et al.: Exploration of spiral progressive software engineering practice teaching system. Exp. Technol. Manage. **35**(02), 174–178 (2018)
15. Wang, H.: Case study of online and offline teaching practice in higher education. In: Proceedings of 3rd International Workshop on Education Reform and Social Sciences (ERSS 2020), pp. 297–300. BCP (2020)

Learning Behavior-Aware Cognitive Diagnosis for Online Education Systems

Yiming Mao[1(✉)], Bin Xu[1,2,3], Jifan Yu[2], Yifan Fang[2], Jie Yuan[2], Juanzi Li[2,3], and Lei Hou[2,3]

[1] Global Innovation Exchange (GIX) Institute, Tsinghua University, Beijing 100084, China
mym19@mails.tsinghua.edu.cn

[2] Department of Computer Science and Technology, BNRist, Tsinghua University, Beijing 100084, China

[3] KIRC, Institute for Artificial Intelligence, Tsinghua University, Beijing 100084, China

Abstract. Cognitive diagnosis, which aims to diagnose students' knowledge proficiency, is crucial for numerous online education applications, such as personalized exercise recommendation. Existing methods in this area mainly exploit students' exercising records, which ignores students' full learning process in online education systems. Besides, the latent relation of exercises with course structure and texts is still underexplored. In this paper, a learning behavior-aware cognitive diagnosis (LCD) framework is proposed for students' cognitive modeling with both learning behavior records and exercising records. The concept of LCD was first introduced to characterize students' knowledge proficiency more completely. Second, a course graph was designed to explore rich information existed in course texts and structures. Third, an interaction function was put forward to explore complex relationships between students, exercises and videos. Extensive experiments on a real-world dataset prove that LCD predicts student performance more effectively, the output of LCD is also interpretable.

Keywords: Cognitive diagnosis · Intelligent education · Graph neural network

1 Introduction

Online education systems, such as Massive Open Online Course (MOOC), Intelligent Tutoring System (ITS), and Mobile Autonomous School (MAS), can provide a series of computer-aided applications for better tutoring, such as computer adaptive test [12] and knowledge tracing [16,23]. Among these applications, one of the critical applications is to diagnose states of student knowledge, which is called **cognitive diagnosis**. In intelligent education systems, cognitive diagnosis aims to discover students' states from history learning records of students, such as their proficiencies on specific knowledge concepts.

J. Zeng et al. (Eds.): ICPCSEE 2021, CCIS 1452, pp. 385–398, 2021.
https://doi.org/10.1007/978-981-16-5943-0_31

Online courses
V1:Data Struture: Red-black Tree
V2:Operation System:Memory Management
V3:Operation System:File System
V4:Algorithm:Quick Sort

Exercises	Knowledge Concepts	Records
E1:How to use the heap to sort	Heap	✔
E2:Operating system belongs to	Operation System	✖
E3:Red and black trees are equivalent to__ Rank B tree	Red–black Tree、 B Tree	✔
E4:The results of 3 & 5 are__	Programming	✔

Fig. 1. An example of student learning records in online education system.

Massive efforts have been made in the cognitive diagnosis area. Among them, item response theory (IRT) [17], as a typical unidimensional model, considered each student as a single proficiency variable (i.e., latent trait). Comparatively, multidimensional models, such as Deterministic Inputs, Noisy-And gate (DINA) [2] model, characterized each student by a binary latent vector, which described whether or not he had mastered the knowledge concepts with the help of Q-matrix(exercise knowledge concept matrix labeled by educational experts). Furthermore, Neural Cognitive Diagnosis [20] uses artificial neural network to model complex non-linear interactions.

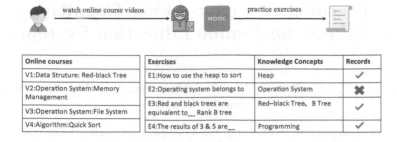

The methods above have been proved effective but still have limitations. A deficiency is that they only consider the interaction between students and exercises and do not explore complete students' learning behaviors in online education systems. Figure 1 shows an example of such a complete learning process of a typical student in Massive Open Online Courses (MOOCs). The student first watches four videos of three courses for his goals and does four exercises as practice. Specifically, he reads the exercise contents, answers them by applying the corresponding knowledge learned from videos, and leaves response records(e.g., right or wrong). The student only answers E_2 incorrectly, which demonstrates that he has a high mastery on knowledge concepts like "Heap", "B Tree" and is weak in "Operation System". We can observe that when students learn in MOOCs, they watch videos first to learn the knowledge they want to master, which helps them practice exercises related to videos.

Moreover, existing methods model each exercise as an independent data instance and do not consider latent relations between exercises. For an online course, it is consists of structured videos and exercises. Besides, exercises are also connected by knowledge relations such as prerequisite [14].

To address these challenges, we propose a learning behavior-aware cognitive diagnosis(LCD) framework to predict student performance by taking advantage of full student learning records and the latent relations of exercises. Specifically, we first build a graph consisting of videos, exercises and construct the structure by utilizing course structures and knowledge concepts. Next, we encode students, exercises, videos as vectors and leverage a multi-layer perceptron (MLP) [6] for modeling the complex interactions of student answering exercises. A graph convolutional network(GCN) [9] is used to automatically refine the representation as feature vectors of exercises and videos. Finally, we conduct extensive experiments on a real-world dataset, which is the first dataset to contain video information and exercise text materials. The results show that our approach outperforms other methods. Detailed parameter analysis and discussion are also discussed. In summary, the contributions are as follows:

- We develop a new method named LCD to model the complex interactions of full student learning behaviors in an end-to-end manner.
- We propose two methods to construct a course graph to build relationships between videos and exercises under the graph neural network framework.
- We construct a novel dataset including students watching video records and exercising records with texts and structure information in MOOC.

2 Related Work

Cognitive Diagnosis. Generally, traditional cognitive diagnostic models (CDM) could be roughly divided into discrete models and continuous models. The fundamental discrete CDM is Deterministic Inputs, Noisy-And gate (DINA) [2], represented each student as a binary vector which denoted whether he mastered or not the knowledge concepts required by exercises with a given Q-matrix prior. The DINA-based models are applied to further specific educational scenarios, such as differential item functioning assessment [7], learning team formation [11] and comprehension test validation. Comparatively, the basic continuous method is item response theory (IRT) models [4], which characterize students by a continuous variable, i.e., latent trait, and use a logistic function to model the probability that a student correctly solves a problem. Many variations of CDMs were proposed by combining learning information. For example, learning factors analysis (LFA) [1] incorporated the time factor into the modeling. In recent years, Wu [10] put forward a fuzzy cognitive diagnosis framework to diagnose students' knowledge state in the examination, which considers both subjective and objective questions. Thai [19] leveraged a multi-relational matrix factorization to map students' knowledge proficiency as a hidden vector. Another typical method is the deep learning method. Piech [16] developed a deep knowledge tracing method based on a recurrent neural network to simulate students' learning process and improve the effectiveness of the task of predicting students' achievement. Huang [8] extended this method, and they used text information for deeper semantic mining of exercises. Wang et al. [20] proposed NeuralCDM that uses artificial neural networks to capture relations between students and exercises.

Graph Neural Networks. The GNN [18] is a type of neural network that can operate on graph-structured data. The graph is a type of data structure representing objects and their relationships as nodes and edges. In recent years, various generalization frameworks and significant operations have been developed with successful results in various applications, such as recommender systems [21] and social network analysis [15]. Very few studies have been done in the field of online learning and education. A recent study on college education by Hiromi proposed a Graph-based Knowledge Tracing (GKT) [13] combines the knowledge tracing with a graph neural network. It encodes a learner's hidden knowledge state as embeddings of graph nodes and updates the state in a knowledge graph. However, GKT aims to predict students' scores and does not distinguish between an exercise and the knowledge concepts it contains. Thus it is unsuitable for cognitive diagnosis.

3 Problem Definition

In this section, we give some basic definitions and formulate the problem of learning behavior-aware cognitive diagnosis in the MOOC platform.

Knowledge component is denoted as $K = \{Kn, Kr\}$, where $Kn = \{k_i\}_{i=1}^{|Kn|}$ is the set of knowledge concepts, which are the subjects taught in courses (e.g., "Red-black Tree" in "Data Structure"). $Kr = [(k_i, k_j)]$ is the relation of knowledge concepts. Here we consider the prerequisite relation, which is crucial for students to learn, organize, and apply knowledge [14]. Specifically, if a pair $(k_i, k_j) \in Kr$, k_i is a prerequisite concept of k_j.

Course corpus is composed by numerous courses, denoted as $M = \{C_i\}_{i=1}^{M}$, where C_i is one course. Each course includes a sequence of videos and exercises, i.e., $C_i = [(c_{ij}, kc_{ij}, t_{ij})]_{j=1}^{|C_i|}$, where the triplet $(c_{ij}, kc_{ij}, t_{ij})$ describes the j-th component of course C_i. t_{ij} refers the type (video or exercise), kc_{ij} (usually labeled by experts) is the set of knowledge concepts the component relates to and c_{ij} indicates its contents (video subtitles or exercise texts). W is the set including all exercises and videos in M and the Q-matrix $Q = \{Q_{ij}\}_{|W| \times |Kn|}$, where $Q_{ij} = 1$ if W_i relates to knowledge concept k_j and $Q_{ij} = 0$ otherwise.

Course graph consists of only one type of node. Each node corresponds to a video or an exercise in W. Formally we represent it as (A, F), where $A \in \{0, 1\}^{|W|*|W|}$ is an adjacency matrix and $F \in \mathbb{R}^{|W|*t}$ is the node feature matrix assuming each node has t features.

Student records consist of two types of behaviors: watching videos and doing exercises. Suppose we have a student set S, where students study individually. Formally, we record the learning behaviors of students as $R = R_1 \cup ... \cup R_{|S|}$, where $R_i = [(w_t, r_t)], w_t \in W$ denotes student S_i learns w_t at step t and r_t denotes the response. For exercises, r_t is the corresponding score the student got on w_i and r_t does not exist otherwise.

Learning behavior-aware cognitive diagnosis is defined as bellows: Given the course corpus M, knowledge component K, student set S, and students' response logs R, the goal of our cognitive diagnosis task is to build a model to mine students' proficiency on each knowledge concept. Since there is no ground truth for diagnosis results, following previous works, we adopt the **student performance prediction task** to validate cognitive diagnosis results' effectiveness.

4 Method

The entire structure of the LCD is shown in Fig. 3. Generally, for a cognitive diagnosis system, three elements need to be considered: student features, exercise features, and the interaction function including them [3]. In this paper, we add video features that can be obtained by course graph representation to enhance interaction function. Specifically, we construct a course graph by course structures, knowledge relations, and text features. Then we learn the representation of nodes by a graph convolutional network (GCN). The results are used as one of the exercise features and video features. We use one-hot vectors of the corresponding student, exercise, and video as input and obtain these three diagnostic features for each response log. The interaction function aggregates different features and outputs predicting score that the student can get on the exercise after training. Finally, we get students' proficiency vectors as diagnosis reports. It is worth mentioning that the GCN and the interaction function are trained in an end-to-end fashion.

4.1 Course Graph Construction

In this section, we introduce details to construct the course graph, which includes the adjacency matrix A and node features F. As shown in Fig. 2, First BERT is used to encode all texts of exercises and videos as the value of F. It is worth mentioning that we only use BERT to extract features of texts and do not fine-tune. Next, we propose two approaches named the structure-based method and the knowledge-based method to assign A. Detailed methods are as follows:

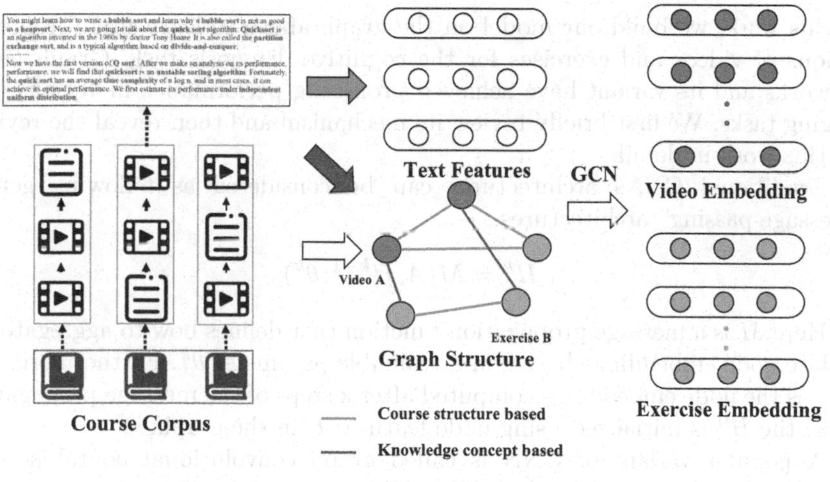

Fig. 2. The framework of course graph learning.

Structure-based Graph is a graph generated by the structured information of courses provided by online education system. We assign the adjacency A^S as:

$$A_{ij}^S = \begin{cases} \frac{MS(w_i,w_j)}{|MC(w_i)-MC(w_j)|}, & |MC(w_i) - MC(w_j)| < \lambda \\ 0, & otherwise \end{cases} \quad (1)$$

where $MS(w_i, w_j) = 1$ if w_i and w_j belong to a same course in M, $MS(w_i, w_j) = 0$ otherwise. $MC(w_i) = n$ indicates w_i is the n-th component of a course in M. That is A_{ij}^S depends on their distance w_i and w_j in the sequence of courses if both them belong to a same course. λ is a parameter to control the threshold.

Knowledge-based Graph is a densely connected graph generated by the knowledge concepts and relations, here we design the weight A^K as:

$$A_{ij}^K = \begin{cases} 1, & kc_i \cap kc_j \neq \emptyset \\ \alpha, & \{(k_x, k_y) | k_x \in kc_i, k_y \in kc_j\} \cap Kr \neq \emptyset \\ 0, & otherwise \end{cases} \quad (2)$$

Where kc_i indicates the set of knowledge concepts of w_i. That is A_{ij}^K is 1 if w_i and w_j contain at least one same knowledge concept; else A_{ij}^K is α if one of knowledge concepts w_i relates to is a prerequisite concept of one which w_j contains; otherwise it is 0. α is a parameter we manually set to control the importance of prerequisite relation.

Finally we assign A as: $A = Normalization(A^S + A^K)$.

4.2 Graph Representation Learning

In this work, we build our model on the graph above to learn useful representations of videos and exercises for the cognitive diagnosis task. Graph neural networks and its variant have achieved promising performance in various challenging tasks. We first briefly review its mechanism and then reveal the revision for this work in detail.

Traditional GNNs architectures can be considered as following general "message-passing" architectures.

$$H^k = M(A, H^{k-1}; \theta^k) \quad (3)$$

Here M is a message propagation function that defines how to aggregate and update node embeddings based on a trainable parameter θ^k. Furthermore, $H \in \mathbb{R}^{n*d}$ is the node embeddings computed after k steps of the message propagation. Here, the H^0 is initialized using node features F in the graph.

A popular variant of GNN is called graph convolutional neural network (GCN), which has demonstrated to be efficient to gather information in a graph through convolution operations. Its message propagation function is as following:

$$H^k = M\left(A, H^{k-1}; \theta^k\right) = \sigma\left(\tilde{D}^{-\frac{1}{2}} \tilde{A} \tilde{D}^{-\frac{1}{2}} H^{k-1} W^{k-1}\right) \quad (4)$$

where $\sigma(\cdot)$ denotes a non-linear activation function such as $ReLU$ function. $\tilde{A} = A + I$ is an adjacency matrix with added self-connections. $\tilde{D}$ is a diagonal degree matrix of $\tilde{A}$ and $\tilde{A}^{k-1}$ is a trainable weight matrix of $k-1$ layer. We divide the set of node embedding H^k into two parts: E^k and V^k, which refers the exercise embedding and video embedding.

4.3 Modeling Process

The goal of the LCD modeling process is to combine different features to predict the probability of a typical student correctly answering a typical exercise. Figure 3 illustrates the structure of the LCD. Details are introduced as below.

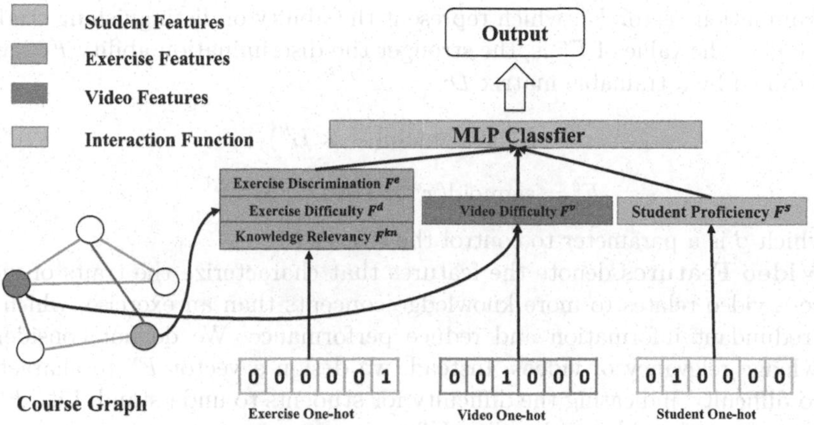

Fig. 3. The architectures of LCD.

Student Features reflect the studying abilities of students in knowledge concepts, which is defined in traditional cognitive diagnosis like DINA [2], IRT [4] and MIRT [19]. One of most important student features is the proficiency on knowledge concepts, which is the goal of cognitive diagnosis task. To better represent knowledge proficiency feature, we design it as explainable vectors similar to DINA and NeuralCDM [20] to guide students' self-assessment rather than the latent trait vectors in IRT and MIRT. Specifically, We use a continuous vector F^s to represent the knowledge proficiency of students, which the size of is the same as the number of knowledge concepts. The formulation of F^s is as below:

$$F^s = \text{sigmoid}(x^s \times B) \tag{5}$$

in which $F^s \in (0,1)^{1\times K}$, $x^s \in \{0,1\}^{1\times N}$ is the one-hot embedding to represent students, $B \in \mathbb{R}^{N\times K}$ is a trainable matrix. sigmoid is the sigmoid function which can control the value of F^s in 0 to 1. Each entry of F^s represents the knowledge level on a specific knowledge concept. For example, $F^s = [0.1, 0.8]$ indicates a low mastery on the first knowledge concept but high mastery on the second.

Exercise Features characterize the trait of exercises. To indicate the relationship between exercises and knowledge concepts, we design the knowledge relevancy vector and denote it as F^{kn}. F^{kn} has the same length of F^s, with each entry represents the relevance of every knowledge concept. In LCD, we use the pre-given Q-matrix Q (exercise knowledge concept matrix labeled by educational experts) to assign F^{kn} for each exercise as:

$$F^{kn} = x^e \times Q \tag{6}$$

where $F^{kn} \in \{0,1\}^{1\times Kn}, x^e \in \{0,1\}^{1\times E^k}$ is the one-hot representation of the exercise. In addition, we design a feature called the exercise difficulty vector F^d to represent the difficulty of each knowledge concept like F^s. F^d directly comes from the exercise embedding E^k. Furthermore, we design the exercise discrimination vector F^e, which represent the ability of distinguishing students. The higher the value of F^e is, the stronger the discrimination ability F^e has. F^e is obtained by a trainable matrix D:

$$F^d = \text{sigmoid}(x^e \times E^k) \tag{7}$$

$$F^e = \text{sigmoid}(x^e \times D), D \in \mathbb{R}^{M\times\beta} \tag{8}$$

In which β is a parameter to control the size of F^e.

Video Features denote the features that characterize the traits of videos. Since a video relates to more knowledge concepts than an exercise, which may get redundant information and reduce performance. We do not consider the knowledge relevancy of videos. Instead, we design a vector F^v to characterize video difficulty, indicating the difficulty for students to understand. Like F^d, F^v is direct from the video embedding V^k:

$$F^v = \text{sigmoid}(x^v \times V^k) \tag{9}$$

where $x^v \in \{0,1\}^{1\times V^K}$ is the one-hot representation of the video which satisfies the following conditions: 1) the knowledge concepts relating to x^e also relate to x^v. 2) Student x^s watched x^v before exercising x^e. It is worth mentioning that if x^v does not exist, we use a zero vector to represent F^v.

Interaction Function. We use Multi-Layer Perceptron (MLP) to obtain the interaction function. The formulations are as below:

$$y = \tau_n\left(\ldots\tau_1\left(F^s, F^{kn}, F^d, F^e, F^v, \lambda_f\right)\right) \tag{10}$$

where τ_n denotes the mapping function of the ith MLP layer and λ_f denotes model parameters of all the interactive layers.

$$\boldsymbol{x} = (F^{kn} \cdot F^s) \oplus F^e \oplus F^d \oplus F^v \tag{11}$$

$$\begin{aligned} \boldsymbol{f}_1 &= \text{sigmoid}\left(\mathbf{W}_1 \times \boldsymbol{x}^T + \boldsymbol{b}_1\right) \\ \boldsymbol{f}_2 &= \text{sigmoid}\left(\mathbf{W}_2 \times \boldsymbol{f}_1 + \boldsymbol{b}_2\right) \\ \boldsymbol{y} &= \text{sigmoid}\left(\mathbf{W}_3 \times \boldsymbol{f}_2 + \boldsymbol{b}_3\right) \end{aligned} \tag{12}$$

where $\oplus$ is the concatenation operation. The loss function of LCD is cross entropy between output y and true label r:

$$loss = -\sum_{i} (r_i \log y_i + (1 - r_i) \log (1 - y_i)) \tag{13}$$

After training, the value of F^s is what we get as the diagnosis result, which denotes the student's knowledge proficiency.

5 Experiments

5.1 Dataset Description

Since there is no open dataset that could provide both watching video records and exercising records with text information, our experimental dataset is derived from MOOCCube [22], a large-scale data repository of over 700 MOOC courses, 100k concepts, 8 million student behaviors with an external resource. We extract 12 computer science courses. Some statistics of the dataset are shown in Table 1. To ensure the experimental results' reliability, we filter the students that both did less than eight exercises and answered less than two exercises incorrectly. We also filter the exercises and videos that no students have done or no contents. In total, we get 271,960 learning behavior records from 2,093 students, 519 exercises, 857 videos with 101 knowledge concepts for diagnostic networks.

Table 1. The statistics of the dataset.

Statistics	Value
# of students	26,969
# of exercises	1,074
# of videos	1,321
# of knowledge concepts	101
# of video records	2,680,833
# of exercise records	116,790

5.2 Experiment Setup

Framework Setting. We now specify the network initializations for LCD models. To extract the semantic representation for exercises and videos, we use the BERT-Base model to get embedding vectors with 768 dimensions. In course graph construction part, we set parameters $\alpha = 0.5$ and $\lambda = 6$. In the video and exercise representation learning part, we set the graph hidden size of 128. The size of the GCN output is 16. In the interaction layer part, The dimensions of the full connection layers are 64, 32, 1 respectively, $\beta = 16$. We initialize the parameters with Xavier initialization [5]. Besides, we set mini-batches as 32 for training and used dropout(with probability 0.6) to prevent overfitting.

Comparison Methods. To compare the performance of our proposed models, we borrow some baselines from various perspectives. Details are as follows:

IRT: A cognitive diagnostic model(CDM) that models student exercising records by a logistic-like function [3].

MIRT: A CDM that extends the latent trait value of each student in IRT to a multi-dimension knowledge proficiency vector with the Q-matrix [17].

NeuralCDM: A CDM that incorporates neural networks to learn the complex exercising interactions [20].

LCD-E: A variant of LCD framework. First, we construct an exercise graph(all nodes are exercises) to replace the course graph. Second, in the modeling process, we use a zero vector to represent video features instead of video embeddings learned by GCN.

In the experiment, we prepare 70% of the data as training data, 10% of the data as validation data, and the remaining 20% as testing data. All models are implemented by PyTorch using Python, and all experiments are run on a Linux server with four 2.4GHz Intel Xeon E5-2640 CPUs and a Geforce 1080Ti GPU.

5.3 Experiment Result

Student Performance Prediction. Here, we conduct experiments on the Student performance prediction task, i.e., predicting students' scores over each subjective or objective problem to indirectly evaluate the effectiveness of LCD, since we cannot obtain the real knowledge proficiency of students. Considering that all the exercises we used in our dataset are objective, we use evaluation metrics from both the classification and regression aspects, including accuracy, RMSE (root mean square error), and AUC (area under the curve).

Table 2. Experimental results on student performance prediction.

Model	Accuracy	RMSE	AUC
IRT	63.1 ± .1	46.7 ± .1	60.3 ± .1
MIRT	77.0 ± .2	39.9 ± .1	74.9 ± .1
NeuralCDM	77.2 ± .2	40.0 ± .1	74.2 ± .1
LCD-E	77.5 ± .1	39.7 ± .1	75.4 ± .1
LCD	**78.0** ± .1	**39.5** ± .1	**75.6** ± .1

Table 2 shows the experimental results of all models on student performance prediction task. The error bars after '±' are the standard deviations of 5 evaluation runs for each model. We observe the performance in the following aspects:

For LCD-E method, it outperforms other baseline methods in terms of accuracy, RMSE, and AUC, which demonstrates that our method can make full use of the relationship between exercises to benefit the prediction performance.

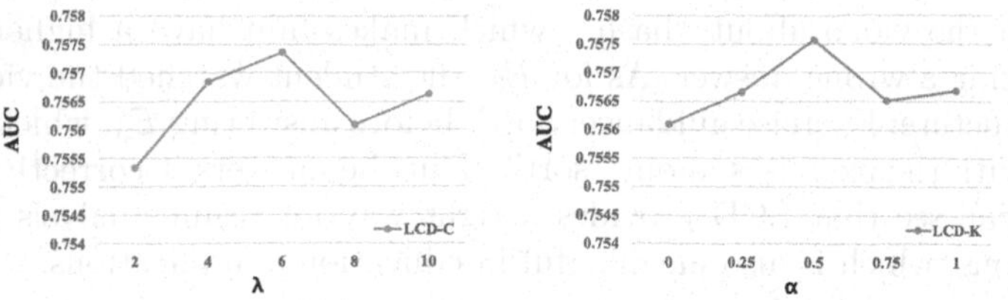

(a) λ (b) α

Fig. 4. The study of different graph methods with parameter influence.

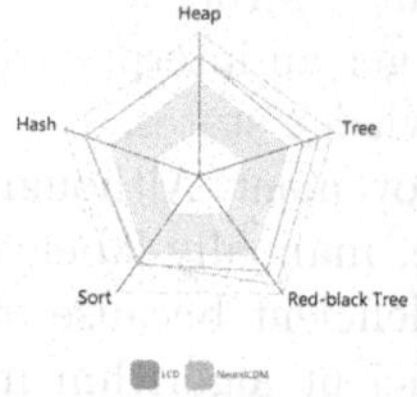

Videos	Exercises	real	NeuralCDM	LCD
V1: Data Structure: IO of B Tree Concepts: IO, B Tree, Tree, heap, computation,	E1: How to use the heap to sort... Concepts: heap	✓	✓	✓
V2: NULL	E2: S is the space of all possible entries, A is the space of all available addresses (|A| < |S|), h is a hash function, then... Concepts: hash	✗	✓	✗
V3: Data Structure: Exercise guidance: Sort Concepts: sort, class, arrange, complexity, computation, sequence, vector	E3: And which sort has the asymptotic time complexity with it? A: Selection sort C: Bubble sort B: Insertion sort D: Merge sort Concepts: sort	✓	✗	✓

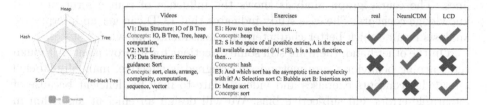

Fig. 5. Visualization of a student's proficiency on knowledge concepts and the responses of three exercises.

For LCD method, it generates better results than LCD-E, indicating the effectiveness of diagnosing student's knowledge states on both watching video behaviors and exercising behaviors than only considering exercising records.

Course Graph Analysis. We investigate the parameters that may influence the performance, including structure-based parameter λ in Eq. 1 and knowledge-based parameter α in Eq. 2.

Figure 4(a) visualizes the performance with the increasing values of λ from 2 to 10. As λ increases, the performances of LCD-C firstly increase but decrease when λ surpasses 6, and it gets the lowest performance while $\lambda = 2$ (each video and exercise only connects with neighbors). Figure 4(b) shows it gets the best performance when $\alpha = 0.5$, which demonstrates that the prerequisite relation has a positive but weak influence.

Case Study. Here, we present an example of the cognitive diagnosis of student knowledge proficiency. As shown in Fig. 5, the radar chart shows a typical student's concept proficiency report diagnosed by NeuralCDM and LCD. The right side is some student records in the online education system. V_i is the video with the same concept as E_i and the student learned before practicing E_i. We can observe that LCD predicts all three exercises correctly, but NeuralCDM gets wrong results on E_2 and E_3 because NeuralCDM does not consider the video information which the student watched. For E_2, although the results of the student diagnosis report by NeuralCDM and LCD are similar, the student does not watch the video about "hash", which makes him have a higher probability of getting a wrong answer. As for E_3, the student watched the video named "Data Structure: Exercise guidance: Sort" before answering E_3, which helps him familiar with the exercises about "sort". Thus he answers it correctly. From the case, we can see that LCD provides a right way for result analysis and model explanations, which is also meaningful in educational applications.

5.4 Discussion

From the experimental results, we can observe that LCD outperforms all baselines. The course graph analysis shows the influence of course structure and knowledge concepts. The case study indicates that LCD can get an interpretive analysis result, which better characterizes the students' knowledge states.

On the other hand, there are still some directions for improvement. Although the learning behavior-aware method we designed is effective, manually-labeled knowledge concepts exercises and videos relate to may be deficient because of inevitable errors and subjective bias. We will design an efficient algorithm in the future to extract knowledge concepts more accurately. Besides, inevitably, students gain and forget the knowledge they learn, especially in online self-learning circumstances. We will make an effort to design reward mechanisms while monitoring the behaviors of students automatically.

6 Conclusion

In this paper, to investigate the influence of student learning behaviors in the cognitive diagnosis task, we propose a learning behavior-aware framework modeling both student's exercising records and watching video records. First, we constructed a graph to build the relationship between videos and exercises and refined the representation by a graph convolutional network (GCN). Next, we propose a novel interaction function to explore complex relationships between students, exercises, and videos. Experimental results on a large scale real-world dataset validate the effectiveness and the interpretation of the proposed method. Promising future directions would be to investigate other important features (guessing features) in the student learning process.

Acknowledgement. This work is supported by the National Key Research and Development Program of China (2018YFB1005100). It also got partial support from National Engineering Laboratory for Cyberlearning and Intelligent Technology, and Beijing Key Lab of Networked Multimedia.

References

1. Cen, H., Koedinger, K., Junker, B.: Learning factors analysis – a general method for cognitive model evaluation and improvement. In: Ikeda, M., Ashley, K.D., Chan, T.-W. (eds.) ITS 2006. LNCS, vol. 4053, pp. 164–175. Springer, Heidelberg (2006). https://doi.org/10.1007/11774303_17

2. De La Torre, J.: Dina model and parameter estimation: a didactic. J. Educ. Behav. Stat. **34**(1), 115–130 (2009)
3. DiBello, L.V., Roussos, L.A., Stout, W.: 31a review of cognitively diagnostic assessment and a summary of psychometric models. Handb. Stat. **26**, 979–1030 (2006)
4. Embretson, S.E., Reise, S.P.: Item Response Theory. Psychology Press (2013)
5. Glorot, X., Bengio, Y.: Understanding the difficulty of training deep feedforward neural networks. In: Proceedings of the Thirteenth International Conference on Artificial Intelligence and Statistics, pp. 249–256 (2010)
6. Hornik, K., Stinchcombe, M., White, H., et al.: Multilayer feedforward networks are universal approximators. Neural Netw. **2**(5), 359–366 (1989)
7. Hou, L., la Torre, J.D., Nandakumar, R.: Differential item functioning assessment in cognitive diagnostic modeling: application of the Wald test to investigate DIF in the DINA model. J. Educ. Measur. **51**(1), 98–125 (2014)
8. Huang, Z., Yin, Y., Chen, E., Xiong, H., Su, Y., Hu, G., et al.: EKT: exercise-aware knowledge tracing for student performance prediction. IEEE Trans. Knowl. Data Eng. **30**, 100–105 (2019)
9. Kipf, T.N., Welling, M.: Semi-supervised classification with graph convolutional networks. arXiv preprint arXiv:1609.02907 (2016)
10. Liu, Q., et al.: Fuzzy cognitive diagnosis for modelling examinee performance. ACM Trans. Intelli. Syst. Technol. (TIST) **9**(4), 1–26 (2018)
11. Liu, Y., et al.: Collaborative learning team formation: a cognitive modeling perspective. In: Navathe, S.B., Wu, W., Shekhar, S., Du, X., Wang, X.S., Xiong, H. (eds.) DASFAA 2016. LNCS, vol. 9643, pp. 383–400. Springer, Cham (2016). https://doi.org/10.1007/978-3-319-32049-6_24
12. Martin, A.J., Lazendic, G.: Computer-adaptive testing: implications for students' achievement, motivation, engagement, and subjective test experience. J. Educ. Psychol. **110**(1), 27 (2018)
13. Nakagawa, H., Iwasawa, Y., Matsuo, Y.: Graph-based knowledge tracing: modeling student proficiency using graph neural network. In: 2019 IEEE/WIC/ACM International Conference on Web Intelligence (WI), pp. 156–163. IEEE (2019)
14. Pan, L., Li, C., Li, J., Tang, J.: Prerequisite relation learning for concepts in MOOCs. In: Proceedings of the 55th Annual Meeting of the Association for Computational Linguistics (Volume 1: Long Papers), pp. 1447–1456 (2017)
15. Peng, H., et al.: Fine-grained event categorization with heterogeneous graph convolutional networks. arXiv preprint arXiv:1906.04580 (2019)
16. Piech, C., et al.: Deep knowledge tracing. In: Advances in Neural Information Processing Systems, pp. 505–513 (2015)
17. Reckase, M.D.: Multidimensional Item Response Theory Models, pp. 79–112. Springer, New York (2009). https://doi.org/10.1007/978-0-387-89976-3
18. Scarselli, F., Gori, M., Tsoi, A.C., Hagenbuchner, M., Monfardini, G.: The graph neural network model. IEEE Trans. Neural Netw. **20**(1), 61–80 (2008)
19. Thai-Nghe, N., Schmidt-Thieme, L.: Multi-relational factorization models for student modeling in intelligent tutoring systems. In: 2015 Seventh International Conference on Knowledge and Systems Engineering (KSE), pp. 61–66. IEEE (2015)
20. Wang, F., et al.: Neural cognitive diagnosis for intelligent education systems. arXiv preprint arXiv:1908.08733 (2019)
21. Ying, R., He, R., Chen, K., Eksombatchai, P., Hamilton, W.L., Leskovec, J.: Graph convolutional neural networks for web-scale recommender systems. In: Proceedings of the 24th ACM SIGKDD International Conference on Knowledge Discovery & Data Mining, pp. 974–983 (2018)

22. Yu, J., et al.: MOOCCube: a large-scale data repository for NLP applications in MOOCs. In: Proceedings of the 58th Annual Meeting of the Association for Computational Linguistics, pp. 3135–3142 (2020)
23. Zhang, J., Shi, X., King, I., Yeung, D.Y.: Dynamic key-value memory networks for knowledge tracing. In: Proceedings of the 26th International Conference on World Wide Web, pp. 765–774 (2017)

Implementation of Online Teaching Behavior Analysis System

Xu Zhao, Changwei Chen(✉), and Yongquan Li

College of Information Engineering, Nanjing Xiaozhuang University, Nanjing 211171, China

Abstract. Online teaching is gradually spreading widely, but there are problems in online teaching evaluation, such as lack of reference index, students' learning state feedback being hard to get in real-time. To improve online teaching quality and evaluation, a classroom behavior analysis and evaluation system based on deep learning face recognition technology is proposed. This system conducts the model training and tests on the deep learning development framework (TensorFlow). It implements the frame image processing through the Mask R-CNN network, extracts the skeleton and the angle direction information to establish the vector, thus carries out the classroom behavior recognition and analysis, and serves as the appraisal important index. The behavior analysis is made and the results show that the recognition rate is high and the learning situation feedback is given in real time. The experiment shows that the system has certain robustness and high accuracy in the real scene. It is convenient for classroom teaching management and implementation, and is helpful to improve the teaching quality.

Keywords: Teaching evaluation · Behavior analysis · Deep learning · Image processing · Face recognition

1 Introduction

Learning behavior analysis is a new technology in the field of education. It uses different analytical methods and techniques to study and analyze the behavior data of learners in the process of learning to find out the learning rules, understand the inherent characteristics of learners, predict the learning effect, and present the results to the teaching stakeholders in a relcvant way. The students can adjust their learning progress and methods and help teachers find out the shortcomings in the teaching process. It is an area of potential for students, teachers, parents, educational managers, educational researchers, educational decision makers, system developers, etc.

In the current application of online teaching, the most similar application to the online teaching behavior analysis system based on deep learning described in this paper is to take pictures of learners at each interval of time during teaching, carry out expression recognition, and finally analyze and get the learning state. This method, which detects students' concentration in class every short period of time, has a certain real-time performance. At the same time, in terms of hardware, TUTOR GROUP LTD [1] proposes to use special monitoring equipment to monitor students' behavior, which has strict

J. Zeng et al. (Eds.): ICPCSEE 2021, CCIS 1452, pp. 399–409, 2021.
https://doi.org/10.1007/978-981-16-5943-0_32

requirements for learners' body posture; McAfee, Inc. [2] proposes to use face recognition combined with biosensors, such as heart rate monitoring instrument to monitor the learning status of learners. In terms of software, Shanghai Fuan Network Information Technology Co., Ltd. [3] proposes a method of combining intelligent terminal system of storage, big data analysis and feedback. But more software uses artificial intelligence technology for homework correction [4], students' academic prediction and early warning, big data analysis of teaching resources [5] and so on. Online teaching behavior analysis system based on deep learning can monitor students' online learning behavior in real time and combine facial expression recognition with body behavior to get students' learning status. In the process of online teaching, the data statistics of students' learning status can make teachers understand students' learning status in front of the computer screen and can also be used as an important indicator of teachers' online teaching evaluation.

2 Research Status

Universities and institutes at home and abroad have invested in many related studies: CMU team has proposed Part Affinity Fields concept to achieve rapid jointing [6]. Megvii Face++ proposed a Cascaded Pyramid Network-based multi-person key point detection framework, with the increase in the number of network stage, the prediction ability of this system gradually improved [7]. A parallel multi-resolution subnet was designed by MSRA (Microsoft Asia Research Institute) to realize multi-scale feature extraction [8]. The "quantitative auxiliary analysis system of 3D human motion simulation technology", "computer simulation and analysis technology of digital 3D human motion" and "video analysis technology" of the Institute of computing technology of Chinese Academy of Sciences, etc. Identification and security technology of Institute of automation, Chinese Academy of Sciences: Microsoft Asia Research Institute and artificial intelligence laboratories of some domestic universities, etc. Although China has made some achievements in the research of human behavior, it is still in the stage of development [9]. The evaluation and analysis of learners' learning behavior in China make less use of computer vision and deep learning. Zhang Zihan uses bone data to analyze [10]. There is still a lack of real-time classroom feedback. Therefore, by mining the video data of the classroom teaching process, extracting the human skeleton and facial key points, and estimating and analyzing the learning state of the learners in the video by computer vision and machine learning technology, the deep learning-based student behavior analysis system is proposed. The main modules include face recognition detection module and human behavior analysis and detection module.

3 System Design and Implementation

This project carries on the model training and tests on the deep learning development framework (TensorFlow). By frame processing of video, the vector is established by extracting the skeleton, direction and other human posture information by using Mask R-CNN [11] technology, so as to identify and analyze the classroom behavior and feedback the information to the third party. The system flow chart is as follows (see Fig. 1).

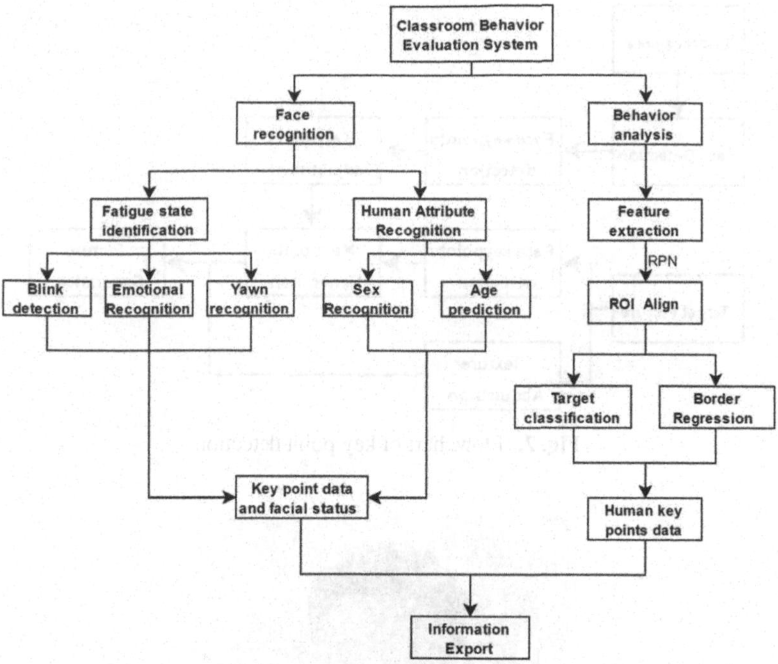

Fig. 1. Flowchart of the system

3.1 Face Recognition Detection Module

Face recognition module uses face key point detection technology to locate the key parts of the face. To further improve the detection accuracy of face key points, we select the dlib library, which has high detection accuracy and more trained key point detection models. The main flow (see Fig. 2) is as follows: firstly, the frame is obtained from the surveillance camera or local video, and 68 feature points are obtained from the face by using the dlib library and saved. Then, three frames are obtained at one time to obtain the variation a between each frame. The key points on the target face image are used as the basic points, and combined with the variation a, they are converted into the key points of the total target, At the same time, the required texture information is obtained from the image, texture mapping is carried out based on the selected target points, and the result is displayed at last (Fig. 3).

3.2 Human Behavior Analysis Module

Image Preprocessing. In this project, the target region is located by image background difference, filtering and computing connected domain, and multiple targets are divided into single targets and then sent to network recognition. By intercepting the original image of the segmented connected domain, it is restored to the target position in the background map of the corresponding empty learning environment, which makes it a complete image of a single target in the learning environment and transforms it into a picture suitable for input Mask R-CNN classification.

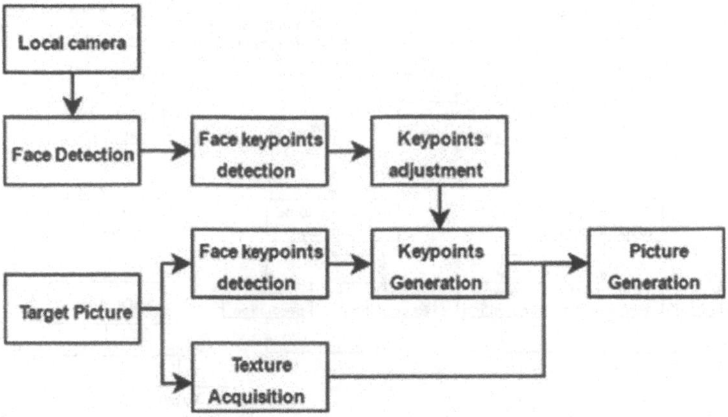

Fig. 2. Flowchart of key point detection

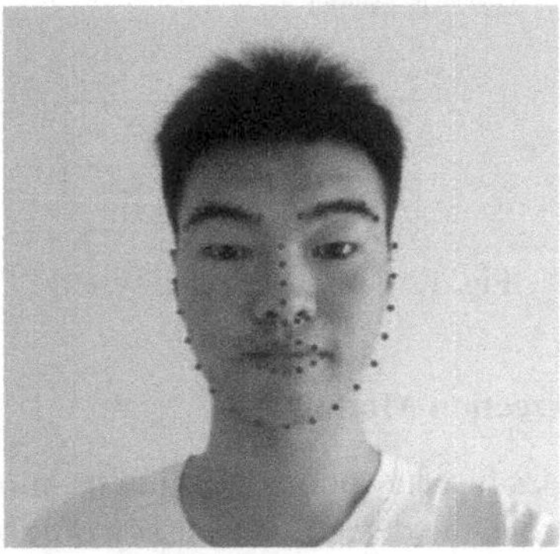

Fig. 3. Effect of face key point detection

Mask R-CNN Algorithm Introduction. Mask R-CNN example segmentation includes three parts: target location, target category classification, segmentation mask prediction. The main framework of MASK R-CNN is the same as that of Faster R-CNN [12], but Mask R-CNN adds a fully connected segmentation subnet after the basic feature network, which changes from two tasks (classification + regression) to three tasks (classification + regression + segmentation). Mask R-CNN is a two-stage framework. In the first stage, the image is scanned and the proposal is generated, namely RPN part. In the second stage, the proposal is classified and the bounding box and mask are generated, namely MRCNN part (Fig. 4).

Firstly, a picture is input to the network, and the feature map of the image is extracted by a series of convolution and pooling operations using the Feature Pyramid Networks [13] (FPN). Secondly, the candidate target is selected on the feature map by RPN network. Softmax classifier is used to judge whether the candidate target belongs to background or foreground. Meanwhile, the candidate target region is generated by using the range

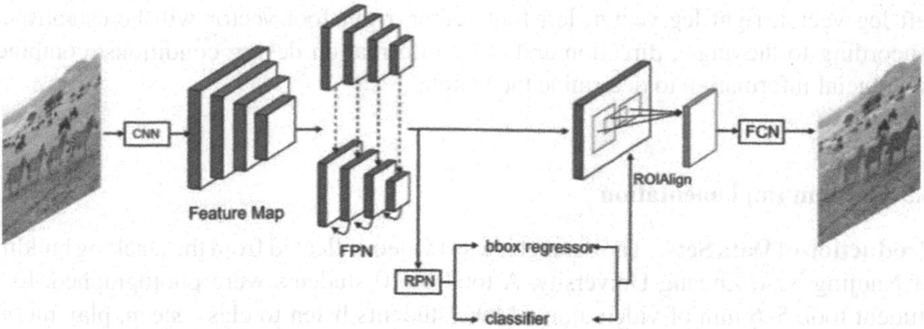

Fig. 4. Mask R-CNN framework content

box regression to modify the position of candidate target. Finally, the target segmentation mask is predicted by Fully Convolutional Networks [14] (FCN). The classification network uses the candidate regions generated by the feature map and RPN network to detect the target category, and FCN uses the feature map to achieve accurate pixel level segmentation of the target. Finally, the coordinates of key points of human body are output.

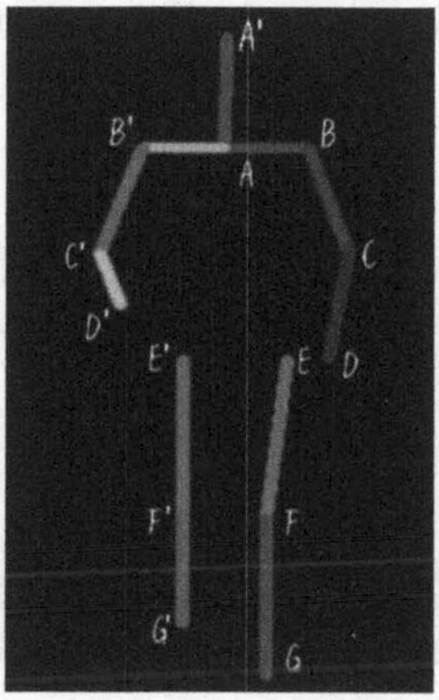

Fig. 5. Information map of key points of human body

Realization of Human Behavior Recognition. Human behavior recognition is mainly divided into computer perspective recognition and motion capture technology recognition. Action recognition in this project mainly extracts the human key point information predicted by the network from the function. If the state is ideal, the $\left\{\overrightarrow{AA'}, \overrightarrow{AB}, \overrightarrow{AB'}, \overrightarrow{BC}, \overrightarrow{B'C'}, \overrightarrow{CD}, \overrightarrow{C'D'}, \overrightarrow{EF}, \overrightarrow{E'F'}, \overrightarrow{FG}, \overrightarrow{F'G'}\right\}$ 11 set of key point information (see Fig. 5) will be obtained. Then the neck vector, left shoulder vector, right shoulder vector, left arm vector, right arm vector, left hand vector, right hand vector, left leg vector, right leg vector, left foot vector, right foot vector will be established. According to the angle, direction and other information design conditions, combined with facial information to determine the action.

3.3 System Implementation

Production of Data Sets. The data set is a test video collected from the teaching building of Nanjing Xiao Zhuang University. A total of 10 students were photographed. Each student took 5–6 min of video alone. Make students listen to class, sleep, play mobile phone action and each action lasts 60–120 s. In each student's behavior mode of each shooting angle, 25 frames of images were randomly selected, and 300 pictures of each student were obtained, totaling 3000 pictures, to form a training set. In the same way, 360 verification sets, and 360 test sets are obtained. In order to avoid overfitting and large generalization error, data enhancement is carried out in two ways:

(1) The experiment collected a variety of postures of 10 students about sleeping, playing mobile phones, and normal listening, and increased the amount of information contained in the experimental data.
(2) Flip horizontally to generate the image, which doubles the database. In order to reduce the correlation between the training samples, the pictures of each student's three behavior patterns were randomly disrupted, and the labels were generated to form the training data set (Figs. 6 and 7).

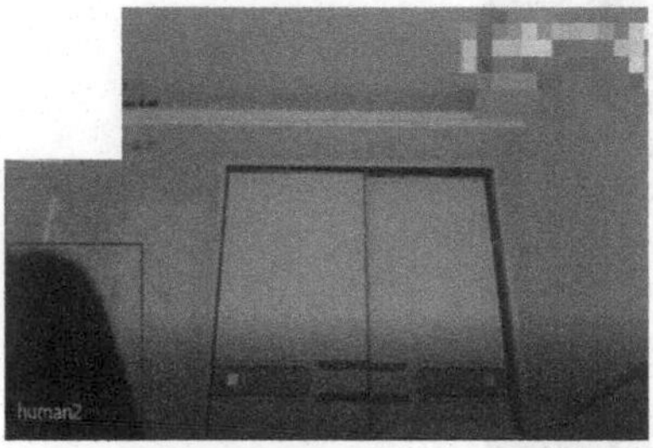

Fig. 6. Effect of the "leave" action

Fig. 7. Effect of yawning

"Leave" Action Detection

(1) Parameters involved: num_point (number of key points in human body), num_face (number of faces), $\overrightarrow{AB}$ (left shoulder vector), $\overrightarrow{AB'}$ (right shoulder vector), $\overrightarrow{\sigma 1}$ (horizontal to right vector), $\overrightarrow{\sigma 2}$ (horizontal to left vector)

(2) Implementation methodology: When both num_point and num_face are less than 1, and maintain more than 3 frames, the system determines it as leaving and records departure time. When the num_point is greater than 5 and the num_face is greater than 1, If $\sin\langle\overrightarrow{AB}, \overrightarrow{\sigma 1}\rangle$ σ1 and $\sin\langle\overrightarrow{AB'}, \overrightarrow{\sigma 2}\rangle$ values are less than 1/2, That is to say, it is judged as coming back, the time of coming back is recorded and the number of departures is increased. When the number of key points is less than 1 and the system can't detect a face, lasting over 3 frames, the system determines it as leaving and records departure time. When the face is detected and the number of key points in the human body is greater than 5, use two - shoulder coordinates to build vectors. If the angle and horizontal direction of the shoulders are within 30°, that is to say, it is judged as coming back, the time of coming back is recorded and the number of times is increased. (departure time, return time, interval time will only appear 2 s time)

"Yawning" Action Detection

(1) Parameters involved: MAR (mouth aspect ratio).
(2) Implementation method: the MAR value in face-to-face recognition is greater than the threshold value and is maintained above 5 frames, which can be judged as yawning, and then the number of yawning increases when the MAR value is less than the threshold value.

"Party Table" Action Test

(1) Parameters involved:
$\overrightarrow{AA'}$, $\overrightarrow{AB}$, $\overrightarrow{AB'}$, $\overrightarrow{BC}$, $\overrightarrow{B'C'}$, $\overrightarrow{CD}$, $\overrightarrow{C'D'}$, $\overrightarrow{EF}$, $\overrightarrow{E'F'}$, $\overrightarrow{FG}$, $\overrightarrow{F'G'}$, $H1$ (screen width)
(2) Implementation: The y axis coordinates of the midpoint of the $\overrightarrow{AA'}$, $\overrightarrow{AB}$, $\overrightarrow{AB'}$, $\overrightarrow{BC}$, $\overrightarrow{B'C'}$, $\overrightarrow{CD}$, $\overrightarrow{C'D'}$, $\overrightarrow{EF}$, $\overrightarrow{E'F'}$, $\overrightarrow{FG}$, $\overrightarrow{F'G'}$ are less than $\frac{1}{3}$H1 for over 10 frames, that is the table action. All the key point coordinates are extracted, and the midpoint coordinates of all line segments are calculated. If all the midpoint coordinates are lower than 1/3 of the relative height of the screen, the maintenance of more than 3 s, can be judged as the table action (Figs. 8 and 9).

"Head Up" Movement Detection

(1) Parameters involved:
$\overrightarrow{AA'}$ (neck vector), $\overrightarrow{AB}$ (left shoulder vector), $\overrightarrow{AB'}$ (right shoulder vector), num_face (face number)

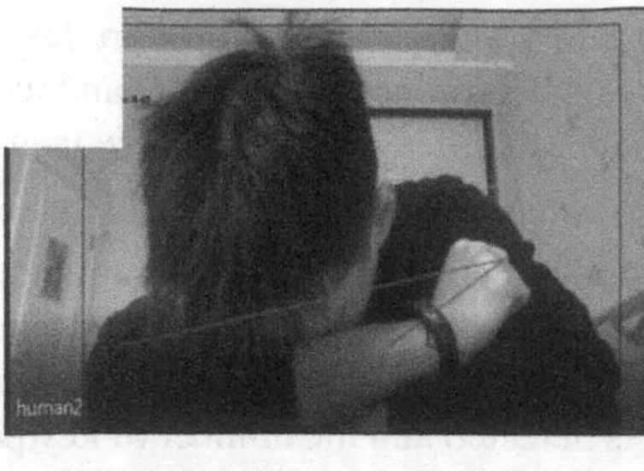

Fig. 8. Effect of the "table" action

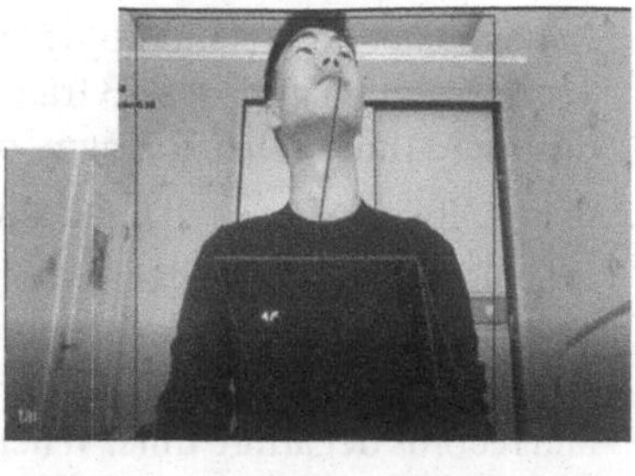

Fig. 9. Effect of "head up" action

(2) Implementation: If the num_face is less than 1, if $\left|\overrightarrow{AA'}\right| > \frac{2\left(\left|\overrightarrow{AB}\right|+\left|\overrightarrow{AB'}\right|\right)}{3}$ and $\cos\left\langle\overrightarrow{AA'}, \overrightarrow{AB}\right\rangle$, $\cos\left\langle\overrightarrow{AA'}, \overrightarrow{AB'}\right\rangle$ are less than 1/2, it is determined to be head. If $\left|\overrightarrow{AA'}\right| > \frac{2\left(\left|\overrightarrow{AB}\right|+\left|\overrightarrow{AB'}\right|\right)}{3}$, $\cos\left\langle\overrightarrow{AA'}, \overrightarrow{AB}\right\rangle$ are greater than 1/2, it is determined to look left. If $\left|\overrightarrow{AA'}\right| > \frac{2\left(\left|\overrightarrow{AB}\right|+\left|\overrightarrow{AB'}\right|\right)}{3}$, $\cos\left\langle\overrightarrow{AA'}, \overrightarrow{AB'}\right\rangle$ are greater than 1/2, it is determined to rise to the right. When the face information is not collected, the key point coordinates of shoulders and neck are extracted, the corresponding vectors are established, and the length of shoulder and neck segments is calculated. If the neck length is more than 2/3 of the sum of shoulders and the angle between neck vector and shoulder is more than 70°, it can be judged as head. If the angle between the neck and the left shoulder is less than 60°, it is determined to look left. If the angle between the neck and the right shoulder is less than 60°, it is determined to look up in the right direction.

"Hand Support Cheek" Action Test

(1) Parameters involved:
$\overrightarrow{AA'}$ (neck vector), $\overrightarrow{AB}$ (left shoulder vector), $\overrightarrow{AB'}$ (right shoulder vector), $\overrightarrow{CD}$ (left hand vector), $\overrightarrow{C'D'}$ (right hand vector), D (Dx, Dy), D′(D′ x, D′ y), AA′: ax + by + c = 0(linear expression)

(2) Implementation: If $\left|\frac{aD'x+bD'y+c}{\sqrt{a^2+b^2}}\right| < \frac{\left(\left|\overrightarrow{AB}\right|+\left|\overrightarrow{AB'}\right|\right)}{4}$, and $\cos\left\langle\overrightarrow{AA'}, \overrightarrow{CD}\right\rangle$ is greater than 1/2 and maintains more than 5 frames, it is determined as left gills. If it $\left|\frac{aDx+bDy+c}{\sqrt{a^2+b^2}}\right| < \frac{\left(\left|\overrightarrow{AB}\right|+\left|\overrightarrow{AB'}\right|\right)}{4}$, and $\cos\left\langle\overrightarrow{AA'}, \overrightarrow{C'D'}\right\rangle$ is greater than 1/2 and maintains more than 5 frames, it is determined as right gills.

Realization method: the coordinates of the key points between the two pairs of hands and the neck are extracted, and the vertical distance from the end point of the hand to the neck segment is calculated. If the distance is less than 1/4 shoulder length and this position keeps more than 3 s, it is determined as supporting gills behavior (Figs. 10 and 11).

Fig. 10. Effect of "hand support gills" action

Fig. 11. Effect of "bow" action

"Bow" Action Detection

(1) Parameters involved:
$\overrightarrow{AA'}$ (neck vector), $\overrightarrow{AB}$ (left shoulder vector), $\overrightarrow{AB'}$ (right shoulder vector), num_face (face number)

(2) Implementation: If the num_face is less than 1, if $\left|\overrightarrow{AA'}\right| < \frac{\left(\left|\overrightarrow{AB}\right|+\left|\overrightarrow{AB'}\right|\right)}{3}$ and $\cos\left\langle\overrightarrow{AA'}, \overrightarrow{AB}\right\rangle$, $\cos\left\langle\overrightarrow{AA'}, \overrightarrow{AB'}\right\rangle$ less than 1/2, it is determined to be straight down. If $\left|\overrightarrow{AA'}\right| < \frac{\left(\left|\overrightarrow{AB}\right|+\left|\overrightarrow{AB'}\right|\right)}{3}$, $\cos\left\langle\overrightarrow{AA'}, \overrightarrow{AB}\right\rangle$ are greater than 1/2, it is determined to bow to the left. If $\left|\overrightarrow{AA'}\right| < \frac{\left(\left|\overrightarrow{AB}\right|+\left|\overrightarrow{AB'}\right|\right)}{3}$, $\cos\left\langle\overrightarrow{AA'}, \overrightarrow{AB'}\right\rangle$ are greater than 1/2, it is determined to bow to the right. When the face information is not collected, the key point coordinates of shoulder and neck are extracted, the corresponding vector is established and the length of line segment between shoulder and neck is calculated. If the length of neck is less than 1/3 of the sum of shoulders, it can be judged as bow. If the angle between neck and shoulder is more than 70°, it is determined to be positive direction. If the angle between the neck and the left shoulder is less than 60°, it is determined to bow to the left. If the angle between the neck and the right shoulder is less than 60°, it is determined to bow to the right (Fig. 12).

3.4 Classroom Evaluation and SMS Push Module

The system will summarize the students' class behavior and calculate the different behaviors according to the number of detected behaviors, so as to evaluate the students' learning state synthetically. The system automatically sends a short message reminder when the student leaves, and then pops up a prompt window to remind the student to leave (Figs. 13 and 14).

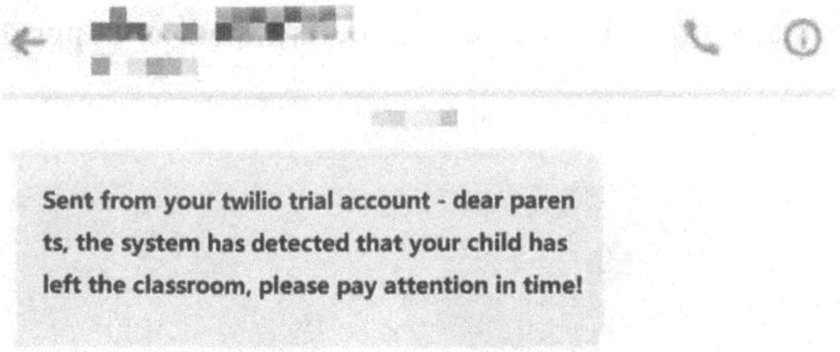

Fig. 12. SMS push test diagrams

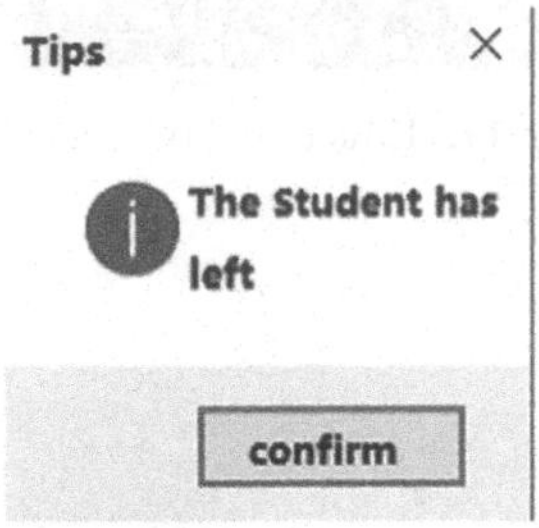

Fig. 13. Test diagram for the prompt window

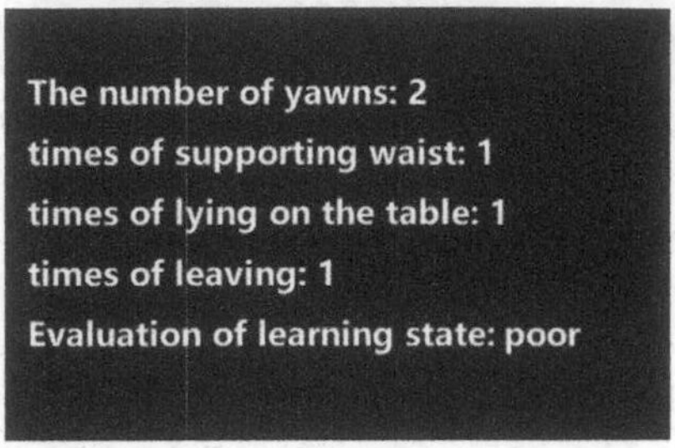

Fig. 14. Student status assessment test chart

4 Conclusion

The classroom behavior analysis and evaluation system based on deep learning expounds and summarizes the overall framework of learners' learning behavior analysis. Its key elements mainly include face recognition detection and human behavior analysis, so as to provide a theoretical basis for the study of learners' behavior analysis. The overall framework of learning behavior analysis is designed. From the perspective of analysis requirements, the behavior characteristics of classroom actions, learning expressions, blinking times and so on issued by learners are analyzed from multiple perspectives. The key technologies are studied to achieve the target of behavior analysis, which provides theoretical and practical basis for the practical application of online learning behavior analysis.

5 Thanks

This article is supported by the Key Project of Innovation Training of Jiangsu University Students' Innovation and Entrepreneurship Plan in 2020 (Project Name: online teaching behavior analysis system based on deep learning, project number: 202011460006Z).

This article is supported by the National Natural Science Foundation of China Youth Science Foundation project (Project name: Research on Deep Discriminant Spares Representation Learning Method for Feature Extraction, No. 61806098).

This article is supported by Scientific Research Project of Nanjing Xiaozhuang University (Project name: Multi-robot collaborative system, No. 2017NXY16).

References

1. TUTOR GROUP LTD.: Method of Real-Time Supervision of Interactive Online Education: JP2019179235A[P], 17 Oct 2019
2. MCAFEE INC.: Systems and Methods for Real-Time User Verification in Online Education: EP3238167A4[P], 16 Jan 2019
3. Shanghai Fuan Network Information Technology Co., Ltd.: An online teaching interactive effect evaluation method and device, China, cn109949189a [P], 28 Jun 2019
4. Anhui Medical College: An online teaching supervision and management method and system: China, cn111626372a [P], 04 Sept 2020
5. Langfang Zhenyugu Education Technology Co., Ltd.: An online education information teaching system based on big data cloud platform, China, cn111681476a [P], 18 Sept 2020
6. Cao, Z., Simon, T., Wei, S.E., et al.: Realtime multi-person 2D pose estimation using part affinity fields. arXiv e-prints (2016)
7. Chen, Y., Wang, Z., Peng, Y., et al.: Cascaded pyramid network for multi-person pose estimation
8. Wang, J., Sun, K., Cheng, T., et al.: Deep high-resolution representation learning for visual recognition (2019)
9. Liang, W., Wei, H., Tieniu, T.: A review of visual analysis of human motion and a review of visual analysis of human motion **25**(3), 225–237 (2002)
10. Zihan, Z.: A study and application of classroom behavior recognition method based on data (2020)
11. He, K., Gkioxari, G., Dollár, P., et al.: Mask R-CNN. IEEE Trans. Pattern Anal. Mach. Intell. (2017)
12. Ren, S., He, K., Girshick, R., Sun, J.: Faster R-CNN: towards real-time object detection with region proposal networks. IEEE Trans. Pattern Anal. Mach. Intell. **39**(6), 1137–1149 (2017)
13. Lin, T.Y., Dollar, P., Girshick, R., et al.: Feature pyramid networks for object detection (2016)
14. Long, J., Shelhamer, E., Darrell, T.: Fully convolutional networks for semantic segmentation. IEEE Trans. Pattern Anal. Mach. Intell. **39**(4), 640–651 (2015)

Teaching Reform and Research of Data Structure Course Based on BOPPPS Model and Rain Classroom

Yongfen Wu(✉), Zhigang Li, Yong Li, and Yin Liu

School of Command and Control Engineering, Army Engineering University of PLA, Nanjing 210007, Jiangsu, China

Abstract. Data structure is the core course for computer science majors. How to improve their 'computational thinking' ability is crucial and challenging in this course. To optimize the teaching effect, a classroom teaching model is proposed. This model combines Yu Classroom online teaching tools, Chinese university MOOC teaching resources and BOPPPS model. The empirical experiment was implemented and the results show that the comprehensive application of this teaching model helps to cultivate students' high-level cognitive ability of 'analysis, evaluation and creation'. It is proved the validity of this teaching model improve students'.

Keywords: BOPPPS model · Rain classroom · Teaching design · Data structure

1 Introduction

Data structure course is a professional basic course which is an important foundation for courses such as compilers principles, operating systems and database principles [1]. These courses not only use the basic knowledge of data structures, but also students' programming skills and computational thinking skills which are more important. Therefore, the teaching reform based on ability training is an important aspect of the teaching reform of this course [2]. How to reform the teaching of data structure course? Many teachers have tried various popular elements, such as "flipped classroom [3, 4]", "two-point teaching [5, 6]", "BOPPPS model [7, 8]", ", and IT tools such as Rain Classroom [9–11] and MOOC resources of Chinese universities, etc., all of which have made different progress. According to the teaching practice of our school's data structure course, combined with rain class, the paper will introduce the teaching design and application of the BOPPPS model in detail.

2 BOPPPS Model

BOPPPS model from North America is used by many universities and institutions in the world, compared with the traditional teaching model, which emphasizes teaching effect, classroom efficiency and teaching benefits [12]. BOPPPS model divides the teaching

J. Zeng et al. (Eds.): ICPCSEE 2021, CCIS 1452, pp. 410–418, 2021.
https://doi.org/10.1007/978-981-16-5943-0_33

process into six stages. The biggest feature of BOPPPS model is student's participatory learning process and student-centered, emphasizes students' full participation in learning and timely feedback information. BOPPPS teaching model is shown in Table 1.

Table 1. BOPPPS teaching model and its content

No	Teaching link	Teaching content
1	Bridge in	Arouse learning motivation and introduce topics
2	Objective	List the learning goals learners should achieve
3	Pre-test	Evaluate the knowledge and ability of learners
4	Participatory learning	Encourage learners to actively learn and participate in teaching activities
5	Post-assessment	Use effective methods and evaluate teaching effects
6	Summary	Summarize

BOPPPS model emphasizes participatory learning, which is in line with the "OBE" achievement-orient and "student-centered" teaching philosophy. The introduction of the BOPPPS model framework into the data structure allows students to not only "retain" more knowledge through participatory learning, but also to enhance high-level cognitive abilities through the process of thinking, thereby enhancing computational thinking skills.

BOPPPS model emphasizes participatory learning, which is in line with the "OBE" result-oriented and "student-centered" teaching philosophy. The introduction of the BOPPPS model framework into the data structure allows students to not only "retain" more knowledge through participatory learning, but also to enhance high-level cognitive abilities through the process of thinking, which enhances computational thinking skills.

Taking 45 min as an example, Table 2 shows BOPPPS framework for one lesson in data structure courses. Among them, each link can be used flexibly according to actual needs.

Table 2. BOPPPS framework for a class of about 45 min

No	BOPPPS link	Suggested duration
1	Bridge in	3–5 min
2	Objective	2–3 min
3	Pre-test	3–5 min
4	Participatory learning	25–30 min
5	Post-assessment	3–5 min
6	Summary	2–5 min

3 Rain Classroom

Rain Classroom is a smart teaching tool which is jointly developed by Xuetang Online and Tsinghua University Online Education, implements the student-centered educational concept through functions such as scan code sign-in, Q&A barrage, and data analysis.

Rain Classroom improves students' sense of participation and the interaction between teachers and students through answering questions on everyone's mobile phone. At the same time, real-time feedback, statistical student data analysis and other functions can quickly collect student feedback, which can be used in the pre-test and post-test of BOPPPS model and is very convenient and flexible (Fig. 1).

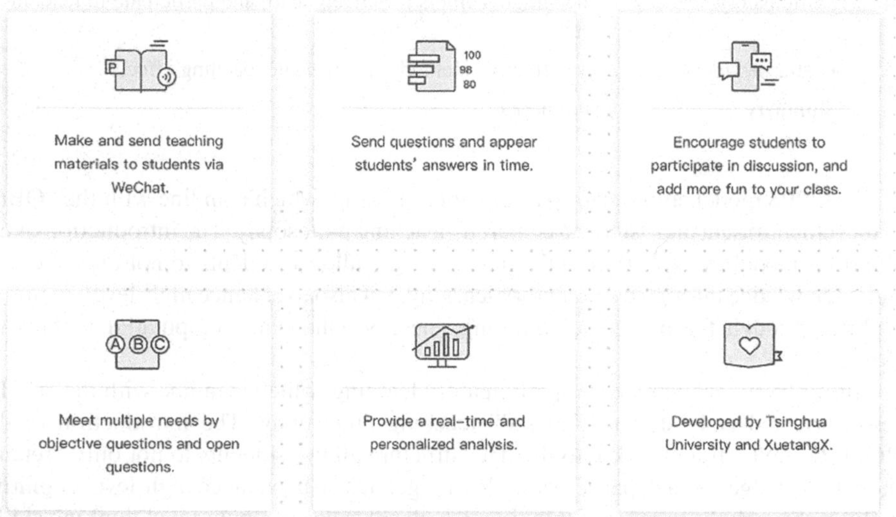

Fig. 1. Features of rain classroom

4 The Teaching Reform of Data Structure Course Based on BOPPPS Model and Rain Classroom

Taking the learning of Huffman tree (Lecture 15) in Data Structure Course as an example, the paper discuss the practical application of BOPPPS teaching mode in the teaching of this course. In order to improve students' enthusiasm for participatory learning, and to reproduce the innovative process of scientific thinking, the teaching process of BOPPPS is designed, as shown in Fig. 2 below.

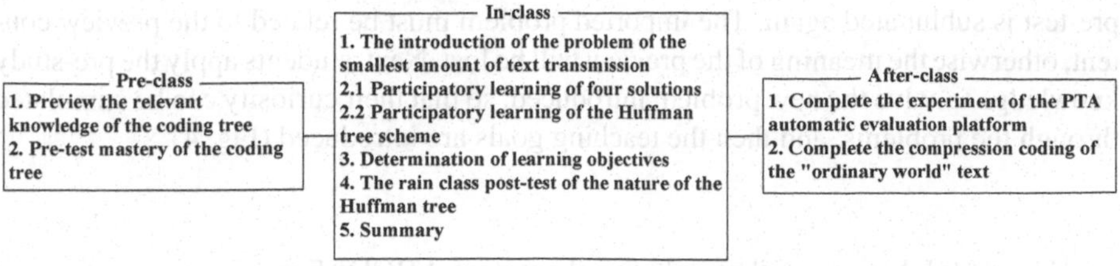
Pre-class
1. Preview the relevant knowledge of the coding tree
2. Pre-test mastery of the coding tree
In-class
1. The introduction of the problem of the smallest amount of text transmission
2.1 Participatory learning of four solutions
2.2 Participatory learning of the Huffman tree scheme
3. Determination of learning objectives
4. The rain class post-test of the nature of the Huffman tree
5. Summary
After-class
1. Complete the experiment of the PTA automatic evaluation platform
2. Complete the compression coding of the "ordinary world" text

Fig. 2. The teaching process of BOPPPS

The steps of BOPPPS model should not be fixed, and the steps of participatory learning should be designed reasonably according to the teaching needs. An instructional design that can induce students to think and guide students to actively participate in learning should be tried. BOPPPS model only gives a way to implement participatory active learning. In Huffman's lecture, our initial step is to preview before class.

4.1 Pre-class

Preview Before Class. In this step, the teacher posts learning tasks through the WeChat group, and asks students to submit mind maps and pre-test results. The pre-study step is very important, and the time should not be too long, otherwise the enthusiasm for pre-study will be discouraged. While cultivating students' self-learning ability, we enable students to listen to lectures in the classroom with questions and in-depth thinking, which is more effective.

Pre-test Before Class. We believe that students must take pre-tests after their preparation, otherwise it is not known whether the students have completed the preparation goals we set (Fig. 3).

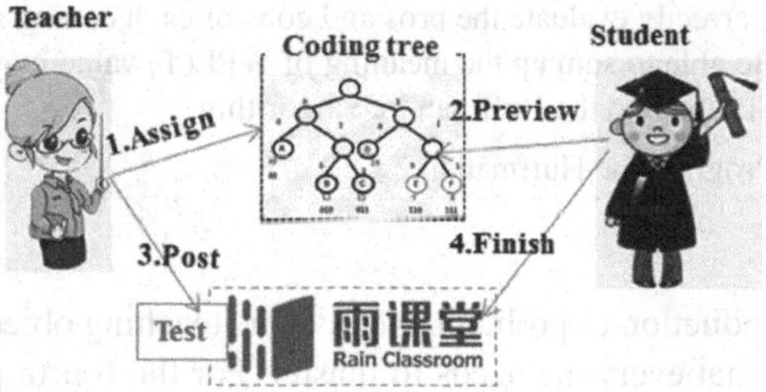

Fig. 3. Examples of pre-class preview and pre-test

4.2 In-class

Bridge in. Before introducing the problem, we must comment on the pre-test assignments for the preview. Through the commentary, the learning effect of the students' pre-test is sublimated again. The imported problem must be related to the preview content, otherwise the meaning of the preview will be lost. Next, students apply the pre-study knowledge to solve the new problem introduced, so that their curiosity can be stimulated through the problems, and then the teaching goals are introduced (Fig. 4).

[Example] Assuming that only six characters ABCDEF may appear in the communication, then a file "ABC......" with one hundred thousand characters needs to be transmitted. In order to improve transmission efficiency, you can try to select or design a coding scheme that minimizes the amount of transmitted data after coding.

Fig. 4. Bridge in

Objectives. BOPPPS model emphasizes the need for detailed and measurable teaching goals. The establishment of learning goals must be based on four elements [6]: Who? What to learn? Under what premise? How much did you learn? Follow-up instructional design should be developed around the teaching objectives. In addition, specific and measurable learning goals need to be written using bloom verbs [6], and there should be no unmeasurable words such as "understanding" and "mastering". The teaching objectives of the Huffman tree lecture are set as follows (Table 3):

Table 3. Learning objectives for the lecture of the Huffman tree

Learn link	Learning objectives
Pre-class	1. Describe the characteristics of the coding tree
In-class	2. Correctly evaluate the pros and cons of each coding scheme 3. Be able to sum up the meaning of WPL(T) value by yourself 4. Summarize the Huffman tree algorithm
After-class	5. Program the Huffman tree

Through the introduction of problem, we give the teaching objectives of the 2nd and 3rd points in Fig. 5 that everyone needs to master. For the fourth point in Fig. 5, after completing the 4 solutions in the next step of participatory learning, when introducing the Huffman tree, we will give the fourth point of teaching goal.

Participatory Learning. This is the core stage of classroom implementation, where students take the initiative to learn the core content of this lecture. And there are 4 kinds of solutions. By proposing solution 1, the teacher guides students to optimize to solution 2, and then from solution 2 to solution 3. At this point, the students' potential has been basically tapped. Then, the teacher proposes option 4 for students to evaluate. In the next step, according to all the 4 schemes, let the students summarize the optimization goal, and the teacher leads the structure of the Huffman tree. Before the construction of the Huffman tree, the teacher gave the teaching objective 4 in Fig. 5 (summarize the algorithm of the Huffman tree).

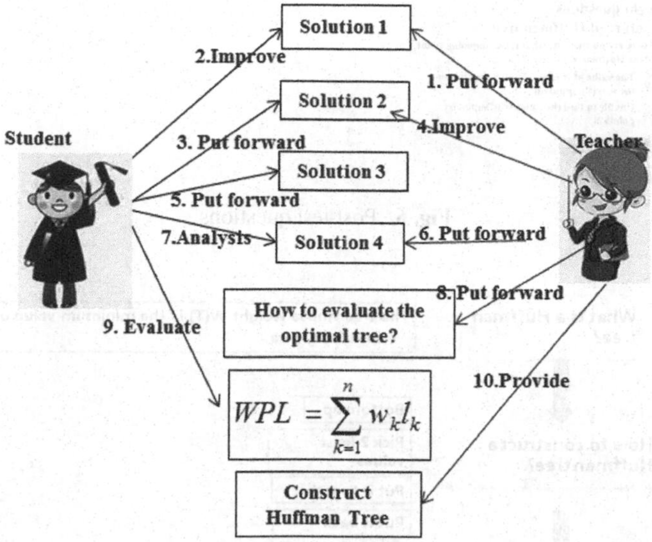

Fig. 5. The process of active participatory learning in this lecture of the Huffman tree

Post-test. After the main body of teaching is completed, we test whether students have achieved our learning goals. We use Rain Classroom to implement post-testing. The specific method is: use Rain Classroom to publish the post-test questions; use Rain Classroom statistics to grasp everyone's learning level, and trigger students to think again; if there is no time, the post-test can also be completed by the students after class (Fig. 6).

Summary. Next, we use the way of asking questions to summarize, what is the Huffman tree? How to construct the Huffman tree? How to implement the Huffman tree? The specific process is shown in Fig. 7.

4.3 After-Class

Assign after-class tasks: After class, you program to implement the coding and decoding of the Huffman tree. You can choose offline experiment (submit the experiment report) or online experiment (PTA automatic evaluation) (Fig. 8).

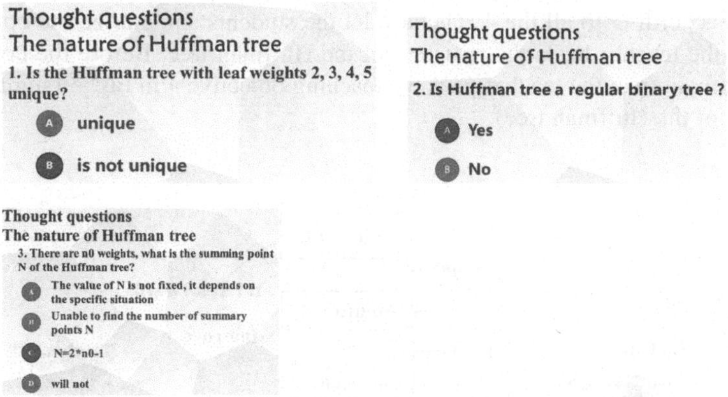

Fig. 6. Post-test questions

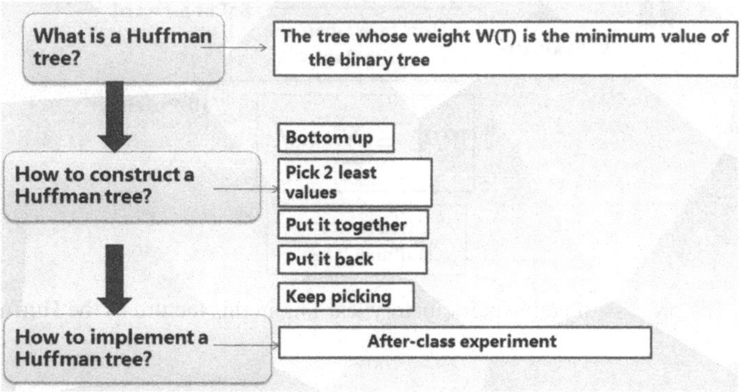

Fig. 7. Summary

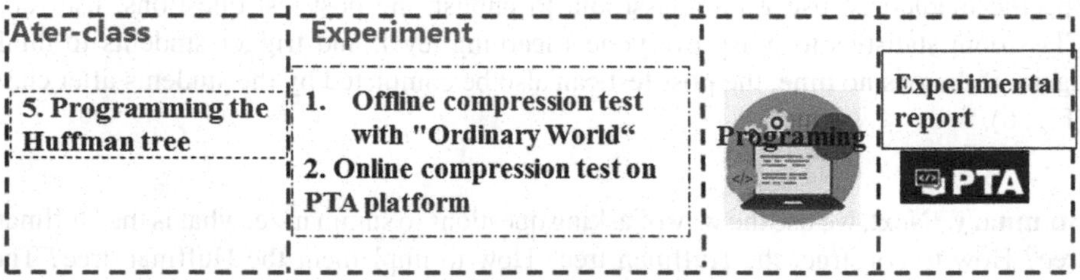

Fig. 8. After-class tasks

4.4 Implementation Effect Evaluation

Through the questionnaire survey of student satisfaction, the completion of the PTA online program, the statistics of the failure rate, etc., the data are better than the data of previous years. Therefore, we believe that the combined use of BOPPPS model and Rain Classroom tools is effective in data structure courses (Fig. 9).

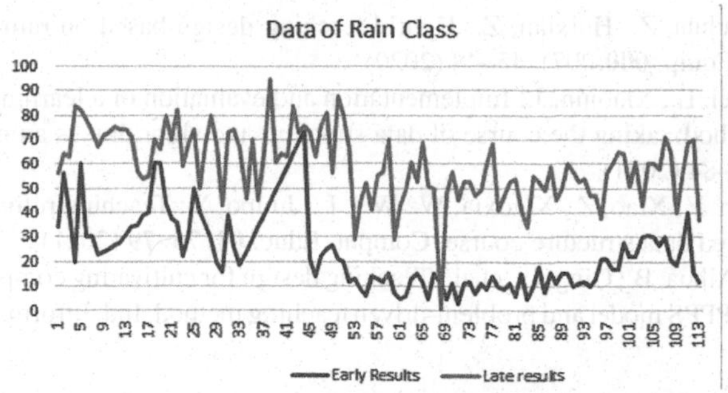

Fig. 9. Before and after comparison of Rain Classroom results of 199 students of grade

5 Conclusion

Taking the Huffman tree as an example, this paper proposes a teaching design scheme based on the integration of BOPPPS and Rain Classroom. Emphasizing participatory learning not only allows students to retain as much knowledge as possible, but also trains students in scientific thinking. The part of learning objectives helps students to take questions into the next part of participatory learning; the pre-test part helps teachers to understand students' level of preparation for this lecture; active participation in learning activities is the core which will be carried out, through a series of thinking, questions, and discussions; the post-test link will deepen and consolidate the learning of this lecture; the summary part will stimulate students to think further while reviewing this lecture. Teaching practice shows that the comprehensive application of the above teaching methods helps to cultivate students' high-order cognitive ability of "analysis, evaluation, and creation", and then cultivate their computational thinking ability.

References

1. Guowen, T.: Data Structure Curriculum Design, pp. 1–18. Tsinghua University Press, Beijing (2019)
2. Hongyi, R., Wanjuan, S.: Data Structure Curriculum Design: C Language Description. 2nd edn, pp. 1–34. Electronics Industry Press, Beijing (2019)
3. Wei, W., Bai, H.-T., Liu, W., Jiang, N.: The research on teaching mode of combining BOPPPS model with flipped classroom. Educ. Teach. Forum **10**, 1674–9324 (2018)
4. Ying, C., Mingzhu, W., Li, L.: The application of "half-flip" mode in the teaching of "data structure" course in higher vocational education. Comput. Educ. **02**, 163–166 (2019)
5. Divide the classroom: a new exploration of college classroom teaching reform. Fudan Educ. Forum **12**(5), 5–10 (2014)
6. Ruiling, F.: BOPPPS and "Diversification" hybrid teaching design and offline and online implementation. Lectures on Higher Education and National Training (2020)
7. Weiwei, C., Aihua, B., Qing, L., et al.: Teaching design based on BOPPPS model and problem-driven teaching method to cultivate computational thinking. Ind. Inform. Educ. **000**(006), 8–11 (2014)
8. Xin, L., Bin, Z., Xuzhou, L., Bo, Z.: Teaching reform practice of "Data Structure Experiment" based on BOPPPS model. High Educ. J. **18**, 64–66+70 (2020)

9. Aihua, L., Yuhua, Z., Huixian, Z.: Hybrid teaching design based on rain classroom. Chin. Mod. Educ. Equip. **000**(007), 45–48 (2020)
10. Jiana, M., Wei, L., Xiaojun, J.: Implementation and evaluation of a learning-centered hybrid teaching method: taking the course of data structure and algorithm as an example. Comput. Educ. **02**, 80–84 (2021)
11. Sun Lianshan, F., Xiao, Z., Xiaoxia, W., Wei, L., Junpo, Y.: Teaching reform and practice of output-oriented data structure course. Comput. Educ. **01**, 75–79 (2021)
12. Weiwei, C., Aihua, B., Qing, L., et al.: Teaching design for cultivating computational thinking based on BOPPPS model and problem-driven teaching method. Ind. Inform. Educ. **6**, 8–11+18 (2014)

An Empirical Study on Teachers' Informationized Teaching Ability in Higher Vocational Colleges

Liya Liu(✉), Yuanyuan Zhang, and Min Li

Sanya Aviation and Tourism College, Sanya 572000, Hainan, China

Abstract. Information-based teaching plays a positive role in enhancing the interaction between teachers and students in vocational colleges, motivating students' learning interests and improving teaching efficiency. This paper aims at analyzing the reliability, validity and factors of the questionnaire designed for evaluating teachers' ability of educational informationization in higher vocational colleges. The research results show that the Cronbach's Alpha coefficient of the designed questionnaire is 0.899, which indicates that the reliability test results are good and can be generalized. KMO value is 0.886, Bartlett's sphericity test value is 2492.773, and the significance level is less than 0.05. It is shown the overall level of validity is good, while the local reliability and validity need to be strengthened. In aspect of educational informationization practice, male teachers and female teachers behave different. Moreover, most teachers give low scores to informationization environment. Finally, analytical results show information-based teaching is important and the classroom teaching efficiency is improved with it.

Keywords: Higher vocational colleges · Informatization · Teaching ability · Empirical research

1 Introduction

China's educational informatization has been developing since the 1990s, when it focused on the construction of school infrastructure. Information technology has been continuously developing in depth in the teaching practice, research and teaching management in colleges and universities. The Circular of the State Council on the Printing and Distribution of the Implementation Plan of the National Vocational Education Reform points out that we should adapt to the development needs of "Internet + vocational education", use modern information technology to improve teaching methods, and promote the construction and widespread application of virtual factories and other online learning Spaces. The Ministry of Education and the Ministry of Finance jointly issued the Opinions on the Implementation of the Plan for the Construction of High-level Higher Vocational Schools and Majors with Chinese Characteristics, which proposed to build smart classrooms and virtual factories, and widely apply a mix of online and offline teaching. It can be seen that the integration of information technology and education is

J. Zeng et al. (Eds.): ICPCSEE 2021, CCIS 1452, pp. 419–433, 2021.
https://doi.org/10.1007/978-981-16-5943-0_34

an urgent task of teaching reform, and higher requirements are required for teachers' information teaching ability in higher vocational colleges.

Foreign university teachers in the classroom implementation of information teaching level is relatively high, more representative are the UK and Japan. In 1998, the United Kingdom took the lead in setting the most basic requirements for teachers to apply ICT to teaching. In 2007, Japan put forward requirements for teachers' informationized teaching ability from five dimensions.

In the China Journal Full-Text Database (CNKI), the research literature on "information-based teaching" was retrieved from 1998 and entered the research peak in 2019 according to the law of Bradford discrete literature. In 2020, 5,176 papers were published, with Central China Normal University and other institutions playing a leading role in the research.

Wenjun Su et al. analyzes the specific connotation of the flipped classroom model and the specific connotation of teachers, and explains the training strategies of higher vocational teachers from five aspects [1]. GE Wenshuang et al. combined a factor model with four dimensions including "awareness, literacy, strategy, and research" through exploring and testing factor analysis methods based on testing and sampling data investigation [2]. Liu Lei et al. designs the training model of pre-service teachers' information-based teaching ability, and carried out practical exploration in combination with specific subject contents so as to improve the information-based teaching ability of pre-service teachers [3]. Gong Lihua starts with the specific elements of information-based teaching ability in the ero of educatonal informatization 2.0 [4].

Our society has an increasing demand for skilled personnel with professional ability nowadays. Vocational education is an important part of education in China. Exploring the factors that influence the development of teachers' informationization teaching ability in higher vocational colleges plays an important role in promoting the further development and improvement of this ability comprehensively and efficiently. It should be the core content in the reform and development of informationized teaching in vocational colleges.

2 Research Method and Process

2.1 Investigation Object, Method and Content

Psychologist Bandura put forward the three element interaction model, that is, individual cognition, behavior and environment interact with each other [5]. Based on this model, this study conducted a questionnaire survey on the informationization ability of teachers in higher vocational colleges in Hainan Province through random sampling. A total of 205 questionnaires were sent out and 201 were effectively received with an effective recovery rate of 98.05%. The basic situation of the respondents is most of them are female teachers, accounting for 64.68%. Most of them are aged 31–40 years old, accounting for 53.73%. Most of them had master's degree, accounting for 55.72%.

On the basis of analyzing the informationization level of teachers in higher vocational colleges, the questionnaire adopts attitude measurement Likert self-leveling 5-point scale method. Each item is scored as 5 points, 4 points, 3 points, 2 points and 1 point respectively according to "completely consistent", "consistent", "generally consistent", "not quite consistent" and "completely inconsistent", of which 3 is a critical value, and the higher the score is, the higher the ability and consciousness of information level is [6]. Frequency are shown in Table 1.

Table 1. Frequency table.

Variables	Values				
Item	1	2	3	4	5
V1	1	14	38	111	37
V2	0	1	17	98	85
V3	0	4	15	126	56
V4	1	9	25	107	59
V5	0	2	27	137	35
V6	0	8	17	133	43
V7	1	26	86	73	15
V8	0	8	70	106	17
V9	0	11	83	90	17
V10	1	13	81	88	18
V11	1	8	63	104	25
V12	1	15	82	90	13
V13	0	21	84	84	12
V14	1	1	25	89	85
V15	7	21	91	61	21
V16	5	16	86	70	24
V17	8	30	108	42	13
V18	6	19	116	47	13
V19	10	24	97	58	12
V20	10	20	110	51	10

V1: Paying attention to the development of information-based teaching.
V2: Informatization is very important to teachers' teaching.
V3: Information literacy is a necessary component of teachers' professional literacy.
V4: The importance of using information teaching methods in teaching.
V5: Being conscious of applying information teaching means in classroom teaching.
V6: Being able to screen and integrate online teaching resources.
V7: Being clear about the the related theory of information teaching.
V8: Being able to use information means to improve classroom activity.
V9: Being able to fully explore mixed teaching.
V10: Being able to select the appropriate technology in teaching process.
V11: Being able to communicate and cooperate with classmates and teachers.
V12: Master of teaching intervention in the information teaching environment.
V13: Good command of information technology in the aspects of knowledge.
V14: The need to attend the training that a few promote information- based teaching ability.
V15: The degree of satisfaction with the campus network construction.
V16: The degree of satisfaction with multimedia teaching equipment.
V17: Funding support for information-based teaching reform.
V18: Information-based teaching related incentive measures.
V19: Technical support and service.
V20: The existing ways to improve teachers' information-based teaching.

In content, the project design is measured from the angle of two level and three dimensions. Namely, from the level of informatization of teachers' teaching ability and the informationization teaching ability to motivate teachers to improve the external conditions of two level to measure teachers' information awareness, information of teaching in higher vocational colleges teaching ability and the university informatization external condition judgment. The three dimensions relate to the importance of teachers' understanding of informationized teaching, their attention to new ideas of informationized teaching, their design of informationized teaching, their acquisition and processing of teaching resources, their operation of teaching tools and software, campus network construction, multimedia teaching equipment, and their support for informationized teaching. In content, this paper measures the teachers' awareness of informatization teaching in higher vocational colleges, the ability to implement informatization teaching and the judgment of the external conditions of informatization teaching in colleges and universities from two aspects: teachers' own informatization teaching ability level and the external conditions that promote the improvement of teachers' informatization teaching ability.

As can be seen from Table 1, 91.04% of the interviewees believe that informatization is very important to teachers' teaching and that they can use informatization in teaching to improve the teaching effect. Most of the interviewees have a low evaluation on the informationized teaching environment of the school.

The correlation between the original variables is analyzed by using two correlation coefficient matrices to investigate their relevance. In order to facilitate typesetting, only the correlation matrix of the first six variables is given.Correlations are shown in Table 2.

Table 2. Correlations.

		V7	V8	V9	V10	V11	V12
V7	Pearson correlation	1	0.530**	0.489**	0.282**	0.470**	0.203**
	Sig. (2-tailed)		0	0	0	0	0.004
	N	201	201	201	201	201	201
V8	Pearson correlation	0.530**	1	0.617**	0.411**	0.491**	0.168*
	Sig. (2-tailed)	0		0	0	0	0.017
	N	201	201	201	201	201	201
V9	Pearson correlation	0.489**	0.617**	1	0.576**	0.600**	0.282**
	Sig. (2-tailed)	0	0		0	0	0
	N	201	201	201	201	201	201
V10	Pearson correlation	0.282**	0.411**	0.576**	1	0.525**	0.243**
	Sig. (2-tailed)	0	0	0		0	0
	N	201	201	201	201	201	201

(*continued*)

Table 2. (*continued*)

		V7	V8	V9	V10	V11	V12
V11	Pearson correlation	0.470**	0.491**	0.600**	0.525**	1	0.373**
	Sig. (2-tailed)	0	0	0	0		0
	N	201	201	201	201	201	201
V12	Pearson correlation	0.203**	0.168*	0.282**	0.243**	0.373**	1
	Sig. (2-tailed)	0.004	0.017	0	0	0	
	N	201	201	201	201	201	201

** Correlation is significant at the 0.01 level (2-tailed).
* Correlation is significant at the 0.05 level (2-tailed).

The results show that there is a strong or weak correlation between these variables, which indicates that there is information overlap between these variables, and it is necessary to conduct factor analysis on the data.

2.2 Reliability Analysis

Reliability is the extent to which Random error is excluded from the survey value. If the Random error is 0, the survey is completely credible. In academia, Cronbach's α coefficient is usually used to judge the internal consistency reliability of a questionnaire, and then to evaluate the reliability of a questionnaire. Its calculation formula is as follows:

$$\alpha = \frac{k}{k-1}\left(1 - \frac{\sum S_i^2}{\sum S_X^2}\right) \tag{1}$$

α is reliability coefficient. S_i^2 is the variance of the answer to the question i of all respondents. S_X^2 is the variance of all respondents and answers to all questions and k is the number of questions.

The higher the reliability coefficient, the better the reliability of the questionnaire. If the reliability coefficient is above 0.8, the questionnaire is considered to have a high internal consistency. In practical application, Cronbach's α value should be at least greater than 0.5, preferably greater than 0.7 [6].

The reliability analysis of this scale shows that there are no invalid missing values in the Case Processing Summary. The reliability statistics show that Cronbach's αcoefficient is 0.899, and the coefficient is above 0.8. The reliability of the overall scale is good, close to 0.9, indicating that the reliability test results are good and can be popularized. The results of sensitivity analysis are shown in Table 3. The results show that the Cronbach's α coefficient does not increase after deleting any item, which indicates that all the constituent items can be retained.

Table 3. Item-total statistics.

Item	Scale mean if item deleted	Scale variance if item deleted	Corrected item-total correlation	Cronbach's alpha if item deleted
V1	69.13	75.397	0.487	0.896
V2	68.65	76.8	0.509	0.895
V3	68.81	76.424	0.554	0.894
V4	68.91	76.012	0.456	0.896
V5	68.96	75.943	0.652	0.892
V6	68.93	78.029	0.379	0.898
V7	69.6	74.891	0.524	0.895
V8	69.32	75.118	0.62	0.892
V9	69.41	74.994	0.596	0.893
V10	69.43	74.947	0.563	0.893
V11	69.26	75.803	0.508	0.895
V12	69.48	74.821	0.589	0.893
V13	69.54	74.609	0.597	0.893
V14	68.7	80.43	0.155	0.904
V15	69.64	74.082	0.507	0.895
V16	69.52	74.821	0.478	0.896
V17	69.87	72.797	0.632	0.891
V18	69.77	74.04	0.592	0.893
V19	69.79	72.749	0.614	0.892
V20	69.82	73.648	0.588	0.893

2.3 Information Enrichment Using Factor Analysis

The Basic Mathematical Model of Factor Analysis. Factor analysis is to find out a few random variables that can synthesize all variables by studying the internal dependence of the covariance matrix among multiple variables [7]. These random variables are not measurable and are usually called factors. Factor analysis is to replace all the original variables with a few random variables to solve the original problem.

Set N samples and P indicators as random vectors, and the common factor to be sought is, then the model is

$$X_1 = a_{11}F_1 + a_{12}F_2 + \cdots a_{1m}F_m + \varepsilon_1 \tag{2}$$

$$X_2 = a_{21}F_1 + a_{22}F_2 + \cdots a_{2m}F_m + \varepsilon_2 \tag{3}$$

$$\vdots$$

$$X_3 = a_{m1}F_1 + a_{12}F_2 + \cdots a_{1m}F_m + \varepsilon_m \quad (4)$$

It's called the factor model. The matrix $A = (a_{ij})$ is called factor load matrix, and the load of the variable i on the factor j, a_{ij} is called factor load. In essence, it is the correlation coefficient between the variable X_i and the common factor F_j, which indicates the degree to which the variable X_i depends on the common factor F_j and reflects the importance of the variable X_i to the common factor F_j. Special factors ε represent influencing factors other than common factors, which are ignored in practical analysis.

The Factor Analysis. The structure validity analysis of 20 questions was conducted to explore the structure of the questionnaire. The method of structure validity analysis is factor analysis. Before factor analysis, it is necessary to check whether the KMO value is greater than 0.7 and whether Bartlett's sphericity test is significant. As can be seen from Table 4, KMO value is 0.886, greater than 0.5. It can be seen that this sample is suitable for factor analysis.

Table 4. KMO and Bartlett's test.

Kaiser-Meyer-Olkin measure of sampling adequacy		0.886
Bartlett's Test of Sphericity	Approx. Chi-Square	2492.773
	df	190
	Sig.	0.000

Principal component analysis method is used to carry out factor analysis on the topic. Based on the principle that the eigenvalue is greater than 1, three principal components are proposed systematically. As shown in Table 5, it includes extracted factors, factor loading amount, characteristic value and cumulative variance contribution rate, etc.

Table 5. Total variance explained.

Component	Initial eigenvalues			Extraction sums of squared loadings		
	Total	Variance%	Cumulative %	Total	Variance%	Cumulative %
1	7.2	35.999	35.999	7.2	35.999	35.999
2	3.589	17.947	53.947	3.589	17.947	53.947
3	1.695	8.477	62.424	1.695	8.477	62.424
4	0.983	4.914	67.337			
5	0.832	4.158	71.495			

(*continued*)

Table 5. *(continued)*

Component	Initial eigenvalues			Extraction sums of squared loadings		
	Total	Variance%	Cumulative %	Total	Variance%	Cumulative %
6	0.769	3.846	75.341			
7	0.612	3.059	78.399			
8	0.576	2.878	81.277			
9	0.513	2.565	83.842			
10	0.474	2.369	86.211			
11	0.433	2.167	88.378			
12	0.39	1.948	90.325			
13	0.38	1.902	92.228			
14	0.349	1.747	93.974			
15	0.296	1.481	95.456			
16	0.26	1.299	96.755			
17	0.206	1.029	97.783			
18	0.194	0.971	98.754			
19	0.154	0.772	99.527			
20	0.095	0.473	100			

If Scree Plot is selected in the "Extraction" sub-dialog box, the relationship between factors in the above analysis results and characteristic roots will be shown in the form of Fig. 1 [8].

As can be seen from Fig. 1, although the characteristic root sizes of the fourth, fifth and sixth common factors are less than 1, they are very close to each other and significantly higher than the 7–20 common factors, indicating that the importance of the fourth, fifth and sixth common factors is slightly higher than that of the 7–20 common factors.

The load graph is plotted based on the component matrix coefficients, as shown in Fig. 2. Table 6 is Rotated Component Matrix. Coefficients below 0.5 can be ignored basically.

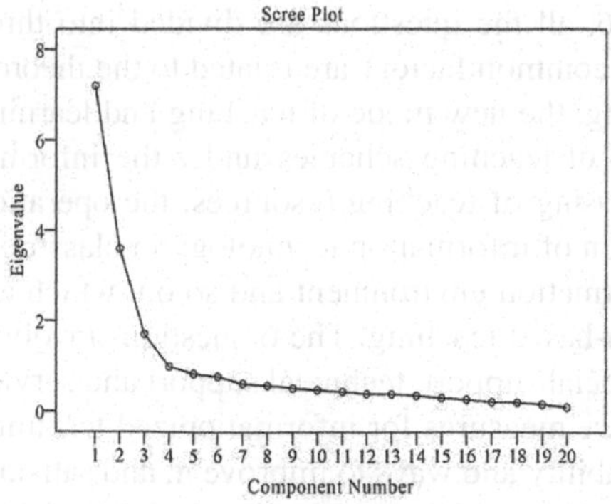

Fig. 1. Scree plot

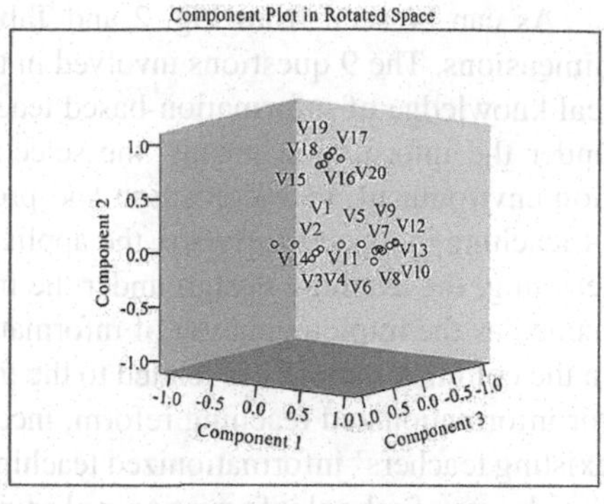

Fig. 2. Component plot in rotated space

Table 6. Rotated component matrix[a].

Title	Component		
	1	2	3
V13	**0.825**	0.139	0.052
V12	**0.813**	0.149	0.040
V9	**0.798**	0.121	0.116
V10	**0.756**	0.076	0.169
V8	**0.748**	0.089	0.273
V7	**0.702**	0.085	0.146
V6	**0.616**	−0.049	0.106
V11	**0.567**	0.127	0.239
V5	**0.556**	0.169	0.541
V17	0.138	**0.924**	0.090
V19	0.121	**0.888**	0.147
V18	0.109	**0.878**	0.104
V20	0.189	**0.853**	0.028
V15	0.035	**0.794**	0.162
V16	0.032	**0.794**	0.084
V2	0.194	0.166	**0.771**
V3	0.375	0.047	**0.756**
V4	0.318	0.075	**0.609**
V1	0.347	0.107	**0.580**
V14	-0.180	0.095	**0.571**

Extraction Method: Principal Component Analysis.
Rotation Method: Varimax with Kaiser Normalization.
[a]Rotation converged in 5 iterations.

As can be seen from Fig. 2 and Table 6, all the questions are divided into three dimensions. The 9 questions involved in the common factor1 are related to the theoretical knowledge of information-based teaching, the new mode of teaching and learning under the information means, the selection of teaching schemes under the information environment, the acquisition and processing of teaching resources, the operation of teaching tools and software, the application of information technology in classroom teaching, the teaching design under the information environment and so on, which are named as the implementation of information-based teaching. The 6 questions involved in the common factor2 are related to the financial support, technical support and service for informationized teaching reform, incentive measures for informationized teaching, existing teachers' informationized teaching ability and ways to improve it, and satisfaction degree of school informationized environment, which are named as the information-based teaching conditions. The 5 questions involved in the common factor3 are related to the the importance of informatization in teachers' opinion, whether teachers use informatization means to improve information literacy, reduce the teaching burden, pay attention to the new theory of informatization teaching, and whether they need to participate in the training of informatization ability improvement, which are named as the awareness of informationization teaching.

The Information Teaching Ability Analysis Based on the Results of Factor Analysis. Based on the characteristics of teachers in different higher vocational colleges, the differences in the scores of each factor are analyzed, and the influencing factors of the information teaching ability of teachers with different characteristics in higher vocational colleges are finally identified, so as to improve the information ability of teachers in a targeted way.

According to the gender of the interviewees, the t-test is used to analyze whether female and male teachers have different scores in various factors.

Table 7. Group statistics.

Common factor	V1	N	Mean	Std. Deviation	Std. Error mean
Implementation of information-based teaching	Female	130	−0.167	1.024	0.090
	Male	71	0.306	0.882	0.105
Conditions of information-based teaching	Female	130	0.028	0.906	0.079
	Male	71	0.051	1.158	0.137
Awareness of information-based teaching	Female	130	0.069	0.937	0.082
	Male	71	−0.127	1.102	0.131

Table7 and Table8 are the mean values and test results of each factor under different gender conditions. $T = -3.279$, $df = 199$, $Sig. = 0.001 < 0.05$, it can be considered that female teachers and male teachers in vocational colleges have significant differences

Table 8. Independent sample test.

Common factor		Levene's test for equality of variances		T-test for equality of means				
		F	Sig.	t	df	Sig. (2-tailed)	Mean difference	Std. Error difference
Implementa-tion of information-based teaching	Equal variances assumed	3.073	.081	−3.279	199	0.001	−0.47254	0.14409775
	Equal variances not assumed			−3.426	163.075	0.001	−0.47254	0.13792418
Conditions of information-based teaching	Equal variances assumed	5.292	0.022	0.538	199	0.591	0.07957	0.14783243
	Equal variances not assumed			0.501	117.477	0.617	0.07957	0.15874175
Awareness of information-based teaching	Equal variances assumed	2.906	0.090	1.331	199	0.185	0.19607	0.14728565
	Equal variances not assumed			1.270	125.624	0.207	0.19607	0.15442391

in information teaching implementation factors. The average value of informationized teaching implementation factors of males is much higher than that of females, and the data are relatively concentrated. However, there is no difference in the two factors of the information teaching condition and the knowledge of the information teaching. There is little difference between female and male in informationization condition factor and informationization teaching cognition factor.

3 Conclusions

The results show that informationized teaching has become a very important teaching method which is welcomed by higher vocational students. Teachers will often use the information platform to carry out teaching, management, assessment and independent learning in course teaching. In the implementation of information teaching, there are obvious differences between male and female teachers, male teachers are better than female teachers. Most teachers have a low evaluation of the information environment. A large number of teachers have a higher awareness of information teaching. In the future, vocational colleges should continue to take information-based teaching as the breakthrough point to improve the effectiveness of classroom teaching. On the premise of optimizing the informatization teaching environment, they should focus on realizing the deep integration of information technology and courses, strengthen the top-level design and improve the system and mechanism as the guarantee, so as to effectively improve the teaching effect of classroom. However, there are some problems in curriculum informationization teaching in higher vocational colleges, such as limited network teaching environment, insufficient construction of informationization platform and teaching resources, and insufficient integration of information technology and teaching [9]. In the future, we should continue to take curriculum informationization teaching as the breakthrough point to improve the effectiveness of the curriculum teaching. On the premise of optimizing the informationization teaching environment, we should focus on realizing the deep integration of information technology and teaching, strengthen the top-level design and improve the system and mechanism as the guarantee to effectively improve the education effect of the teaching.

3.1 We Should Continue to Take Informationization Teaching as a Breakthrough to Improve the Effectiveness of Teaching

After more than 30 years of development in China, information technology has been increasingly prominent in its status and role, from its initial use as an "auxiliary means" of teaching to "active use" and then to its use as an "important means" of education and teaching. The Acation Plan for Educational Informatization 2.0 (2018–2022) issued by the Ministry of Education in 2018 regards Teaching application covers all teachers, learning application covers all school-age students, digital campus construction covers all schools, informatization application level is generally improved, information literacy of teachers and students is generally improved, and "Internet + education" platform is built [10].

To promote the in-depth development of curriculum informationization teaching in colleges and universities is not only an active response to the development of The Times, but also a need to solve the development dilemma of the curriculum. For a long time, many colleges and universities have been faced with the "three low" problems of low attendance rate, low head-raising rate and low nodding rate. Higher vocational colleges generally have low admission scores, poor students' basic knowledge, poor self-management ability, difficulties in learning supervision and low retention of interest in learning, so it is more difficult to carry out teaching for students [11]. In the new era, college students have active thinking and strong sense of innovation. Traditional teaching mode and teaching means can hardly meet their psychological needs of pursuing fashion and individuality. The combination of information technology and education in teaching objectives, teaching means and other aspects makes the information-based teaching possible. With the help of information technology, it is beneficial to optimize the environment of ideological and political education and enrich the content of education. Flexible and diverse teaching methods can meet the psychological needs of college students in the new era, to seek innovation and change and pursue individuality, which is conducive to enhancing the effect of teaching.

3.2 We Should Optimize the Information Teaching Environment as the Premise

By optimizing the informationization teaching environment, teachers' awareness of informationization teaching is enhanced, and teachers are encouraged to actively learn and implement the informationization skills. Informationized teaching environment refers to the synthesis of all kinds of circumstances and conditions that affect informationized teaching. The generalized informationized teaching environment includes teaching concept, teaching skills, software and hardware environment, policy, system and atmosphere, etc. In the narrow sense, the informationized teaching environment mainly refers to the software and hardware environment. According to the results of field investigation, the problems in the course information teaching in higher vocational colleges are mainly manifested as: slow network speed, congestion of information platform and other phenomena. The function of information platform is limited, and some application functions are not given full play. While the colleges and universities in the network infrastructure construction, multimedia classrooms, such as digital campus informationization construction has done a lot of work. Because curriculum informationization teaching is still a new thing, it is still in the exploration stage. There are still many shortcomings in the selection of information platform, the mutual cooperation of various information platforms and the creation of network teaching environment [12].

Higher vocational colleges usually adopt the form of "large class teaching". Whether the classroom is equipped with the necessary facilities and equipment to carry out informationized teaching, whether the wireless network is opened, the speed and stability of the network, and the configuration of wired and wireless facilities with the same screen will directly affect the teaching quality and teaching effect. Higher vocational colleges should choose some platforms with large scale, good reputation, stable performance and complete functions to meet the needs of curriculum informationization teaching, improve teaching facilities and create a good environment and atmosphere for informationization teaching according to their own reality.

3.3 We Should Focus on The establishment of Curriculum Informationization Teaching Team

The construction of curriculum informationization teaching team should establish and innovate the team cooperation mechanism, formulate clear development goals, establish effective operation mechanism, and improve the overall informationization teaching ability of the team by means of mutual help and mutual learning, mutual supervision, research and discussion [13].

Curriculum information teaching can not only stay in the general application of information technology in curriculum teaching. The key to enhance the effectiveness of information-based teaching lies in the deep integration of information technology and curriculum. "Deep integration" emphasizes the two-way integration and mutual promotion of education and information technology. It requires more than "application", "combination" and "integration". In the "integration" stage, information technology is fully and deeply integrated into education and triggers the overall and systematic reform of teaching ecology. We should build a new model in the field of classroom teaching, management, student learning, assessment and so on, and realize the deep integration of the two. It involves the reform of teaching idea, teaching management mode, examination and evaluation mode, etc. In the process of reform, we must give full consideration to the school situation, learning situation and teaching situation of higher vocational colleges, give consideration to the nature and characteristics of courses, and pay attention to the details and highlight the key points as well as the overall planning and long-term prospect.

3.4 We Need to Strengthen Top-Level Design and Improve Systems and Mechanisms

At present, the application of information technology to promote curriculum reform in most vocational colleges is still in the spontaneous stage, and there is a lack of top-level design and overall planning. Vocational colleges should formulate relevant policies to improve teachers' informatization teaching, carry out rich and effective informatization training, organize and encourage teachers to participate in informatization teaching ability competitions, and enhance teachers' enthusiasm and thinking on scientific research [14]. The "information research type" teacher is a new requirement for the role orientation of college teachers.

There is not a close connection between vocational colleges, between schools and government agencies and enterprises, and lack of joint construction and sharing mechanism. Thus is specifically reflected in the following aspects: First, the cooperation between vocational colleges, government institutions and enterprises in the aspects of teaching service platform, network construction and information construction standard system is not close enough. There is a lack of clear division of roles between each other, and the role of each institution in the informationized teaching has not been fully played. Second, vocational colleges generally lack of course co-construction and sharing mechanism. For example, in terms of teaching resources, teachers of each course have their own teaching resource package, but the school lacks an overall curriculum "resource pool". There is a lack of system and mechanism of resource co-construction and sharing among vocational colleges. The reasons are not only related to the degree of attention paid by higher vocational colleges, but also related to the ability and level of mutual cooperation. Higher vocational colleges lack unified standards for the construction of digital teaching resources. It involves technical standards, management norms, resource evaluation and review system, etc., which requires the participation and cooperation of various subjects [15]. Most higher vocational colleges are still exploring on their own, and are in the state of "fighting alone" in the course information construction.

Information technology is a double-edged sword. The purpose of applying modern information technology is to promote students' learning, but some students with poor self-control ability will also use mobile information technology to engage in activities unrelated to teaching in class. Information-based platforms such as cloud class and wisdom tree can help teachers quickly understand the learning progress of students, but some students attach importance to acquiring "experience value" instead of caring about the learning content itself. In addition, how to combine quantitative assessment with qualitative evaluation through information platform, and how to better play the role of information platform in students' independent learning and deep learning, these are the contents that need to be further explored in the future.

Acknowledgments. This research was financially supported by the General Project of Higher Education and Teaching Reform Research in Hainan Province of China (Hnjg2020–161), Sanya Aviation and Tourism College Field project in Hainan Province of China (SATC2020JG-07), "Double Leaders" Teacher Party Branch Secretary Studio Construction Project of Hainan University.

References

1. Su, W.: Design of information teaching in higher vocational education. In: Proceedings of 2nd International Conference on Social Science and Higher Education (ICSSHE 2016 V53), pp. 64–66. Atlantis Press (2016)
2. Ge, W.: Developing a validated questionnaire to measure teaching competency for university teachers in digital age **38**(06), 123–128 (2017)
3. Liu, L.: Research on training of pre-service teachers' information-based teaching ability under TPACK theory. Softw. Guide **18**(12), 267–270+276 (2019)
4. Gong, L.: Cultivation and improvement of college teachers' informatization teaching ability in the era of education Informatization 2.0. Lifelong Educ. **9**(5) (2020)
5. Albert, B.: Social Learning Theory, 1st edn. China Renmin University Press, Beijing (2015)
6. Wu, S.: SPSS and Statistical Thinking, 1st edn. Tsinghua University Press, Beijing (2019)
7. Zhang, W.: Advanced Course in SPSS Statistical Analysis, 2nd edn. Higher Education Press, Beijing (2013)
8. Xia, Y.: Essentials and Examples of SPSS Statistical Analysis, 1st edn. Electronic Industry Press, Beijing (2010)
9. Niu, C.: An analysis of the information-based teaching in higher vocational colleges. Theor. Pract. Educ. **40**(15), 24–26 (2020)
10. The Acation Plan for Educational Informatization 2.0 The Ministry of Education. http://www.moe.gov.cn/srcsite/A16/s3342/201804/t20180425_334188.html. Accessed 15 Jan 2021
11. Zhao, L.: Teachers' informationization teaching leadership: connotation, influencing factors and improving path. Chongqing High. Educ. Res. **7**(3), 86–97 (2019)
12. Lu, A.: Promotion paths of information-based teaching in applied colleges and universities under the background of digital campus construction. Educ. Vocat. **9**, 54–58 (2019)
13. Cong, B.: Promotion of informationized teaching ability of physical education teachers in colleges and universities in the Information 2.0 Era. J. Shenyang Sport Univ. **40**(1), 40–48 (2021)
14. Sui, X.: Practical research on influence factors of ICT teaching competence of faculty—a case study of some colleges in Hunan province. Chin. Educ. Technol. **5**, 128–134 (2020)
15. Diao, J.: A strategy analysis of teachers' ICT competence training during COVID-19: based on cases from 28 vocational colleges. E-educ. Res. **42**(1), 115–121 (2021)

Research on Teaching Evaluation of Courses Based on Computer System Ability Training

Wenyu Zhang(✉), Run Wang, Yanqin Tang, En Yuan, Yongfen Wu, and Zhen Wang

Army Engineering University of PLA, Nanjing, China

Abstract. The courses based on computer system ability training has its particularity of training objective. To achieve objective feedback of teaching, a multi-subject and multi-dimensional hierarchical teaching evaluation system is proposed based on the analysis of computer system ability training objectives. Then, the teaching content, implementation, learning experience and teaching effect were evaluated by supervisors, colleague teachers and students. Finally, the multi-dimensional evaluation results show that the evaluation system proposed is effective and provides a reference for teaching evaluation.

Keywords: Multiple evaluation subject · Hierarchical evaluation index · Multi-dimensional evaluation · Computer system ability

1 Introduction

In the new network era characterized by mobile Internet, big data, cloud computing and Internet of things, the demand for computer professionals' ability has been transformed from program development ability to computer system design ability [1, 2]. The course group of computer has been constructed with the guidance of system ability training. However, the teaching evaluation still follows the previous general evaluation index [3], which is not suitable to give effective evaluations about how to reflect system ability training in teaching content [4], teaching implementation and teaching effect of the course. Therefore, it is necessary to discusses the teaching evaluation of computer specialty courses based on system ability training.

2 Problems in Traditional Teaching Evaluation

The purpose of teaching evaluation is to form a reasonable and scientific incentive mechanism and a fair competition environment, ultimately to improve the quality of college education and teaching. Through analysis of traditional teaching evaluation, the following problems are found.

- Unclear evaluation orientation
  The previous teaching evaluation is used for every course regardless of different classification and different target-oriented [5]. However, evaluation indexes should vary

J. Zeng et al. (Eds.): ICPCSEE 2021, CCIS 1452, pp. 434–442, 2021.
https://doi.org/10.1007/978-981-16-5943-0_35

from course to course. In the process of teaching reform and implementation, many courses have adopted new teaching ideas to organize and construct courses according to new training objectives and characteristics [6]. However, teaching evaluation has not been adjusted accordingly [7]. Transformation of teaching concept and ability training which are important to improve teaching quality has been neglected. There is not much difference between course reform and non-reform in teaching evaluation.

- Mismatched evaluation subjects and evaluation indexes
  Most teaching evaluations take supervisors and students as evaluation subjects, without considering the applicability of the evaluation index to the evaluation subject [8, 9]. For example, it is impossible for the supervisor to have a clear understanding of the content organization and ability training objectives of each course. Students' knowledge structure is limited and they are not clear about the course knowledge system. Thus the evaluation results are biased rather than objective and fair.
- Evaluation process is not student-centered
  Teaching evaluation focuses on the evaluation of teachers' teaching and ignores students' learning experience [10]. Some teaching evaluations set students' evaluation of teaching, but only rough statistics of teacher satisfaction and course satisfaction, without specific quantitative statistics of students' learning experience [11].
- Evaluation results focus on score rather than improvement
  The teaching evaluation is scored according to each evaluation index, and the total score is finally given based on the comprehensive scores [12]. Such a single evaluation result turns into grading, reward or punishment for teachers. There is no multi-directional evaluation and lack of closed-loop feedback on the teaching process [13]. Teachers do not know the advantages and disadvantages of their teaching from the evaluation. Therefore, teaching evaluation cannot play a guiding and motivating role.

3 Teaching Evaluation Orientation and Subjects Based on Computer System Ability Training

The course group based on system ability training has its own uniqueness and should not continue to use the traditional teaching evaluation.

3.1 Evaluation Orientation

Teaching evaluation orientation should be consistent with the training objectives of the course. The general idea based on computer system ability training is to take computer system design and implementation as the means, and integrate the courses of computer composition and design, digital logic, operating system and compilation principle [14]. Based on the unified view of system design, a hierarchical, progressive and open course system and practical goal are established to enable students to design a simple but complete computer system in a progressive manner.

The training of systematic ability is progressive in the course group. According to this, the course group of computer major can be divided into levels (see Fig. 1). Courses of different levels correspond to different training objectives and evaluation guidance.

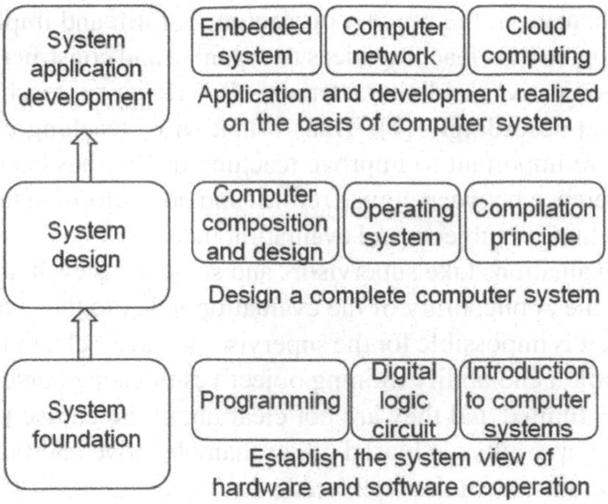

Fig. 1. Curriculum hierarchy division and training objectives based on systematic ability training

3.2 Multiple Evaluation Subjects

The multiple evaluation subjects are formed by not only supervisors and students, but also colleague teachers. The evaluation indexes are multi-dimensional and diversified, thus evaluation results will not be one-sided and lack of objectivity and fairness.

- Colleague teachers
  Colleague teachers are those who are most familiar with the course content be-sides the teacher. They are experts in the organizational structure and key points of the course content. The content of courses based on computer system ability should reflect the advanced, complete, software and hardware integrated features. While the experienced colleague teachers in the same field have more ad-vantages in evaluation of teaching content. For classroom teaching, colleague teachers can make evaluation by attending lectures. For online courses in the form of MOOC [15], colleague teachers can visit the network platform at any time to browse the content for evaluation.
- Supervisors
  From the perspective of objectivity and authority, supervisors should evaluate the implementation process of teaching by attending lectures in class, and can also impart experience such as organization and teaching methods to the teacher. In order to improve the pertinence and effectiveness of teaching evaluation, supervisors can learn about students' pre-class learning activities by communicating with students or visiting the network platform of online courses.
- Students
  Students have direct feelings and experience of teachers' teaching methods, effects and class organization. They can also feedback teaching satisfaction. Questionnaire survey is a common form for students to evaluate the teaching of teachers, and online tools can be used to conduct online voting. The design of questions is of vital importance. Reasonable and effective questions can feed back students' demand for teaching. Thus teachers could improve teaching design and quality purposefully.

4 Hierarchical Teaching Evaluation Evaluation Index

Traditional teaching evaluation indexes are unified. It considers the convenience of management, but ignores the individuality of teaching and characteristics of course. For the courses based on system ability training, teaching evaluation indexes should reflect the characteristics of the course consistent with training objectives.

Quantitative evaluation and qualitative evaluation are combined in the course evaluation index based on system ability training. Two-level evaluation index system is set up. Each level 1 index is subdivided into several level 2 indexes. In level 2 evaluation indexes, evaluation result is evaluation conclusion or evaluation grade instead of scores, such as very satisfied, satisfied, dissatisfied, very dissatisfied or excellent, good, medium and poor.

Table 1 shows level 1 evaluation indexes. There are four level 1 indexes in the evaluation system: (1) teaching contents; (2) teaching implementation; (3) learning experience; (4) teaching effect. The evaluation subject of teaching content is colleague teachers, and that of teaching implementation is supervisors. The evaluation subject of learning experience is students. Teaching effect is evaluated by both supervisors and students.

Table 1. Teaching evaluation level 1 indexes

Number	Evaluation index	Evaluation subject
1	Teaching contents	Colleague teachers
2	Teaching implementation	Supervisors
3	Learning experience	Students
4	Teaching effect	Both supervisors and students

4.1 Level 2 Indexes of Teaching Content Evaluation

Table 2 shows level 2 indexes of teaching content evaluation. Applicability, rigor, advancement, completeness, rationality of structure and relevance are the key points of teaching content evaluation. Evaluation results are recorded in the form of evaluation grade, evaluation conclusion, and yes or no.

Teaching effect varies greatly whether teaching content reflects the relevance or not. For example, processor is the main part of the course "computer composition and design". Traditional textbooks focus on the organization and working principles of CPU. The textbook written by professors Patterson and Hennery related CPU performance to program execution. When students learn this part of the content, they will be able to establish a correlation with program design, operating system and other courses and establish system view of hardware and software cooperation.

Table 2. Level 2 indexes of teaching content evaluation

Level 2 indexes	Description	Types of evaluation results
Applicability	Whether teaching content is applicable to teaching object	Yes or no
Rigor	Whether teaching content is scientific and rigorous	Yes or no
Advancement	Teaching content include forefront knowledge of subject. Advanced teaching content and cases are selected. Obsolete content is eliminated	Evaluation grade
Completeness	Teaching content is complete and reflect the basic knowledge system	Evaluation grade
Rationality of structure	The organization of teaching content conforms to the cognitive rule formed by students' systematic view	Evaluation grade
Relevance	Teaching content reflects the connection with other subjects, including the knowledge of hardware and software	Evaluation conclusion

4.2 Level 2 Indexes of Teaching Implementation

The evaluation of teaching implementation focuses on teaching ability, methods, organization and coordination ability, etc. The evaluation indexes are shown in Table 3.

Table 3. Level 2 indexes of teaching implementation

Level 2 indexes	Description	Types of evaluation results
Order	On time start and finish class, comply with the teaching system, good manners, normal order	Yes or no
Activity	Interact with students, students take the initiative to think and answer questions	Evaluation grade
Applicability and novelty of teaching methods	Reasonable use of heuristic, inquiry-based, seminar, flipped classroom, and other teaching methods	Evaluation conclusion

(*continued*)

Table 3. (*continued*)

Level 2 indexes	Description	Types of evaluation results
Arrangement of teaching activities	Reasonably arrange pre-class preview, classroom teaching, homework and other teaching activities	Evaluation conclusion
Teaching tools	Flexible use of blackboard, multimedia and other teaching tools	Evaluation conclusion

4.3 Level 2 Indexes of Learning Experience

The evaluation of learning experience reflects students' intuitive feelings and experience of teaching content, methods, organization and participation, etc. The evaluation indexes are shown in Table 4.

Table 4. Level 2 indexes of learning experience

Level 2 indexes	Description	Types of evaluation results
Class attraction	Be interested in courses and willing to learn	Evaluation conclusion
Class activity	Actively participate in the discussion and answer questions in class	Evaluation conclusion
Difficulty level of learning	The course is difficult to learn	Evaluation conclusion
Interest in learning	Take the initiative to study and think after class	Evaluation conclusion
Problem solving	After asking questions, whether the teacher answer in time and give solutions	Evaluation conclusion
Novelty	The novelty of the new technology of computer development introduced by the teacher	Evaluation conclusion

4.4 Level 2 Indexes of Teaching Effect

Teaching effect evaluation is to evaluate whether teaching effect is consistent with training objective from the perspective of students and supervisors. The evaluation indexes of students are shown in Table 5, and that of supervisors are shown in Table 6.

Table 5. Level 2 indexes of teaching effect by students

Level 2 indexes	Description	Types of evaluation results
Mastery	Proficiency in course knowledge	
	Evaluation conclusion	
Completion	Assignments or experiments can be completed on time with high accuracy	Evaluation conclusion
System cognition	Understand the knowledge of hardware and software collaboration	Evaluation conclusion
System design	Completion of computer system design	Evaluation conclusion

Table 6. Level 2 indexes of teaching effect by supervisors

Level 2 indexes	Description	Types of evaluation results
Class reaction	Students are focused and active in class	Evaluation grade
Achievement of teaching objectives	Students' mastery of the content	Evaluation conclusion
Knowledge expansion	Students can master new knowledge	Evaluation conclusion
Ability development	Computer system designs by students are well done	Evaluation conclusion
Team collaboration	Students collaborate with each other to complete group discussions, presentations and experiments	Evaluation conclusion

5 Multi-dimensional Evaluation Results

The multi-dimensional evaluation results are presented in the form of radar chart. The final evaluation results are no longer a single overall evaluation score or the grades of excellent, good, medium and poor, but the evaluation of each dimension. The evaluation conclusions of all dimensions are independent of each other. What aspects should be improved and suggestions for improvement are reflected in the evaluation conclusions. In Fig. 2, two courses are taken for example to show multi-dimensional evaluation results. Course 1 reflect better students' learning experience, but lower evaluation of teaching content and implementation. The reason might be that teaching content is not accord with the goal of system ability training. Or the content is too simple, thus students don't have to spend too much energy on it. The evaluation of course 2 is balanced in all dimensions, but still have room for improvement in learning experience and teaching effect.

Multi-dimensional evaluation is oriented to curriculum improvement because students' evaluation results also contain specific suggestions and constructive feedback to improve teaching. For example, for course 1, students' feedback is that the learning content is easy to understand, the homework and experiment are low difficulty. It is suggested that teacher can increase the difficulty of learning content. For course 2, students' feedback is that the learning content is from simple to deep, they need to invest energy in studying teaching materials and reviewing learning content, and the homework and experiment are challenging. It is suggested that teacher should guide students to learn more.

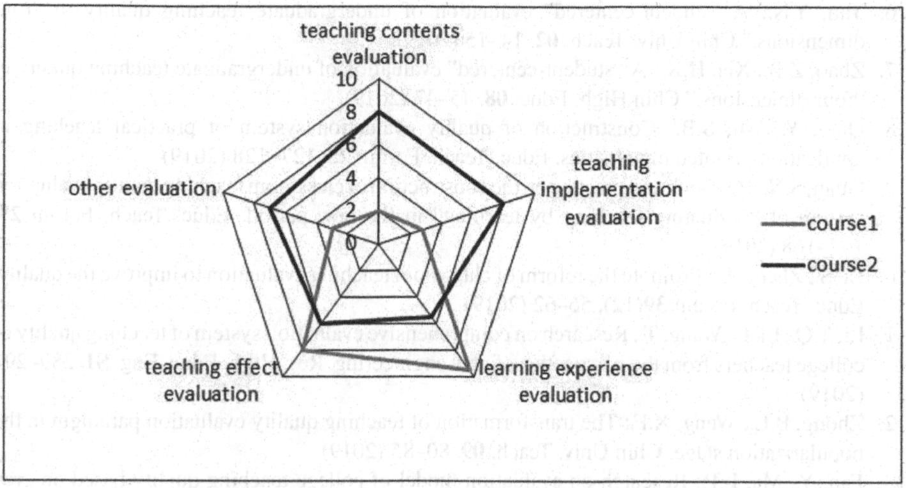

Fig. 2. Multi-dimensional course evaluation represented by radar chart

6 Conclusion

The courses based on computer system ability training has its particularity of training objective. Therefore, the traditional teaching evaluation should not be used to measure the teaching. The evaluation results should be objective and fair, and reflect the characteristics of diagnostic, encouraging and feedback of teaching evaluation. Thus a hierarchical teaching evaluation index system is proposed to evaluate the teaching content, implementation, learning experience and teaching effect. The evaluation subjects are multiple, consisting of supervisors, col-league teachers and students. The evaluation results are multi-dimensional in the form of radar chart. Teaching evaluation is an extremely complex system, and the teaching evaluation method is a positive attempt to provide a reference for teaching evaluation.

References

1. Guan, G.X., Ren, L.L., Wang, J.: Exploration and practice of curriculum reform based on computer system ability cultivation. Exp. Technol. Manage. **36**(05), 239–241 (2019)

2. Wang, L., Gao, N.: Construction of evaluation indicators for undergraduate course teaching quality – based on the perspective of student experience. Res. High. Educ. Eng. **02**, 195–200 (2021)
3. Cao, C.M., Sun, F., Qin, K.: On the construction and practice of teaching quality evaluation system of theory course in colleges and universities. Educ. Teach. Forum **04**, 98–102 (2021)
4. Zhang, X., Liu, J.N., Liu, Q.: Online evaluation system of teacher course teaching quality under the concept of OBE. Educ. Teach. Forum **13**, 128–129 (2020)
5. Chen, X., Wang, X.L., Wang, Y., Zhang, H.Y.: Reconstruction of curriculum teaching quality evaluation system and construction of "golden course." Chin. Univ. Teach. **05**, 43–48 (2019)
6. Yan, Y.N.: A "student-centered" evaluation of undergraduate teaching quality in "four dimensions." Chin Univ. Teach. **02**, 14–15 (2019)
7. Zhao, Z.B., Xia, H.X.: A "student-centered" evaluation of undergraduate teaching quality in "four dimensions." Chin High. Educ. **08**, 45–47 (2019)
8. Qiao, Y., Wu, S.B.: Construction of quality evaluation system of practical teaching in application-oriented universities. Educ. Teach. Forum **25**, 127–128 (2019)
9. Guan, S.N.: Research and practice on the construction of classroom teaching quality evaluation system of "evaluating teaching by learning" in the new period. Educ. Teach. Forum **29**, 167–168 (2019)
10. Li, S., Zheng, L.: Promote the reform of classroom teaching evaluation to improve the quality. Educ. Teach. Forum **39**(12), 56–62 (2019)
11. Li, Y.Q., Li, F., Wang, T.: Research on comprehensive evaluation system of teaching quality of college teachers from the perspective of new engineering. Res. High. Educ. Eng. **S1**, 289–291 (2019)
12. Zhong, B.L., Wang, X.F.: The transformation of teaching quality evaluation paradigm in the popularization stage. Chin Univ. Teach. **09**, 80–85 (2019)
13. Fan, Y., Ma, L.P.: Research on evaluation model of college teaching quality based on grey system theory. Univ. Educ. **02**, 185–188 (2019)
14. Xiong, L.S., et al.: Design of computer composition principle experiment board based on CPLD. In: International Conference on Mechanical, Electronic and Information Technology Engineering (ICMEIE 2016), Barcelona, pp. 122–127, ISBN: 978-1-60595-340-3 (2016)
15. Bing, X., Kefen, M., Gang, F.: The influence of MOOC on the computer teaching courses in colleges. In: Joint International Information Technology, Mechanical and Electronic Engineering Conference (JIMEC 2016), pp. 359–362. Atlantis Press, Xi'an (2016)

Study on Vocational College Students' Communicative Competence of Intercultural Communication

Dan Ren(✉) and Dan Xian

Sanya Aviation and Tourism College, Sanya, Hainan, China

Abstract. Under the background of development of Hainan Free Trade Zone (port) to meet its demand of English language talents with good intercultural communication competence, an empirical study is conducted to evaluate vocational college students' communicative competence of intercultural communication. The results indicate that among the five factors of communicative competence of intercultural communication, students' listening competence is the highest, followed successively by oral competence, strategic competence, sociolinguistic competence, discourse competence. The analytical results show that each factor among the five factors are positively correlated with the other four factors. Moreover, students' scores in college English Test (Grade A) have a positive correlation with their oral English competence, discourse competence and sociolinguistic competence, respectively.

Keywords: Communicative competence of intercultural communication · Listening competence · Oral competence · Strategic competence · Sociolinguistic competence · Discourse competence · Empirical study

1 Research Background

Since 2018, China has rolled out a plan for building its southern island province of Hainan into a pilot free trade zone [1]. Developing Hainan Free Trade Zone has granted Hainan more autonomy to reform, and has brought Hainan more opportunities to further open up and promote economic globalization. However, it also poses a challenge for talents, especially English talents with great intercultural communication competence [2]. In the press conference of *Introducing One Million Talents to Hainan* in June, 2018, Hainan Provincial Government points out that in order to solve the problem of lacking high quality talents, Hainan should not only introduce talents from all over the world, but also cultivate local talents within the province [3]. The building of Hainan Free Trade Port is in urgent need of the following five levels of English talents: First, the English talents who is well versed in international trade rules and regulations and be able to conduct business negotiations in English; Second, senior English translators who are qualified for the translation and interpretation work of the government, enterprise, and public institutions; Third, English teachers and trainers who are able to cultivate new language force and contribute to improving the English quality of the whole people;

J. Zeng et al. (Eds.): ICPCSEE 2021, CCIS 1452, pp. 443–455, 2021.
https://doi.org/10.1007/978-981-16-5943-0_36

Fourth, English service personnel who can provide considerate service for foreigners in various government agencies, enterprises and service windows; Fifth, English volunteers who can provide concierge service, escort service and help for foreigners [4]. There are altogether 20 universities and colleges in Hainan, including 7 undergraduate colleges and 13 vocational colleges. Vocational colleges undoubtedly play a very important role in cultivating local English talents for Hainan Free Trade Zone, especially the last two levels of English talents.

Intercultural communication refers to the communication between native speakers and non-native speakers. It also refers to the communication between people with different language and cultural backgrounds [5]. Since the concept of communication competence was put forward by Hymes in 1972 [6], scholars at home and abroad have done a great deal of researches on it. It can be mainly divided into two stages. In the early stage from 1972 to 1997, foreign scholars mainly focused on studying its connotation and application. Many scholars, such as Byram, Lustig and Koester [7], try to define and distinguish intercultural communication competence from intercultural competence. In China, intercultural communication competence was firstly introduced by Professor He Daokuan from Shenzhen University in 1983 [8]. In the early stage, domestic researches on intercultural communication lay particular stress on language teaching, seldom involving psychology, which is quite different from USA [8, 9]. In 2020, Maxin, Sumin and Lijie analyzed the co-cited references of Web of Science journal articles and proceeding papers published between 1998 and 2017 based on a bibliometric analysis of relevant core literature. The findings indicate: international IC/CC (intercultural communication/cross-cultural communication) researches are mainly distributed into five disciplines, including education & educational research, social science (interdisciplinary), linguistics, language & linguistics, communication; The year 2010 became the turning point in the increasing number of IC/CC published articles; Physician-patient communication, translation, teaching and Sino-American comparison are the major four theme clusters in this field; New emerging hot fields in the past five years include translation, intercultural communicative competence, and language teaching acquisition [10]. In recent years, domestic scholars, such as Jing Yuan, Liu Dan, etc. do research on how to build a model to measure intercultural communication competence [11].

In this paper, an empirical study is conducted to study the current situation of vocational college students' communicative competence of intercultural communication and the correlations between each internal factors and College English Test (Grade A) scores.

2 Research Design

2.1 Variables

In 1980s, Canale and Swine defined communication competence in four levels: grammatical competence, sociolinguistic competence, discourse competence and strategic competence [12]. Intercultural communication competence is a concept with wide connotation, including global mentality, knowledge, cultural adaption and communicative competence [5, 6, 10], as it is shown in Fig. 1. However, in this paper, Canale and Swain's definition of communication competence is adopted as communicative competence of intercultural communication, as it is shown in Fig. 2.

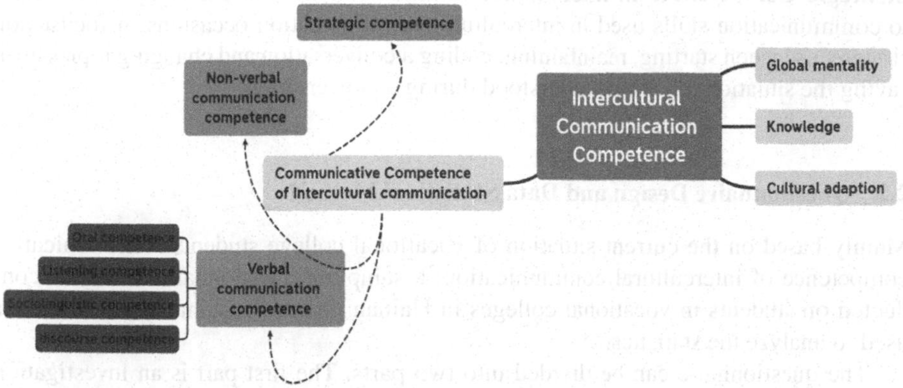

Fig. 1. The definition of intercultural communication competence

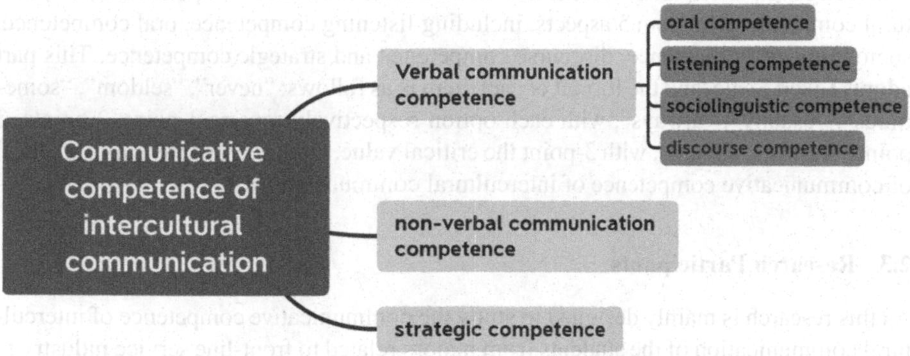

Fig. 2. Research variables in this paper

Oral Competence and Listening Competence. In intercultural communication, oral and listening competence here not only refers to communicators' oral and listening skills and abilities, it can also reflects communicators' proficiency of other skills, such as translation. Here, it refers to the communicators' language knowledge and skills.

Sociolinguistic Competence. In intercultural communication, sociolinguistic competence refers to pragmatic competence. Two levels of competence are included in its connotation. First, the ability to correctly use foreign language to express and understand verbal behaviors; Second, the ability to properly use social language conforming to a particular culture.

Discourse Competence. In intercultural communication, discourse competence refers to the competence to organize sentence groups in verbal or written forms, as well as the competence to analyse paragraphs and utterance. In different culture or different contexts, there are different ways for people to choose a topic, to start a conversation, to take turns, to change a topic, even to end a conversation. The mastery of these rules belongs to discourse competence.

Strategic Competence. In intercultural communication, strategic competence refers to communication skills used in intercultural communication occasions. It focuses on strategy use when starting, maintaining, ending a conversation and changing topics even saving the situation when misunderstood during a conversation.

2.2 Questionnaire Design and Data Statistical Method

Mainly based on the current situation of vocational college students' communicative competence of intercultural communication, a sampling questionnaire survey is conducted on students in vocational colleges in Hainan province, China, and SPSS 24 is used to analyze the statistics.

The questionnaire can be divided into two parts. The first part is an investigation on demographic information of questionnaire participants, including their gender, age, major, grade, score of English Level A Examination. The second part includes 32 items. It investigates participants' current situation of communicative competence of intercultural communication from 5 aspects, including listening competence, oral competence, sociolinguistic competence, discourse competence and strategic competence. This part adopts Likert Scale, and the format of each item is as follows: "never", "seldom", "sometimes", "usually", "always", with each option respectively scoring 1 point, 2 points, 3 points, 4 points, 5 points, with 3-point the critical value. High score represents high level of communicative competence of intercultural communication.

2.3 Research Participants

As this research is mainly designed to study the communicative competence of intercultural communication of the students from majors related to front-line service industry in Hainan, the questionnaire is mainly delivered to students majoring in hotel management, English, Russian, tourism management and international cruise steward management, etc. The research participants include both the hotel and tourism industry staff who need to upgrade their educational background and the vocational college students. They are the reserve of talents for the future front-line service industry in Hainan. 250 students participate in the survey, and 239 questionnaires are valid.

In the 239 participants, there are 135 females, accounting for 56.5%. The participants are mainly freshmen and sophomores, accounting for 87.9% and 12.1% respectively. They study in different majors, and their age mainly range from 18 to 23. Through the frequency analysis of demographic variables, the result is shown in Table 1.

Table 1. Frequency statistic (N = 239)

Variables	Attribute	Frequency	Percentage (%)
Gender	Male	104	43.5
	Female	135	56.5
Age	18–23	239	100
Grade	Freshman	210	87.9
	Sophomore	29	12.1
Major	English	33	13.8
	Hotel management	57	23.8
	Tourism management	74	31
	Russian	30	12.6
	International Cruise Steward management	45	18.8

3 Empirical Research Procedure

3.1 Reliability and Validity Test

Reliability Test. Cronbach's α coefficient, as the most frequently-used method to analyse the reliability by scholars and researchers, is used here as an index to test the reliability. SPSS 24 is used to analyze the statistics of the survey, and Table 2 reports the reliability statistics.

As it is shown in Table 2, the Cronbach's α coefficient of the whole questionnaire is 0.956, which is greater than 0.6, indicating that the reliability of the questionnaire is very good.

Table 2. Reliability statistics

	Cronbach's Alpha	Case number	Number of items
Communicative competence of intercultural communication	0.956	239	32

Validity Test. Construct validity is used here as an evaluation index to test the validity of the questionnaire. As it is shown in Table 3, the KMO value is 0.925; the chi-square value of Bartlett's test is 6696.003; the degree of freedom is 496; and its significance level is about 0, suggesting that this sample has good validity.

Table 3. KMO and Bartlett's test

Sampling sufficient KMO measures		0.925
Bartlett's test	Approximate chi-square	6696.003
	df	496
	Sig.	0.000

3.2 Description Statistics

Description statistics is used to analyse the current situation of college students' communicative competence of intercultural communication. The mean value and standard deviation can be seen in Table 4, which can be used to judge students' communicative competence of intercultural communication. The higher mean value represents better communicative competence of intercultural communication. As it is shown in Table 4, vocational college students' communicative competence of intercultural communication is not so good. Their listening and oral competence are just above average, with the mean value 3.1140 and 3.0883 respectively. Their strategic competence is below average, successively followed by sociolinguistic competence and discourse competence.

Table 4. Description statistics

	Number of case	Minimum value	Maximum value	Mean value	Standard deviation
Listening competence	239	1.00	5.00	3.1140	0.99200
Oral competence	239	1.00	5.00	3.0883	0.93610
Discourse competence	239	1.00	5.00	2.7331	0.82759
Strategic competence	239	1.00	5.00	2.9833	0.79898
Sociolinguistic competence	239	1.00	5.00	2.8696	0.91852

3.3 Student's t Test

Student's t test is conducted to test if there is significant difference among participants of different genders, grades or majors.

Gender. As it is shown in Table 5, there is significant difference between male students and female students on listening competence ($t = -2.713$, $p < 0.05$), oral competence ($t = -3.019$, $p < 0.05$), and sociolinguistic competence ($t = -3$, $p < 0.05$). And the mean value of these three competence of female students are significantly higher than that of male students. Moreover, there is no significant difference on discourse competence and strategic competence between male students and female students.

Table 5. Student's t test on gender

	Gender	Number of case	Mean value	Standard deviation	t	p
Listening competence	Male	104	2.9183	1.01233	−2.713	0.007
	Female	135	3.2648	0.95259		
Oral competence	Male	104	2.8835	0.9267	−3.019	0.003
	Female	135	3.2461	0.91578		
Discourse competence	Male	104	2.675	0.77544	−0.952	0.342
	Female	135	2.7778	0.86582		
Strategic competence	Male	104	2.8726	0.77955	−1.89	0.06
	Female	135	3.0685	0.80616		
Sociolinguistic competence	Male	104	2.6699	0.91079	−3	0.003
	Female	135	3.0235	0.8978		

Grade. As it is shown in Table 6, vocational college students from different grades has significant difference on sociolinguistic competence ($t = 2.111$, $p < 0.05$), and the mean value of sociolinguistic competence of freshmen are obviously higher than that of sophomores. Moreover, there is no significant difference between freshmen and sophomores on listening, oral, discourse and strategic competence.

Table 6. Student's t test of grade

	Grade	Number of case	Mean value	Standard deviation	t	p
Listening competence	Freshman	210	3.1190	0.97154	0.211	0.833
	Sophomore	29	3.0776	1.14779		
Oral competence	Freshman	210	3.0910	0.91718	0.119	0.906
	Sophomore	29	3.0690	1.08111		
Discourse competence	Freshman	210	2.7276	0.81593	−0.273	0.785

(*continued*)

Table 6. (*continued*)

	Grade	Number of case	Mean value	Standard deviation	t	p
	Sophomore	29	2.7724	0.92230		
Strategic competence	Freshman	210	3.0024	0.78086	0.995	0.321
	Sophomore	29	2.8448	0.92320		
Sociolinguistic competence	Freshman	210	2.9159	0.90266	2.111	0.036
	Sophomore	29	2.5345	0.97832		

Major. As it is shown in Table 7, no significant difference exists between every two dimensions of listening, oral, discourse, strategic and sociolinguistic competence for vocational college students from different majors ($P > 0.05$).

Table 7. Student's t test of major

		Number of case	Mean value	Standard deviation	F	p
Listening competence	Russian	30	3.1333	0.99076	1.008	0.404
	Hotel management	57	2.9956	1.11252		
	Tourism management	74	3.0743	0.92226		
	English	33	3.4167	1.01871		
	International Cruise Steward management	45	3.0944	0.91901		
	Total	239	3.1140	0.99200		
Oral competence	Russian	30	3.0259	0.92627	1.537	0.192
	Hotel management	57	2.9220	1.09007		
	Tourism management	74	3.0616	0.89177		
	English	33	3.4108	0.79421		
	International Cruise Steward management	45	3.1481	0.87328		
	Total	239	3.0883	0.93610		
Discourse competence	Russian	30	3.0000	0.74649	1.772	0.135
	Hotel management	57	2.6491	0.93372		
	Tourism management	74	2.7162	0.86135		
	English	33	2.9030	0.89878		
	International Cruise Steward management	45	2.5644	0.54820		

(*continued*)

Table 7. *(continued)*

		Number of case	Mean value	Standard deviation	F	p
	Total	239	2.7331	0.82759		
Strategic competence	Russian	30	2.9958	0.78406	1.806	0.128
	Hotel management	57	2.8355	0.91251		
	Tourism management	74	2.9105	0.77540		
	English	33	3.2576	0.78806		
	International Cruise Steward management	45	3.0806	0.65991		
	Total	239	2.9833	0.79898		
Sociolinguistic competence	Russian	30	2.9889	0.90437	1.385	0.24
	Hotel management	57	2.6637	0.97717		
	Tourism management	74	2.8986	0.87055		
	English	33	3.0960	0.92799		
	International Cruise Steward management	45	2.8370	0.90557		
	Total	239	2.8696	0.91852		

3.4 Correlation Analysis

The structure of the dimensions and corresponding items are determined through reliability and validity test mentioned before. Pearson correlation coefficient is used to analyse the correlation between each two dimensions by analyse the mean value of the items in each dimension. The value range of correlation coefficient is between -1 and 1, denoted by r. If the correlation coefficient r is positive, it means positive correlation exists between the variables. If r is negative, it means negative correlation exists between the variables. Moreover, higher absolute value represents closer correlation between the variables, and vice versa. To be specific, if $|r| = 1$, complete correlation exists between the variables. If $0.7 \leq |r| < 1$, high correlation exists between the variables. If $0.4 \leq |r| < 0.7$, moderate correlation exists between variables. If $0.1 \leq |r| < 0.4$, low correlation exists between variables. If $|r| < 0.1$, it can be regarded as no correlation existing between the variables. The conclusions mentioned above can be true only when the significant value $P < 0.05$.

Correlation Analysis Between Each Variable Within Communicative Competence of Intercultural Communication. As it is shown in Table 8, positive correlation exists between each two variables of communicative competence of intercultural communication. Among them, high positive correlation exists between listening competence and oral competence ($r = 0.803$, $p < 0.05$), and between oral competence and strategic competence ($r = 0.713$, $p < 0.05$). Moderate positive correlation exists between the following groups of variables, including listening competence and discourse competence ($r = 0.445$, $p < 0.05$), listening competence and strategic competence ($r = 0.659$, $p < 0.05$), oral competence and discourse competence ($r = 0.361$, $p < 0.05$), discourse competence and strategic competence ($r = 0.623$, $p < 0.05$), discourse competence and sociolinguistic competence ($r = 0.642$, $p < 0.05$), as well as strategic competence and sociolinguistic competence ($r = 0.577$, $p < 0.05$). Low positive correlation exists between the following two groups of variables, including listening competence and sociolinguistic competence ($r = 0.347$, $p < 0.05$), oral competence and sociolinguistic competence ($r = 0.361$, $p < 0.05$).

Table 8. Correlation analysis between each variable with communicative competence of intercultural communication

		Listening competence	Oral competence	Discourse competence	Strategic competence	Sociolinguistic competence
Listening competence	Pearson correlation	1	0.803**	0.445**	0.659**	0.347**
	Significance (two-tailed test)		0	0	0	0
	Number of case	239	239	239	239	239
Oral competence	Pearson correlation	0.803**	1	0.457**	0.713**	0.361**
	Significance (two-tailed test)	0		0	0	0
	Number of case	239	239	239	239	239
Discourse competence	Pearson correlation	0.445**	0.457**	1	0.623**	0.642**
	Significance (two-tailed test)	0	0		0	0
	Number of case	239	239	239	239	239
Strategic competence	Pearson correlation	0.659**	0.713**	0.623**	1	0.577**
	Significance (two-tailed test)	0	0	0		0
	Number of case	239	239	239	239	239
Sociolinguistic competence	Pearson correlation	0.347**	0.361**	0.642**	0.577**	1
	Significance (two-tailed test)	0	0	0	0	
	Number of case	239	239	239	239	239

Correlation Between Each Variable and College English Test (Grade A) Scores. College English Test (Grade A) is national English test specially designed for vocational college students. It tests vocational college students' competence in listening, grammar, reading, discourse analysis, translation and writing. In this questionnaire, participants are required to write down their scores of this test. Pearson correlation coefficient is also used to test the correlation between each variable in communicative competence of intercultural communication and College English Test (Grade A) scores, and the results are shown in Table 9.

As it is shown in Table 9, positive correlation exists respectively between college English test scores and the following variables, including oral competence ($r = 0.132$, $p < 0.05$), discourse competence ($r = 0.14$, $p < 0.05$), strategic competence ($r = 0.2$, $p < 0.05$), sociolinguistic competence ($r = 0.248$, $p < 0.05$).

Table 9. Correlation analysis between communicative competence of intercultural communication and college English test (Grade A) scores

		Listening competence	Oral competence	Discourse competence	Strategic competence	Sociolinguistic competence	College English Test (Grade A) scores
College English Test (Grade A) scores	Pearson Correlation	0.116	0.132*	0.140*	0.200**	0.248**	1
	Significance (two-tailed test)	0.081	0.047	0.036	0.002	0.000	
	Number of case	229	229	229	229	229	229

4 Some Suggestions to Improve Students' Communicative Competence of Intercultural Communication

The development of Hainan Free Trade Zone needs talents with international vision, solid professional knowledge and skills, as well as excellent intercultural communication competence. In order to cultivate excellent local English talents to meet the needs of developing Hainan Free Trade Zone, some suggestions are given based on the results of this research.

4.1 Make Full Use of the International Competitions and Conferences in Hainan

From the survey, it is obvious that vocational college students' scores on oral competence, listening competence, sociolinguistic competence, strategic competence are moderate or even below moderate. As we all know, Hainan is one of the favorite choices for international organizations, companies, groups to host their conference, competitions and matches. All of these international conferences and competitions provide the students in

Hainan lots of opportunities to communicate with foreigners. Thus, vocational colleges should take full advantage of these opportunities to organize and encourage their students to actively take part in the volunteer activities, and vocational college students should make full use of these opportunities to practice their oral English, listening skills as well as cross cultural communication ability.

4.2 English Teaching Should Adapt Teaching Content and Methods to Cultivating Students' Communicative Competence of Intercultural Communication

Traditional English Teaching usually attaches great importance to improving students' five basic language skills [13], including listening, speaking, reading, writing and translation, but to some extent neglects the cultivation of students' global mentality, cultural adaption, cross-cultural attitude and sensitivity. From this survey, it is reasonable for teachers to adjust their teaching contents and methods to make it suitable for cultivating students' communicative competence of intercultural communication, especially discourse competence and strategic competence [14]. Cultural knowledge and cultural difference between China and western countries can be integrated into the language course, and students are encouraged to discover the deep reasons behind the difference, thus to form cross-cultural awareness [15].

5 Conclusion

Through this empirical study, it can be concluded that vocational college students' communicative competence of intercultural communication is not so good, especially discourse competence, strategic competence and sociolinguistic competence. Significant difference exists between male students and female students on listening competence, oral competence, and sociolinguistic competence, and these three competence of female students are significantly higher than male students. Significant difference exist between freshmen and sophomores on sociolinguistic competence, and the score of freshmen are significantly higher than sophomores. There is no significant difference between students from different majors on each competence. By Pearson correlation analysis, it comes to conclusion that positive correlation exists between each two dimensions of communicative competence of intercultural communication, and positive correlation exists between college English test (Grade A) scores and each of the following variable, including oral competence, discourse competence, strategic competence, sociolinguistic competence.

Cultivating and improving students' intercultural communication competence is a long journey. Vocational college students should take full advantage of the international conferences and competitions held in Hainan Free Trade Zone to practice their listening and oral competence as well as cross-cultural communication skills. Moreover, in addition to basic language skills, English teaching should attach more importance to cultivate students' global mentality, cultural adaption and enrich students' knowledge, as well as improve their communicative competence of intercultural communication, especially discourse competence and strategic competence.

Acknowledgment. R. X. thanks the Administration Office of Hainan Province Education Science Planning Leading Group for their support in the research project "The Research on Approach

Exploration and Strategies for All People Learning Foreign Language in Hainan Free Trade Zone" (QJC20191011).

References

1. Chinadaily Homepage. http://www.chinadaily.com.cn/a/201810/16/WS5bc58ef6a310eff303282aaa.html. Accessed 16 Oct 2018
2. Dan, R.: Research on strategy exploration for cultivating versatile English talents in higher vocational college——under the background of developing Hainan free trade zone. J. Shijiazhuang Univ. Appl. Technol. **31**(1), 22–25 (2019)
3. Chinadaily Homepage. http://www.chinadaily.com.cn/a/201805/04/WS5aeba607a3105cdcf651be48.html. Accessed 04 May 2018
4. Xiaoping, W.: The development of free trade zone and optimization of international language. J. Guangdong Inst. Admin. **28**, 89–92 (2016)
5. Dan, L.: A study of intercultural communicative competence model and development. Foreign Lang. Res. **6**, 127–131 (2015)
6. Ying, Y., Enping, Z.: Frame construction of intercultural communication competence in foreign language teaching. Foreign Lang. World **4**(121), 13–21 (2007)
7. Lustig, M.W., Koester, J.: Intercultural Competence: Interpersonal Communication across Cultures, 3rd edn. Addison Westerly Longman Inc, New York (1999)
8. Wenzhong, H.: The current situation and future of intercultural communication in China from the perspective of discipline construction. J. Foreign Lang. **6**, 28–32 (2010)
9. Yan, H.: Research on college students' intercultural communication sensitiveness. Foreign Lang. World **3**(144), 68–73 (2011)
10. Xin, M., Min, S., Jie, L.: The research on current situation of intercultural communication based on a bibliometric analysis. Foreign Lang. Educ. **1**(41), 59–64 (2020)
11. Jing, Y.: The construction of theoretical model of intercultural communicative competence scale for college students. Foreign Lang. Res. **1**, 74–78 (2021)
12. Canale, M., Swain, M.: Theoretical bases of communicative approaches to second language teaching. Appl. Linguis. **1**, 1–47 (1980)
13. Feng, Y.: Study and strategy on cultivation of intercultural communication ability in the college english teaching of ethnic minorities region in Guizhou Province. Guizhou Ethnic Stud. **38**(196), 235–238 (2017)
14. Lei, L.: Exploration on the cultivation of students' intercultural communication ability in higher vocational english teaching under the background of the "Belt and Road." J. Heilongjiang Inst. Teach. Develop. **39**(2), 56–58 (2020)
15. Min, C.: The cultivation of cross-cultural awareness in English teaching. J. Chin. Soc. Educ. **S1**(93), 93–94 (2019)

Exploration and Practice for the Cultivation Mode of College Students' Innovation Ability

Yinglun Xi[1(✉)], Xiang Chen[1], and Yang Li[2]

[1] School of Computer Science and Technology, Beijing Institute of Technology, Beijing, China
xiyinglun@bit.edu.cn

[2] Information Resources Department, China North Industries Corporation, Beijing, China

Abstract. Science and technology innovation has become a major driving force for economic growth and social development. Engineers are the primary group that carry out technology innovation in industrial production. And universities are the major institutions for cultivating engineers. However, with the rapid updating of technology, engineering education in universities failed to cope with the actual requirements of enterprises. In order to bridge the gap between university education and industry's demands for talents, an innovating engineer education center is proposed in this paper. It was jointly developed by universities and enterprises. In the innovating center, enterprises were incorporated into universities' teaching systems. The longevity of university-enterprise cooperation was guaranteed in terms of funding, personnel and supervision. It shows that through the joint training of university and enterprises, students' engineering ability of collaborative development for on-the-ground projects is significantly strengthened.

Keywords: Engineering education · Innovation · University

1 Introduction

The fundamental source of science and technology innovativeness originates from talents, in which universities are the primary institutions for cultivating talents. Over recent years, colleges and universities have placed an increasing emphasis on fostering students' innovative consciousness and innovation competence. Nevertheless, there are still plenty of issues in the practice of innovation education. To a certain extent, the university curriculums are not entirely oriented to the demands of enterprises, but are decoupled from industrial needs. The innovative approaches taken by students are not in line with the practical work of enterprises, which require a long period of retraining for them to be qualified for positions. Moreover, due to limited resources, colleges and universities have

J. Zeng et al. (Eds.): ICPCSEE 2021, CCIS 1452, pp. 456–464, 2021.
https://doi.org/10.1007/978-981-16-5943-0_37

failed to effectively integrate enterprise resources to satisfy students' needs for innovation and practice, thus leading to a lack of students' innovative and practical abilities. In spite of numerous internships available in many enterprises, students suffer from high academic pressure, limited spare time, and difficulty in getting connected with enterprises. On the other hand, there is a greater demand for practical work and less innovative guidance from enterprises, which do not meet the development demands of students.

To address the contradiction between university student training and industry demand, we proposed an innovative talent cultivation method which is co-conducted by universities and enterprises. And we practice the method in the School of Computer Science & Technology of Beijing Institute of Technology (BIT). First of all, the School of Computer Science & Technology of BIT works with top enterprises in the industry and jointly establish an Innovation Center, namely *Undergraduate Innovation Center and Incubator of BIT*. Based on the Innovation Center, enterprises participate in engineering education through courses, scholarships, competitions, internships and practices, as detailed in Sect. 2.1. The Innovation Center has developed a relatively mature and dynamically updated curriculum system, which enables students to acquire knowledge oriented to the needs of the industry and technical problems encountered in front-line practice, see Sect. 2.2 for details. A mechanism has been formulated to facilitate the Innovation Center's continued role in fostering students' innovative capacity, including financial security, personnel management and supervision, see Sect. 2.3 for details.

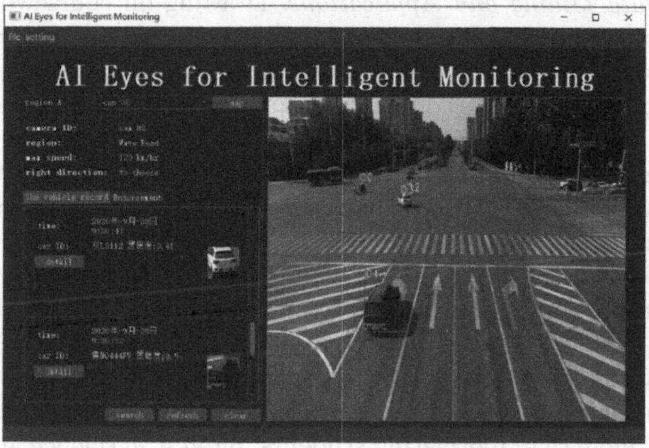

Fig. 1. An intelligent traffic application based on computer vision. The product is developed by a team of juniors, which relies on a hands-on student project of the Innovation Center. The product has the functions of vehicle detection, tracking, license plate number recognition and alarm for violation of regulations.

Through years of practice, the Innovation Center program established by universities and enterprises has achieved noticeable results. The students of

Innovation Center have won a total of 1,596 science and technology competitions and published 146 academic papers. The teachers of Innovation Center have published 22 papers about engineering course. Many student science and innovation projects have formed software modules that can be implemented into industry or project frameworks that can serve social governance, such as Fig. 1.

Briefly, we have made the following contributions to our work:

1. We establish the university-enterprise cooperation platform to address the issue of low enterprise participation in engineering education.
2. We establish an enterprise teaching system to tackle the issue of disconnecting between teaching and demand.
3. We establish an operating mechanism to avoid the issue of unsustainable cooperation.

2 Method and Experiments

2.1 University-Enterprise Cooperation Platform

Exploring New Paths for Cooperation. We actively explore the school-enterprise cooperation model and broaden the cooperation channels through courses, scholarships, competitions, internships and practices.

In collaboration with enterprises, the Innovation Center has carried out academic work and produced courses including *IOS Application Development Practice Course, Android Application Development Practice Course, Speech Recognition and Data Mining and Applications.* Drawing on coursework, it encourages students to continue their exploration and research, which results in their graduation design projects. Furthermore, internships are offered which combine graduation design and professional internships, with thesis written on industrial products.

Based the Innovation Center, the school of computer science and technology has jointly established an entrepreneurship alliance with enterprises, which offers students with support such as innovation scholarships, incubation funds and guidance for entrepreneurship projects. As of 2020, there is a total of RMB 2.4 million in corporate funds invested in engineering education at the Innovation Center.

The Innovation Center hold a "Programming Competition" with Lssec Tech and an "Artificial Intelligence Competition" with Byte Dance. Through these competitions, more than motivating students' interest in science and technology, enterprises select outstanding talents to engage in the core technology research of enterprises.

Ensuring Win-Win Cooperation. The Innovation Center regularly listens to the opinions of University Enterprises and students, timely responds to the needs of all parties. The Innovation Center seriously consider their interests and ensure win-win results.

From the perspective of students, the new mechanism of university-enterprise cooperation addresses the decoupling of classroom knowledge from enterprise demand, through the joint development of practical courses by universities and enterprises. It enables enterprise R&D to align with students' science and technology innovation, which fulfills students' needs for internship and practice with the goal of enhancing their employment competitiveness.

In terms of enterprises, it is to fulfill the need to promote enterprises' brands by offering scholarships and conducting science and technology competitions, thus boosting the recognition of enterprises in the student community. Meanwhile, enterprises can be facilitated to overcome technical difficulties through the intelligence of university students. Enterprises can acquire a realistic understanding of students' actual abilities through the courses, thereby getting opportunities for early talent selection without complex recruitment processes. In particular, enterprises involving in the Innovation Center can "pinch the tip" earlier than the recruitment season.

In the viewpoint of university, it expands the education funding and faculty by consolidating enterprise resources. By exchanging with front-line engineers, university teachers can keep abreast of the development direction of the industry in a timely manner so as to reform the teaching contents and approaches. Moreover, the employment rate can be raised through direct recruitment by enterprises. It accelerates the career development of the alumni by delivering talents to leading enterprises, which increases the universities' voice and influence in the technology sector.

2.2 Curriculums

Curriculum is a key component in our exploration of university-enterprise cooperation for the cultivation of innovative and high-tech talents. The School of Computer Science & Technology has introduced a four-credit course, namely "innovation and Entrepreneurship", which are incorporated into the training programs of academic disciplines. It should be emphasized that there is no fixed and specific content for the course, but enterprises prepare the course content, and students choose freely. As a result, it is a great opportunity for students to choose course content based on their career development plans and interests. Over a period of years, we have created a substantial curriculum offering for students to choose from, as shown in Table 1.

Table 1. Courses offered at the Innovation Center.

Course name	Teaching organization	Teaching hours	Programming hours
Video processing software development on IOS system	Beijing Byte Dance Technology Co., Ltd.	32	32
Video processing software development on Android system	Beijing Byte Dance Technology Co., Ltd.	32	32
Speech recognition	Huawei Technology Co., Ltd.	32	32
Practical game development	Beijing Yunchang Game Technology Co., Ltd.	32	32
Data mining and applications	Shui Qian Technology Co., Ltd.	32	32
Artificial intelligence basics	Artificial Intelligence Club	32	32
Web front and back-end development fundamentals	Mobile Government Innovation Lab	32	32
Basic of algorithms and programming	ACM Club	32	32
Chess gaming vs. human-machine gaming	Gaming and Intelligent Projection Student Innovation Lab	32	32
Simulation for national key projects	Digital Media Simulation Lab	32	32

2.3 Operational Mechanism

Funding and Facilities. Various departments, such as the Fangshan District Government, Academic Affairs Department of BIT and the Youth League Committee of BIT, have provided significant support for the construction of the Innovation Center. Over the last five years, there has been a steady allocation of funding amounting to approximately RMB 500,000 annually from various departments. Apart from that, the Innovation Center has raised RMB 2.4 million for enterprises and alumni, which has been used to finance scholarships and grants for students' innovating projects and to launch university-level innovation competitions. The university has equipped the Innovation Center with classrooms and laboratories with a construction area of 500 m^2 at the Haidian Campus of BIT, where students can carry out science and innovation work. There has also been a 40 m^2 club house available at the Fangshan Campus for the ACM Club and Artificial Intelligence Club to conduct activities. The Fangshan District Government has provided the Innovation Center with a 70 m^2 complimentary office in the Fangshan Entrepreneurship Park to allow students to incubate their business projects.

Personnel Management. In terms of the personnel management, there are 20 full-time instructors, 5 full-time managers and 20 student secretaries at the Innovation Center. On the basis of the strategic cooperation agreement between the base and enterprises, the "Enterprise, University and Innovation" alliance has been formulated, which is based on win-win cooperation and guaranteed by legal regulations, so as to ensure the long-term effectiveness and stability of the curriculum construction. As a result, an innovation guiding team, primarily composed of technical personnel from enterprises and supplemented by university faculty and doctoral students, has been assembled. Of these, the technical personnel of enterprises are responsible for training in cutting-edge technologies and guiding students in the general direction of innovation and entrepreneurship; university faculty members take charge of the operation of basic projects and curriculum connection; and the student Innovation clubs handle training and practice of basic professional knowledge. With a rational structure and a clear division of labor, a stable multi-level faculty has been formed.

Supervision and Evaluation Mechanisms. In conjunction with enterprises, the Innovation Center has offered menu driven services for students with a curriculum including artificial intelligence, information security, and digital art. It allows students to make their own choices and "vote with their feet" to discard courses that do not meet industry development and student needs. Simultaneously, lecturers will evaluate students' learning outcomes on a grading scale and issue certificates of completion. Thus, the Innovation Center is able to assess performance based on student learning and awards for credit granting.

3 Performance

3.1 Student Satisfaction

Fig. 2. Game of the Drawing BIT. This game is developed by university seniors on the basis of "Practical Game Development" course which is conducted by Beijing Yunchang Game Technology Co. Ltd. It takes the history of BIT as the background for the level making, while opening the "layout mode" to allow players to adjust the school scene freely. This game has as many as 1500 players online at the same time. As an important element of BIT's 80th anniversary celebration, the game has received unanimous praise from students, teachers and alumni of the university.

The Innovation Center enables students to earn course credits and acquire practical experience without stepping out of school, as a result of killing two birds with one stone. It is very beneficial to address the issue of students' difficulties in innovation activity and enterprise internships due to course pressure. A number of students have continued their development based on coursework, which has resulted in excellent software products, such as Fig. 2. Satisfaction survey results have revealed that students are over 95% satisfied with the courses.

3.2 Enterprise Satisfaction

With direct involvement in front-line university teaching, enterprises have direct contact with students. During the lecture process, enterprises can design assignments based on the main business of the enterprise. In addition, the enterprise's human resource department can establish contact with outstanding students in advance, after recognizing the solid skills and innovative potential of students. Over the last three years, Baidu has recruited 47 candidates from the School of Computer Science & Technology, While Byte Jump has recruited 15. These are some of the most talented individuals in the student body, who have quickly become an intermediate force in the company's core technology initiatives. Enterprises have expressed great satisfaction with the cooperation.

3.3 Competition Results

Since 2015, there have been more than 4,200 students enrolled in the Innovation Center, with a cumulative number of science and technology competition awards for 1596 students, including 230 awards at the international level and 974 awards at the national level. The students of the Innovation Center have successively won major international and domestic science and technology awards, including the Gold Prize of the International Computer Gaming Competition, the Fourth Prize of the International Robotic Football Competition, the Gold Prize of the "Challenge Cup" University Student Entrepreneurship Competition, the First Prize of the National University Information Security Competition and the First Prize of the China Software Cup. These competition results have also inspired students to keep up their participation in technology and innovation, creating a virtuous cycle.

3.4 Academic Achievements

The university-enterprise cooperation of the Innovation Center has cultivated the academic ability of students to uncover and solve problems. As of 2019, there have been 146 academic papers published by the Innovation Center students. In which, there are 109 papers indexed by SCI, such as references [3,4,7,9], and 37 papers indexed by EI, such as references [2,8,11]. In particular, the paper by the Innovation Center student Ma et al. which relied on the Robotic Football Project [1] has been named the best paper of the 2013 IEEE International Conference on Granular Computing; drawing on the Data Intelligence Project, the academic paper presented by the Innovation Center student Xu et al. at the IEEE 14th Intl Conf on Dependable, Autonomic and Secure Computing [10] has been rated as the best paper of the conference.

3.5 Teaching Material Achievements

Since 2015, on the basis of this project, the faculty has published 22 papers about education reform, as listed in references [5,6,12]. In total, there have been 16 textbooks and 2 national high-quality online open courses published. The Innovation Center has been honored by the Beijing Municipal Education Commission in 2015, as an on-campus model innovation and practice base for higher education institutions in Beijing.

4 Conclusion

Focusing on the training goal of practical innovative talents, aiming at the problems of low participation of enterprises in traditional innovative ability training, we have constructed the mode of university research team enterprise joint participation in innovative ability training and established the Undergraduate Innovation Center and Incubator. Based the Innovation Center, the curriculum system of enterprise teaching mechanism is integrated into teaching system of universities, while teaching management and operation guarantee mechanism ensure the quality of teaching. Our practice has demonstrated that this model can effectively improve the innovation ability of college students.

References

1. Chen, J., Ma, C., Yan, Z., Chen, B., Shen, Y., Liang, Y.: Defensive strategy of the goalkeeper based on the 3D vision and field division for the middle-size league of robocup. In: 2013 IEEE International Conference on Granular Computing (GrC), pp. 49–52. IEEE (2013)
2. Chen, J., Zhao, Z., Shi, J., Zhao, C.: A top-N recommendation model based on feedback learning to rank and online learning. In: 2017 IEEE 2nd Advanced Information Technology, Electronic and Automation Control Conference (IAEAC), pp. 384–389. IEEE (2017)
3. Guo, F., et al.: Video saliency detection using object proposals. IEEE Trans. Cybern. **48**(11), 3159–3170 (2017)
4. Li, D., Wang, S., Yuan, H., Li, D.: Software and applications of spatial data mining. Wiley Interdiscip. Rev. Data Min. Knowl. Discovery **6**(3), 84–114 (2016)
5. Li, L., Chen, Y., Li, Z., Li, D., Li, F., Huang, H.: Online virtual experiment teaching platform for database technology and application. In: 2018 13th International Conference on Computer Science & Education (ICCSE), pp. 1–5. IEEE (2018)
6. Shi, S., Huang, H., Li, K.: The construction and practice of integration cultivation mechanism for innovative talents in CS discipline. In: Proceedings of ACM Turing Celebration Conference-China, pp. 110–114 (2018)
7. Song, H., Wu, X., Yu, W., Jia, Y.: Extracting key segments of videos for event detection by learning from web sources. IEEE Trans. Multimedia **20**(5), 1088–1100 (2017)
8. Song, H., Du, W., Zhao, Q.: Automatic depression discrimination on FNIRS by using FastICA/WPD and SVM. In: Deng, Z., Li, H. (eds.) Proceedings of the 2015 Chinese Intelligent Automation Conference. LNEE, vol. 336, pp. 257–265. Springer, Heidelberg (2015). https://doi.org/10.1007/978-3-662-46469-4_27
9. Wang, W., Shen, J., Yang, R., Porikli, F.: Saliency-aware video object segmentation. IEEE Trans. Pattern Anal. Mach. Intell. **40**(1), 20–33 (2017)

10. Xu, J., Shi, M., Chen, C., Zhang, Z., Fu, J., Liu, C.H.: ZQL: a unified middleware bridging both relational and NoSQL databases. In: 2016 IEEE 14th International Conference on Dependable, Autonomic and Secure Computing, 14th International Conference on Pervasive Intelligence and Computing, 2nd International Conference on Big Data Intelligence and Computing and Cyber Science and Technology Congress (DASC/PiCom/DataCom/CyberSciTech), pp. 730–737. IEEE (2016)
11. Zhang, F., Mao, Z., Ding, G., Xu, L.: Design of Chinese natural language in fuzzy boundary determination algorithm based on big data. J. Inf. Hiding Multimedia Signal Process. **8**(2), 423–434 (2017)
12. Zhao, S., Li, F., Huang, H., Chen, Y.: Design and implementation of virtual experiment for stack operation via visualization method. In: 2018 13th International Conference on Computer Science & Education (ICCSE), pp. 1–5. IEEE (2018)

An Analysis and Validation Toolkit to Support the Undergraduate Course of Computer Organization and Architecture

Yanjun Shu, Zhuangyu Ma, Hongwei Liu(✉), Zhan Zhang, Dongxin Wen, Bing Xu, Danyan Luo, and Decheng Zuo

Harbin Institute of Technology, Harbin 150001, China
liuhw@hit.edu.cn

Abstract. As an important part of the computer organization and architecture (COA) course, the experiment teaching is generally about the computer system design. Students use the hardware description languages (HDLs) tools to implement the computer system on the Field Programmable Gate Array (FPGA) based platform. However, the HDLs tools are made for expert hardware engineers and the computer system is a very complex hardware project. It is hard for students to implement their computer system design in the limited lab hours. How to help students get the design validation and find the failure root is important in COA experiment teaching. To this end, an analysis and validation toolkit which is special for COA experiment teaching is designed. For two main steps of FPGA-based hardware design, waveform simulation and on-board testing, two packages were implemented for them respectively. The comparison results of using and not using our toolkit show it improves the effectiveness of experiment teaching greatly.

Keywords: Computer organization and architecture · Experiment teaching · Hardware design · Analysis and validation

1 Introduction

Computer organization and architecture is an essential topic in undergraduate IEEE/ACM Computer Science (CS) curricula [1]. Teaching COA courses to CS students can be challenging, as the concepts are on a high abstraction level and not easy to grasp for students [2,13]. Experiment assignments are very important in COA teaching. Kehagias [9] surveyed assignments collected from 40 undergraduate computer architecture courses which are selected from 40 CS departments of universities listed among the 120 top north American universities. This

Supported by 2019 Heilongjiang province higher education and teaching research reformation fund (No. SJGY20190214) and Harbin Institute of Technology "Smart Base" project.

J. Zeng et al. (Eds.): ICPCSEE 2021, CCIS 1452, pp. 465–474, 2021.
https://doi.org/10.1007/978-981-16-5943-0_38

survey summarizes four categories of assignments in COA course: problem sets, assembly language programming, the Computer architecture design and test using HDLs-based tools, exploring computer architecture topics using high-level programming and instrumentation tools. In these four categories assignments, the third category is the most important part for COA experiment teaching. Students use HDLs tools to design components of the computer system on FPGA-based hardware platforms [6,7,10].

Although existing HDLs tools (e.g. Vivado [17] or Quartus [8]) facilitate hardware design greatly, they are all designed for expert hardware engineers. Compared to other programming software, students need more patient and effort for HDLs programming and debugging. It is very difficult for students to validate their hardware design and analysis errors, especially for the pipelined processor or cache hierarchy. How to help students get the hardware design validation and find the failure root is very important in COA experiments. To this end, we implement an analysis and validation toolkit for COA experiments.

For every COA experiment, we decompose its design scheme and abstract the most important signals in correctness determination. Based on these signals, we further design the test cases for every experiment. The test cases are completed in HDLs. In order to help students better debugging errors, we summarize the failure root and insert the corresponding references into the test cases when the experiments test case is not passed. As FPGA-based hardware design commonly have two steps: waveform simulation and on-board testing. The test cases are also designed for these two steps respectively. Two kinds of test cases are implemented as two packages and compose the analysis and validation toolkit. It needs to mention that our toolkit can be compiled as an IP (intellectual property) core and becomes a plug-in component for the HDLs tool.

The organization of the paper is as follows: Sect. 2 explains the background and challenges in computer architecture courses. Section 3 introduces our validation and analysis toolkit in detail. Section 4 presents the comparison results for using and not using the toolkit in the COA course. Finally, Sect. 5 is the conclusion.

2 Background and Challenges

2.1 The COA Lab Projects

In 2018, our computer science and technology school has finished the innovation of the major curriculum according to the building requirements of "China's double first class" discipline. The computer organization and architecture course also reformed in the new major curriculum which follows the "bottom-up" teaching mode.

The principles of bottom-up teaching mode can be concluded as [3,4,15]: (1) from simple to complex, (2) from easy to difficult, (3) from concrete to abstract. Based on these principles, the experimental teaching of computer organization and architecture course is revised. We decompose a prototype computer system, which is a MIPS (Microprocessor without interlocked pipelined stages) CPU based computer system [12], into 10 series progressive lab projects. Figure 1 shows the COA lab projects. The 10 lab projects are demonstrated as an engineering schedule control diagram. A real hardware development board, Digilent Nexys3 [5], is the hardware platform for lab projects.

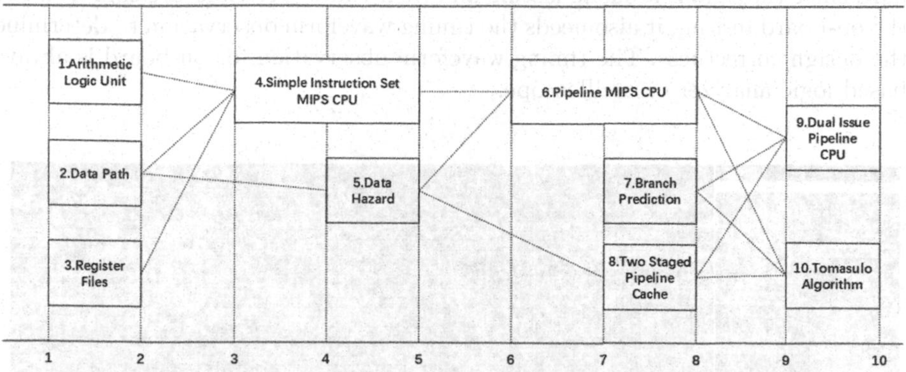

Fig. 1. COA experiment course

The 10 projects in Fig. 1 can be divided into three levels: (1) Register transfer level (RTL), (2) component level, and (3) system level. The RTL lab projects, lab 1, lab 2, and lab 3, they are basic experiments, mainly to help students get familiar with the Verilog programming environment and hardware design methods. The component level lab project, lab 4, they are the realization of the hardware components of the computer, mainly the simple instruction set CPU. Based on MIPS instruction set, we combine RTL level experimental projects to complete a simple CPU design. The system level experiment projects, lab 5 to lab 8, they complete a pipelined MIPS CPU, including the implementation of data conflict, branch prediction and two-level pipelined cache. In addition, based on this pipeline CPU, we can further complete the projects of dual issue pipeline (lab 9) and Tomsulo algorithm (lab 10) which realize the performance optimization at the computer system level.

In all 10 COA lab projects, lab 1 to lab 8 are compulsory. Lab 9 and lab 10 are additional projects for the students who want extra points. As lab 1 to lab 4 are very basic, we don't implement the test cases for them. For lab 9 and lab 10, they are prepared for few excellent students. We also don't implement the test cases for them. Thus, we just implement the analysis and validation test cases for lab 5 to lab 8, which are data hazard, pipeline MIPS CPU, branch prediction, and two stage pipelined cache, in our toolkit.

2.2 Challenges for COA Labs

In general, there are two steps in FPGA-based hardware design: waveform simulation and on-board testing [11]. For determining the design correctness, both two steps need to observe the timing waveform of signals [14,16]. Figure 2 shows the simulated waveform of a pipeline MIPS CPU. We can find there are a lot of signals in this waveform diagram. In order to check the correctness of the MIPS CPU design, students need to observe the timing of signals and determine whether the relations between signals are accurate. However, this work is very hard since the signals are long lasting and their timing relations are complex. For the on-board testing, it also needs the timing waveform observation to determine the design correctness. The timing waveform observation for on-board is always based logic analyzer or oscilloscope.

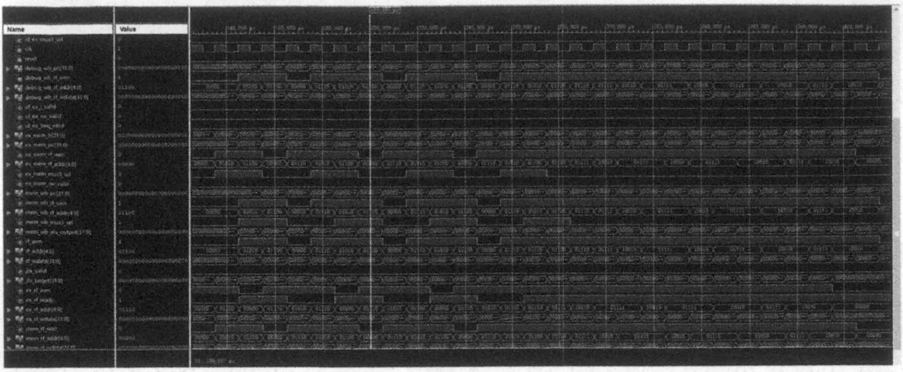

Fig. 2. Simulated waveform of a pipeline CPU

In the above discussion, we show students need massive effort to observe the timing waveform of signals. If we want to make the waveform observation easily, the test cases must be simplified. However, the simple test cases will lead to an incomplete validation for hardware design. More serious, for on-board testing, it needs the logic analyzers or oscilloscope to observe signals. A limited number of signals can be observed. When the design is complex and the number of signals is large, we can't observe all the key signals.

3 The Analysis and Validation Toolkit

In this section, we introduce the analysis and validation toolkit for the COA experiment teaching. The purpose of the toolkit is to replace signal timing observation in waveform simulation and on-board test by a series of test cases for determining the correctness of the COA labs. The test cases design goals includes:

(1) verifies the correctness: verifies the correctness of the experiment design and improves the detection efficiency

(2) Locates design errors: locates design errors, infers the cause of the error, and helps students standardize design

(3) Reduces the randomness: Expands the number of signal detection, increases the size of test cases, and reduces the chance and randomness of detection

(4) Avoids on-board debugging: Avoids using logic analyzers to capture signals for on-board debugging.

3.1 Basic Ideas

Students' design schemes are diverse when they do hardware design experiments. It is impossible to design test cases for the experiments of each student. Moreover, the number of signals used in hardware design is huge. It is impossible to detect all signals. We hope to design unified test cases to verify the lab correctness by using the most critical signals.

For example, suppose we need to test the memory access function, addition function and subtraction function of a pipeline MIPS CPU, we can use the benchmark in Fig. 3 to test. This benchmark has six instructions. The first three instructions write 1, 2, and 3 to the registers R1, R2, and R3. The 4th subtraction instruction writes −1 into register R6. The 5th and 6th add instructions write 4 and 3 into registers R4 and R5 respectively. After this benchmark execution, the data in register R1 to R6 is changed. In general, in MIPS CPU, the execution result commonly writes data back to a specific general-purpose register during the instruction execution.

```
R0 = 0           mem[i] = i
0x00000000       LW    R1, 1(R0)     ; mem[R0(1)] → R1
0x00000004       LW    R2, 2(R0)     ; mem[R0(2)] → R2
0x00000008       LW    R3, 3(R0)     ; mem[R0(3)] → R3
0x0000000c       SUB   R6, R1, R2    ; R1  -  R2 → R6
0x00000010       ADD   R4, R1, R3    ; R1  +  R3 → R4
0x00000014       ADD   R5, R1, R2    ; R1  +  R2 → R5
```

Fig. 3. Benchmark of pipeline MIPS CPU

Figure 4 shows the simplified structure of a five stage pipelined MIPS CPU (i.e., lab 6). It is a typical MIPS CPU which includes the instruction fetch (IF) stage, the instruction decoding stage (ID) stage, the execution (EX) stage, the memory access (MEM) stage, and the write-back (WB) stage. When the first instruction in Fig. 3 executes, the program counter PC (wb_pc) is 0×00000000, the enable signal written back to the register (rf_wen) is 1, the general register number (rf_waddr) is 1, and the data (rf_wdata) is 1. These four signals are shown in Fig. 4. We can only detect these 4 write-back related signals to verify the correctness of the pipelined MIPS CPU.

Similar to this five stage pipelined MIPS CPU, we analyze the design scheme of other lab projects and their benchmarks. The key signals of these lab projects' benchmarks are abstracted for generating the corresponding test cases.

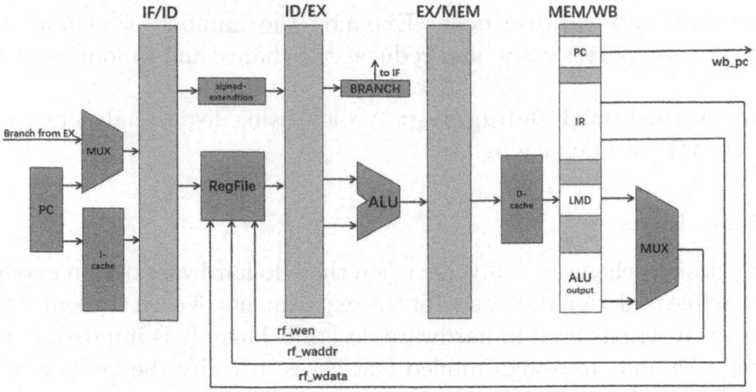

Fig. 4. Verification method

3.2 Test Cases for Waveform Simulation Analysis and Validation

The simulation test cases use the verilog programming language which is a widely applied HDL. The test cases are embedded in the testbench of HDLs tools. For every experiment, we abstract the key simulation incentive signals. Then, based on these simulation incentive signals, we predefined their standard results. Taking the five stage pipelined MIPS CPU as an example, the benchmark in Fig. 3 is a test case and the content in the instruction memory is the simulation incentives. If the MIPS CPU experiment runs, its program counter (PC) and register write-back information are written into a comparison file. The standard comparison results for simulation verification are generated.

Figure 5 shows a code fragment of the MIPS CPU test case. The red box is the write-back results comparison part. The signal debug_rf_wen is the register write-back enable signal captured by the HDLs tool. When this signal is 1, it means that the current instruction's operation result will be written back to a register. The PC signal debug_wb_pc, the write-back address debug_rf_waddr, and the written-back data debug_rf_wdata will be compared with the correct result PC signal ref_wb_pc, the standard write-back address ref_rf_waddr, and the standard write-back data ref_rf_data.

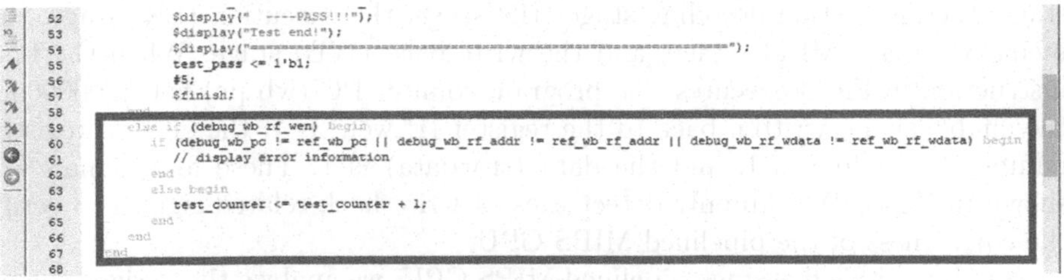

Fig. 5. Comparison mechanism of simulation test case

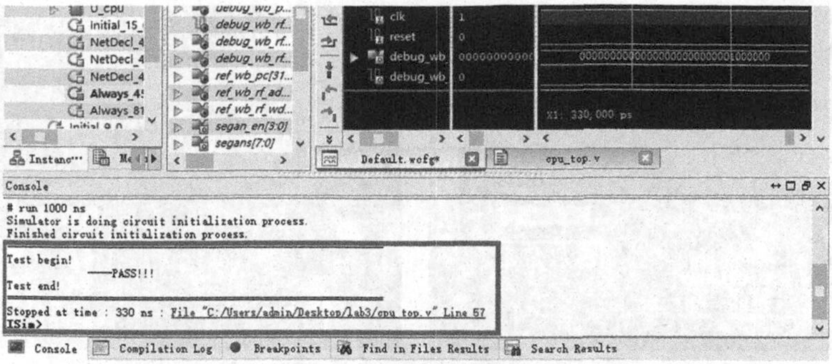

Fig. 6. CPU waveform simulation success validation

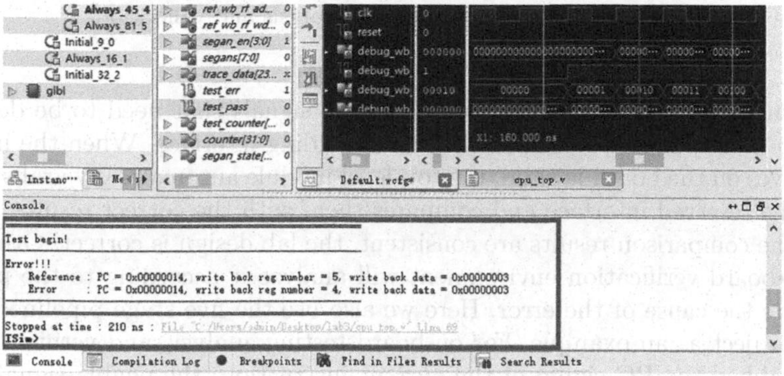

Fig. 7. Cause analysis for CPU waveform simulation

If all write-back results are the same as the standard results, the MIPS CPU runs correctly. Then, the console will display "PASS" to inform students that this design is correct. Figure 6 shows the screenshot of the correct validation of the MIPS CPU. With the "PASS" information, students can get the lab validation rapidly and continue other test cases.

When the comparison results are inconsistent, it means the design is incorrect. To help students find the error, we define some references for error analysis in the test cases. The references include the error location, the detected error signals, and correct signals. Figure 7 shows an example of the error in the "ADD R5, R1, R2" instruction. The console prints out the captured signals and the correct results. Students can find the error register number in the console.

3.3 Test Cases for On-Board Testing Analysis and Verification

When a lab design passes all waveform simulation test cases, it will carry out the on-board testing. Here we adopted the module interface for implement on-board test cases. We realize the top-level module to control the I/O interface and

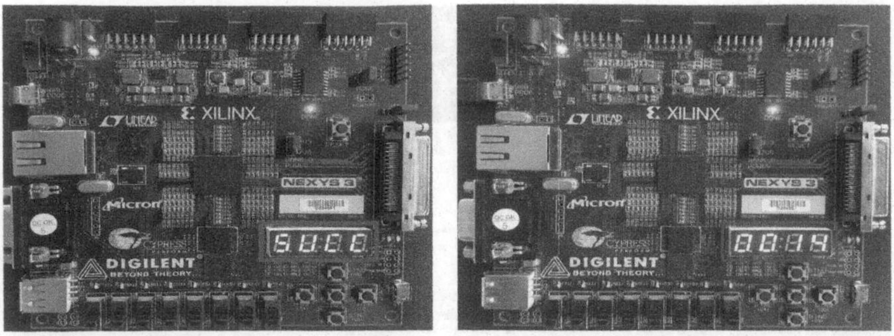

(a) CPU on-board testing verification success

(b) CPU on-board testing analysis

Fig. 8. CPU on-board testing analysis and validation

the comparison mechanism. According to the signals that need to be detected, the top-level module predefined the corresponding interface. When the input of incentives on the board is valid, the top-level module automatically grabs signals from the reserved interface and compares them with the correct results.

If the comparison results are consistent, the lab design is correct. Otherwise, the on-board verification environment will analyze it according to the position and give the cause of the error. Here we also use the five stage pipelined MIPS CPU project as an example. For on-board testing analysis and verification, we also grabbed the PC signal of the current instruction, the enable signal of the register write-back, the address signal of the write-back, and the data signal of the write-back. The difference between waveform simulation and on-board testing is that we achieve write-back results comparison by using the top module of the CPU and the I/O interface on the board.

Figure 8(a) shows the on-board test passed. "SUCC" is displayed on digital tubes and indicates that the on-board verification is passed. Figure 8(b) is the scene when there is an error in the five stage pipelined MIPS CPU experiment. Digital tubes display the PC when the error occurred. The current PC with the error is 0×0014, which means the instruction at 0×0014 is failed. It is "ADD R5, R1, R2" instruction. Error analysis helps students determine the incorrect instruction. The on-board testing analysis and validation test cases provide a new way for hardware correctness analysis. Students have not to use logic analyzers and other tools to re-grab signals, which greatly reduces the complexity of the hardware design.

For one lab project, we designed a series of test cases for analysis and validation. The test cases of lab 5 to lab 8 for waveform simulation and on-board testing are composed two packages respectively. Then, the waveform simulation testing package and the on-board testing package are integrated into the analysis and validation toolkit. This toolkit can be compiled as a IP core which is easy to plug into the HDLs tools. Students can invoke it during the different hardware design steps of COA experiments.

4 Results

In 2020, we complete the analysis and validation toolkit for the COA course and use it in our experimental teaching. Table 1 shows the comparison results with using and not using the toolkit. From the table, we can find that students spend the time on labs is reduced. The implement ratio and the assessment for the experiment part of the COA course are also increased. But for the mean score, the average score points of lab 6 and lab 8 dropped slightly because we use complex test cases in these two lab projects. Students need to pass more difficult test cases to get higher score points. In summary, our analysis and validation toolkit assistants students in their COA lab design and improves the experiment teaching of the COA course.

Table 1. The toolkit effectiveness in COA labs

	2019				2020			
	lab 5	lab 6	lab 7	lab 8	lab 5	lab 6	lab 7	lab 8
Mean time (hours)	2.36	6.82	3.49	1.73	2.18	6.27	3.21	1.57
Achieved ratio	94.87%	84.62%	89.74%	97.44%	95.74%	91.49%	93.61%	97.87%
Average score point	81.79	75.32	77.37	91.89	89.97	74.67	82.11	90.07
Experiment assessment	87.98	80.78	91.25	93.57	90.60	85.63	92.22	98.29

5 Conclusion

In this paper, we present the development of an analysis and validation toolkit for the computer organization and architecture course. Applying the validation and analysis package in computer architecture courses allows students not only to obtain the correct feedback rapidly but also to determine the root of error quickly when their experiments fail. The comparison results of using and not using our toolkit demonstrate that it improves the teaching quality of COA experiments.

References

1. ACM, IEEE: Computer engineering curricula (2016). https://dx.doi.org/10.1145/3025098. Accessed Mar 2019
2. Aydogan, T., Ergun, S.: A study to determine the contribution made by concept maps to a computer architecture and organization course. Eur. J. Contemp. Educ. **15**(1), 76–85 (2016)
3. Brailas, A., Koskinas, K., Alexias, G.: Teaching to emerge: toward a bottom-up pedagogy. Cogent Educ. **4**(1), 137–145 (2017)
4. Cummings, R., Phillips, R., Tilbrook, R., Lowe, K.: Middle-out approaches to reform of university teaching and learning: champions striding between the top-down and bottom-up approaches. Int. Rev. Res. Open Distrib. Learn. **6**(1) (2005)

5. Digilent: Nexys 3 spartan-6 fpga trainer board (2021). https://store.digilentinc.com/nexys-3-spartan-6-fpga-trainer-board-limited-time-see-nexys4-ddr/. Accessed Mar 2021
6. Gao, Z., Lu, H., Guo, H., Luo, Y., Xie, Y., Fang, Q.: An analogous teaching method for computer organization course design. In: 2016 8th International Conference on Information Technology in Medicine and Education (ITME), pp. 414–418 (2016). https://doi.org/10.1109/ITME.2016.0099
7. Harris, S.L., et al.: MIPSfpga: using a commercial MIPS soft-core in computer architecture education. IET Circ. Dev. Syst. **11**(4), 283–291 (2017). https://doi.org/10.1049/iet-cds.2016.0383
8. Intel: Quartus prime software suite (2021). https://www.intel.com/content/www/us/en/software/programmable/quartus-prime/overview.html?wapkw=%20Quartus. Accessed Mar 2021
9. Kehagias, D.: A survey of assignments in undergraduate computer architecture courses. iJET **11**(6), 68–72 (2016)
10. McGrew, T., Schonauer, E., Jamieson, P.: Framework and tools for undergraduates designing RISC-V processors on an FPGA in computer architecture education. In: 2019 International Conference on Computational Science and Computational Intelligence (CSCI), pp. 778–781 (2019). https://doi.org/10.1109/CSCI49370.2019.00148
11. Palnitkar, S.: Verilog HDL: a guide to digital design and synthesis. Verilog HDL: a guide to digital design and synthesis (1996)
12. Patterson, D.A., Hennessy, J.L.: Computer Organization and Design, Fifth Edition: The Hardware/Software Interface. The Hardware/Software Interface, Computer Organization and Design, Fifth Edition (2013)
13. Qiao, B.Y., Zhao, X.G., Yuan, Y.: Teaching reform practice of computer architecture. Educ. Teach. Forum **51**, 85–86 (2019)
14. Qin, G., Hu, Y., Huang, L., Guo, Y.: Design and performance analysis on static and dynamic pipelined CPU in course experiment of computer architecture. In: 2018 13th International Conference on Computer Science Education (ICCSE), pp. 111–116 (2018). https://doi.org/10.1109/ICCSE.2018.8468729
15. Shu, Y., et al.: Bottom-up teaching reformation for the undergraduate course of computer organization and architecture. In: Mao, R., Wang, H., Xie, X., Lu, Z. (eds.) ICPCSEE 2019. CCIS, vol. 1059, pp. 303–312. Springer, Singapore (2019). https://doi.org/10.1007/978-981-15-0121-0_23
16. Wang, L., Yu, Z., Zhang, D., Qin, G.: Research on multi-cycle CPU design method of computer organization principle experiment. In: 2018 13th International Conference on Computer Science Education (ICCSE), pp. 760–765 (2018). https://doi.org/10.1109/ICCSE.2018.8468694
17. Xilinx: Vivado design suite (2021). https://www.xilinx.com/products/design-tools/vivado.html. Accessed Mar 2021

Evaluation and Prognostics of the Higher Education Based on Neural Network and AHP-PLS Structural Equations

Fanyueyang Zhang[1], Xunbo Zhang[2], Zirui Tang[3], and Xianhua Song[1(✉)]

[1] School of Science, Harbin University of Science and Technology, Harbin, China
songxianhua@hrbust.edu.cn
[2] School of Economics and Management, Harbin University of Science and Technology, Harbin, China
[3] School of Measurement and Control Technology and Communication Engineering, Harbin University of Science and Technology, Harbin, China

Abstract. As a subsystem of society, higher education is an inevitable choice to meet the political and economic development of a country. Research of the higher education evaluation system is of great significance to the development of society. In this paper, a backpropagation (BP) neural network model is established to predict the future development scale of higher education. The analytic hierarchy process (AHP) method and partial least squares regression (PLS) structural equations were used to verify the scientificity and feasibility of the model. BP neural network has strong nonlinear mapping capabilities, and it is capable of the prognostics. It performed well on issues with more complicated internal mechanisms. Through experimental simulations, it is found that the BP neural network model has a good fit when making predictions and the relative error is less than 3%, which shows that the prediction results obtained with this model have high reliability.

Keywords: AHP-PLS structural equations · BP neural network · Higher education system

1 Introduction

The evaluation system of higher education is extremely important for the development of a country, and the health of the higher education system determines the future development of a country and the sustainable and healthy development of the national culture. The health of the higher education system determines the sustainable and healthy development of national culture. To evaluate the quality of higher education development, we must first clarify the connotation of the quality of higher education development [1]. Taking China as an example, this paper introduces the development of higher education and its scale forecast.

Using China as an example, it can be seen from the reference [2] that the development of higher education has roughly gone through four stages. The exploratory period

J. Zeng et al. (Eds.): ICPCSEE 2021, CCIS 1452, pp. 475–490, 2021.
https://doi.org/10.1007/978-981-16-5943-0_39

before the founding of New China (192 ~1948) formed the basic area education thought and formed the "Yan'an Model", which accumulated valuable experience and laid the foundation for the development of higher education. From 1966 to 1976, China higher education was severely damaged, leading to a dramatic improvement in the quality of higher education. China higher education did not return to normal until the unification of the college entrance examination system was restored in 1977. The optimization period of reform and development is from 1978 to 2015. In that period, Chinese higher education has accelerated development, the scale of higher education has continued to expand, the quality has been greatly improved, the reform of the management system has achieved qualitative development, and the development of higher education has achieved unprecedented achievements. This stage of Chinese comprehensive strength of higher education has been greatly improved since 2016: the scale of higher education continues to expand, the legal system governing education according to law has initially formed, and the education evaluation system with the participation of multiple subjects has been constructed.

Judging from the current actual situation, the academic circles have not formed a unified understanding of the quality of higher education development. In terms of social development as a whole, development quality refers to the sum total of endogenous, coordination, symbiosis and efficiency embodied by social organisms in the process of development. It reflects the characteristics of social organisms that meet their internal needs with their total resources, and the optimized state of each component in a certain time and space during its operation. On this basis, we can define the quality of higher education development as: talent training, scientific research and social development in the process of higher education movement from small to large, from weak to strong.

The development of higher education is an important part of social development. Continuously improving the quality of higher education development has become a prominent problem facing the development of higher education in the popular era. But so far, the quality of higher education development has not really attracted the attention of experts, scholars and education administrators [3]. Research on the development of higher education in the theoretical field is still mainly limited to development concepts, development goals, development models, development mechanisms, development motivation, development laws and other aspects. In reference [4], the Yangtze River Delta is used as the research area to determine the Yangtze River Delta The development goal of the high-quality development of higher education integration is an active growth pole for regional higher education development, a sample area and demonstration area for the high-quality development of regional higher education integration, and a comprehensive reform pioneer area for the coordinated development of regional higher education. Reference [5] pointed out that, as far as the current stage of development is concerned, higher education has gradually begun to develop in a new direction. It should adopt the "Internet+" platform method and use network technology and information technology as the basis to promote early development. Teaching methods have changed to ensure that educational activities are more dynamic and to promote the completion of information transformation in higher education. Regarding the development momentum of higher education quality, reference [6] pointed out that innovation and entrepreneurship education promotes the high-quality development of higher education. The quality of research and development, especially the development of discussion and evaluation quality, is particularly rare [7]. On this basis, this article discusses the evaluation of the development quality of higher education and predicts the future development, with a view to improving the quality of international higher education development.

In summary, this article has made the following contributions: (1) The Delphi method and the analytic hierarchy process are used to establish a higher education evaluation system and calculate the weights to provide basis and ideas for the evaluation of the health rating indicators of the higher education system. (2) Establish the PLS structural equation model, select the development of different regions in China as the research object, sort and classify the development level of higher education in different regions, and verify the feasibility of the model. (3) Regarding regional economic development as an influencing factor affecting the regional development of higher education, and promoting the coordinated development of regional economy and society as a long-term important policy, according to the regional differences in China's economic and social development, the coordinated development of higher education in China suggestions. (4) Established a BP network model to predict the future development scale of higher education.

2 The Preparatory Work

2.1 AHP—Satty Scale Method

Analytic Hierarchy Process (AHP) provides a simple and scientific decision-making method for complex problems by hierarchizing, quantifying and then analyzing them with mathematics [8]. It has been widely used in society, management, economy and education.

2.1.1 Construction of Reciprocal Judgment Matrix

When using the analytic hierarchy process to solve specific problems, what are the judgment matrices and what elements in each judgment matrix are determined by the hierarchical model structure of the problem, and the structure of the hierarchical model is determined by the characteristics of the specific problem and the business involved decided. When the hierarchical model is finalized, the judgment matrix is finalized. The process of constructing the hierarchical model itself is also an important link in the comprehensive evaluation or decision-making using the analytic hierarchy process.

In this project, experts are selected and sent to them a weighted opinion consultation table for each dimension to explain the method and standard of value assignment. Experts are asked to compare one by one through pair comparison among indicators. 1–9 and its reciprocal are used as the scale for value assignment to judge the difference in the importance of two elements. The thinking judgment is quantified, as shown in Table 1.

The above-mentioned Satty scale method is used to conclude and sort out the consultation results through expert consultation, and the following discrimination matrix is obtained [9].

Table 1. 1–9 scale principle

Scale	Meaning
1	Indicates that two elements are of equal importance compared to each other
3	Indicates that one element is slightly more important than the other
5	Indicates that one element is significantly more important than the other
7	Indicates that one element is more important than the other
9	Indicates that one element is extremely important compared to the other
2,4,6,8	2, 4, 6, and 8 are the median values of the above adjacent judgments
Reciprocal	Represents the importance of comparing the exchange order of the corresponding two factors

2.1.2 The Weight Calculation Process

The calculation process of single-layer weight of judgment matrix is shown as follows:

(1) Adopt the multiplication root method to calculate the geometric mean value of each line of the judgment matrix ($\overline{W_i}$) [10].

$$\overline{W_i} = \left(\prod_{j=1}^{n} a_{ij}\right)^{\frac{1}{n}} \quad \text{i, j} = 1, 2, \ldots, \text{n} \qquad (2.1.1)$$

Where, a_{ij} represents the elements in row i and column j in the original judgment matrix, n represents the number of indicators, and $\overline{W_i}$ represents the geometric average value in row i of the original judgment matrix.

(2) The feature vectors are obtained by normalizing the geometric mean values of each line:

$$W_i = \frac{\overline{W_i}}{\sum_{j=1}^{n} \overline{W_j}} \quad \text{i, j} = 1, 2, \ldots, \text{n} \qquad (2.1.2)$$

Where, W_i is the weight of the ith index. n is the number of indices, W_i is the geometric average of row i of the original judgment matrix. The calculation results of the weight coefficient of the first level index are obtained.

(3) The maximum eigenvalue of the judgment matrix was calculated λ_{max}[11].

$$\lambda_{max} = \frac{1}{n}\sum_{i=1}^{n} \frac{\left(\sum_{j=1}^{n} a_{ij} W_j\right)}{W_i} \quad \text{i, j} = 1, 2, \ldots, \text{n} \qquad (2.1.3)$$

Where, a_{ij} is the element in row i and column j in the original judgment matrix, n is the number of indicators, W_i is the weight of index i, and λ_{max} is the maximum eigenvalue of the judgment matrix. From this, the maximum eigenvalue can be calculated.

2.1.3 The Weight Calculation

MCE software is used to calculate the discriminant matrix. MCE (Modern Comprehensive Evaluation) is software used to deal with complex Comprehensive Evaluation problems. MCE provides three comprehensive evaluation methods which are widely used in modern times: AHP (Analytic Hierarchy Process), Fuzzy (Fuzzy) and Gray (Gray).

2.1.4 Consistency Test

In order to judge whether the above weight calculation results are scientific, consistency test must be carried out. The so-called consistency test is to determine whether the Cr value is within the allowable range of inconsistency through scientific calculation [12].The eigenvector corresponding to the maximum eigenvalue is used as the weight vector of the influence degree of the factors being compared on the upper layer of a factor. The greater the degree of inconsistency, the greater the judgment error will be. In this project, the consistency test module of MCE software was used to conduct consistency test on the results. The standard for whether the consistency test is passed is: CR represents the consistency ratio, that is, to judge whether the CR value is less than 0.1.When Cr is less than 0.1, the calculation results of the matrix pass the consistency test, and the weight calculation results are reliable. On the contrary, the weight calculation result is not reliable. According to the above AHP analysis results, the Cr of all the discriminant matrices is less than 0.1, so the judgment of each criterion layer is consistent. Through consistency test, the calculated results are scientific and effective.

2.2 PLS Structural Equations

The structural equation model (SEM) can fit, modify and evaluate the subjectively constructed index system based on the collected objective data to obtain a relatively simple and accurate index, and determine the size and weight of the index system through the fitting factor load. The PLS structural equation model is a statistical method proposed by the Swedish statistician H. Wold to construct a predictive structure model, which combines principal component analysis and normative correlation analysis. The Bootstrap method is used for parameter estimation and importance testing. The Bootstrap method is one of the resampling techniques [13]. This method obtains the sample distribution of statistical information from the original sample. The process is as follows: ①Resampling. First, select a random sample from the research population as the original sample. In the original sample, re-sampling is performed using alternative sampling methods. ②Calculate the Bootstrap distribution. Calculate the statistics of each Bootstrap sample, and the partition formed by these statistics is the Bootstrap partition. The Bootstrap distribution approximates a statistical sampling distribution. (3) Estimate and infer overall parameters [14]. Use the Bootstrap distribution to obtain information about the shape, center, and ductility of the statistical sample distribution, and then estimate and infer overall parameters. In order to overcome the small sample size and the robustness of parameter estimation, the sample size of Bootstrap is generally set to 500 or greater than 1000. Compared with the covariance structural equation model such as LISREL, the PLS structural equation model has the following advantages: (1) The premise is relatively loose There is no strict requirement on the normal distribution of variables, and the latent variable score can be directly calculated; (2) In the case of small samples, the accuracy and robustness of parameter estimation is very good, and the measurement of reflection type and formation type can be processed at the same time model. (3) It can be used when LISREL cannot be used. For example, when the number of indicators is greater than the number of samples, the sample covariance matrix is not positive definite. In this case, the model estimated by LISREL usually does not converge, but PLS can still be used in this case. (4) Good performance in overcoming the problem of multicollinearity. An-Anderson et al. have applied this method to a study by the International Association for Evaluation of Educational Achievement, which examined the impact of factors such as school environment, teacher and student personality traits on student academic performance in nine countries. In addition, C. Guinot et al. proposed a comprehensive index construction method based on the PLS structural equation model. The comprehensive indicators constructed in this way can not only reflect the information contained in the latent variables, but also have the strongest correlation with all specific indicators. Wang Huiwen and others used this method to establish a comprehensive urban evaluation index, and conducted a comprehensive evaluation of the city, and achieved satisfactory results.

2.3 BP Neural Network

BP (Back Propagation) neural network refers to the multi-layer feed forward neural network with error Back propagation. BP neural network is a multi-level acyclic network which adopts the minimum error mean square criterion and nonlinear differentiable function to train or learn the weight [15]. The basic idea of BP algorithm is gradient descent (GD) method. GD method uses gradient search technique in order to minimize the mean squared error between the actual output value and the desired output value of the network. The structure of BP neural network is composed of input layer (1 layer), hidden layer (1 layer or multiple layers) and output layer (1 layer), as shown in Fig. 1:

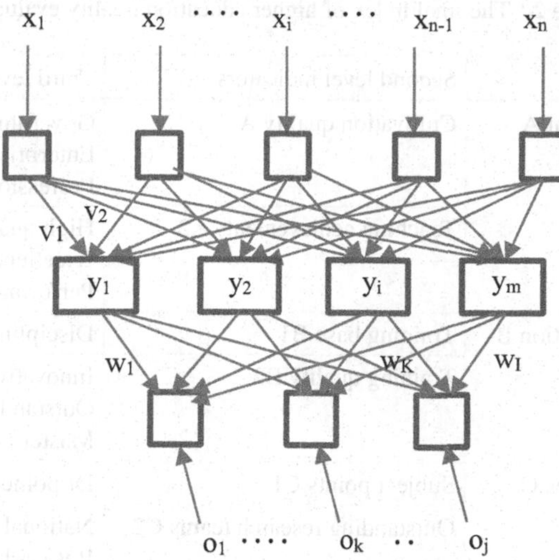

Fig. 1. BP neural network structure diagram

The number of nodes in the hidden layer of the BP neural network is one of the important parameters that affect the performance of the neural network. If the number of nodes is too small, the fault-tolerant performance of the network model is poor. Although increasing the number of nodes can enhance analysis capabilities and improve the convergence performance of the network, it will also complicate network training and prolong the learning time. The number of nodes in the hidden layer is directly related to the actual needs of the problem and the number of nodes in the input and output layers.

3 Evaluation and Prediction of the Health Status of the Higher Education System

3.1 The Evaluation Basis and Ideas

This part adopts the Delphi method and proposes the Analytic Hierarchy Process (AHP), which analyzes and simulates the index forms of different dimensions. In the first round of index evaluation, there are seven experts to evaluate the importance and necessity of each aspect. , Feasibility, and based on the professional knowledge of the evaluation, provide basis and ideas for the evaluation of the health rating indicators of the higher education system.

Use the average value and the coefficient of variation to assess the importance of setting indicators. Use SPSS software to establish a database, record the scores of experts and perform statistical processing. Thereby retaining, merging and eliminating some indicators. After fully absorbing expert suggestions, four first-level indicators, and twelve second-level indicators and 30 third-level indicators were determined, as shown in Table 2.

Table 2. The final index of higher education quality evaluation

First level indicators	Second level indicators	Third level indicators
Personnel recruitment A	Cultivation quality A1	Government management Enterprise management Professional and technical
	Teaching achievements A2	High quality courses Excellent teaching material Performance results
The quality of education B	Training base B1	Discipline construction
	Training quality B2	Innovative talents Outstanding talents Master points
Educational resources C	Subject points C1	Dr points
	Outstanding research teams C2	National Innovation Research Group Academician Number of doctoral supervisors Number of Master Supervisors
Scientific research D	Paper publishing D1	Domestic Citation Database Papers and Applications Foreign Citation Database Papers Citations of academic works
	Scientific payoffs D2	National Science Award International Social Science Award Authorized invention patent Advance in science and technology
	Scientific research project D3	National Natural Science Program National Social Science Program National Basic Research Program
	Scientific research base D4	National Scientific Research Institution International-level research Science and technology transformation mechanism

(*continued*)

Table 2. (*continued*)

First level indicators	Second level indicators	Third level indicators
	Scientific research funds D5	National Natural Science Foundation National Social Science Fund

In this project, experts are selected and sent to them a weighted opinion consultation form for each dimension to explain the method and standard of value distribution. Experts are required to make a one-to-one comparison between indicators. We can get the following discriminant matrix: (Table 3)

Table 3. Classification index judgment matrix

Indicators	A1	A2	A3	A4
A1	1	2	2	3
A2	1/2	1	1	2
A3	1/2	1	1	2
A4	1/3	½	1/2	1
Single weight	0.4231	0.2274	0.2274	0.1222

In this paper, the MCE software is used to calculate the discriminant matrix. The following formula is used to calculate the comprehensive weight [16]: $W = W1 \times W2 \times W3$, and the calculation results are shown in the following table:

Table 4. The weight of quality evaluation index of higher education system

First level indicators	W1	Second level indicators	W2	Third level indicators	W3	W
A1	0.4231	B1	0.3333	C1	0.4126	0.0582
				C2	0.3275	0.0462
				C3	0.2599	0.0367
		B2	0.6667	C4	0.4000	0.1128
				C5	0.2000	0.0564
				C6	0.4000	0.1128

(*continued*)

Table 4. (*continued*)

First level indicators	W1	Second level indicators	W2	Third level indicators	W3	W
A2	0.2274	B3	0.7500	C7	1.0000	0.1706
		B4	0.2500	C8	0.6667	0.0379
				C9	0.3333	0.0189
A3	0.2274	B5	0.2500	C10	0.2500	0.0142
				C11	0.7500	0.0426
		B6	0.7500	C12	0.4179	0.0713
				C13	0.2485	0.0424
				C14	0.1090	0.0186
				C15	0.2245	0.0383
A4	0.1222	B7	0.2055	C16	0.7500	0.0188
				C17	0.2500	0.0063
		B8	0.2055	C18	0.2500	0.0063
				C19	0.7500	0.0188
		B9	0.1831	C20	0.4000	0.0089
				C21	0.2000	0.0045
				C22	0.4000	0.0089
		B10	0.1153	C23	0.2500	0.0035
				C24	0.7500	0.0106
		B11	0.1453	C25	0.0393	0.0007
				C26	0.5651	0.0100
				C27	0.2696	0.0048
				C28	0.1260	0.0022
		B12	0.1453	C29	0.2500	0.0044
				C30	0.7500	0.0133

3.2 PLS Structural Equation

Since this study compares the development level of provincial higher education, the sample size is only 31, which cannot meet the requirement that the sample size of the covariance structural equation model is not less than 200, so the PLS (Partial Least Squares) structural equation model is used [17]. In order to overcome the small sample size and the robustness of parameter estimation, the sample size of Bootstrap is generally set to 500 or more than 1000, and this article is set to 1000. This paper uses the PLS structural equation model to screen the indicators and verify the system model of higher education development indicators. Calculate and compare latent variables. The data was processed using SPSS 16.0 and the software SmartPLS 2.0 developed by Ringle, Wende and Will of the University of Hamburg in Germany.

By calculating the scores of potential variables in the five aspects of the comprehensive level of higher education development and the needs of higher education in each province, the pros and cons of the comprehensive level of higher education in each province and the development of various aspects are obtained. Regions can be checked and ranked. The results are shown in Table 5. According to the scores of the latent variables in Table 4, the SPSS software is used for cluster analysis, and the development of higher education in 31 provinces and regions in China is divided into five categories. The first category includes Beijing, which is far ahead of other provinces in terms of score. The second category includes Shanghai and Tianjin. The third category is Jiangsu, Zhejiang, Liaoning, Guangdong and Shandong in the east, Hubei, Jilin, Heilongjiang and Hunan in the middle, and Shaanxi, Sichuan and Chongqing in the west. The fourth category includes Fujian, Hebei and Hainan in the east. Shanxi, Anhui, Jiangxi and Henan; the west is Gansu, Ningxia, Guangxi, Inner Mongolia, Qinghai and Xinjiang; the fifth category includes the western provinces of Yunnan, Tibet and Guizhou.

Table 5. The comprehensive score and ranking of each province and region

Region	Personnel recruitment		The quality of education		Educational resources		Scientific research	
	Score	Rank	Score	Rank	Score	Rank	Score	Rank
Beijing	3.608	1	1.926	3	3.794	1	2.169	2
Shanghai	−2.341	2	2.054	1	2.043	2	2.546	1
Tianjin	1.112	3	2.048	2	0.244	11	2.143	3
Jiangsu	0.983	4	0.77	6	1.077	3	0.999	6
Zhejiang	0.776	5	0.077	15	0.734	4	1.019	5
Hubei	0.679	6	0.793	5	0.722	5	−0.058	12
Shaanxi	0.679	7	0.875	4	0.693	6	−0.323	17
Liaoning	0.594	8	0.203	13	0.684	7	1.093	4
Ji Lin	0.356	9	0.477	8	0.217	12	0.287	9
Heilongjiang	0.167	10	−0.093	16	0.49	8	0.718	8
Guangdong	0.162	11	0.24	12	0.395	9	0.095	10
Shandong	−0.056	12	0.242	11	0.016	13	−0.179	14
Hunan	−0.078	13	0.293	10	−0.038	14	−0.901	27
Sichuan	−0.112	14	−0.856	28	0.294	10	−0.385	20
Chongqing	−0.157	15	−0.363	23	−0.163	15	−0.341	19
Fujian	−0.462	16	0.133	14	−0.506	19	−0.386	21
Hebei	−0.507	17	−0.152	20	−0.547	21	−0.782	26

(continued)

Table 5. (*continued*)

Region	Personnel recruitment		The quality of education		Educational resources		Scientific research	
	Score	Rank	Score	Rank	Score	Rank	Score	Rank
Shanxi	−0.508	18	0.55	7	−0.503	18	−0.599	23
Gansu	−0.53	19	−0.28	22	−0.642	22	−1.371	30
Anhui	−0.534	20	−0.536	25	−0.36	17	−0.625	24
Ningxia	−0.535	21	−0.126	18	−0.804	26	−0.269	15
Jiangxi	−0.623	22	−0.146	19	−0.351	16	−0.332	18
Hainan	−0.644	23	−0.27	21	−1.024	31	0.9	7
Henan	−0.676	24	−0.125	17	−0.703	25	−0.747	25
Guangxi	−0.697	25	−0.556	26	−0.545	20	−0.948	28
Inner Mongolia	−0.749	26	0.325	9	−0.982	29	−0.305	16
Yunnan	−0.76	27	−1.827	30	−0.658	23	−1.08	29
Qinghai	−0.868	28	−0.728	27	−1.01	30	0.045	11
Xinjiang	−0.888	29	−0.523	24	−0.963	28	−0.593	22
Tibet	−0.911	30	−2.632	31	−0.658	24	−0.15	13
Guizhou	1.157	31	−1.791	29	−0.944	27	−1.637	31

According to the scores and clustering results of 31 provinces and autonomous regions, the following characteristics are obtained: (1) Beijing's higher education development level is much higher than other provinces and autonomous regions, and is in a leading position. In terms of the supply of higher education, it shows its advantages as a political, cultural and educational center; Shanghai and Tianjin rank second, third and third respectively. (2) Some economically developed provinces, such as Zhejiang, have developed rapidly in higher education. Although the number of universities in the university layout is relatively small, due to its huge investment, the development level of higher education in Zhejiang surpassed that of Shaanxi, Hubei and the three northeastern provinces, and its university layout is relatively good (thanks to the supply of higher education in Zhejiang Province Ranked fourth) (behind Beijing, Shanghai and Jiangsu). However, the development performance of higher education in the two major economic provinces of Guangdong and Shandong lags far behind their economic performance, and the comprehensive level of higher education is at the middle level. (3) The level of higher education in developing western provinces is generally low. Except for Shaanxi, Sichuan and Chongqing, all other western provinces are in the fourth and fifth levels, while Yunnan, Tibet and Guizhou are the only provinces and regions in the fifth level [18]. The development level of higher education in the western region urgently needs to be improved. (4) In terms of higher education performance (the main indicator is the per capita resource conditions for running schools), the central provinces performed poorly. Jiangxi, Shanxi, Anhui, Henan and other provinces are ranked last, and the central provinces should strive to improve the conditions of educational resources for students. (5) Although Tibet is at the fifth level of higher education development, its per capita educational resources are relatively good (ranked fifth in higher education performance), and its higher education enrollment and participation rates are also higher than in some western provinces. This is mainly due to the continuous special support of the central government and eastern provinces.

The higher education development index system constructed according to the UNESCO Education Index System framework has been verified by the PLS structural equation model, meets the requirements, and can be used as an effective tool for analyzing the development level of higher education in China. This paper establishes a comprehensive evaluation model based on the index system of higher education development, compares the development level of higher education in 31 provinces and cities, and finds that the relationship between supply and demand of higher education is the two most influential factors. Among them, the indicators closely related to the overall level are: the ratio of higher education workers to educators, the number of patent applications per thousand teachers, the number of equipment per student, and the proportion of the population with higher education. It is necessary to focus on these aspects in order to improve the development level of provincial higher education. According to scores in terms of comprehensive level and higher education demand, the development level of higher education in China's 31 provinces, autonomous regions, and municipalities is divided into five categories. The level of higher education development in Beijing, Shanghai and Tianjin is ahead of other provinces. The level of higher education development in the eastern provinces is higher than that of the central and western provinces. The performance of developed provinces is different. Jiangsu and Zhejiang are higher than the other eastern provinces except Shanghai, while Guangdong and Shandong are in the middle level, lagging behind Shaanxi, Hubei and Northeast provinces [19]. In terms of educational resources for each student, the central provinces performed poorly.

3.3 Theoretical Basis for Using BP Neural Network to Predict the Future Scale of the Higher Education System

In this paper, through the number of training and the principle of overall minimum error, combined with the change of training error, finally determine the appropriate number of hidden layer elements to 8. The input variables are the total population, GDP, the proportion of tertiary industry in GDP, the Engel coefficient of rural residents and thc disposable income of urban residents. The output variable is the number of undergraduates and college students; it contains a hidden layer, and the number of hidden layer units is 8. Use the data from 2009 to 2020 as the training sample to train the neural network. The global error E of network training is 10–4, and the maximum number of training times is 300,000. BP neural network is used to predict the scale of higher education in China, as shown in Table 6.

It can be seen from the above table that the model fits well, and the relative error is less than 3%. Therefore, it can be said that the BP neural network model has more accurate prediction capabilities than traditional prediction models such as linear regression models. Therefore, it should be considered in education Application and promotion in forecasting and planning. Facts have proved that in the field of higher education resource quality sharing, higher education resources can be deeply explored, in the construction of high-quality, intensive management, effective utilization, supplementary supply, formation of a good higher education ecosystem, and enhancement of education in universities in the same region Overall quality, promote the development of the connotation of higher education.

Table 6. Actual value and predicted value of the number of college students. (Unit: 10,000)

Year	Actual value	Predictive value	Relative error
2009	3015	3021.46	0.21%
2010	3105	3142.586	1.20%
2011	3201	3226.748	0.80%
2012	3325	3335.713	0.32%
2013	3425	3449.068	0.70%
2014	3524	3564.628	1.14%
2015	3647	3676.484	0.80%
2016	3699	3676.892	−0.60%
2017	3779	3777.163	−0.05%
2018	3833	3878.571	1.17%
2019	3996	3981.754	−0.36%
2020	4106	4162.838	1.37%

4 Case Analysis

Based on the BP neural network model of the scale of China's higher education, the development speed and trend of the scale of higher education in Jiangxi Province in the next few years can be predicted, as shown in Table 7.

Table 7. Forecast of scale trend of higher education in 2021–2025

Year	2021	2022	2023	2024	2025
Scale of higher education	4285.347	4386.794	4467.746	4538.355	c
Annual growth rate	2.94%	2.37%	1.85%	1.58%	1.55%

As can be seen from the above table, after the implementation of the new policy, China's higher education scale will continue to expand from 2021 to 2025, and the annual growth rate will gradually decline.

5 Conclusion

The model used in this paper is applicable to all cases, and only takes China as an example in this paper. In this article, firstly, using AHP to analyze and simulate the indicator forms of different dimensions. In the first round of indicator evaluation, there are seven experts to evaluate the importance, necessity, and feasibility of each aspect. Based on the professional knowledge of the evaluation, it provides the basis and ideas for the evaluation of the health rating indicators of the higher education system. Secondly, the PLS structural equation model was established. The comprehensive indicators constructed in this way can not only reflect the information contained in the latent variables, but also have the strongest correlation with all specific indicators. Finally, the BP neural network model is used to evaluate the level of higher education in different regions of China and verify the practicability of the model. So far, the research on the quality evaluation of the higher education system is not comprehensive. The future research work can be carried out from several perspectives: First, the expert consultation sample used in the AHP method in this paper is 7, in order to improve the objectivity of the results and eliminate The influence of individual factors can expand the sample size of expert consultation; second, the results obtained in this paper are carried out in an ideal state, that is, data errors caused by accidental events are not increased. In order to improve the practicality of the model, we can collect some results Data caused by accidental events, and put them into the model to test the stability of the model. If the stability of the model is not enough, some parameters can be adjusted according to actual needs.

References

1. Li, Z., Shihuang, S., Xianhui, Z.: Development of standard examination system of special course for remote education. J. Donghua Univ. **19**(01), 99–102 (2002)
2. Qingquan, Z., Kangkai, Y.: The hundred years, achievements and prospects of the communist party of China in the development of higher education. Univ. Edu. Sci. **39**(02), 34–41 (2021)
3. Chen, S., Zhu, X., Androzzi, J.: Evaluation of a concept-based physical education unit for energy balance education. J. Sport Health Sci. **7**(03), 353–362 (2018)
4. Yijiang, L.: High-quality development goals and action paths for the integration of higher education in the Yangtze River Delta. J. Soochow Univ. (Edu. Sci. Edn.) **8**(4), 37–45 (2020)
5. Gan, L.: Discussion on the development of higher education under the background of Internet+. Educ. Teach. Forum **13**(1), 58–61 (2021)
6. Dandan, X., Feng, S., Ming, L., Jia, S.: Research on the driving mechanism and trend of innovation and entrepreneurship education leading the high-quality development of higher education. Educ. Teach. Forum **12**(19), 5–6 (2020)
7. Yijian, S., Rufu, H., Dilin, C.: Fuzzy set-based risk evaluation model for real estate projects. Tsinghua Sci. Technol. **13**(S1), 158–164 (2008)
8. Dong, Q., Saaty, T.L.: An analytic hierarchy process model of group consensus. J. Syst. Sci. Syst. Eng. **23**(03), 363–375 (2014)
9. Weihua, S.: Research on the Theory and Method of Multi-index Comprehensive Evaluation. Xiamen University (2000)
10. Jianbin, L.: Research on priority of analytic hierarchy process and fuzzy multi-attribute decision-making. Southwest Jiaotong University (2006)
11. Chen, W., Fang, T., Jiang, X.: Research and application of group decision making based on Delphi and AHP method. Comput. Electr. Comput. Eng. (05), 18–20 (2003)

12. Hongyan, L., Hui, L.: An improved MCE training algorithm for speaker recognition. Comput. Syst. Appl. **24**(06), 143–147 (2015)
13. Chunfu, L., Jie, Z., Guizeng, W.: Adaptive batch process quality prediction based on PLS model. J. Tsinghua Univ. (Sci. Technol.) **90**(10), 1363–1370 (2004)
14. Sheng, L., Jinlan, L., Wenxiu, H.: Customer satisfaction evaluation method based on PLS-structural equation. J. Syst. Eng. **21**(06), 653–656 (2005)
15. Minsheng, L., Bin,L.: Improvement and application of BP learning algorithm. Beijing Inst. Technol. **44**(06), 721–724 (1999)
16. Lam, B.K., Lee, E.W.: Estimating liver weight of adults by body weight and gender. World J. Gastroenterol. **12**(14), 2217–2222 (2006)
17. Qu, H.: Modeling method based on PLS. Zhejiang Univ. (Eng. Sci.) **44**(05), 19–22 (1999)
18. Yuhong, D.: An empirical study on the difference of higher education development in China. High. Educ. Res. **21**(03), 44–48 (2000)
19. Xingyun, L.: On the economic function of higher education and the development of regional higher education. Jiangsu High. Edu. **22**(05), 27–29 (2006)

Construction of Multiple Linear Regression Prediction Model of PRETCO-A Scores and Its Positive Backwash Effect on Teaching

Haiyun Han and Yuanhui Li(✉)

Sanya Aviation & Tourism College, Sanya Hainan 572000, China

Abstract. PRETCO-A is a standardized English proficiency test set up to evaluate the English application ability of the students in higher vocational college. In order to improve the passing rate, a multiple linear regression prediction model is constructed in this paper. A significance test was first performed on the regression model and the regression coefficient to verify a high correlation among the variables. The confirmed model was then put into application to predict the students' scores and identify the students who may fail the exam, leading to targeted tutoring assistance given to those students in advance. Finally, 60 students with predicted scores lower than 60 points were selected as research samples, and randomly divided into the control group and the experimental group, 30 students in each group. Finally, the experimental group students were given 40 teaching hours of precision assistance and targeted training, while the control group did not engage in any teaching intervention. The experimental results indicate that the pass rate of experimental group is 20% higher than the control group, which means the backwash effect of the test prediction is positive. The prediction model is proved to be scientific and reliable for teaching.

Keywords: PRETCO-A · Multiple linear regression · Prediction model · Significance test · Backwash effect · Targeted teaching assistance

1 Introduction

Practical English Test for College Students (PRETCO) was officially implemented in 2000, which has been adopted by more than 20 provinces, municipalities and autonomous regions in China. It has played an important role in promoting the application-oriented teaching reform of English curriculum in higher vocational colleges and has gradually been recognized by the talent market. PRETCO is divided into two levels, A and B. PRETCO-A is the standard requirement for higher vocational college students, which become an important means to measure the teaching quality of English course, a "benchmark" to evaluate students' English proficiency, and a career pass to ensure the students' employment after graduation in higher vocational colleges.

The backwash effect refers to the impact of the test on the teacher's teaching and students' learning, which is divided into positive effects and negative effects. Language tests, especially high-stakes large-scale language tests (such as CET-4/6, TEM4/8, CEE,

J. Zeng et al. (Eds.): ICPCSEE 2021, CCIS 1452, pp. 491–502, 2021.
https://doi.org/10.1007/978-981-16-5943-0_40

etc.) will undoubtedly affect teaching content, methods, progress and other aspects and also have a practical impact on students' learning attitude, motivation, effect, etc. The purpose of the research is to explore ways to reduce negative backwash effects and to study improve positive backwash effects.

Predicting test results and verifying its positive backwash effect help to improve the test passing rate. Many scholars have predicted the test results by constructing various models. For example, Li Nan and Hao Wenjia (2021) used logistic algorithm to predict the performance variables of the postgraduate entrance examination. [1]; Han Lina (2018) used the Bayesian classification model based on the results of a course's computer test, combined with the student's usual grades and students' survey data to predict the test results. The accuracy rate has reached more than 80% [2]; Chen Yong (2016) used genetic algorithm to optimize BP neural network, and then designed genetic neural network model which had high accuracy in test results prediction [3]; Cao Xinyu (2018) proposed a residual data interpolation method based on K-nearest neighbor local optimal reconstruction, combined with random forest model to achieve effective performance prediction [4]; Li Mengying and the other three coauthors (2020) proposed a two-way attention (TWA) model to predict the students' performance [5]; Wu Xiaoqian, Quan Lili, etc. (2020) proposes a student performance analysis and prediction model based on big data decision tree algorithm and its accuracy reached 94% [6]; Yang Donghai (2019) constructed a model to predict the students' exam results in the higher vocational college by using the orthogonal kernel least squares algorithm (OKLSA) [7]; Zhang Chao (2017) proposed that prediction models based on firefly optimization neural network could improve the accuracy of sports performance prediction [8]. However, from the existing research, there is little about the prediction of the scores of PRETCO-A test.

Since Hughes (1989) put forward the concept "washback effect" [9], research on the washback effect has received more and more attention. Bailey (1996) made a review of the washback concept in language testing [10]. Domestic research on the washback effect of language testing is mostly concentrated in large-scale national tests. For example, Gong yuhuan (2015) revealed the backwash effect of English testing on English teaching in the adult college [11]. Li Xu (2019) studied how to use the backwash effect of translation on cultures in CET-4 to introduce Chinese culture into college English classrooms to promote cultural output [12]; Wan Shuxia of Huaiyin Normal University studied the washback effects of the CET listening reform on college English teaching [13]. Xu Qian and Liu Jun adopts a multi-method and multi-phase approach to investigate the washback effects of Test for English Majors (TEM), shedding new light on TEM reform and the reform of English teaching and learning in China [14]. Han feifei reviewed the key research on the washback of CET-4, which shows a mixture of positive and negative effects of washback so some solutions to reduce the negative effects of the washback were proposed [15].

The backwash effect of language testing on teaching has been widely recognized, but there are few studies on how much the predictions on testing itself influenced the teaching. This paper intends to construct a set of prediction models for the PRETCO-A test results by collecting data of some students, so as to provide a reference for the prediction and evaluation of PRETCO-A test results and try to prove its positive backwash has a forward-looking guiding significance for teaching.

2 Data and Methods

2.1 Data Collection and Description

In this study, the scores data of 106 freshmen of the same major in the college SATC including scores of college English final exams, learning process evaluation, College Entrance Exam (CEE) and PRETCO-A in the first semester, were extracted from the database of educational administration system for research, as is shown in Table 1.

The students were enrolled after CEE, and attended the PRTETCO-A for the first time at the end of the first semester, which means they have had learned College English for almost a whole term before the PRETCO-A. All the examinations are organized by different examination committees and scores are distributed directly into the database of academic affairs department of the college. So all the data collected are comparatively objective and reliable.

Table 1. English scores of the 106 students

No.	PRETCO-A	CEE	Learning process evaluation	Final examination
1	60	73	99	53
2	34	64	81	29
3	60.5	57	90	48
4	44	43	89	43
5	32.5	38	92	39
……	……	……	……	……
……	……	……	……	……
104	61	91	100	58
105	52	49	70	46
106	52	50	88	33

2.2 Modeling Procedures

The modelling process is as follows: (1) selecting dependent variables; (2) determining the explanatory power of independent variables on dependent variables; (3) eliminating multiple correlations of independent variables; (4) fitting linear regression equations; (5) testing equations; (6) analyzing Residual; (7) confirm the model and use it for prediction.

Selection Dependent Variable. The goal of this study was to predict PRETCO-A test scores, so the scores of PRETCO-A in the samples were chosen as the dependent variables.

Determine the Explanatory Power of the Independent Variable to the Dependent Variable. The scatter plots figures of English scores in college entrance examination, final exam, learning process evaluation and PRETCO-A in the first semester are shown in Fig. 1, Fig. 2, and Fig. 3. It can be seen from the figures that there is no obvious linear dependence between the scores of PRETCO-A and the three variables, so it is not suitable to be fitted with the linear regression equation. Now try to introduce multiple linear regressions.

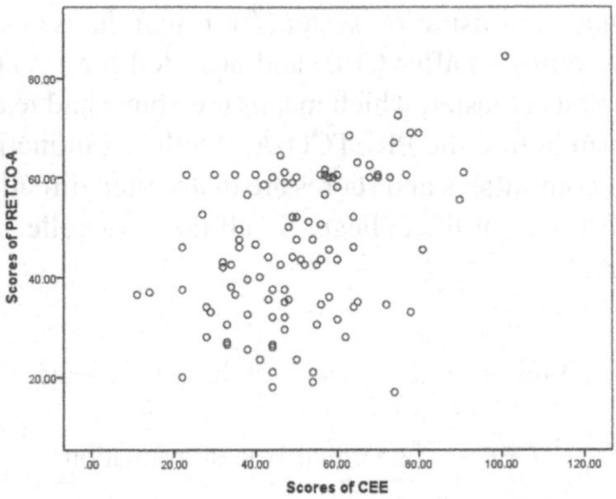

Fig. 1. Scatter plot of scores of PRETCO-A and CEE

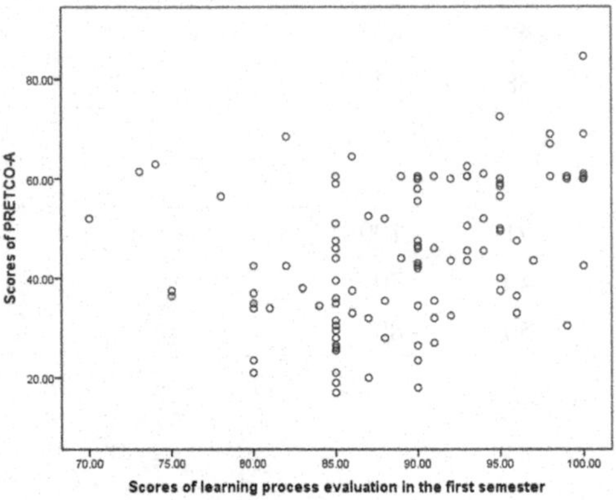

Fig. 2. Scatter plot of scores of PRETCO-A and learning process evaluation in the first semester

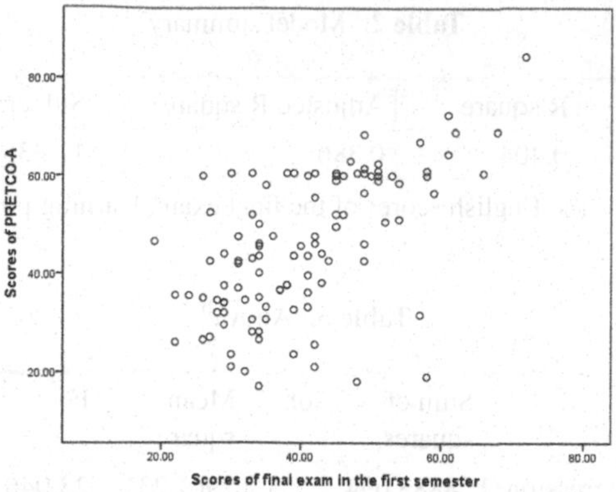

Fig. 3. Scatter plot of scores of PRETCO-A and final exam of english course in the first semester

The idea and hypothesis of multiple linear regression: multiple linear regression is a statistical method for studying the dependence of multiple variables. Among the multiple variables, the dependent variable is represented by $\hat{y}$, the independent variable $x_1, x_2, \cdots, x_n$, and the linear regression equation is as Eq. (1)

$$\hat{y} = \beta_0 + \beta_1 x_1 + \beta_2 x_2 + \cdots + \beta_n x_n + \varepsilon \quad (1)$$

$\beta_0, \beta_1, \beta_2, \cdots, \beta_n$ are the regression coefficient value, and ε is the random error of the Eq. (1).

Multivariate linear regression needs to satisfy the following five conditions, none of which is indispensable [16]:

Linearity is reasonable. There is only a linear relationship between dependent variables and independent variables, and there is no non-linear relationship.

The mean of random errors is zero, independent of each other, and the variances are equal, which satisfies the following Eqs. (2):

$$E(\varepsilon_i) = 0;\ var(\varepsilon_i) = \sigma^2;\ cov(\varepsilon_i, \varepsilon_j) = 0 \qquad i, j = 1, 2, \cdots, n, i \neq j. \quad (2)$$

Random errors have the same distribution and are subject to a normal distribution, as Eq. (3):

$$\varepsilon_i \sim N\left(0, \sigma^2\right),\ i = 1, 2, \cdots, n. \quad (3)$$

Random errors and independent variables are independent of each other in Eq. (4):

$$cov\left(\varepsilon_i, \varepsilon_j\right) = 0, \qquad i, j = 1, 2, \cdots, n. \quad (4)$$

The independent variables are not correlated with each other.

The SPSS software is used to substitute the three independent variables, such as the English scores of the CEE, final exam, learning process evaluation, into the multiple linear regression model. The results are shown in Table 2, Table 3 and Table 4 below.

It can be seen from Table 3 that the significance test of the equation can pass, and the regression coefficient is not all 0; however, in Table 4, if the significance level is

Table 2. Model summary

Model	R	R square	Adjusted R square	Std. error of the estimate
1	0.636[a]	0.404	0.386	11.33537

a. Predictors:(Constant), the English scores of the final exam, learning process evaluation, CEE

Table 3. Anova[b]

Model		Sum of squares	df	Mean square	F	Sig.
1	Regression	8883.694	3	2961.231	23.046	.000[a]
	Residuals	13106.035	102	128.491		
	Total	21989.729	105			

a. Predictors:(Constant), the English scores of the final exam, learning process evaluation, CEE.
b. Dependent Variable: scores of PRETCO-A

Table 4. Coefficients[a]

Model		Unstandardized coefficients		Standardized coefficients	t	Sig.
		B	Std. error	Beta		
1	(Constant)	−26.119	15.006		−1.741	0.085
	CEE scores	0.115	0.076	0.133	1.509	0.134
	Scores of learning process evaluation in the first semester	0.461	0.175	0.21	2.631	0.01
	Scores of the final exam in the first semester	0.61	0.117	0.466	5.227	0

a. Dependent Variable: scores of PRETCO-A

assumed to be 0.05, then the regression coefficients of the two variables, (the "first semester scores of learning process evaluation" and "the first semester volume score") are not significantly zero, and the regression coefficients of the remaining variables are not significant. The reason for this is that there is multiple collinearity between the independent variables in multiple linear regression.

Eliminating Multiple Correlations of Independent Variables. Multicollinearity refers to the strong linear correlation between variables, which destroys the fifth condition of multiple linear regression: independent variables are irrelevant. The stepwise regression method is used to allow SPSS to automatically select the appropriate independent variables to establish the regression equation. The variables with multiple collinearity with the selected variables will be excluded from the regression model, and a linear multi-collinearity fitting equation with smaller collinearity will be obtained. Table 5 shows the results after eliminating multicollinearity.

Table 5. Model summary

Model	R	R square	Adjusted R square	Std. error of the estimate	Durbin-Watson
1	0.587[a]	.344	.338	11.77338	
2	0.625[b]	.391	.379	11.40550	2.195

a. Predictors:(Constant), the English scores of the final exam, CEE.
b. Predictors:(Constant), the English scores of the final exam, scores of learning process evaluation

As can be seen from the goodness of fit shown in Table 5, the second model is improved over the first model, indicating that the second model is superior to the first model. Then look at the significance test results of the equation in Table 6.

Table 6. ANOVA[c]

Model		Sum of squares	df	Mean square	F	Sig.
1	Regression	7574.023	1	7574.023	54.642	.000[a]
	Residuals	14415.705	104	138.613		
	Total	21989.729	105			
2	Regression	8590.923	2	4295.462	33.020	.000[b]
	Residuals	13398.806	103	130.085		
	Total	21989.729	105			

a. Predictors:(Constant), the English scores of the final exam.
b. Predictors:(Constant), the English scores of the final exam, learning process evaluation.
c. Dependent Variable: scores of PRETCO-A

It can be seen from Table 6 that both models have passed the equation significance test, indicating that the regression coefficients of the two models are not significantly zero.

Table 7 shows the results of the coefficient significance test. It can be seen from Table 7 that the coefficients of the two models are significant. The second model adds the variable "scores of learning process evaluation" to the first model, and the goodness of fit increases significantly, so this variable should be increased.

Table 7. Coefficients[a]

Model		Unstandardized coefficients		Standardized coefficient	t	Sig.
		B	Std. error	Beta		
1	(Constant)	14.371	4.362		3.295	.001
	Final exam	.768	.104	.587	7.392	.000
2	(Constant)	−26.156	15.099		−1.732	.086
	Final exam	.690	.104	.528	6.613	.000
	Learning process evaluation	.490	.175	.223	2.796	.006

a. Dependent Variable: scores of PRETCO-A

Fitting Linear Regression Equation. In summary, the multiple linear regression fitting equation for PRETCO-A scores can be obtained, as shown in as Eq. (5):

$$\hat{y}_{\mathrm{i}} = -26.156 + 0.690x_{\mathrm{i}1} + 0.490x_{\mathrm{i}2} \tag{5}$$

Testing the Equations. The regression coefficient in the equations indicates that for every 10 scores increased in the final exam, the score of PRETCO-A is expected to increase by 6.9 points; for every 10 points increased in the learning process evaluation, the PRETCO-A score is expected to increase by 4.9 points.

Analysis of Residuals. Make a normal P-P diagram, as shown in Fig. 4.

As can be seen from Fig. 4, the residual data is basically subject to a normal distribution. The autocorrelation analysis of the residuals is then performed by the Durbin-Watson statistic. The Durbin-Watson statistic is defined as as Eq. (6):

$$D.W. = \frac{\sum_{i=2}^{n}(e_i - e_{i-1})^2}{\sum_{i=1}^{n} e_i^2} \tag{6}$$

The Durbin-Watson statistic is between 0 ~ 4. When the residuals are positively correlated, that is, e_i, e_{i-1} are close, the Durbin-Watson statistic is close to 0. Similarly, when the residuals are negatively correlated, the Durbin-Watson statistic is close to 4. When the Durbin-Watson statistic value is close to 2, it indicates that there is no autocorrelation of the random error.

As can be seen from Table 5, the Durbin-Watson statistic has a value of 2.195, which is close to 2, indicating that there is no autocorrelation in the sequence.

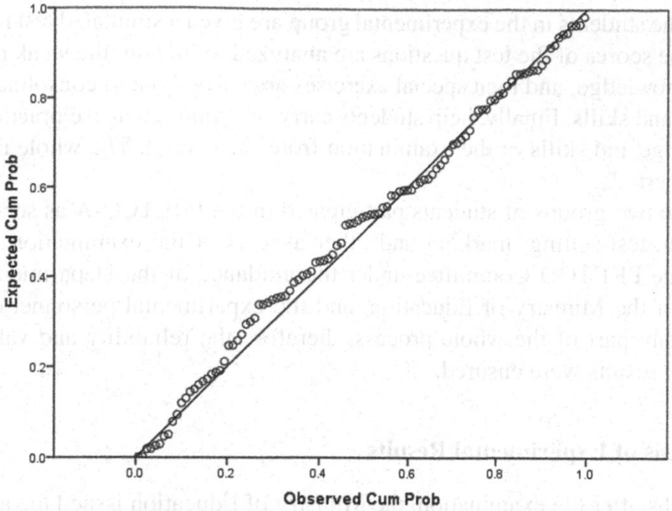

Fig. 4. Normal P-P plot of unstandardized residual

Confirm the Model and Use it for Prediction. The prediction model for the scores of PRETCO-A is now confirmed as Eq. (7):

$$\hat{y}_i = -26.156 + 0.690x_{i1} + 0.490x_{i2} \tag{7}$$

Then this model was applied to predict the 106 samples' PRETCO-A scores.

3 Positive Backwash Effect of the Model

3.1 Determination of Experimental Samples

Base on the predicted results, 60 students whose prediction score is lower than 60 points (excluding 60 points) as the experimental samples, and then they were randomly divided into control group and experimental group, with 30 samples in each group. The average predicted score of the control group was 51.07, and the average predicted score of the experimental group was 51.17. In addition, the two groups of students are from the same grade, the same major, so they are almost the same in the quality of students, curriculum, learning time and other aspects. Thus it ensures that there is no significant difference between the two groups of students before the experiment, and it reduces the interference factors of the experimental results as far as possible.

3.2 Experimental Procedures

On condition that the predicted scores and other aspects of the two groups were basically the same, the experimental group was given targeted tutoring assistance, while the control group was given no teaching intervention. Targeted assistance measures are as follows:

first of all, the students in the experimental group are given a simulated test in their spare time, and the scores of the test questions are analyzed to find out the weak points of the students' knowledge, and then special exercises are carried out to consolidate the basic knowledge and skills. Finally, help students carry out comprehensive practice to master the knowledge and skills of the examination from the overall. The whole helping time is 40 class periods.

Then the two groups of students participated in the PRETCO-A as scheduled. The organization, test-setting, marking and other aspects of the examination were in the charge of the PRETCO Committee under the guidance of the Department of Higher Education of the Ministry of Education, and the experimental personnel did not participate in any part of the whole process, therefore the reliability and validity of the examination results were ensured.

3.3 Analysis of Experimental Results

Three months after the examination, the Ministry of Education issued the actual scores of the two groups of students in the PRETCO-A examination, and a comparison analysis are performed on the scores. The results are shown in the following table (Table 8):

Table 8. Comparison of PRETCO-A scores of experimental group and control group

Items analysed	Control group	Experimental group
Average score	47.83	50.37
Standard deviation	11.25	11.73
Standard error of mean	2.05	2.14
Number of passing students	7	13
Passing rate	23.33%	43.33%

As is shown in the table, the average score of the experimental group is 2.54 points higher than that of the control group, and the pass rate of the experimental group is 20% higher than that of the control group. Therefore, the significant difference of the two group students' scores can verify that the pre-test assistance provided based on the prediction of model is effective, and the backwash effect of performance prediction on English teaching is positive.

4 Conclusions

By establishing a reasonable data collection method and prediction model to scientifically and effectively predicting PRETCO-A test scores can provide important decision-making reference for educational management and teaching activities. Predicting students' PRETCO-A scores through the establishment of a scientific mathematical prediction model can help teachers identify objects that need to be focused in advance, and then provide targeted assistance to them, and conduct personalized teaching interventions, thereby improving students' academic performance and effectively reduce the probability of failing the exam. It can be seen that the application of prediction model to teaching practice has a positive backwash effect on teaching. Compared with the backwash effect of tests on teaching, the backwash effect of tests prediction on teaching can turn the traditional post-remedial teaching reform into a pre-warning assistance, which provides a foresight guidance for teaching.

The drawback is that when using this model to predict PRETCO-A results, sometimes there is a large difference between the predicted and actual scores of individual students. In the follow-up research, we should take more factors that have great impacts on the prediction model into consideration, improve the collection of various performance data and optimize the prediction model.

Acknowledgments. I would like to extend my sincere gratitude to all those who have offered cordial support in writing this paper.

First and foremost, I am extremely grateful to my colleague Li Yuanhui, who has contributed significantly to improving the methods of construction of the prediction model. Thanks to his expertise, patience and encouragement, I finally finish this paper.

Secondly, I am much obliged to my friend, Wu Han, who has assisted me with the revising work.

My thanks also go to my colleagues Li Chao, Liu Xia, Zhan Man and Xian Dan who has helped me a lot during the research process. It's their generous help and great support make this paper possible.

Last but not least, I'd like to express my heartfelt appreciation to my family members, especially my mother and my husband who has motivated, encouraged, and supported me a lot, which makes me concentrate on the writing of this paper.

References

1. Nan, L., Wenjia, H.: Method of predicting performance variables of postgraduate entrance examination based on logistic algorithm. J. Jilin Univ. **39**(1), 114–120 (2021)
2. Lina, H.: Application of the Bayesian classification model in the forecast of student achievement. Comput. Digital Eng. **46**(10), 2039–2041 (2018)
3. Yong, C.: Research and implementation of result prediction based on genetic neural network. Mod. Electron. Tech. **39**(5), 96–100 (2016)
4. Xinyu, C.: Prediction method of college students' scores toward uncertain missing data. Mod. Electron. Tech. **41**(6), 145–149 (2018)
5. Mengying, L., Xiaodong, W., Shulan, R.: Student performance prediction model based on two-way attention mechanism. J. Comput. Res. Dev. **57**(8), 1729–1740 (2020)
6. Xiaoqian, W., Lili, Q., Cheng, C.: Analysis of student performance and simulation of prediction model based on big data decision tree algorithm. Electron. Des. Eng. **28**(24), 138–141 (2020)
7. Donghai, Y., Yuerui, W.: Using orthogonal kernel least squares algorithm for predicting students' exam results in higher vocational college. Softw. Guide **18**(6), 143–146 (2019)
8. Chao, Z.: Sports performance prediction model based on glowworm algorithm optimizing neural network. Mod. Electron. Tech. **40**(15), 94–100 (2017)
9. Hughes, A.: Testing for Language Teachers. Cambridge University Press, Cambridge (1989)

10. Bailey, K.: Working for washback: a review of the washback concept in language testing. Lang. Test. **13**(3), 257–279 (1996)
11. Gong, Y.: Backwash effect of english testing on english teaching for adults. In: Zhang, Q., Yang, H. (eds.) Pacific Rim Objective Measurement Symposium (PROMS) 2014 Conference Proceedings, pp. 263–270. Springer, Heidelberg (2015). https://doi.org/10.1007/978-3-662-47490-7_20
12. Xu, L.: Introducing Chinese culture into the translation class——the washback effects of the CET reform on language teaching for university students. Educ. Teach. Forum **21**(3), 48–50, (2018)
13. Shuxia, W., Wei, B.: The washback effects of the CET listening reform on college english teaching. Educ. Teach. Forum **18**(1), 91–94 (2019)
14. Qian, X., Jun, L.: A Study on the Washback Effects of the Test for English Majors (TEM). Springer, Singapore (2018). https://doi.org/10.1007/978-981-13-1963-1
15. Han, F.: Washback of the reformed college english test band 4 (CET-4) in english learning and teaching in china, and possible solutions. In: Lanteigne, B., Coombe, C., Brown, J.D. (eds.) Challenges in Language Testing Around the World, pp. 35–46. Springer, Singapore (2021). https://doi.org/10.1007/978-981-33-4232-3_4
16. Yifan, X.: Essentials and Examples of SPSS Statistical Analysis, 1st edn. Electronic Industry Press, Beijing (2010)

Research Demo

Demos of Passing Turing Test Successfully

Shengyuan Wu(✉)

Shandong University, Jinan, China
wsy@sdu.edu.cn

Abstract. Recently, a new kind of machine intelligence was born, called as UI (Ubit intelligence). The basic difference between UI and AI is encoding; UI is based on word encoding; but AI is based on character encoding. UI machine can learn from human, remember the characters, pronunciation, and meaning of a word like human. UI machine can think among the character, pronunciation, and meaning of words like human. Turing Test is similar to a teacher testing a student; Before Test, tester must teach the content of the test questions to UI machine first; after UI machine learning, tester asks testee questions; to check testee has remembered what he taught; to check testee can think among character, pronunciation, and meaning of words. This paper demonstrates that testee can remember what testee taught; and answer all 6 questions correctly by thinking. UI machine passes Turing Test easily and successfully with score 100. Following on, the works related to this study is briefly introduced. At last, this paper concludes that UI machine is based on word encoding, can form word, form concept, can possess brain like intelligence, also can possess human like Intelligence; therefore UI machine passes Turing Test easily and successfully. On the contrary, AI machine is based on character encoding; can't form word; can't form concept, AI machine can't possess brain like intelligence, nor possess human like Intelligence. Therefore, AI machine can't pass Turing Test.

Keywords: Human like intelligence · Machine learning · Machine thinking · Turing test · Word encoding · Concept · Character encoding

1 Introduction

In the paper "Make Machine Humanlike Intelligence", I present a new kind of machine intelligence, called as Ubit intelligence, for short, UI. UI machine can learn like human; and remember the characters, pronunciation, and meaning of words like human. UI machine can think among character, pronunciation, and meaning of words like human. UI can make machine to possess human like intelligence; but AI can't [1–4]. What makes human like intelligence realized? The basic difference between UI and AI is encoding. UI is based on word encoding; it is the word encoding makes human like intelligence realized. On the contrary, AI is based on character encoding; it is the character encoding makes human like intelligence can't be realized.

J. Zeng et al. (Eds.): ICPCSEE 2021, CCIS 1452, pp. 505–514, 2021.
https://doi.org/10.1007/978-981-16-5943-0_41

2 Making Machine Remember the Pronunciation and Meaning of Singular Character Word by Word Encoding

For example, there is a famous Chinese couplet, which consists of 18 characters, all are same character.

The left line: 长长长长长长长

The right line: 长长长长长长长

The top line: 长长长长

Like human, before machine learning, UI machine can't understand the couplet, doesn't know how to read, what meaning. However, after learning, machine to know there are two words of the couplet, and remembers each word as the following three parts:

Character: 长, pronunciation: "chang ", meaning is long

Character: 长, pronunciation: "zhang ", meaning is grow

After UI machine learning, the three parts are grouped as a basic unit, a word, a concept; then it can read the couplet by machine thinking as following:

The left line: "chang zhang chang zhang chang chang zhang"

The right line: "zhang chang zhang chang zhang zhang chang"

The top line: "zhang chang chang zhang"

Machine can express the meaning of the couplet as following:

Often grow, often grow, often and often grow;

Grow longer, grow longer, grow and grow longer

Grow loner and often grow. [1]

(http://www.ubit.hk/uai/duilian.mp4)

This is because UI machine is word encoding, the same character 长 is encoded in two different word codes, one 长 associates with pronunciation "chang ", the meaning is long; the other associates with pronunciation "zhang", the meaning is to grow.

Human can read the couplet correctly by learning; but AI machine can't read the couplet correctly by machine learning, Chinese comprehensions algorithm, or "The DeepMind AI, a so-called neural network made up of layered algorithms"; both can't transform characters into words, into concepts [1–4].

3 Making Machine Remember the Boundary of Each Word

Another example, the meaning of "乒乓球拍卖完了" is ambiguous,

If segmented as "乒乓球" "拍卖" "完" "了";

The meaning is: Ping pong balls/auction/finished/already.

If segmented as"乒乓" "球拍" "卖" "完" "了",

The meaning is: Ping pong/ rackets/sell/finished/already.

Human can read this sentence in two different meanings by learning, but AI machine can't. Before machine learning, UI machine nor can express this sentence in two different meanings. But after machine leaning, UI machine can read this sentence in two different meaning [1]. (http://www.ubit.hk/uai/pingpang-read.mp4).

Because In UI machine,"乒乓球" "拍卖" "乒乓" "球拍" "卖" "完" "了"are coded as words by machine learning, every word consists of three parts: characters, pronunciation and meaning. Each word is like a concept in human brain [1–4].

UI machine can select different words to express different meaning like human.

UI machine can easily distinguish what is the boundary of words by word codes.

However, there are two basic problems must be solved in order to realize word encoding.

First, different length of word codes must compatible one another; for examples: "乒乓球"is 6 bytes, "拍卖"is 4 bytes, "卖" is 2 bytes; 2, 4, 6 byte word codes must compatible one another.

Second, the word code must be compatible with any existed character code system. That is to say, UI machine must be compatible with all existed machines.These two problems have been totally solved by Ucode (Ubit code) [1–5].Turing Test is used to judge if a machine possessing human intelligence [6].

The aim at this paper is furthered to proof UI machine does possess human like intelligence by Turin Test.

4 UI Machine Learning Before Turing Test

Turing Test is described as the following:

The tester and testee (a person and a machine) is isolation from each other, asks random questions on the testee through some device (such as a keyboard). After a few questions, if more than 30 percent of the responses failed to identify which was a person and which was a machine, the machine passed the test and was deemed to have human intelligence [6].

All questions on the next Turing Test are about comprehensions of the following Chinese sentence:

会计师骑自行车去银行开会

Turing Test is similar to a teacher testing a student; before test, tester must teach UI machine learning first.

5 UI Machine Learning to Remember the Pronunciation of Each Word

First, Tester tells testee.

The sentence consists of 6 words as follows.

会计师, 骑, 自行车, 去, 银行, 开会

The meaning of the sentence is:

Accountant rides bicycles to bank for a meeting. Then, teach testee.

Then, Tester teaches the pinyin of each character by human machine interface.

I have developed a number of UI machines learning methods, which are divided into three categories: manual, half automatic and automatic. To make more clearly, manual machine learning method is adopted here as shown in Fig. 1.

Figure 1 hints how Chinese character transforms into singular character word in UI machine learning. The pinyin of会 in word 会计师 is kuai4 by tester selecting 2; the

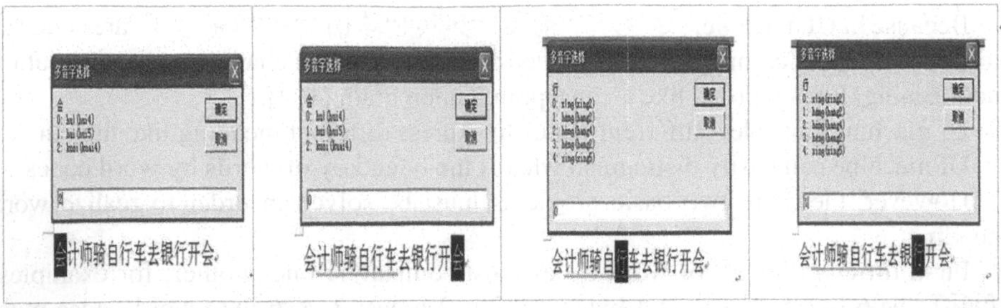

Fig. 1. Screen shot of tester teaching pinyin of words

pinyin of会in word 开会 is hui4 by tester selecting 0; the pinyin of 行 in word自行车 is xing2 by tester selecting 0; the pinyin of 行 in word银行 is hang2 by tester selecting 1. The digit at the last of the pinyin is for tones of Chinese character pronunciation; there are 1, 2, 3, 4, 5 tones; digit 5 is for gentle voice.

After learning, characters are transformed into word, which associated with its pronunciation and meaning.

6 UI Machine Learning to Remember the Characters of Each Chinese Word

The machine learning is adopted a half automatic learning method. Tester first divides the sentence into words as the following:

会计师 骑 自行车 去 银行 开会

The words are separated by space.

Then, UI machines to transforms the 12 singular character words of 6 words automatically.

After machine learning, the 6 words are coded by 6 Ucodes. Each Ucode is associated with the characters, pronunciation, and meaning of each word.

The characters of each Chinese word must be determinate; otherwise, word would

For example:

The meaning of word自行车 is bicycles.

If the word is divided into two words: 自行 车; The meaning of word自行 is voluntarily, the meaning of word 车 is a vehicle.

7 Demos of Passing Turing Test

In Turing test, tester asks testee questions about the sentence; to check if testee has remembered what he taught; if testee can think along associations in each word and between words.

Test 1

Question.

Tester uses the keyboard asking the testee to mark pinyin of the sentence and 4 characters; the characters, 会会行行, are taken from word: 会计师, 开会, 自行车和银行 respectively (Fig. 2).

Testee's answer.

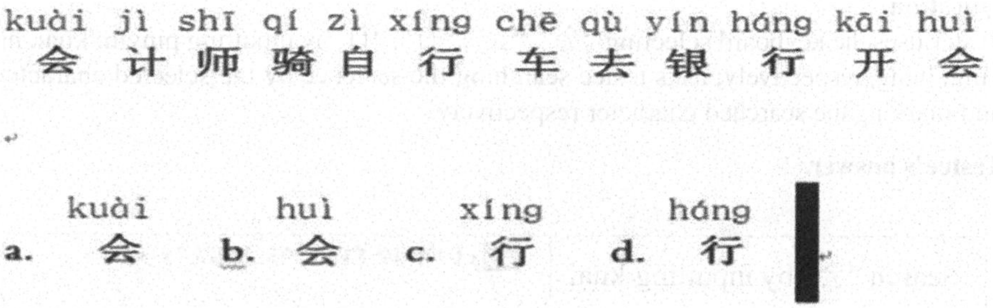
kuài jì shī qí zì xíng chē qù yín háng kāi huì
会 计 师 骑 自 行 车 去 银 行 开 会
kuài huì xíng háng
a. 会 b. 会 c. 行 d. 行

Fig. 2. Testee's answer for marking pingyin

(The video link: http://ubit.hk/kuaiPingyin.mp4)

Tester's confirming.

Testee can remember the pinyin of each singular character word, and can think from character to pinyin correctly, as shown in Fig. 1.

The thinking procedure of UI machine.

Scan each singular character word, thinking from character to associated pinyin in memory, get the pinyin, and put the pinyin on screen to display.

On the contrary, AI can mark pinyin of the sentence correctly, but can't mark the four characters correctly; the two 会 are marked with pinyin: hui; the two 行 are marked with pinyin: hang.

Test 2

Question.

Tester uses a keyboard asking the testee to read the sentence and characters; the characters, 会会行行, are taken from: 会计师, 开会, 自行车和银行 respectively.

Testee's answer.

(audio link: ubit.hk/UIreadZi.mp3).

(video link: http://ubit.hk/kuaiYuyin.mp4).

Tester's confirming.

Testee can remember the speech of each singular character word, and can think from word to speech correctly; can read the sentence correctly; also read the 4 singular character word correctly as shown in audio and video.

The thinking procedure of UI machine.

Scan each of the 16 singular character word, thinking from word to associated audio in memory, get the audio, put the audio to speaker and pronounce.

On the contrary, AI can read the sentence correctly, but can't read the four characters correctly. the two 会 are pronounced with same voice: hui; the two 行 are pronounced with same voice: xing.

Test 3

Question.

Tester uses the keyboard selecting "会", "会", "行", "行"by inputting pinyin: kuai, hui, xing, hang, respectively; asks testee searching the sentence by the selected characters, and marking the searched character respectively.

Testee's answer.

Search "会" by inputting kuai	会计师骑自行车去银行开会
Search "会" by inputting hui	会计师骑自行车去银行开会
Search "行" by inputting xing	会计师骑自行车去银行开会
Search"行" by inputting hang	会计师骑自行车去银行开会

Fig. 3. Testee's answer for marking words by inputted word

(video link: http://ubit.hk/kuaijiZichaxun.mp4)

Tester's confirming.

Testee can remember the character; can associate the pinyin of each character, and can search by the character and its pinyin as a whole; the answers are correctly as shown in Fig. 3.

The thinking procedure of UI machine.

Scan each of the 12 singular character words, compare each word with the inputted word; if equal, then mark the word.

On the contrary, AI can't search the four characters correctly; it marks two 会in the sentence now matter by inputting kuai or hui; and marks two 行in the sentence now matter by inputting xing or hang.

Test 4

Question.

Tester uses the keyboard inputting pinyin: kuai, hui, xing, hang respectively; and asks testee searching the sentence by the inputted pinyin respectively, then marking a vertical line before the searched character.

Testee's answer.

Search by kuai	会计师骑自行车去银行开会
Search by hui	会计师骑自行车去银行开会
Search by xing	会计师骑自行车去银行开会
Search by hang	会计师骑自行车去银行开会

Fig. 4. Testee's answer for marking words by inputted pinyin

(video link: http://ubit.hk/kuaiZiPinyinChaxun.mp4)

Tester's confirming.

Testee can remember the pinyin of each singular character word; and can think from pinyin to character correctly as shown in Fig. 4.

The thinking procedure of UI machine.

Scan each of the 12 singular character word, thinking from word to associated pinyin in memory, get the pinyin, and check if the pinyin contains the inputted pinyin, if containing, then mark the word with a vertical line.

On the contrary, AI can't search the characters correctly by pinyin; because the character is not associated with pinyin.

Test 5

Question.

Tester uses the keyboard to input pinyin: kuai, hui, xing, hang, respectively; asks testee searching words from the sentence by inputted pinyin; and list searched words.

Testee's answer.

Fig. 5. Testee's answer for listing words by inputted pinyin

Tester's confirming.

Testee can remember pinyin of each word, and think from pinyin to word correctly as shown in Fig. 5.

The thinking procedure of UI machine.

Scan each of the words coded by Ucode in the sentence, thinking from word to associated pinyin in memory, get the pinyin, and check if the pinyin contains the inputted pinyin, if containing, then put the characters of the word on screen to display.

On the contrary, AI machine is based on character encoding; it doesn't know which characters are word, nor pinyin contained in word, it is impossible to list the characters of words as in Fig. 5.

Test 6
Question.

Tester controls the keyboard, and asking the testee to read the sentence word by word.

Testee's answer.

(ubit.hk/UIreadCi.mp3).

(video link: http://ubit.hk/kuaijiCiYuyin.mp4);

Tester's confirming.

Testee can distinguish each word, and think from word to speech of word; and read one by one correctly.

The thinking procedure of UI machine.

Scan words coded by Ucode in the sentence, distinguish the 1st word by word code in the sentence, thinking from the word with associated audio of each character in memory, get the audio, put the audio to speaker and pronounce; then, pause; after tester push the next button, then repeat the steps above for the next word; until the end of the sentence.

On the contrary, AI can't read the sentence word by word controlled by tester.

In a short, all answers by UI machine are correctly; UI machine passes Turing Test successfully with score 100.

On the contrary, AI can't pass Turing Test with score 0.

8 The Works Related to This Study

Chinese segmentation algorithm is the bottleneck problem of Chinese language processing. In 90's, I tried to solve the problem by developing new algorithms; but, I found it is impossible to solve by this way.

I set up a new way; making machine intelligence based on word encoding, beginning with multilevel machine code. [7–12].

In 00's, multilevel machine code is improved as multilevel mark code, which can make all character code systems compatible; including Unicode. [13–22].

In 10's, multilevel machine code is further improved on Ucode (Ubit code); and new theory, called as Ubit semantic computing; which makes semantic data understandable to machine; and makes semantic data processing without artificial algorithm and parser. In practice, Ubit semantic computing is about human like intelligence.[1–5].

In 2017, my experimental videos of a scientist dreamed web, awarded by Internet society of China; in practice, it is demos of brain like intelligence web. [23].

The demos here can be extended to various areas; such as:

- Demos of brain like intelligence web, the webpage with words, or concepts;
- Various interface system; such as machine learning method by keyboard inputting, machine can learn natural language, lerning various objects: image, audio, video etc.
- Chinese language processing systems;
- Database processing systems and etc.

9 Conclusion

UI machine is based on word encoding, can form word, form concept, can possess brain like intelligence, also can possess human like Intelligence; therefore UI machine passes Turing Test easily and successfully.

On contrary, AI machine is based on character encoding; can't form word; can't form concept, AI machine can't possess brain like intelligence, nor possess human like Intelligence. Therefore, AI machine can't pass Turing Test.

Acknowledgment. Thank Bin Wu, my wife, for her great contributions and fully supporting. During the three decades, she helps me to do everything about thinking machine; such as experiment, prepare material, discuss problems with me; as difficulties met, she gives me confidence.

References

1. Shengyuan, Wu.: Make machine humanlike intelligence, Apr 2020. https://ieeexplore.ieee.org/document/9071113
2. Shengyuan, W.: Introduction to Ubit semantic computing. In: Proceedings of the 2014 International Conference on Semantic Web and Web Services of Computer Science, July 2014
3. Shengyuan, Wu.: Methods and apparatus of digital data processing. Chinese Patent Application, Nov 2013
4. Shengyuan, Wu.: Methods and apparatus of digital data processing. US Patent, US 10637643, Apr 2020
5. Shengyuan, Wu.: Introduction to multilevel mark coding theory. Proceedings of the 2007 International Conference on Foundations of Computer Science (FCS'07), June 2007
6. Turing, T.A.M.: Computing Machinery and Intelligence. Oxford University Press on behalf of the Mind Association (1950)
7. Shengyuan, Wu.: A Chinese parallel word segmentation method based on muletilevel coding technology. Comput. Res. Dev. (1997)
8. Shengyuan, Wu.: A Chinese word segmentation method, Comput. Res. Dev. (1996)
9. Shengyuan, W.: Muletilevel coding text telepohone system. Comput. Bus. Inf. **12**(23), P20 (1996)
10. Shengyuan, Wu.: Methods and apparatus of character information processing based on muletilevel coding technology. Chinese State Intellectual Property Office, Patent, Oct 1996
11. Shengyuan, Wu.: The applicatiion effects of muletilevel coding technology. PC World China, pp. 68–69, Sept 1995
12. Shengyuan, Wu.: The muletilevel coding theory and the muletilevel coding computer system. J. Shandong Ind. Univ. (1995)
13. Shengyuan, Wu.: A method to realize alphabetization of Chinese characters based on muletilevel coding technology. China Technol. Online (2000)

14. Shengyuan, Wu.: Object representing and processing method and apparatus, China. Patent and Trademark Office, CN1567174 (2003)
15. Shengyuan, Wu.: Object representing methods and apparatus, PCTIB2003060369, Nov 2005
16. Shengyuan, Wu.: Object representing methods and apparatus, Patent No. US 7724158, Dec 2004
17. Shengyuan, Wu: Object integrated processing method based on 1-0 UTF-8. In: Proceedings of the 2007 International Conference on Multimedia Systems and Applications (MSA 2007), June 2007
18. Shengyuan, Wu.: Compression and improvement of web page based on 1-0 UTF-8. In: Proceeding of the 2007 International Conference on Internet Computing (ICOMP 2007), June 2007
19. Shengyuan, Wu.: 1-0 form of UTF-8. In: Proceedings of the 20th Pacific Asia Conference on Language, Information and Computation, Qinghua University Press, Nov 2006, pp. 354–359
20. Shengyuan, Wu.: A search method for text and multimedia, the fourth symposium of search engine and web mining of China. J. Shandong Univ. (natural science), **41** (2006)
21. Shengyuan, Wu.: Object representing methods and apparatus. Chinese Patent Application, June 2003
22. Shengyuan, Wu.: The muletilevel coding technology is the hope of Chinese software enterprize. Computer World, pp. C1–C2, 14 Aug 2000
23. Shengyuan, Wu.: Award certificate for machine thinking internet expariment video, Internet society of China, July 2017. http://www.ubit.hk/uai/award.jpg

Author Index

Printed in the United States
by Baker & Taylor Publisher Services